3D Bioprinting Revolution

Dr. Sabrie Soloman
Ph.D., Sc.D., MBA, PE

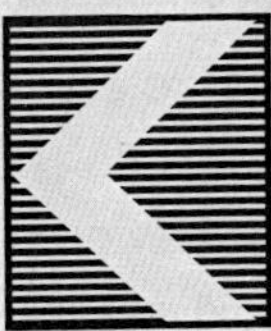

KHANNA BOOK PUBLISHING CO. (P) LTD.

Publisher of Engineering and Computer Books

4C/4344, Ansari Road, Darya Ganj, New Delhi-110002
Phone : 011-23244447-48 **Mobile:** +91-9910909320
E-mail : contact@khannabooks.com
Website : www.khannabooks.com

Dear Readers,

To prevent the piracy, this book is secured with HIGH SECURITY HOLOGRAM on the front title cover. In case you don't find the hologram on the front cover title, please write us to at contact@khannabooks.com or whatsapp us at +91-99109 09320 and avail special gift voucher for yourself.

Specimen of Hologram on front Cover title:

Moreover, there is a SPECIAL DISCOUNT COUPON for you with EVERY HOLOGRAM.

How to avail this SPECIAL DISCOUNT:

Step 1: Scratch the hologram
Step 2: Under the scratch area, your "coupon code" is available
Step 3: Logon to www.khannabooks.com
Step 4: Use your "coupon code" in the shopping cart and get your copy at a special discount
Step 5: Enjoy your reading!

Indian Price: ₹ 2995.00
International Price: $ 164.99

ISBN: 978-93-89139-16-7

3D Bioprinting Revolution
by Dr. Sabrie Soloman

First Edition: 2020
Reprint: 2021, 2022 2023

To view complete list of books,
Please scan the QR Code:

Published by:
Khanna Book Publishing Co. (P) Ltd.
CIN: U22110DL1998PTC095547

Visit us at: **www.khannabooks.com**
Write us at: **contact@khannabooks.com**

Printed in India by: Vaibhav Products, Delhi.

DEDICATION

To My Beloved …!

The Virtuous and Gifted One

Wakes up ahead of dawn - Opens her soft eyes reflecting my morning golden rays
Flying on the wing of her soaring soul – Joyfully hurdling to my loving arms
With warm scarlet robe she bundles me in mid-winter sending me to my daily paths
Her gentle hands carry my desired hot coffee preparing me for my morning chores
The singing birds join the melodies of my contented heart overflowing with gratitude
Enriched my life when she planted the seeds of her greatness deep in my heart

Only for her – I dedicate the humble work of this book!

To my beloved wife - Julie

Preface

"Rising costs. Shorter delivery time. Complex organ specifications. Intricate organ design. Organ shortage from across the street--and around the world."

The Business of organ bio-manufacturing today faces an ever-increasing number of challenges. The bio-manufactures that develop more effective and efficient forms of production, development, and faster time to patients, will be the ones who meet these challenges.

The use of *Additive Bio-Manufacturing* – 3D bioprinting makes a fundamental commitment to rapid organ manufacturing solutions based on simple and affordable 3D bioprinting biotechnology. With this biotechnology, one can integrate *Additive Bio-Manufacturing* – 3D bioprinting processes, react to rapidly changing organ performance conditions, help management to react to rapidly changing organ conditions, help personnel to react more effectively to complex qualitative decisions, and lower the cost of and improve producing organ quality throughout the bio-manufacturing enterprise.

The first step in achieving such flexibility is to establish a rapid entity system that can be reshaped whenever necessary, thus enabling it to quickly respond to the changing organ requirements of the enterprise-and the environment. This reshaping must be accomplished with minimal cost and disruption to the ongoing organ production operation.

Additive Bio-Manufacturing--3D bioprinting will play a key role in achieving flexibility in the organs performance and bio-manufacturing system. However, this biotechnology alone cannot shorten lead time, increase organ inventories, and minimize excess capacity to the extent required by today's bio-manufacturing operation. This can be accomplished only by integrating various types of biomaterials presented in different innovative forms with appropriate control means throughout the bio-manufacturing process within *computer integrated digital bio-manufacturing* strategy. The result is that individual organ design and bio-manufacturing processes will be able to flow–communicate –and respond together as a unified cell, well-structured to face rapid efficient organ-product to market demands.

In order to develop a Computer Aided Design & Bio-Manufacturing CAD/CAM and control information system that will achieve these objectives, the medical enterprise must start with a specific long-range strategy, one providing a foundation that accommodates today's organ needs as well as taking those of tomorrow's – including the support of new bio-manufacturing processes, incorporating new data functions, and establishing new data bases and distributed channels-into account. The tools for this control and integration are available today through the implementation of *Additive Bio-Manufacturing Technology*.

The United States leads the world in inventing new biological products; however, many of these new products, ultimately, are manufactured by other countries. The inability of U.S. manufacturers to compete globally cannot be only blamed on low-cost labor in other countries; more than one- half the trade deficit comes from foreign industries that pay higher wages. The inability to apply rapid affordable design and manufacturing systems for production can be a contributing factor to this dilemma. *Additive Bio-Manufacturing* – Rapid Prototyping – 3D Bio-Printing is the game changer.

A rapid and affordable organ bio-manufacturing system is, simply, a system that contains a variety of reliable biological materials, harmoniously merged together to generate a specific product that will achieve a particular bio-manufacturing operation, directed and controlled by simple and effective biosensors and control systems. Modern rapid bio-manufacturing technology is prevalent in the rapid change in CAD, and in applying the bio-sensory fast-feedback and control biotechnology in the rapid iteration of bio-manufacturing improvement systems to simplify organ fabrication, bio-assembly, and performance. This is greatly enabled through the fast iterations in 3D bioprinting technology.

When considering the Computer Aided Design aspect of a system the following must be taken into account: dynamics, kinematics, statics, and even styling of biological components. All of these play a vital role in forming optimum bio-manufacturing parameters in organ design. The generation and control through various iterations of rapid prototyping provides a review of manufacturing bio-engineering concepts and performance from a system's point of view, directed towards solving the bio-manufacturing and mass production problems.

Additive Bio-Manufacturing – 3D printing plays a key role in translating new organ-product specifications from design bioengineering into bio-manufacturing process plans which are then used to manufacture the final organ. As the organ-product is being deigned and reiterated, bio-manufacturing technical evaluators work with design bioengineers to ascertain if biomaterial strength and organ specifications for bio-manufacturing can be integrated into the design of a total organ-product, and at what cost. Tolerances, biomaterials, clearances, appropriate handling of bio-components, acceptable types and orientation of biomaterial and organ-product assembly times are particularly important factors in this evaluation because they directly affect productivity, the guidelines of which are essential to this analysis.

Cost estimates are equally important. If a new biomaterial or process is needed because, for example, existing biomaterials or processes are deemed too expensive or incapable of producing the desired organ, process bioengineers would be asked to either develop a new economical bio-component part or process or to change organ-product design. Choosing the best alternative could be very difficult because such a decision is based on many conflicting objectives, e.g., cost, feasibility, timeliness, market share, organ's available biomaterial, standard biomaterial, availability of organ scaffold, biosensors locations, and the like. A well CAD- for a bio-component incorporating simple strategy for the rapid generation of a new organ to improve the performance of a bio-manufacturing system plays a fundamental role in the new bio-manufacturing thinking.

Advanced 3D bioprinting technology is more than an implementation of new biotechnologies. It is a long-range strategy that allows bio-components of the entire bio-manufacturing operation to work together to achieve the qualitative and quantitative goals of organ manufacturing business. It must have top management commitment and may entail changing the mind–set of people in the organization and managing that change. However, the rewards are great since successful implementation of this technology is, in large measure, responsible for the success of rapid prototyping through *Additive Bio-Manufacturing*–3D bioprinting strategy today.

In the last few years, *Additive Bio-Manufacturing*–3D bioprinting technology has taken the world by storm, revolutionizing applications in every field from education to astrophysics. The bio-manufacturing and medical world has grasped numerous benefits as 3D bioprinting becomes more affordable and more versatile. Everything from microelectronics to personalized organs and from implants to medical prostheses can be 3D printed. The future of medicine certainly contains 3D bioprinters. Whether medical professionals are using them to help students practice and research new treatments and procedures, or patients are receiving new organs and prosthetics, 3D bioprinting has hundreds of possible applications.

The future of industries and medicine certainly contains countless 3D bioprinters. Whether entrepreneurs creating novel products or medical professionals using them to help students practice and research new treatments and procedures, or patients are receiving new organs and prosthetics, *Additive Bio-Manufacturing* --3D bioprinting technology has countless of applications. It is a game changer!

Sabrie Soloman

Foreword

The ongoing revolution in Additive Bio-Manufacturing– 3D Printing Biotechnology, well into its third decade, has now (albeit somewhat belatedly) become recognized and joined by government, industry, health and academe in USA and around the world. It is based on a number of concepts which have made their way into the professional jargon and have been brought to the public's awareness by technical and business writers: concurrent bio-engineering, Additive Bio-manufacturing, digital manufacturing, flexible manufacturing, just-in-time production and inventory, automation, and bio-manufacturing quality. Each of these are ingredients that contribute to the ultimate goal, which, simply stated, is to achieve the highest quality products at the lowest possible cost, and to do so in a timely fashion. A tall order, this, but one on which depend the welfare of a host of individual bio-medical companies and, even more importantly, the economic health of entire countries, with political and social implications beyond overstatement.

A principal ingredient in the process, perhaps the most important one, is the achievement and implementation of error-free 3D bioprinting production, at one and the same time a guarantor of performance and a minimizer of waste of biomaterials and labor. At first impression, the term "error-free" will sound like a pious ideal, to be striven for but impossible to attain. A moment of reflection will persuade, however, that the aim need not be a philosophical abstraction. In the final analysis, it is the end 3D bioprinting product alone that must fall within the range of prescribed acceptance limits, not each of the many steps in the bioprinting process. That is to say, given within the context of computer-aided-design integrated 3D bioprinting a sufficient array of monitors distributed throughout the bioprinting envelope- i.e., biomedical sensors measuring layer thickness, temperature, speed, biomaterials (and appropriate means to feedback and respond to, in real time, the information gathered by them) and control systems which can identify, rectify, or remove defects in the course of part bioprinting- every item that reaches the end of the part bioprinting will be, ipso facto, acceptable.

The bioprinting book before the reader contains not only an exposition of 3D bioprinting organs and controls, but a host of invaluable asides, comments, and extended discourses on key topics of modern Additive Bio-Manufacturing. The author, Dr. Sabrie Soloman is an active practitioner of advanced bio-manufacturing techniques and a highly regarded teacher of the subject. In this volume, he brings error-free 3D bioprinting a step nearer to realization; the world of additive bio-manufacturing – 3D bioprinting, digital bio-manufacturing and design engineering owes him a debt of gratitude.

H. Deresiewicz
Chairman
Applied Science and Engineering
Columbia University

About the Author

DR. SABRIE SOLOMAN
Ph.D., Sc.D., MBA, PE

England: Dr. Sabrie Soloman awarded his Doctor of Science & Technology Degree- D. Sc., from New Castle Upon-Tyne University, England due to his discovery of "Light Energy Phenomena" in the Near Infrared Domain - named "Sabrie's Index." Earlier on, he was also awarded his Doctor of Philosophy Degree- Ph. D., from the same University in "Dynamic Linkage Mechanisms." Dr. Soloman was appointed as an Assistance Professor once he completed his Bachelor of Science Degree B.SC., in Engineering & Science Technology.

USA: Dr. Soloman was employed by Pfizer Pharmaceuticals, Inc., USA as the Director of Advanced Manufacturing Technology, NY, USA. He was advanced to take the position of Vice President Pharmaceutical Manufacturing Technology, NY-USA.

Columbia University: Dr. Soloman was appointed Adunct-Professor of Science and Engineering at Columbia University. He Founder the Post Graduate Studies of Advanced Manufacturing Technology at Columbia University. Dr. Soloman Lectured on "Computer Integrated Manufacturing, "Flexible Manufacturing Systems," "Sensors and Control Systems in Manufacturing," "Affordable Automation," and "Machine Design."

Chairman & CEO: Dr. Soloman founded American SensoRx, Inc., based on several new patented proprietary technologies; US Patent 8,146,331 B2, US Patent 8,413,797 B2, and US Patent 5,679,954, as well as a Trademark 5,665,657.

Patents & Trade Marks: The Corporate fields of expertise are: Pharmaceuticals; Medical; Near Infrared, Immunotherapy Products; Passive Immunotherapy Product Line includes: Antibody-Rich Products and Product Candidates for the Treatment of Immune Deficiencies, Infectious Diseases and Potential Biodefense Applications; Protein Purification Technology, Innovative Development Capabilities, and Manufacturing "Know-How" to Develop various Product Lines; Cosmetics; Chemicals; Active Ingredients; Food: Pork, Beef, Poultry, Microelectronics; Automotive; Gas; Liquids; Solids; Consumers Products; Bio-surgical Products for the Control of Bleeding, or Hemostasis; Passive Immunotherapy Products includes Antibody-Rich Products for the Treatment of Immune Deficiencies and Infectious Diseases, and for

Potential Biodefense Applications; and Highly Advanced "3D Printing" and 3D Bioprinting" Technology. Participated in launching numerous products worldwide.

Author: Dr. Soloman Authored 17 Books and 2 Handbooks Published by McGraw-Hill Publishing Co., USA: The following books are among the published: Sensors & Control Systems in Manufacturing, 1st and 2nd edition, Sensors Handbook, 1st and 2nd edition, Affordable Automation, Introduction to Electromechanical Engineering Modern Welding Technology; 3D Printing & Design book, and 3D Bioprinting Revolution book - both books published by "Khanna Books Publishing Company, India.

Contents

1 3D Bio-Printing Technology

UNDERSTANDING BIOPRINTING TECHNOLOGY

Bioprinting technology is the manufacturing applications that combine two basic facets of creativity; science and art to make a unique biological product to help mankind. The science of Bioprinting is geared to deploy a technique cultivating living biological assay in a form of ink-like drops. While, the art of Bioprinting is articulated to deliver ink-like drops through innovative devices similar to traditional printer depositing, spraying, or abating its ink.

Bio-Manufacturing can be defined as the production of complex living and non-living biological products from raw materials such as living cells, molecules, extracellular matrices, and biomaterials.

Nomenclatures

DNA: Deoxyribonucleic acid, a self-replicating material which is present in nearly all living organisms as the main constituent of chromosomes. It is the carrier of genetic information.

ECM: Extracellular matrix – It is a three-dimensional network of extracellular macromolecules, such as collagen, enzymes, and glycoproteins, that provide structural and biochemical support of surrounding cells.

Cytokines: Cytokines are a large group of proteins, peptides or glycoproteins that are secreted by specific cells of immune system. Also, Cytokines are a category of signaling molecules that mediate and regulate immunity, inflammation and hematopoiesis.

Calvaria: Calavaria or skullcap is the top part of the skull. It is the upper part of the neurocranium and covers the cranial cavity containing the brain. It forms the main component of the skull roof.

Creative Bio-Ink

Bioprinting is then defined as means to deposit "bio-inks" of biologically active components. The biologically active components encompass several biological substances such as proteins, peptides, DNA, cells, ECM molecules, native or synthetic biopolymers hormones: including cytokines, growth factors and synthetic hormonal signaling peptides. The scale of bioprinting applications varies size and shape. The may be as small as molecules such as DNA and as large as implantable scaffolds.

Scaffolds

Scaffolds are the structural frame where the organs can be erected. Eventually, scaffolds will be dissolved soon after the organ is constructed, or more precisely is printed. Several types of printers can be used.

The printer dispensing device, which deposits ink-like drops, may drop its ink substance in two dimensional X–Y coordinates, while other printers may deposit their substances in X-Y-Z coordinates. The advancement of depositing each step is considered a printer step following the X-Y coordinates in one plane forming a layer.

Depositing Successive Layers

A successive layer can be constructed, or more precisely deposited, when the printer dispensing nozzle or orifice steps in the Z coordinate to generate a new layer of ink deposits. This stepping process in X-Y and Z coordinates is repeated unto the entire intended structure is formed.

Debates – Printing Organs

There has been much too much debate around 3D printing organs; however, many are left pondering how does the 3D bioprinting process work? Can we actually use the technology to print actual printed organs? Surprisingly, the process for printing living and viable organs is closely similar to most printing machines.

Printing Head - Biological Living Cells

The actual printer head is a special orifice designed to dispense liquefied ink-like substance made of biological living cells. The printed head moves in an X-Y plane to create a printed layer. The layer thickness is equal to thickness of the deposited drop out of the printer head. Nonetheless, the surface where the print is deposited may move down in the Z direction to create another layer of deposited living cells. The process is repeated until all the layers are printed creating the intended biological configuration. In some applications the bio-printing head moves in the vertical detections instead of the printer bed. The biological material is deposited through a small nozzle and built up layer by layer.

In industrial traditional 3D printer the dispensed material is an extruding plastic passing through a heated nozzle. 3D bioprinter, dispense small amounts of tissues and different chemical compounds through a small orifice.

Dispensing Multiple Bio-Materials Simultaneously

3D bioprinters dispense multiple bio-materials simultaneously, similar of traditional industrial printing machines equipped with multiple nozzles. Scientists were able to successfully develop several extensive types of bio-material options to successfully print tissues from complex organs.

Organ Donors - Less than Organ Demands

As organ donors are less than the organ demands to meet a growing need, the demand for printing organs is at an all-time high. Therefore, producing synthetic organs through bioprinters is literally a lifesaver to provide reliable printed organs. These 3D bioprinting breakthroughs, which have recently occurred, allowed the large-scale production of synthetic organs to expediently take place.

3D Bioprinting - Revolutionary Technology

3D Bioprinting has gained public recognition and overall acceptance of its value added criteria. It is no longer considered as infant technology for prototyping and modest design. 3D Bioprinting is a revolutionary technology rapidly reaching the tipping point of a full-scale bio-manufacturing competence.

In recent past the entrepreneur understanding of 3D Bioprinting may have been restricted to simple tissue growth and dispensing, tooth forming, or printing a simplistic cast of a fractured arm, like school science projects or biology hobbyists. Recently, 3D Bioprinting has evolved from top to bottom. Apparently, day by day, the raw bio-materials become more varied, the bio-engineering more refined, and the output more precise, consistent, and repeatable—indicating that the vast potential for complex bioprinting industrial use substantially expands every year.

Unique Organ Absent of Organ Donor

It is unimaginable to receive an organ absent of the organ donor! A no waiting list and no waiting at all could be a dream comes true to thousands of patients in their daily life. An unimaginable world where a patient is able to be fitted with a suitable custom made organ printed on-the-fly by a bio-printer. Bio-medical 3D Bioprinting corporations are heavily competing to release many printers to allow users to Bioprint their own 3D printed human organs. As stated earlier, bio printer work with a liquefied substance of different cultivation. This biological liquefied substance is referred to as "Bio-ink." This liquefied biological substance is pushed through a nozzle and bonded together on the printer bed using blue light as curing agent.

3D Bio-Printing On-Demand

An object that has blood vessels and organ tissue could be printed on demand. This has successfully facilitated the objective to use this to manufacture liver-like organ for drug testing and skins cosmetic research and testing. Accordingly, the need for testing on humans is eliminated.

The versatilities of the 3D Bioprinting operations together with their advancements in printing speed and dispensing finer resolution enabled the creation of tissue engineering, organs construction, customizing prosthetics, anatomical models, and pharmaceutical research. 3D Bio-Print tissue has led to the intensive research, which is currently being conducted on artificial hearts, liver structures, and kidneys. Printing other major organs is underway.

Cell Survival

The survival of the cells must be maintained in a certain environment. This environment is maintained in gel scaffolds, often stated as bio-paper in simple. There are a number of means of 3D bioprinting. One of these means is to print a layer of gel and then a layer of cell in a continuous way until the anticipated structure is achieved. Another way of 3-D bioprinting is to initially print a gel scaffold and then enable cell growth in the gel scaffold by inserting in cells, Figure 1.1.

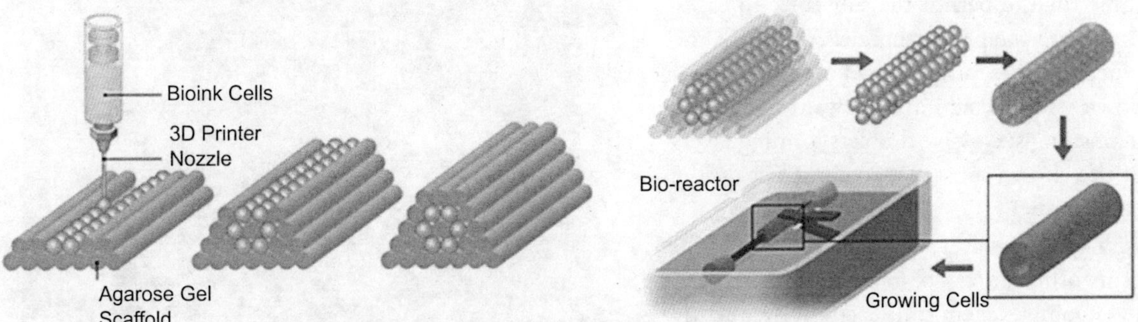

Figure 1.1: Cell 3D Printing Operation
Courtesy: Modern Meadow

Droplets - Organic Bionic Substance

Conventional inkjet printing methodology is used to generate 3D Bioprinting applications. In this technology Pico-liter droplets of "Organic Bionic Substance" are precisely deposited by the aid of a well-developed computer controlled routine. The droplets are deposited onto a culture dish or hydrogel substrate, Figure 1.1. The ink droplets are formed by a thermoplastic method that heats up an inflated air bubble,

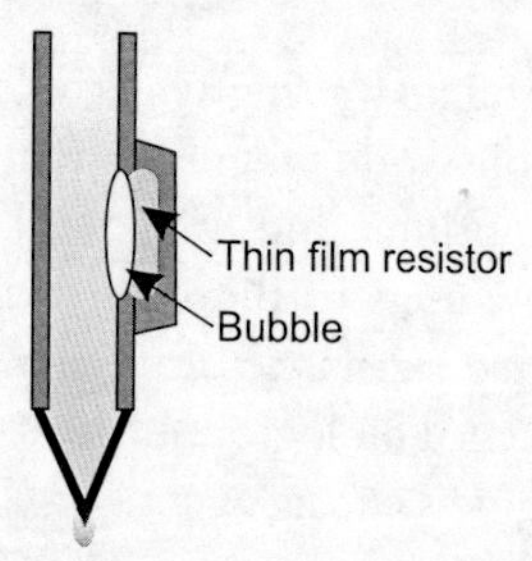

Figure 1.2: An Ink Droplets Formed by a Thermoplastic Method

which pushed the ink droplet out of a narrow nozzle, Figure 1.2. A few microseconds are only what it takes to reach a high temperature to generate the necessary pressure to push out the droplet.

While this technology is highly economical the deposited droplets size may not consistent. The nozzle is frequently clocked to maintain precise and smooth Bio-printing. The organic bio ink substance is greatly affected by the thermal stress, causing substantial viability in the cellular and protein characteristics. To avoid such problem in printing quality a use of piezoelectric method is used to prevent orifice clogging, and allow droplets to remain equal and consistent size, Figure 1.3.

Figure 1.3: Equal and Consistent Droplet Sizethrough Piezoelectric Method

Piezoelectric technology, if used repeatedly may damage the cell membrane and cell lysis. Nonetheless, a high success rate can be achieved using piezoelectric droplets deposit in human osteoblasts, fibroblasts, and bovine chondrocytes.

Creating 3D Bioprint Object

To create a 3D object, you primarily need a blueprint—that is, a digital file generated utilizing modelling software. Once generated, the computer-created model is directed to the printer. Your selected material, such as cell tissue is placed into the 3D Bioprinter, prepared to be subjected to heat for a very short time to permit the substance to easily flow from the printer nozzle. As it reads the blueprint, the printer head moves up and down, side to side and forward and back, depositing successive droplets forming layers of the chosen substance to construct up the end product.

As each layer is 3D Bioprinted it transforms into a compacted form, either by cooling, chemical reaction, frequently often induced light, or by the mixing of two diverse solutions dispensed by the Bioprinter head. New layers stick to the previous one to generate a stable, cohesive object. Nearly, any configuration can be generated in this manner, as well as objects of intricate layers.

Figure 1.4: 3D Bioprinting Machine -Multiple, Photo-curing, Electromagnetic and Extrusion - Materials: Hydrogel, and Organoids

Courtesy: FABION Technology

An extensive variety of objects are already being generated utilizing 3D printers, including skins, bones, nose, ears, prosthetics, dentures, hearing aids, and high-end Bioprinted organs.

BIOPRINTING

Bioprinters work in nearly the meticulous similar way as 3D printers, with one important variance. As an alternative of bringing materials such as plastic, ceramic, metal or food, they deposit layers of Biomaterial that may contain living cells, to construct intricate configurations like blood vessels or skin tissue.

How to Generate Living Cells

In the beginning, since creation; the creation of every tissue in human body is naturally made up of different cell kinds. Therefore, the needed cell, such as kidney cells, skin cells and many other cells are acquired from a patient and then cultured until there are sufficient cell to create the "bio-ink." The Bioink is placed into the Bioprinter. However, this may not br feasible all the times. Accordingly, some tissues, such as adult stem cells—which can develop to form the cells required in different tissues—can be utilized.

Subsequent comprehensive detailed computer designs and models often based on 3D scans taken directly from a patient. The 3D scan could be needed nose, ear, skin, or an orthopedic part, or a fractured bone. A precise printer heads deposit cells exactly where they are needed and, over the course of several hours, an organic object is constructed utilizing a great number of very thin layers.

Obviously, one may require more than just cells. Therefore, most 3D Bioprinters also deposit certain types of organic or synthetic "adhesive"—a dissolvable gel, collagen scaffold or other kind of support structure that the cells can cling to and cultivate. This assists the cells to form and stabilize into the precise configuration. Astonishingly, countless cells can assume the precise setting by themselves with no scaffolding.

Inherit Cells Creative Properties

Scientists and highly educated scholars have consistently wondered the creative inherent properties that the cell has maintained throughout ages to restructure itself to assume the precise settings without any external manipulation allowing it to reform itself autonomously to become a functioning organ, such as an ear, an eye, a nose, or even a heart. How do they know where to go? They use their inherent given properties from the beginning of creation to seek out similar cells to join with. They innately know where they are required, as the manner cells in an embryo grow in the womb, or tissue in an adult changes to repair damage. Scientists are then capable to mechanize the configuration in which cells achieve that purpose, with the Bioprinter constructing the end structure.

Versatile 3D Bioprinters

There are many diverse Bioprinters on the market and enduring testing, and consequently methods for carrying these cells and delivering biomaterials also vary. Some of the more commonly utilized methods at the present time include:

The 3D Bioprinting Extrusion Type: The Extrusion Bioprinting is a technique where continuous filaments of a material are enforced through a nozzle in a precise manner to build a 3D structure, Figure 1.5.

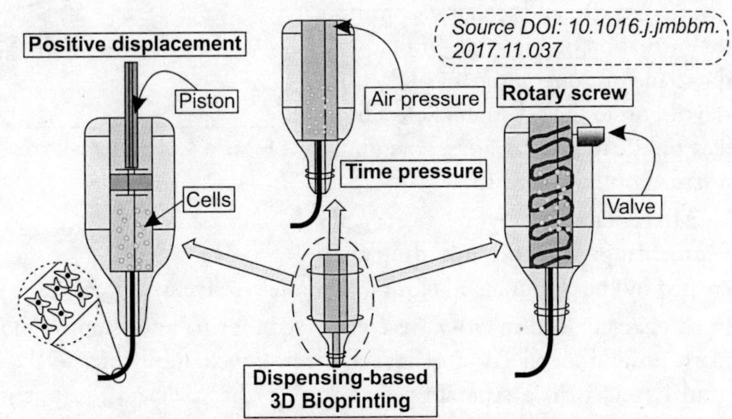

Figure 1.5: Extrusion 3D Bioprinting Dispensing Type
Courtesy: Biogelx, Ltd

Extrusion-Construct Bioprinting Techniques have been utilized in current years to provide scientists with cost-effective alternate approaches for scaffold construction. The acceptance of these approaches depends on distinctive handing approaches leading to a procedure which offers ease, variety, and expectedness. The extensive variety of printable biomaterials and affordable equipment are also amongst the benefits of this printing practice.

The extrusion-construct bioprinting technique is classified into two classifications:

- pneumatic, piston-driven Figure 1.5, and
- screw-driven dispensing, Figure 1.5.

In pneumatic dispensing, air pressure delivers the essential driving force, Figure 1.5, though in piston and screw-driven dispensing, vertical and rotating motorized forces start printing, correspondingly. All of these approaches needed a bio-ink which requires being in the liquefied phase to evade nozzle blockage, but gummy enough that it holds the printed configuration, shields cells during extrusion and delivers the cells with an "*in vivo*"-like environment.

Scientists favor to utilize biomaterials beforehand confirmed to be well-matched with cells to lessening the variables in their projects. Though, several biomaterials are not printable deprived of making extensive variations to their mechanical and potentially, consequently, their biological characteristics. Therefore, it is significant to select a material for your research which does not only deliver you with a 3D cell culture atmosphere but is also printable.

Laser Type of 3D Bioprinting: In laser-guided direct writing technique, photons from a laser beam trap and guide cells by manipulating the variances in refractive indexes of cells and cell media.

A characteristic **Laser Assisted Bioprinting (LAB)**" arrangement encompasses a pulsed laser beam, a focusing system, a ribbon (a translucent glass slide, layered with a laser-absorbing layer of metal, onto which a thin layer of bio-ink is spread, and a delivery substrate facing the ribbon. The physical principle of **LAB** is based on the creation of a cavitation-like bubble, into the depth of the bio-ink film, whose growth and collapse prompts the formation of a jet and, in that way, the transmission of the bio-ink from the ribbon to the substrate, here a bone flaw on the mouse calvarias, emerging a micro-droplet, Figure 1.6.

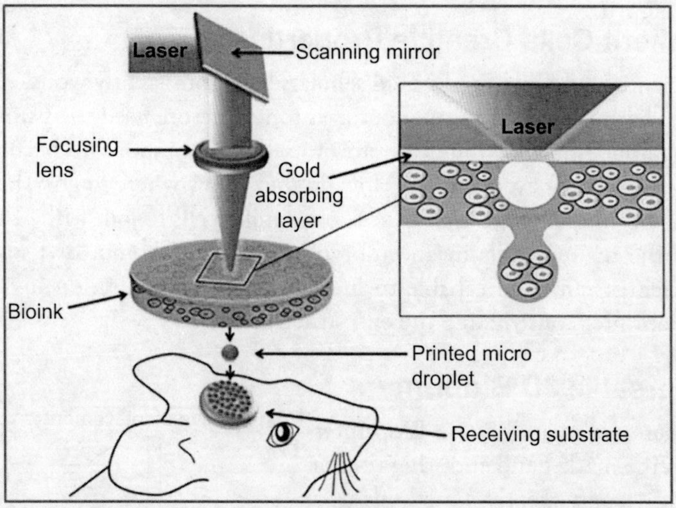

Figure 1.6: Schematic diagram of the "Laser Assisted Bioprinting (LAB)" approach

Courtesy: University of Bordeaux, Tissue Bioengineering, U1026, F-33076, Bordeaux, France

Micro-valves Type of 3D Bioprinting: The bio-ink droplet is created by the opening and closing of a micro-valve under constant pneumatic pressure, Figure 1.7.

A characteristic micro-valve-construct bioprinting system contains a three-axis transportable robotic platform and a collection of several electromechanical micro-valve print-heads. Each micro-valve print-head is linked to a separate gas controller that delivers the pneumatic positive pressure and the valve opening time, which is 0.1 milli-second minimum. It is regulated by the movement of both the plunger

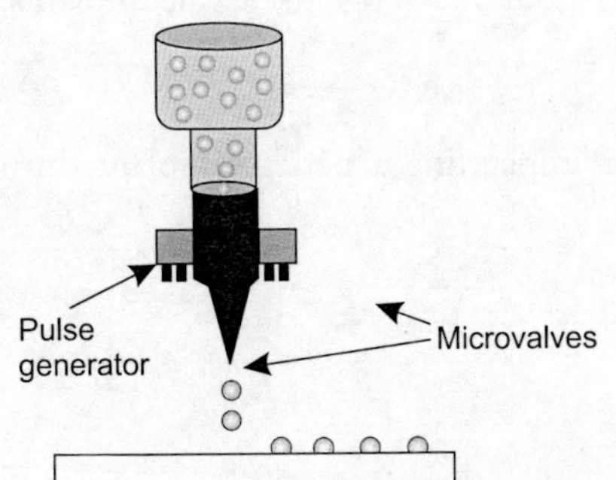

Figure 1.7: Bioprinting Deposit through Micro-valve Technique
Courtesy: Singapore Centre for 3D Printing (SC3DP), School of Mechanical and Aerospace Engineering, Nanyang Technological University

and the solenoid coil. The supplied voltage pulse prompts a magnetic field in the solenoid coil, which opens the nozzle orifice by dragging the plunger up in a climbing motion. The bio-ink is deposited when the air pressure overpowers the fluid viscosity and surface tension at the opened orifice. The material deposition procedure is exceedingly reliant on the nozzle diameter, the viscosity and surface tension of the bio-ink, the air pressure and the valve opening time.

It provides regulated deposition of materials via a layer-by-layer manufacturing method; the benefits of micro-valve-based bioprinting are the coordinated ejection of biomaterials and cells from diverse print-heads, deposition of a thin material layer of a thickness of 1–2 μm, accurate cellular positioning with a feasibility greater than 86% and high amount printing, reaching about 1000 printed droplets per second.

Nevertheless, it is only likely to print hydrogels within a restricted range of viscosities, which is about 1 to 200 mPa s, and cell concentrations of up to106 cells/ml, according to the blocking problems in the small nozzle orifice of a diameter between100 to 250 μm. The cells incline to have a residue over time, affecting the overall cell homogeneity within the bio-inks.

Inkjet Type of 3D Bioprinting: Drop-based bioprinting, inkjet generates cellular build utilizing individual droplets of a selected material, which may also be integrated with a cell line.

Thermal-construct inkjet printing utilizes a heated component to nucleate a bubble. The bubble allows a build-up pressure within the print head. This leads to the discharge of a droplet. The thermal component can reach temperatures between 100 °C to 300 °C. Originally there have been anxieties that such high temperatures would harm the cells, though investigation has revealed that the high temperatures are contained and are only exhibited for a very short span of time, Figure 1.8.

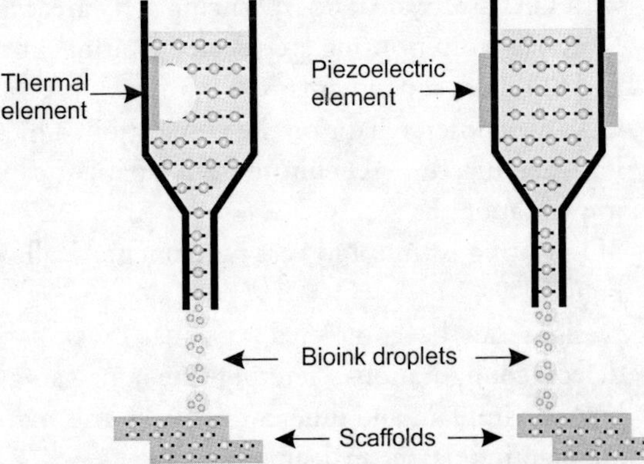

Figure 1.8: Schematic of Inkjet-based Bioprinting. Thermal inkjet uses heat-induced bubble nucleation that propels the Bioink through the micro-nozzle. Piezoelectric actuator produces acoustic waves that propel the Bioink through the micro-nozzle.
Courtesy: Faculty of Medicine and Health, University of Osijek, 31000 Osijek, Croatia

Piezoelectric-construct device utilizes acoustic waves to expel the bio-ink. This mechanism restricts the utilization of highly concentrated and viscous bio-inks as such viscosity reduces the supplied acoustic/pressure waves, hampering the ejection of a droplet. A low viscosity is attained by utilizing low concentration solutions, a restrictive factor for creating 3D structures.

Inkjet printing provides a high resolution of up to 50 μm. Majority inkjet bioprinters deliver high cell capability, and while there is the prospect for prompted sheer stresses to harm the cells, most investigation shows that this is not the case. The rewards of inkjet-construct bioprinting embrace high print speeds, affordable and an extensive accessibility; nevertheless difficulties consist of low droplet directionality and undependable cell encapsulation due to the low concentration of the ink.

The ultimate aim of bioprinting is to provide an alternative to autologous and allogeneic tissue implants.

3D Printing Unique Patient Implant

Figure 1.9, representing a schematic of Bioprinting Scaffolds for implanting a patient implant during clinical utilization. The steps are summarized as follows:

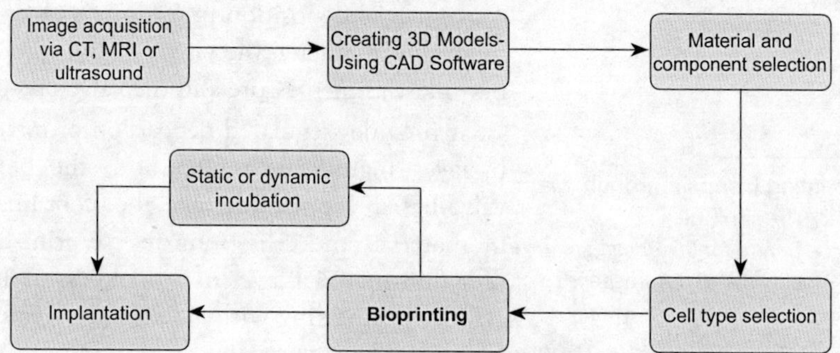

Figure 1.9: 3D Bioprinting Procedure for Unique Patient Implant

1. Digital 3D images gained from CT, MRI or ultrasound, are utilized to design a appropriate scaffold with CAD software, and 3D slicing software, Figure 1.9;
2. Materials from printing are selected contingent upon the request, and can contain polymers, ceramics, and bioactive constituents;
3. Cells are chosen reliant on the application, a bio-ink can comprise of individual or several cell kinds;
4. Post-manufacture 3D culture from a patient can be utilized for classification, valuation and eventually implantation.
5. 3D printing is time and cost fundamentals, allowing rapid alterations and implementation of designs, Figure 1.9.
6. Designs can be completed to equal exact patient defect configurations, refining the amalgamation between implant and inherent tissue, in this manner augmenting tissue amalgamation.
7. Additive manufactured scaffolds have revealed reasonable precision matching the designs.
8. Different kinds of tissues and organs have been fashioned utilizing bioprinting, for example; blood vessels, heart tissue, skin, liver tissue, neural tissue, cartilage and bone, Figure 1.10.

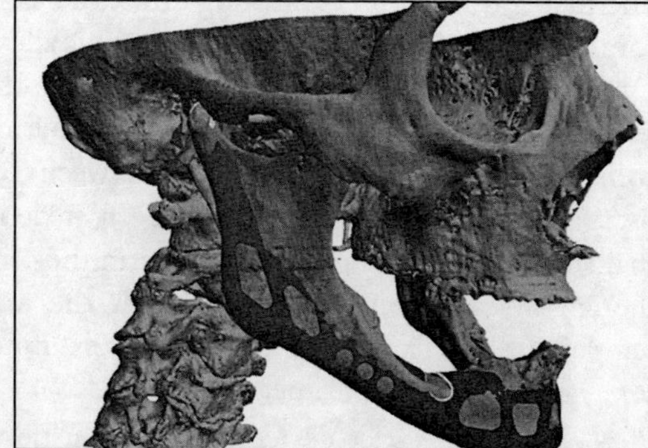

Figure 1.10: The Lower Jaw implant Bioprinted Using Titanium Powder

Courtesy: Biomedical Research Institute at Hasselt University in Belgium

3D bioprinting is a comparatively novel facet to tissue engineering and has unlocked the opportunity of generating an unparalleled biomimicry, which could eventually substitute the current gold standard of autografts. Biomimicry, in procedure and purpose, has countless implication in regenerative medicine, drug screening and sympathetic pathology. *In vitro* requests have been utilized to evaluate pathological and toxicological circumstances, as well as implant amalgamation, and compromise a methodology with a high-throughput, Figure 1.9. Biomimetic microfluidic chips have great potential in substituting animal research for drug and material screening.

Preparing Cells for 3D Bioprinting

Acquiring cells any type of cells according the need of a patient can be accomplished by any one of the stated techniques illustrated in Figure 1.11.

The stem cells cultures may be taken to be cultivated to grow and multiply in preparation to 3D Bioprinting operation, Figure 1.12.

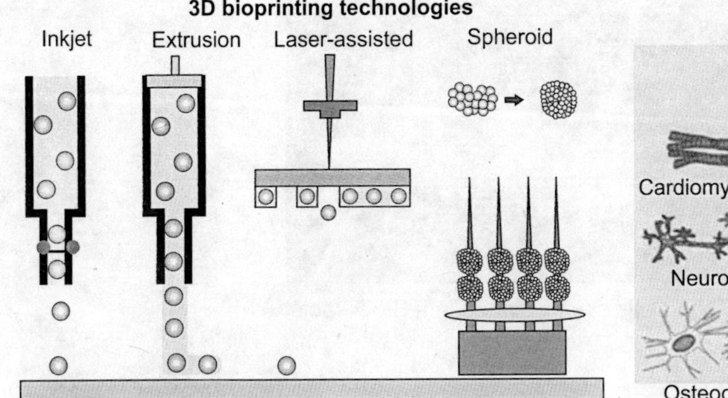

Figure 1.11: Four Different Types of 3D Bioprinting

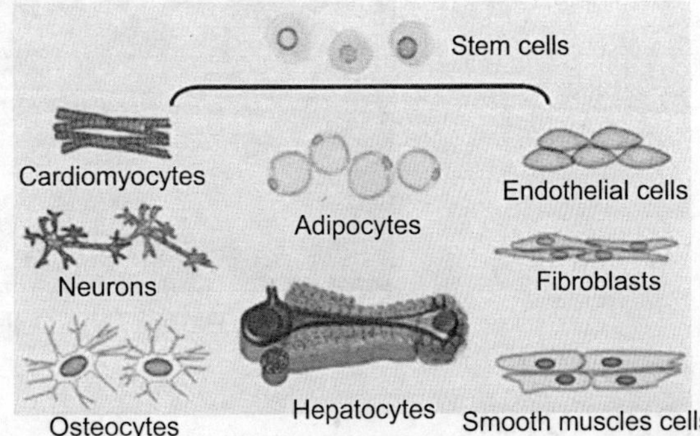

Figure 1.12: Various Types of Stem Cells in the Body

While stem cells is capable to be differentiated to take the form of any type of tissues of the body to create the necessary body organ, using specialized cells also can be cultured and grown to be used successfully in 3D Bioprinters, Figure 1.13.

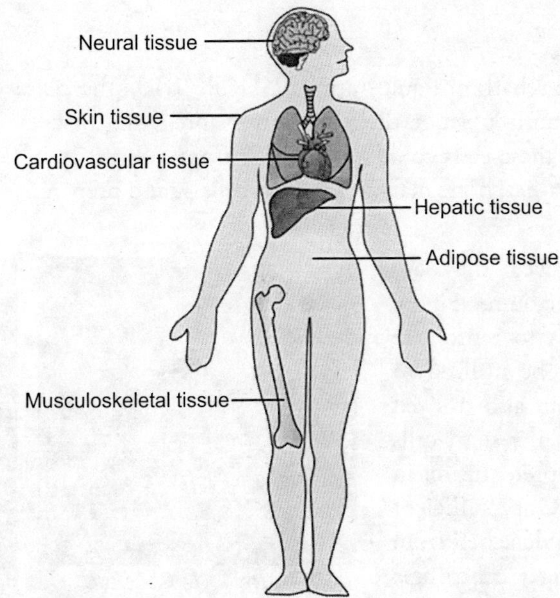

Figure 1.13: Body Tissues Required for 3D Bioprinting Operation

Most organs in the body are filled with veins, so the ability to print vascular tissue is a critical building block for complete organs. The printed veins are about to start testing in animal trials, and eventually go through human clinical trials. You may be able to replace a vein that has deteriorated, due to frequent injections of chemo treatment, for example, with custom-printed tissue grown from your own cells, Figure 1.14.

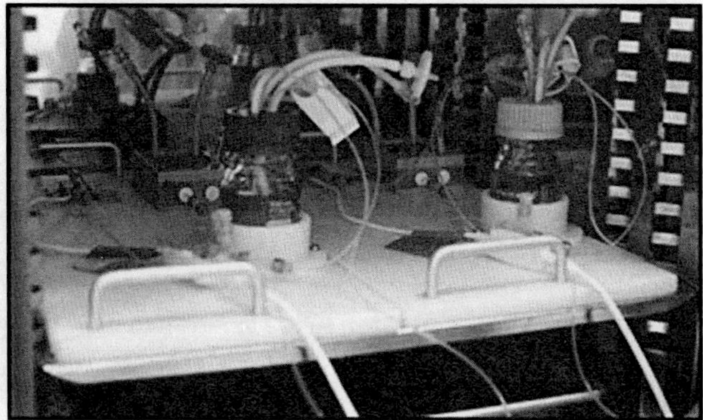

Figure 1.14: Patient Own Cells are Cultivated in a Bio Reactor

BIOREACTOR

It can be positioned inside an incubator anywhere it will be propelled with a growth medium for a few days. The bioreactor utilizes a distinct combination of chemicals, Figure 1.15, that are comparable to what cells would experience when they propagate inside the body, which will aid the cells become robust vascular tissue.

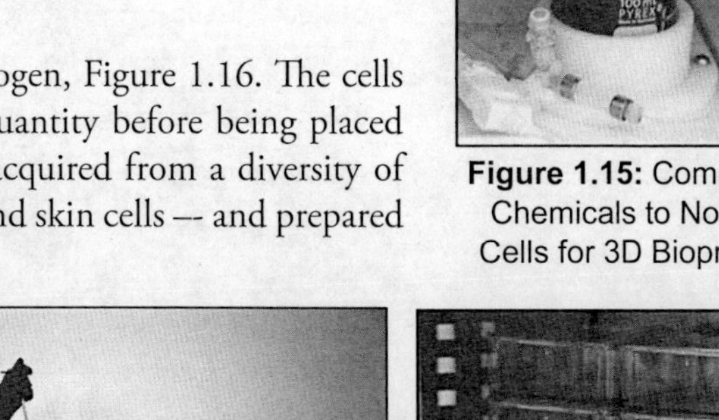

Figure 1.15: Compatible Chemicals to Nourish Cells for 3D Bioprinting

Stem Cells

Stem cells are removed from a bath of liquid nitrogen, Figure 1.16. The cells will be cultivated to significantly upsurge their quantity before being placed into the printer. Ultimately these cells could be acquired from a diversity of locations in a patient's body -- fat, bone marrow and skin cells -- and prepared into a working vein.

After the cells are thawed they are cultured in a growing medium, Figure 1.17. This permits the cells to reproduce and propagate so they can be utilized to configure veins. The medium also utilized special chemicals to express the stem cells to breed into the cell type prerequisite, in this case blood-vessel cells. Once sufficient cells are multiplied, they are detached from the growth medium utilizing a centrifuge, Figure 1.18, and compacted into pellets.

Figure 1.16: Cells are subjected to a Bath of Nitrogen

Figure 1.17: Thawed Cells Cultured in Growth Medium

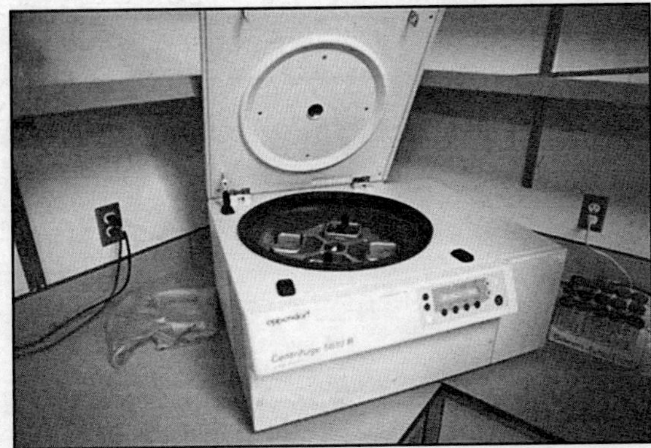

Figure 1.18: Sufficient cells detached from the Growth Medium through a Centrifuge

Hydrogel Scaffolding

The primary step of the printing procedure is to lay down a material called hydrogel, which is utilized as provisional scaffolding to upkeep the vein tissue.

The custom-made printer utilizes two pump heads that squeeze out either the scaffolding structure or the cells into a petri dish. The pump heads are attached on a precise robotic assembly for microscopic precision. The head on the right is dipping into the container of hydrogel in the photo, Figure 1.19.

A chamber compartment called a bioreactor is utilized to excite the vein, Figure 1.20. It is arranged before the vein is printed. The bioreactor is a known standard part of biotech equipment. It is fabricated out of a chunk of aluminum that wraps around a plastic container with several ports. These ports are utilized to propel in chemicals that will nourish the emergent vein, Figure 1.20.

Figure 1.19: Two Pump Heads Attached to Precision Robot

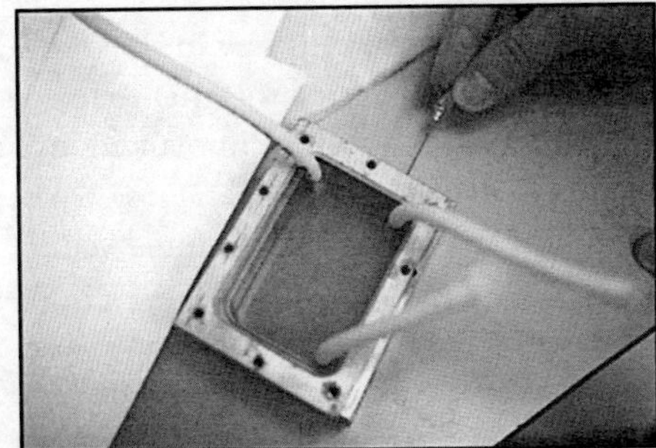

Figure 1.20: Bioreactor Chamber

Before printing the veins, tubes of the cultured cells are loaded into the robotic print head manually, like a biomass print cartridge1.19.

HYDROGEL MOLD FOR BLOOD VESSELS

Lines of the hydrogel are laid down in parallel in a trough shape on the petri dish. Then cylinders of cell pellets are printed into the trough.

One more cylinder of hydrogel is printed into the middle of the cells, which serves to create the hole inside the vein where blood will eventually flow, Figure 1.21.

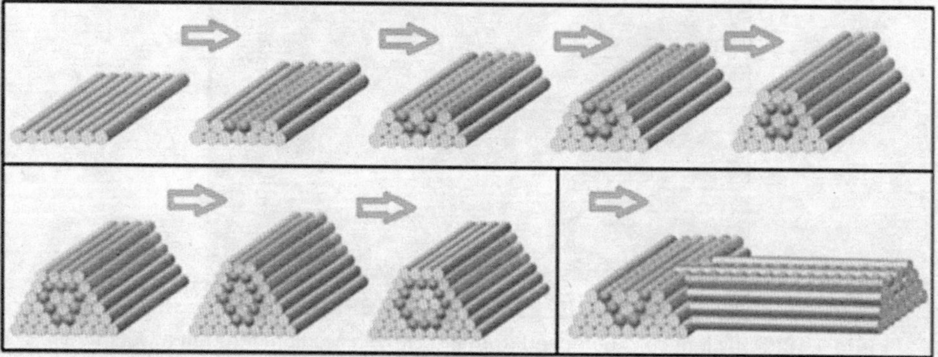

Figure 1.21: Lines of Hydrogel Laid Down, also Hydrogel Layers Printed in the Middle where Blood Flows

The printed veins are then left behind in another growing medium for a number of weeks. The cells rapidly release from the hydrogel, and an empty hollow tube of vascular cells is left behind.

VEINS FORMATION

The printed cells in tube-shaped configurations are then located into the bioreactor, Figure 1.22. The bioreactor propels a special mixture of proteins, and several other chemicals through the already printed vein, to nurture and nourish and maintain healthy survival.

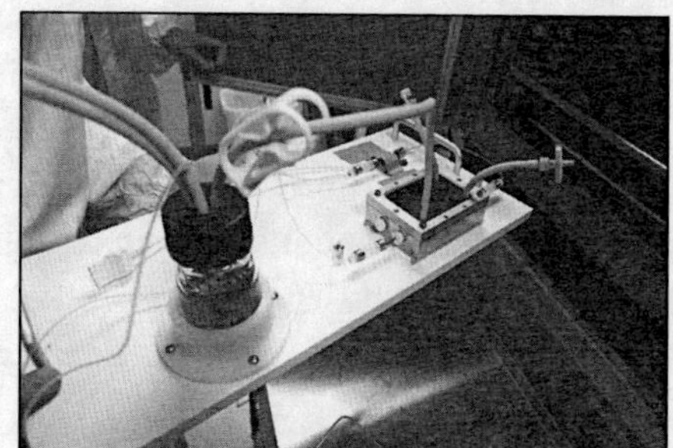

Figure 1.22: The Bioreactor Pumps Proteins and Chemicals to Nurture Printed Veins

Final Product

After the printed tubes of cells stay in the bioreactor, the pellets of cells grow and amalgamated together to form the final unique lengthy veins, which can then be implanted in the patient. Since the veins are developed from the patient's own cells, the patient's body is not likely to reject the implanted vein.

The progression of 3D printing of cells to form the necessary veins are described in figures.

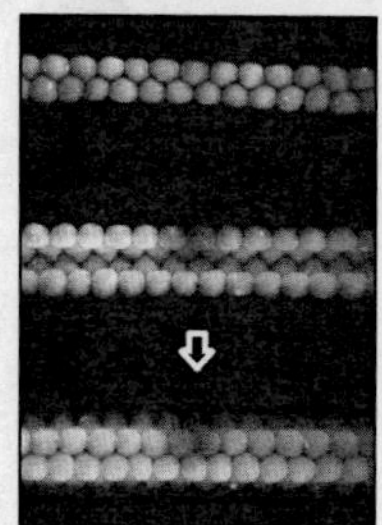

Figure 1.23: A–Layers of Cells Lined Up

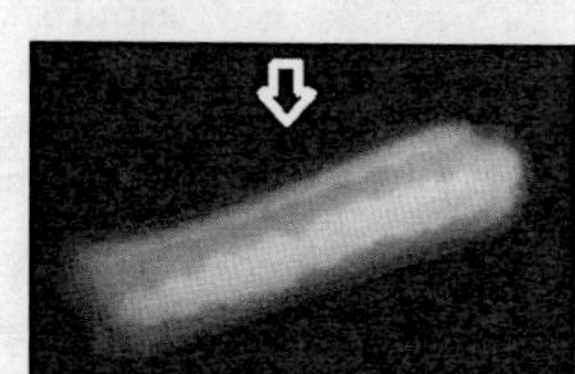

Figure 1.23 B: Lined-up Layers formed Cylindrical Shapes Supported Hydrogel Scaffolds

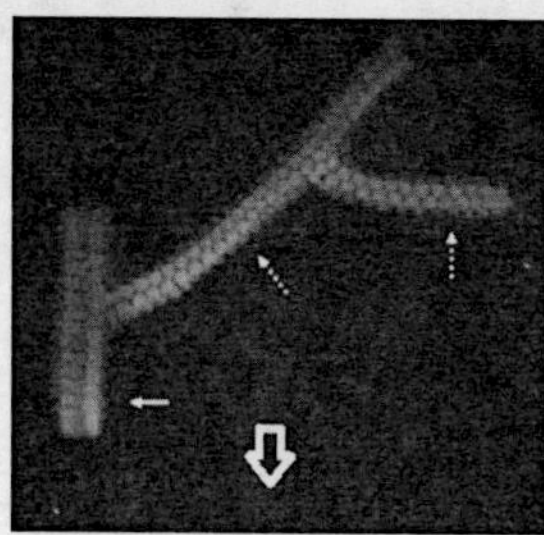

Figure 1.23C: Tree of Veins Amalgamated together

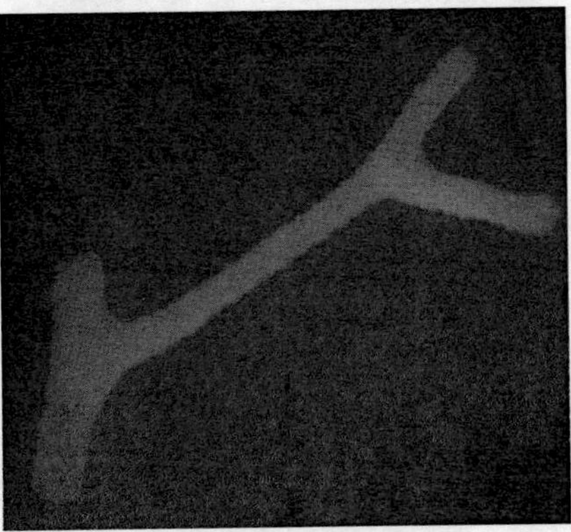

Figure 1.24: Progression of 3D Bioprinting to Form Required Veins.

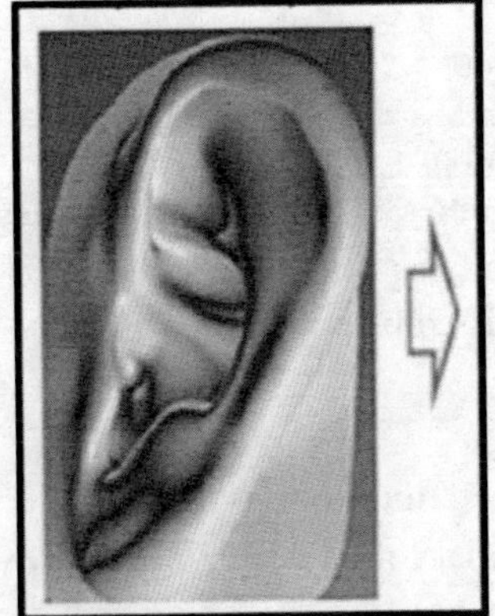

Figure 1.25A: Medical Imaging – CT-MRI – DICOM Format

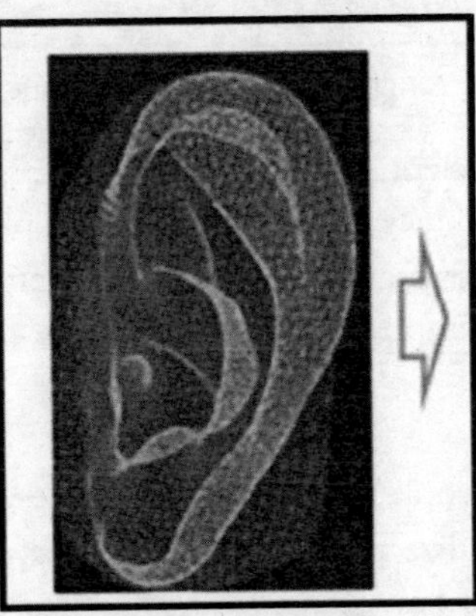

Figure 1.25B: 3D CAD Format – STL Format

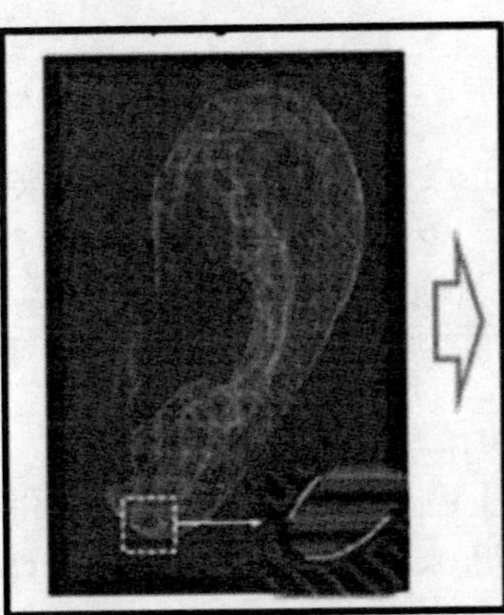

Figure 1.25C: Visualized Motion Program – Text-Based Command

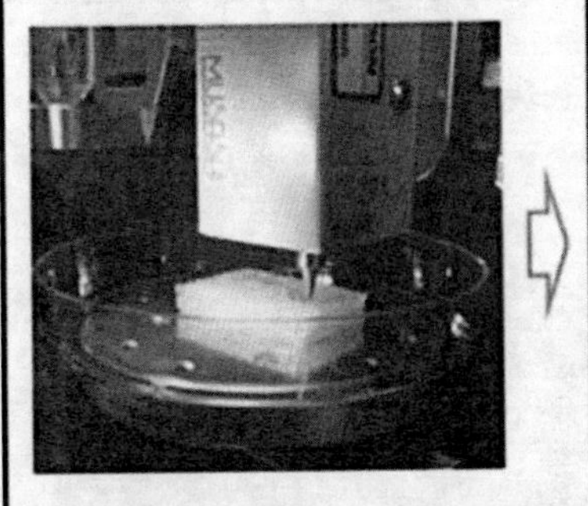

Figure 1.25D: 3D Printing Process

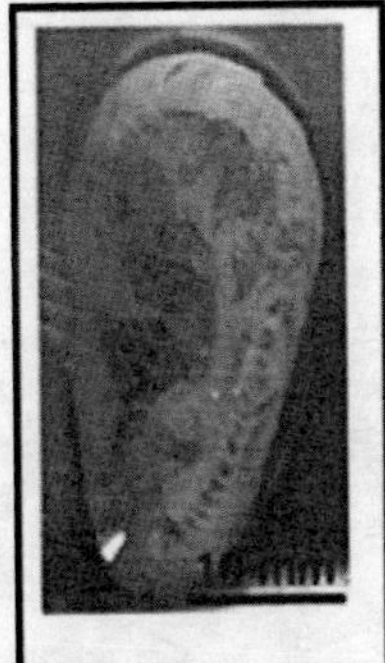

Figure 1.25E: 3D Bioprinted Tissue Product

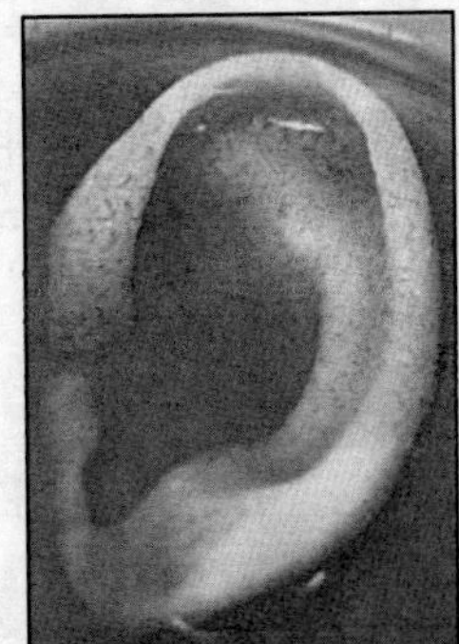

Figure 1.25F: 3D Bioprinted Human Ear

Courtesy: Wake Forest, Nature Biotechnology.

Tissue Fragment Type of 3D Bioprinting: It is the combination of bioprinting techniques with biological self-assembly. This approach exploits the intrinsic capacity of closely spaced tissue fragments to fuse together, otherwise known as tissue fluidity.

Each method has its individual rewards and downsides, and each offers exclusive trials to be overcome. Scientists forecast it will be the mixture of a number of these methods that will deliver the most substantial innovations in bioprinting, Figure 1.26.

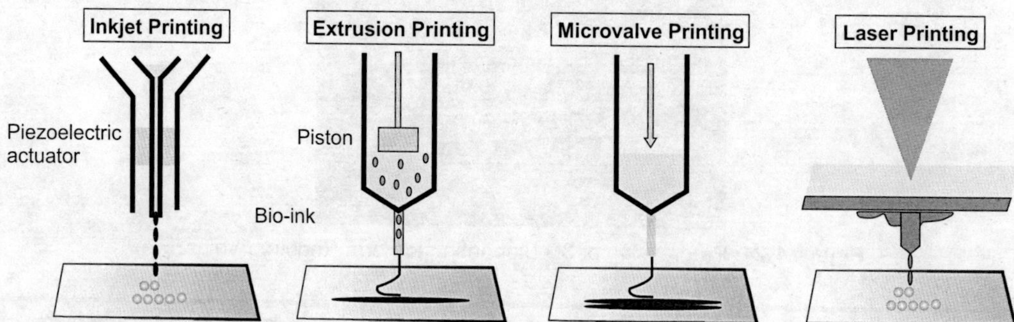

Figure 1.26: Prominent Types of Bioprinting Techniques

Essentially, researchers and medical scientists are utilizing 3D scanners and printers traditionally reserved for auto design, model building and product prototyping to in its place to generate living human tissues.

At present many Universities, scientists and private corporations around the world are all engaged in progressing bioprinting technologies. Samples of the current work are discussed below.

Bone

Each year, additional than 2.2 million people universally need bone-graft operations to restore bone defects.

Present bone implants frequently utilize a synthetic cement-based material mixed with a patient's own bone. Restrictions with these materials have caused in structures that do not have suitable mechanical reliability or permit for the formation of new tissue. Duplicating the naturally occurring bone-cartilage interface can also been challenging.

Though, scientists have advanced a bioprinting procedure that can generate a synthetic bone matrix in the exact shape of the bone needed, utilizing a biocompatible material that is both tough and regenerative. These anew 3D Bioprinted 'bones' are transplanted into the body where, over a period of a number of months, they fuse and eventually substituted by a patient's natural bones with few, if any, impediments, Figure 1.27.

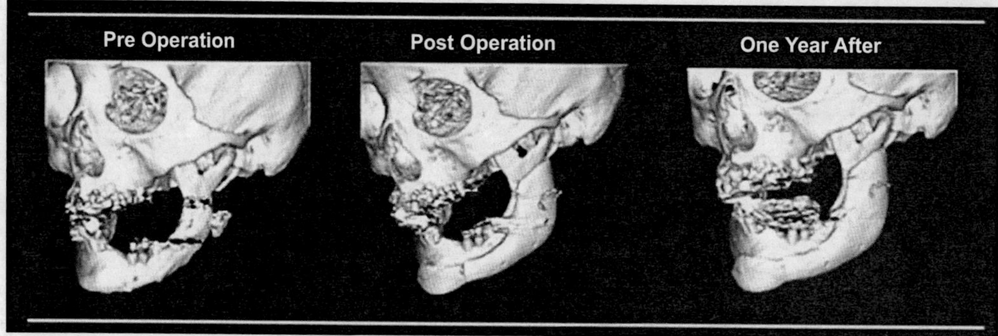

Figure 1.27: 3D Printed Calcium Phosphate Implant Converted to Real Bone by the Patient's Own Body.

To print a small bone with trabecular, which is "Cancellous bone," also known as spongy or trabecular bone is one of the two kinds of bone tissue present in the human body. Cancellous bone is present at the ends of long bones, in addition to the pelvic bones, ribs, skull and the vertebrae in the spinal column. Surgeons may soon be able to Bioprint them in the operating theatre as during the procedure.

Other scientists worldwide are working on comparable bone substitutes. They 3D bioprint a meticulous copy of the portion of the bone they are attempting to substitute—this develops the scaffold. It is then covered with adult human stem cells—able of emerging into almost any different tissue kind. This is joined with bio-ink from the printer—a amalgamation of polylactic acid, which delivers the mechanical power of bone and alginate—a gel-like material that performs as a softening material for the cells. The end product is then implanted into the body, where, within three months, the scaffold will dissolve and be substituted by novel bone.

Investigators anticipate that, in the foreseeable future, bioprinted bones could be generated with enough dependability to reinforce complex spinal restoration, and that the bone material could be additionally enhanced to augment its compatibility with cartilage cells.

Skin

If a person is severely burned, healthy skin can be acquired from another part of the body and utilized to shelter the affected area. Occasionally, though, there is not enough unharmed skin to yield, Figure 1.28.

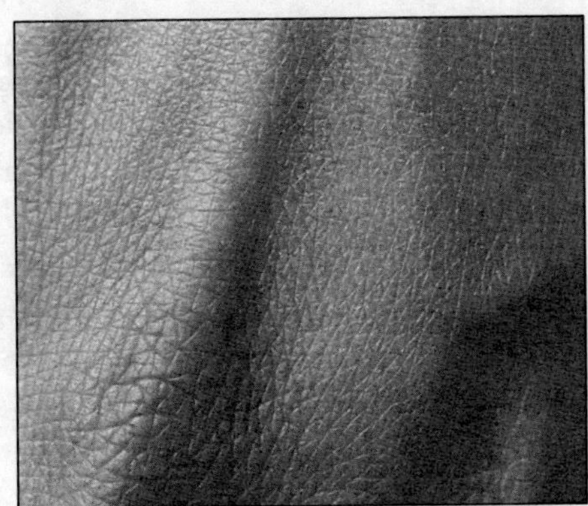

Figure 1.28: A 3D Bioprinted skin Illustrates a Close-up of the skin on a hand *Courtesy: Yale University School of Medicine.*

Several scientists at various medical schools have positively designed, built and tested a Bioprinter that can Bioprint skin cells directly onto a burń wound. A 3D scanner initially defines the dimensions and depth of the wound and this information is delivered to the printer, which, after the applicable cells have been cultured, applies the precise cell kinds at the precise depth to shelter the wound. Dissimilar outdated skin grafts, you only need a cover of skin one-tenth the size of the burn to be able to propagate enough skin cells for skin Bioprinting. Although this technology is still in the tentative phase, scientists anticipate it will be extensively obtainable shortly.

As stated, 3D Bioprinters print in layers, and since skin is a multilayered organ with dissimilar cell kinds, it is well matched to this kind of expertise. Few American corporations have already shaped multilayered skin containing of the dermis and epidermis layers. There are still several contests ahead however, mainly how to stop the heat created by the Bioprinter from harming the cells or their sustainability. Obviously, as with most fragments of the human body, skin is more intricate than it first seems—there are nerves, blood vessels, and a countless of other features that essential to be resolved.

Blood Vessels

Bearing in mind there are hundreds of thousands of kilometers of veins, arteries and capillaries in our bodies, it is worthy to learn that scientists have been occupied on ways to substitute skins should they ever fail. generating practicable blood vessels is likewise vital to lead all those other possible bioprinted body parts to work appropriately, Figure 1.29.

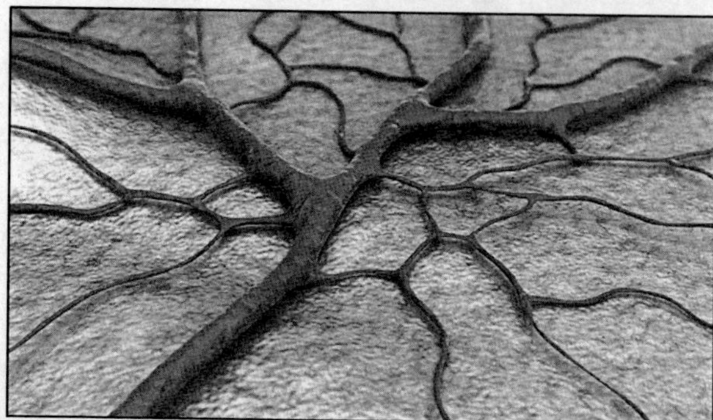

Figure 1.29: 4D Bioprinting Advanced Technology Based on 3D Bioprinting- Assimilates Stimuli-Responsive Biomaterials Fabricating Active Constructs - Alter their Forms upon Stimulation Accomplish Prearranged Functionality.

Scientists are utilizing Bioprinting to generate "living blood vessels." The materials and atmosphere generated by Bioprinters are planned in such a method as to allow small blood vessels—human capillaries—to grow on their own. This growth consumes a while. Accordingly, to aid tubes of cells and other Biomaterials are Bioprinted out to assist distribute essential nutrients to the immediate Bioprinted surroundings. Over a period of time, the self-assembled capillaries join with the Bioprinted tubes, by this means commencement to deliver nutrients to the cells on their particular manner—mimicking the way these structures perform in the human body.

Having positively generated a "disorderly" system of blood vessels, scientist are now concentrating on generating a tiered pyramid such as that which occurs in the human body.

Additional technique of generating blood vessels has been advanced by other scientists around te world. Utilizing an agarose fiber pattern sheltered with hydrogel, the scientists were competent to build microchannel networks which showed numerous architectural structures.

Internal Organs

There are presently about 4,160,000 results (1.05 seconds) on the organ transplant waiting list worldwide. Many scientists anticipate that, within 10 years, these lists will be a thing of the past. They envisage a world in which an organ can be Bioprinted and transplanted in just a few hours, deprived of rejection from the body. These Bioprinted organs will be generated from the actual cells of the patient's body they will re-enter, identical the precise configurations, characteristics and necessities of each separate patient.

At the present time, the Bioprinting of completely functional intricate internal organs, such as hearts, kidneys and livers, is at minimum 10 years away, nonetheless, progresses are going on at a fast rate.

Less intricate organs—such as bladders—are at present possible. At various Universities in the United States, scientists have successfully acquired cells from a patient's unique, ill operative bladder, cultivated them and supplemented extra nutrients. A 3D Bio-mold of the patient's bladder was then Bioprinted and the cultivated cells saturated through it. The Bio-mold was located in an incubator. When the Bioprint came out, it was capable to be transplanted into the patient's body. The Bio-mold eventually breaks down, remaining only the organic material of the bladder behind.

The same team of scientists has also successfully created sustainable urethras. Once again, a mixture of a patient's individual cells and supplementary nutrients is sowed with scaffolding and then embroidered into the correct shape of a urethra. These organized components can then be transplanted back into the patient's body.

Whereas scientists have successfully Bioprinted a very small, 1 millimeter thick x 4 millimeters wide, 3D Bio-functioning liver, it endured for only five days. It is a start, however, there is a long path to further develop.

Kidneys are also extremely challenging as they comprise intricate cellular processes for filtration—It can be effortlessly replicated. Scientists in USA have utilized human stem cells to propagate a kidney organoid that comprises all the compulsory cell types for a kidney. Such cells may deliver a valued preliminary source for Bioprinting a more intricate kidney construction.

Remarkably, hearts may be one of the simplest organs to Bioprint, Figure 1.30. Hearts are fundamentally a pump with tubing. Obviously, It is not as simple as stated, nevertheless, many scientists trust that we will have transplantable Bio-hearts before kidneys or livers. Scientists in the United States have created "organoids"—3D Bioprinted, beating cardiac cells, which are configured utilizing a particular 3D Bioprinter and are, fundamentally, completely functional, Figure 1.30.

Figure 1.30: Israeli Scientists Generated the World's First 3D-Bioprinted Heart

Curtesy - Tel Aviv University (TAU) in Israel

Even though it may be easy to Bioprint a complete duplication of an organ, it is more challenging to make it "factual," with blood vessels, joining tissues, nerves, filtration systems, power, resilience and all the features that involve in enabling our normal organs so complex. These encounters are not deferment the numerous scientists who are occupied to discover answers to these complicated problems.

Human Cartilage

The 3D Bioprinting of human cartilage should, scientists desire, guide to implantable substitutes for trauma sufferers who require reconstructive surgery, eg., ear cut off, and nose shattered—scientists have established a procedure that could allow hospitals to Bioprint a standard human nose implant in under 20 minutes, Figure 1.31. Researchers believe that any cartilage implant possibly will be shaped with their new process.

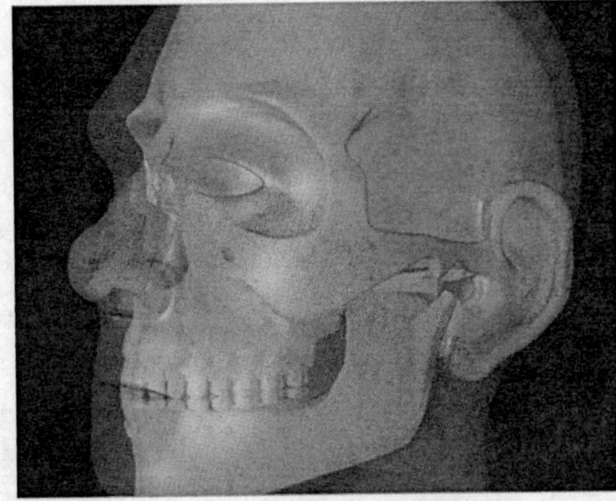

Figure 1.31: 3D Bioprinting Cell-Friendly Bioinks – Cartilage Regeneration and Biomimetic scaffolds.

Courtesy: ETH Zurich

PRACTICAL IMPLEMENTATIONS TECHNIQUE

When a serious car accident outcomes a passenger's nose being crushed. It is conceivable to rebuild this as a 3D Bio-model on the computer. Simultaneously, a biopsy is achieved on the patient, and cartilage cells acquired from the individual own body, for instance from the knee, finger, ear or splinters of the shattered nose.

The cells are deposited in the laboratory and combined with a biopolymer. A toothpaste-like suspension is from. This toothpaste-like suspension, a nose cartilage transplant is generated utilizing the bioprinter, which is implanted in the patient during surgery.

In this procedure, the biopolymer is utilized purely as a procedure of creating mold. Subsequently, the mold is broken-down by the body's own cartilage cells. Within several weeks, it is barely impossible to differentiate between the transplant and the body's own nose cartilage.

As the implant was grownup from the patient body's individual cells, the threat of rejection is far lesser than for an implant constructed of synthetic substance. An additional advantage is that the cellular implant propagates together with the patient, since it is organized by the patient's inherit growth engine, equally is the instance for other body parts. This is predominantly significant for younger implant-recipients.

DRUG TESTING - MEDICAL RESEARCH

One of the important prospective areas for utilizing bioprinted living materials is in the field of medical testing, drug investigation and safety. Many of these bioprinted tissues share mutual structures with innate tissues—such as the existence of multiple cells kinds, cellular density and important architectural properties. In this manner, scientists are capable to learn the special effects of diverse diseases, disease development and potential treatments in an innate microenvironment.

One of the greatest exhilarating current improvements is the impression of a benchtop brain, which has been established by scientists at the ARC Centre of Excellence for Electro-materials Science (ACES). They have advanced 3D Bioprinted, six-layered construction that integrates neural cells that mimic the construction of brain tissue.

This has substantial prospective advantages for scientists, pharmaceutical corporations and private businesses. It will allow them to examine novel products and drugs on tissue that precisely mirrors human brain tissue, as contrasting to animal samples which can yield a very dissimilar response. The benchtop brain could also be utilized to additional investigation into brain disorders such as schizophrenia or Alzheimer's disease.

By letting scientists to work with "real-time" human tissue, it could significantly quickens testing procedures, and yield added convincing, precise outcomes. It could also aid decrease the necessity to utilize lab animals for medical investigations, as well as possibly hazardous experiments on humans.

Mobile 3D Bioprinting

A spin-off from the well-known bioprinter is the Bio-Pen, The pen operates in a comparable manner to a bioprinter, nonetheless is far more mobile. It associates cellular material with shielding gels, which are brought via the pen head as the surgeon spreads with the ink straight on the injured segment of bone or cartilage to patch-up any impairment. As each layer is spread by the surgeon, it is subjected to UV light from a source located by the pen. This solidifies the gel so that added layers can be placed, ultimately generating a 3D framework.

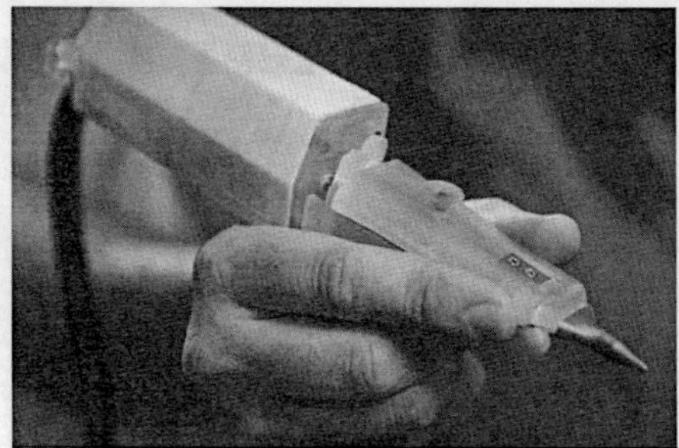

Figure 1.32: Bio-Pen Permits Surgeons to Place Bio-ink on Damaged Bone or Cartilage to repair damage.
Courtesy: University of Wollongong; used with permission.

The pen will provide surgeons a superior degree of precision and control over where they place biomaterial, decrease time in surgery and accelerate the regeneration of bone and cartilage. Once the biomaterial is distributed to the precise location specified by the surgeon, the shielding gels progressively dissolve away, substituted by the new tissue which starts to repair the area.

The Bio-pen has been used pen in clinical trials since 2013, Figure 1.32. Scientists and research teams bring together the discipline of stem cells and polymer chemistry to aid surgeons design and individualize solutions for rebuilding bone and joint flaws in real time.

The human body is a unique entity with all its various elements and much more complex than synchronized mechanisms with unique materials. A human organ has an intricate web of cells, tissues, nerves and structures that necessitate precise positioning for the organ to function appropriately. The positioning the countless of miniscule capillaries in a liver organ, and to literally developing a Bioprinted heart to regularly pulse and contract—it may be a long, exceptionally challenging and very problematic process. Although some parts of the human body are extra intricate than others—an ear or nose is simpler to achieve than a bladder or a urethra, though the higher order organs as kidneys, heart and liver are at the top of the complexity rating—each organ has its individual dedicated necessities and concerns that essential addressing. Additionally, one must address the choice and accessibility of the precise materials, cell kinds, and generating Bioink that encounter the dual burdens of both printability and cell functionality.

Furthermore, there is the necessity to consider in accounting dissimilar growth rates and differentiation issues, as well as emerging the reliable software and hardware to allow it all come together.

Even after scientists trust they have eminent success, bioprinted technologies—like all novel medical treatments—are still necessary to successfully pass safety trials and appropriate regulatory procedures before they can be presented to the public.

Occupied with all these intricacies necessitates the amalgamation and input of technologies from fields together with engineering, biomaterials science, cell biology, physics and medicine, where patience is a virtue.

Assumed the Bioprinting interdisciplinary criteria, 3D Bioprinting is quickening at an ever- accelerating rate. It is a blissful times, though, we essentially must be careful to temper our prospects of this technology with its authenticities. The human body is exceedingly intricate, so as exasperating to duplicate the many elements that it does is cumbersome, stimulating labor. Those employed in the 3D Bioprinting field are achieving developments every day, in both the technology and in their insight of how it may be utilized and enriched. While we have not yet reached the end target of our technological marathon, there is no doubt that the future of medicine will be very different with our advancement in Bioprinting involved technology.

The synthetic generation of human skin, tissue and internal organs may appear like a ahead of its time dream but, unbelievably, much of it is occurring right now, albeit still in the early phases. In research facilities and hospitals around the world, innovations in 3D printing and, more predominantly, Bioprinting are establishing novel preferences for treatment and scientific study. Undeniably, bioprinting has the prospective to be the next big revolution in health care and made to order medicine.

SUMMARY

- Bioprinting is an extension of traditional 3D printing.
- Bioprinting can produce living tissue, bone, blood vessels and, potentially, whole organs for use in medical procedures, training and testing.
- The cellular complexity of the living body has resulted in 3D bioprinting developing more slowly than mainstream 3D printing.
- Bioprinting technology could provide the opportunity to generate patient-specific tissue for the development of accurate, targeted and completely personalized treatments.
- There is still a long way to go before we can create fully functioning and viable organs for human transplant.

REFERENCES

1. Boland T, Wilson WC, Xu T. Ink-jet printing of viable cells. Google Patents 2006. https://www.google.com/patents/US7051654.
2. Murphy SV, Atala A. 3D bioprinting of tissues and organs. Nature biotechnology 2014; 32:773–85.
3. Dimri GP, Lee X, Basile G et al. A biomarker that identifies senescent human cells in culture and in aging skin *in vivo*. Proceedings of the National Academy of Sciences 1995;92:9363–7.
4. Shi Y, Inoue H, Wu JC, Yamanaka S. Induced pluripotent stem cell technology: a decade of progress. Nature reviews Drug discovery 2016; 16:115–130.
5. Tricomi BJ, Dias AD, Corr DT. Stem cell bioprinting for applications in regenerative medicine. Ann N Y Acad Sci 2016; 1383:115–24.
6. Jung JP, Bhuiyan DB, Ogle BM. Solid organ fabrication: comparison of decellularization to 3D bioprinting. Biomater Res 2016;20:27.
7. Jakab K, Norotte C, Marga F, Murphy K, Vunjak-Novakovic G, Forgacs G. Tissue engineering by self-assembly and bio-printing of living cells. Biofabrication 2010;2:022001.
8. Bauwens CL, Peerani R, Niebruegge S et al. Control of Human Embryonic Stem Cell Colony and Aggregate Size Heterogeneity Influences Differentiation Trajectories. STEM CELLS 2008; 26:2300–10.
9. Koch L, Kuhn S, Sorg H et al. Laser printing of skin cells and human stem cells. Tissue Eng Part C Methods 2010; 16:847–54.
10. Gaebel R, Ma N, Liu J et al. Patterning human stem cells and endothelial cells with laser printing for cardiac regeneration. Biomaterials 2011;32:9218–30.
11. Gaetani R, Doevendans PA, Metz CH et al. Cardiac tissue engineering using tissue printing technology and human cardiac progenitor cells. Biomaterials 2012; 33:1782–90.
12. Gaetani R, Feyen DAM, Verhage V et al. Epicardial application of cardiac progenitor cells in a 3D-printed gelatin/hyaluronic acid patch preserves cardiac function after myocardial infarction. Biomaterials 2015; 61:339–48.
13. Zhang YS, Arneri A, Bersini S et al. Bioprinting 3D microfibrous scaffolds for engineering endothelialized myocardium and heart-on-a-chip. Biomaterials 2016;110:45–59.
14. Jang J, Kim TG, Kim BS, Kim SW, Kwon SM, Cho DW. Tailoring mechanical properties of decellularized extracellular matrix bioink by vitamin B2-induced photo-crosslinking. Acta Biomater 2016; 33:88–95.
15. Pati F, Jang J, Ha DH et al. Printing three-dimensional tissue analogues with decellularized extracellular matrix bioink. Nat Commun 2014; 5:3935.
16. Prestwich GD. Hyaluronic acid-based clinical biomaterials derived for cell and molecule delivery in regenerative medicine. J Control Release 2011;155:193–9.
17. Young JL, Engler AJ. Hydrogels with time-dependent material properties enhance cardiomyocyte differentiation *in vitro*. Biomaterials 2011; 32:1002–9.
18. Kang LH, Armstrong PA, Lee LJ, Duan B, Kang KH, Butcher JT. Optimizing Photo-Encapsulation Viability of Heart Valve Cell Types in 3D Printable Composite Hydrogels. Ann Biomed Eng 2017; 45:360–77.

19. Ryu S, Yoo J, Jang Y et al. Nanothin Coculture Membranes with Tunable Pore Architecture and Thermoresponsive Functionality for Transfer-Printable Stem Cell-Derived Cardiac Sheets. ACS Nano 2015; 9:10186–202.
20. Ong CS, Fukunishi T, Nashed A et al. Creation of Cardiac Tissue Exhibiting Mechanical Integration of Spheroids Using 3D Bioprinting. J Vis Exp. 2017 doi: doi:10.3791/55438.
21. Williams SK, Touroo JS, Church KH, Hoying JB . Encapsulation of adipose stromal vascular fraction cells in alginate hydrogel spheroids using a direct-write three-dimensional printing system. Biores Open Access 2013; 2:448–54.
22. Cheung DY, Duan B, Butcher JT. Current progress in tissue engineering of heart valves: multiscale problems, multiscale solutions. Expert Opin Biol Ther 2015; 15:1155–72.
23. Schmidt D, Dijkman PE, Driessen-Mol A et al. Minimally-invasive implantation of living tissue engineered heart valves: a comprehensive approach from autologous vascular cells to stem cells. Journal of the American College of Cardiology 2010; 56:510–20.
24. Dijkman PE, Driessen-Mol A, Frese L, Hoerstrup SP, Baaijens FP. Decellularized homologous tissue-engineered heart valves as off-the-shelf alternatives to xeno- and homografts. Biomaterials 2012; 33:4545–54.
25. Bajaj P, Schweller RM, Khademhosseini A, West JL, Bashir R . 3D biofabrication strategies for tissue engineering and regenerative medicine. Annu Rev Biomed Eng 2014; 16:247–76.
26. Xin Y, Chai G, Zhang T et al. Analysis of multiple types of human cells subsequent to bioprinting with electrospraying technology. Biomed Rep 2016; 5:723–30.
27. Gao G, Cui X. Three-dimensional bioprinting in tissue engineering and regenerative medicine. Biotechnol Lett 2016; 38:203–11.
28. Jana S, Lerman A . Bioprinting a cardiac valve. Biotechnology Advances 2015;33:1503–21.
29. Tian B, Liu J, Dvir T et al. Macroporous nanowire nanoelectronic scaffolds for synthetic tissues. Nat Mater 2012; 11:986–94.
30. Shin SR, Jung SM, Zalabany M et al. Carbon-Nanotube-Embedded Hydrogel Sheets for Engineering Cardiac Constructs and Bioactuators. ACS Nano 2013; 7:2369–80.
31. Soloman, Sabrie "Sensors Handbook, 2nd Edition, McGraw-Hill Publishing Co., NY – ISBN 978-0-07-160570-0
32. Soloman, Sabrie "Sensors and Control Systems in Manufacturing 2nd Edition," McGraw-Hill Publishing Co., NY – ISBN 0-07-059626-3
33. Soloman, Sabrie "Affordable Automation," McGraw-Hill Publishing Co., NY – ISBN 0-07-059633-6
34. Soloman, Sabrie "Sensors Handbook, 1st Edition," McGraw-Hill Publishing Co., NY – ISBN 0-07-059630-1
35. Soloman, Sabrie "Sensors and Control Systems in Manufacturing 1st Edition," McGraw-Hill Publishing Co., NY – ISBN 978-07-10572-4

2 The Bioprinting Revolution

THE ROLE OF STEM CELLS - 3D BIOPRINTING

Stem Cells are the bed-rock of 3D Bioprinting technology, which is tailored to treat multitudes of diseases. Thus, understanding Stem cells will equip the reader with the vast potential for use in therapy.

NOMENCLATURES

Pluripotent: (of an immature or stem cell) capable of giving rise to several different cell types. Many types of cells originate from pluripotent bone marrow stem cells.

Somatic Stem Cell: Somatic cells are any cell in the body that is not sperm or egg

iPSCs: Induced Pluripotent Stem Cells. They are a type of pluripotent stem cell that can be generated directly from adult cells.

Blastocyst: It is a thin-walled hollow structure in early embryonic development that contains a cluster of cells called the inner cell mass from which the embryo arises.

Proliferate: to grow by rapid production of new parts, cells, buds, or offspring; or to increase in number as if by *proliferating* : multiply.

hESCs: *Human embryonic stem cells, hESCs* are derived from inner cell mass of blastocyst-stage embryos, and exhibit unlimited proliferation ability and pluripotency to differentiate into various cell types originated from the three germ layers.

Teratoma: benign tumor - is a tumor made up of several different types of tissue, such as hair, muscle, teeth, or bone. They typically form in the ovary, testicle, or tailbone and less commonly in other areas. Mature teratomas include dermoid cysts and are generally benign. Immature teratomas may be cancerous

Fibroblasts: Cells that yield connective tissues. It plays a critical role in wound-healing. Fibroblasts are the most common cells of connective tissue in animals.

Therapeutic Cloning: Another type of patient-specific stem cell. The production of embryonic stem cells for use in replacing or repairing damaged tissues or organs, achieved by transferring a diploid nucleus from a body cell into an egg whose nucleus has been removed.

Differentiation: It is the process that allows unspecialized stem cells give rise to specialized cells. The production of embryonic stem cells for use in replacing or repairing damaged tissues or organs, achieved by transferring a diploid nucleus from a body cell into an egg whose nucleus has been removed.

HSC: Hematopoietic stem cells

MAPC: Multipotent Adult Progenitor Cell. The therapy Improves Learning after Brain Injury - after the initial responses to Traumatic Brain Injury (TBI).

BMESL: Bone Marrow-Derived Embryonic Stem-Like - Immunotherapies uses the body's own immune system to fight cancer, instead of directly attacking cancer cells.

Riboflavin: It has the ability to create nontoxic effects; it is used particularly in 3D printing of medical grafts

SCNT: Somatic cell nuclear transfer. It is a laboratory strategy for creating a viable embryo from a body cell and an egg cell. The technique consists of taking an enucleated oocyte (egg cell) and implanting a donor nucleus from a somatic (body) cell.

Cell Structure - Tissue Engineering in 3D Bioprinting

At present, the majority of the world populations hardly understand the basics of the types of stem cells. The considerable potential of stem cells therapy may be missed or even ignored altogether. Scientists are deeply occupied on discovering novel ways to use stem cells to cure diseases and heal injuries.

Somatic Cells

Somatic stem cells are often called adult stem cells. They are present naturally in our bodies. They are significant for growing, curative, and substituting cells that are vanished through everyday wear and tear, Figure 2.1.

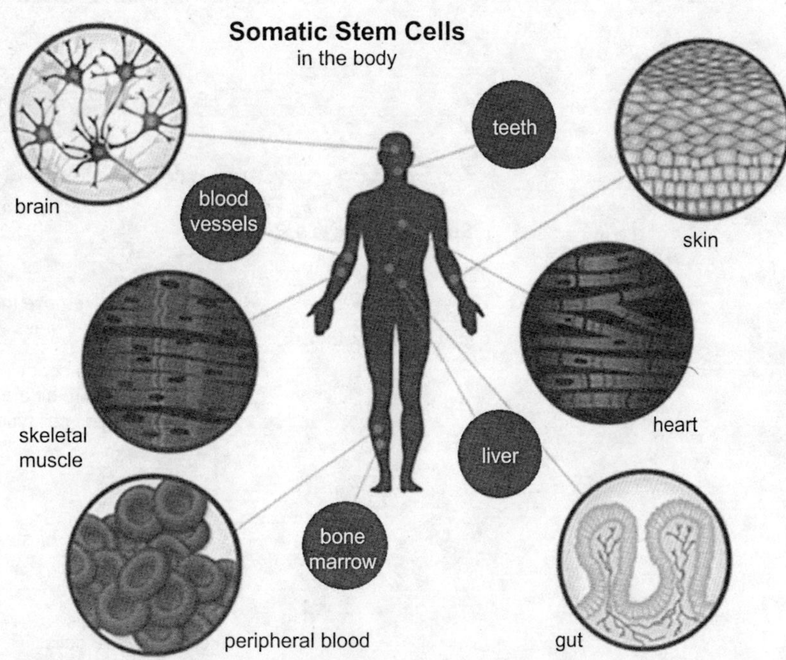

Figure 2.1: Locating Stem Cell in Human Body

Therapeutic Stem Cells

Stem cells generated from the blood and bone marrow are characteristically utilized as a treatment for blood-related illnesses. Though, under normal conditions somatic stem cells can develop only a subcategory of correlated cell types. Bone marrow stem cells, for instance, differentiate mainly into blood cells. This partial differentiation can be a benefit when you need to harvest blood cells; but it is a drawback if you are concerned in generating an unrelated cell type.

Vital Considerations

The majority types of somatic stem cells are problematic to isolate and propagate in culture. Isolation of many kinds could result in substantial tissue or organ destruction, as in the heart or brain. Somatic stem cells can be transplanted from donor to patient, but deprived of drugs that overpower the immune system; a patient's immune system will identify transplanted cells as alien and attack them.

Ethical Therapy

Therapy including somatic stem cells is not contentious; though, it yields to the same ethical attentions that put on to all medical processes.

Embryonic Stem Cells ESc

Embryonic stem (ES) cells are fashioned as a regular part of embryonic growth. They can be isolated from an early embryo and full-grown in a dish, Figure 2.1.

Potential therapy of Embryonic Stem Cells

Embryonic Stem cells have the considerable possibility to become any kind of cell in the human body, enabling them a favorable foundation of cells for healing many sicknesses, Figure 2.2.

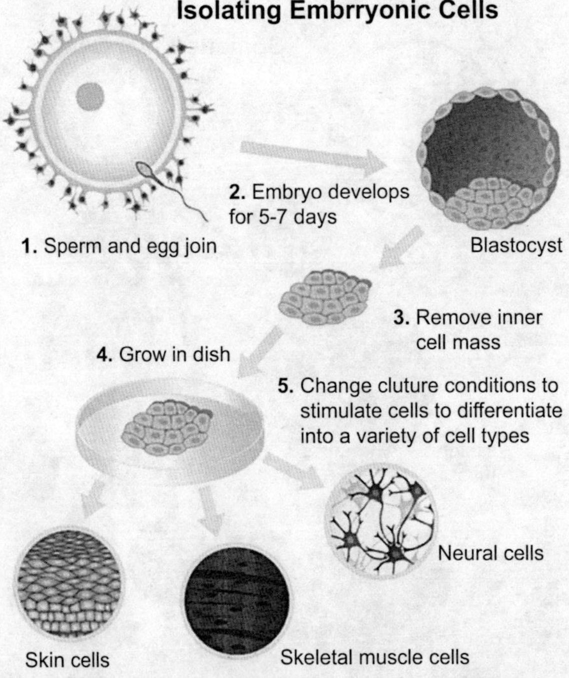

Figure 2.2: Isolating Embryonic Stem Cells

Suppressing Immune System

Deprived of drugs that overpower the immune system, a patient's immune system will identify transplanted cells as foreign alien and attack them.

ETHICS OF EMBRYO DESTRUCTION

Human embryos are living, breathing, functioning, and growing human beings at the moment of inception. As soon as scientists and medical researchers isolate human embryonic stem (hES) cells in the lab, undoubtedly, they destroy the life entity of a human being. The moral and legal consequences of this have led some unwilling individual to sustain research concerning hES cells. In current years, some scientists have concentrated their determinations on generating stem cells that do not necessitate to kill the embryos.

Induced Pluripotent Stem Cells

Induced pluripotent stem (iPS) cells are generated synthetically in the lab by "reprogramming" a patient's own cells. iPS cells can be developed from readily accessible cells counting fat, skin, and fibroblasts, cells that yield connective tissues, Figure 2.3.

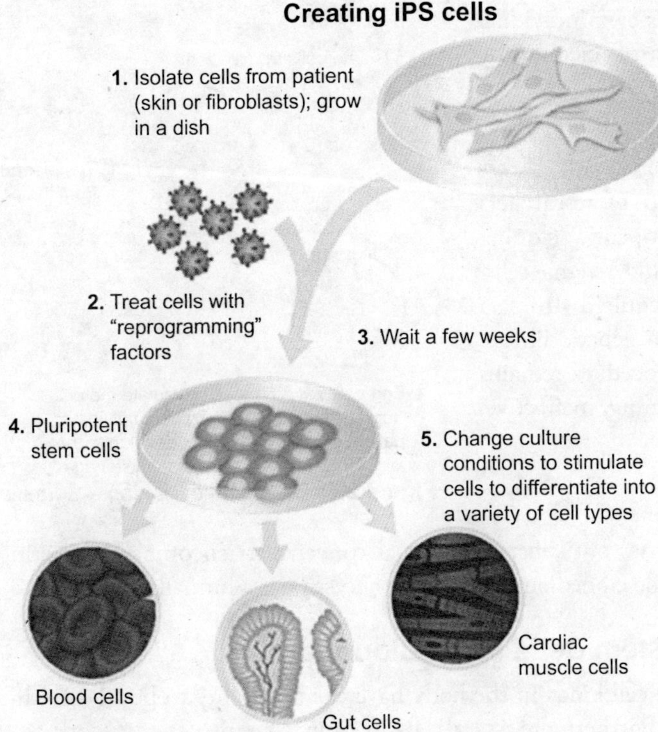

Figure 2.3: Creating Induced Pluripotent Stem Cells

Pluripotent Stem Cells as Therapy

The iPS cells can transform to any cell in the body, or even the whole body. While nuerous analyses are required, the same appears to be true among all human iPS cells, enabling them a favorable source of cells for healing numerous illnesses. Prominently, subsequently iPS cells can be cultivated from a patient's individual own cells, there is no endangerment that their immune system will discard them.

iPS cells are much affordable to generate than ES cells created through therapeutic cloning, another kind of patient-specific stem cell. Though, as the "reprogramming" process presents genetic alterations, the welfare of utilizing iPS cells in patients is indeterminate.

Ethical Therapy

Therapy including iPS cells is not contentious; though, it yields to the same ethical attentions that put on to all medical processes.

Therapeutic Cloning

Therapeutic cloning is a method for creating patient-specific embryonic stem (ES) cells.

Potential as therapy

Therapeutic cloning can, in concept, creates ES cells with the prospective to transform to any sort of cell in the body. Furthermore, meanwhile these cells are acquired from a patient's own individual DNA; there is no harm of rejection by the immune system.

Creating ES cells
Through Theraputic Cloning
Somatic Cell Nuclear Transfer

1 Isolate cells from patient
2 Remove nucleus from an egg cell
3 Transfer nucleus (contains DNA) from the patient's cells to the egg
4 Egg cell "reprograms" the patient's DNA
5 Stimulate the cell to begin dividing; let develop to the blastocyst stage
6 Isolate the inner cell mass from the blastocyst and grow it in a dish

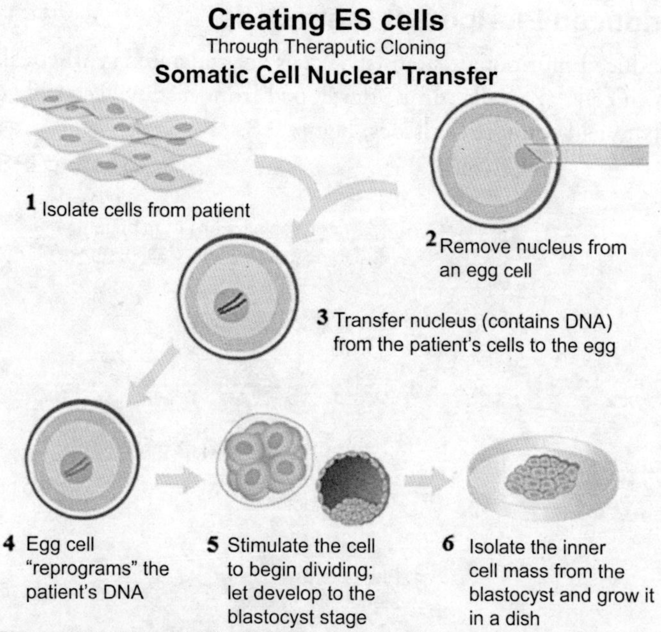

Figure 2.4: Therapeutic Cloning – Somatic Cell Nuclear Transfer

Slow Progress

The first time, a group of researchers in 2013, utilized therapeutic cloning to transform embryonic stem cells. The donor nucleus acquired from a child with a uncommon genetic illness. Though, the cloning procedure remains intense and time consuming, ineffective, and expensive.

Ethical Therapy

Therapeutic cloning carries out substantial moral concerns. It encompasses creating a clone of a human being and terminating the cloned embryo, and it necessitates a human egg donor.

The Vital Role of Stem Cells in 3D Bioprinting

Curiously, many diverse cell kinds in the body have the possibility to be advanced from stem cells during early life and progress. Furthermore, stem cells in numerous tissues can work as an interior restoration mechanism, as they divide deprived of restriction to replace other cells as long as the human or animal continue living. Upon the division of the stem cell, each new cell has the prospective either to develop to another sort of cell with a more specialized purpose or remain a stem cell, such as a brain cell, a muscle cell, or a red blood cell.

Distinguishing Cells

There are two important characteristics to distinguish stem cells from other cell types:

1. They are ***unspecialized*** cells capable of renewing themselves through cell division, sometimes after long periods of inactivity.
2. They can be induced to become ***tissue-*** or ***organ-specific*** cells with special functions under certain experimental or physiologic conditions.

In many instances, stem cells are regularly divided in several organs, such as bones and gut, and bone marrow to repair and replace worn out or damaged tissues. While, in other organs, such as the heart and pancreas, stem cells only divide under special circumstances.

Scientists primarily conducted their research work on two kinds of stem cells from humans and animals:

1. ***Embryonic*** stem cells
2. ***Non-Embryonic "Somatic"*** or ***"Adult"*** Stem cells.

Human Embryonic Stem Cells

In 1981 scientists exposed approaches to develop embryonic stem cells from early mouse embryos. The particulars of this encounter of the mouse biology stem cells directed to a method to develop stem cells from human embryos. These embryos were grown-up in the laboratory. This is the beginning of human embryonic-stem cells.

Adult Cells Becoming Stem Cells

In 2006, researchers accomplished another discovery by recognizing surroundings that would allow some specialized adult cells to be "reprogrammed" genetically to accept the role of a stem cell-like state. The name of this new induced type of stem cell is pluripotent stem cells (iPSCs).

There are many causes making the stem cells to be important for living organisms. The embryo after few days spreads to become a blastocyst, whereas the inner cells progress to be the body of the organism. This contains all of the many specialized cell kinds and organs such as the lung, heart, sperm, skin, eggs, and other tissues. In numerous adult tissues, distinct populations of adult stem cells produce substitutes for cells that are vanished through normal wear and tear, damage, or sickness such as muscle, bone marrow, and brain.

Stem Cells Treat Diseases – 3D Bioprinting Applications

Stem cells provide novel abilities for healing illnesses due to their unique regenerative capacities, such as healing diabetes, Figure 2.5, and heart disease. Similarly, stem cells provide considerable developments in the utilization of 3D Bioprinting applications. Extensive work has been accomplished in the laboratory, as well as many worldwide clinics to utilize stem cell-based therapies to treat disease, which is also referred to as regenerative medicine.

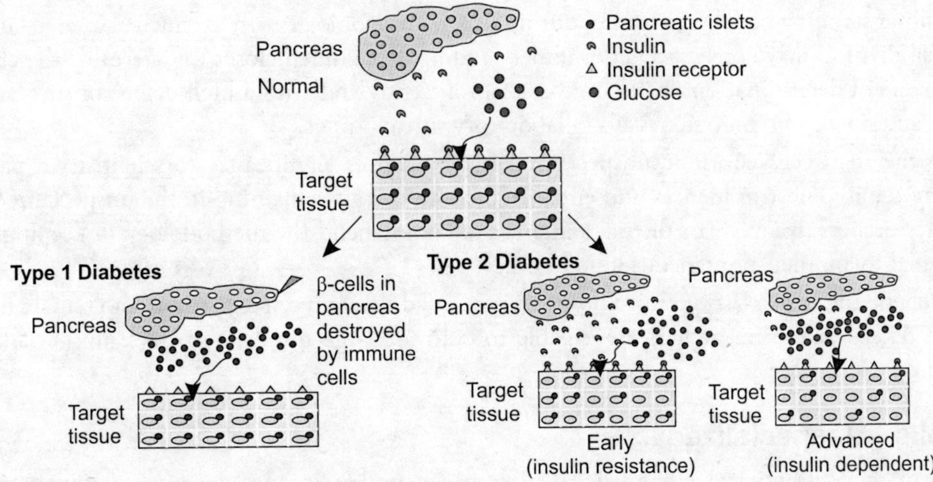

Figure 2.5: Schematic Diagram Differentiating Normal and Two Types of Pancreatic Diabetics.

Stem Cells – Regenerative Medicine

The stem cells' essential properties were extensively studied by many scientists in many clinics, which enabled them to acquire vital information that guided the cells to behave different from specialized cell types. Additionally, scientists have used stem cells shelter new drugs as well as developing model systems leading to further understanding of organs normal growth and identifying the causes of human birth defects.

Scientists continue to study stem cells in depth hoping to discover how an organism may develop from a single cell. Also, and how healthy cells are replacing damaged cells in adult organisms.

Properties of Stem Cells

There are many types of body cells differ than stem cells. There are three primary properties of all stem cells—irrespective of their source—:

1. They are capable of dividing and renewing themselves for long periods;
2. They are unspecialized; and
3. They can develop to be specialized cell types.

Stem cells have the capacity to split and repeat dividing themselves for long periods, whereas muscle cells, blood cells, or nerve cells—do not generally duplicate themselves. Stem cells have the capability to duplicate many times (proliferate). Numerous medical clinics and biological laboratories around the world may colonize stem cells (proliferate) for few months in the laboratory, which may harvest several millions of cells. If the regenerated cells endure to be unspecialized, like their source stem cells, they are recognized to be adept of long-term self-renewal.

There are two important issues of stem cells laboring scientists:

1. Stem cells characteristic of long-term self-renewal. Embryonic stem regenerate cells for few years in the laboratory without differentiating.
2. Non-embryonic stem cells do not have long term self-renewal. The factors of the living organisms to usually regulate stem cell proliferation and self-renewal are still under inquiries.

Closer to Understand Cancer

Profound understanding of cell generation during usual embryonic growth as much as understanding the irregular cell division may convey to better understanding to be much closer to determine the creation of Cancer. The main determination of the study is to proficiently and with a high degree of success produce embryonic and non-embryonic stem cells in laboratory surroundings.

Having the stem cells continue unspecialized is of infallible significance to scientists in particularly determining the foremost influences and circumstances to permit remaining in the unspecialized state. It has taken researchers twenty years of trial and error to comprehend the methodology in keeping the stem cell unaffected to another spontaneous kind of cells.

Comprehend the signals that permit a non-embryonic (adult) stem cell to generate and remain unaffected unspecialized cells before researchers are capable to cultivate large numbers of those unspecialized adult stem cells in the laboratory.

Stem cells - Unspecialized

An significant and essential characteristic of a stem cell is that it does encompass any specific tissue arrangements to permit it to alteration to specialized properties. The stem cell in this event does not cooperate with its adjacent cells to pump blood via the body as in instance of the heart muscle cell. It cannot transport oxygen molecules through the bloodstream as in the instance of red blood cell. However, unspecialized stem cells can be transformed to specialized cells, counting heart muscle cells, blood cells, or nerve cells.

Differentiation

Differentiation is the procedure that permits unspecialized stem cells give rise to specialized cells. There are a number of phases the cells become more specialized whereas differentiating. They are progressed to be more specialized at each step. Researchers are just commencement to comprehend the internal and external cell signals that can trigger each step of the differentiation procedure.

Cells Genes Control Internal Signals

The cell's genes are interspersed across long strands of DNA. This strand of DN carry coded instructions for all cellular structures and functions, while the external signals for cell differentiation allow certain chemicals to secrete by other cells as well as by the physical contact with neighboring cells, combined with certain molecules in the microenvironment.

The cell's DNA acquires epigenetic marks during the differentiation interaction of signals, which restricts DNA expression in the cell. This restriction can be passed on through cell division.

Adult stem cells characteristically generate the same cell types of the tissue in which they belong. Within the bone marrow a blood-forming adult stem cell usually, gives rise to many types of blood cells. It is commonly accepted that a blood-forming cell in the hematopoietic bone marrow—stem cell—cannot give rise to the cells of a very different tissue, e.g., such as nerve cells in the brain. Several scientists claimed that stem cells from one tissue may give rise to cell types of a completely different tissue. Adult stem cells will play a fundamental role in generating necessary sells for organ development and 3D printing as well as the full potential as future therapies.

Embryonic Stem (ES) From Fertilized Eggs

Cells are resulting from embryos fertilized eggs advanced *in vitro* through an *in vitro* fertilization facilities. Embryos are then contributed for research devotions. The embryos in this circumstance are not resulted from eggs fertilized in a woman's body.

Embryonic Stem Cells Grown in Laboratories

Cell culture is developing cells in the laboratory. "Human Creating Embryonic Stem cells (hESCs)" takes place by transferring cells from a preimplantation-stage embryo into a plastic laboratory culture dish. The dish contains a nutrient soup known as culture medium. The surface of the dish contains divided cells spread over the surface. A mouse embryonic skin cells coats the inner surface of the culture dish. The culture dish is treated to prevent them dividing. This coating layer of cells is often called a "feeder layer." In the bottom of the culture dish lays the mouse cells to deliver the cells a gummy clinging surface to so they can attach. Additionally, the feeder cells release nutrients into the culture medium. However, scientists have developed ways to produce embryonic stem cells without mouse feeder cells. Removing the feeder cells removes the threat of viruses to be conveyed to the human cells.

The procedure of generating is somewhat incompetent, so lines are not created each time cells from the preimplantation-stage embryo are located into a culture dish. Though, once the embryonic stem cell line is positively increased enough to pack the dish, they are removed and placed into a number of fresh culture dishes. Sub culturing the cells several times is repetited for several months. Millions of embryonic stem cells are harvested, originated from the original cells.

"Embryonic stem cell line" is embryonic stem cells that are proliferated in cell culture for a certain period of time deprived of differentiating. The embryonic stem cells that converted to pluripotent, and have not exhibited genetic deformity are recognized as an embryonic stem cell line.

Characterization of Embryonic Stem Cells

Scientists examine the cells during the formation of embryonic stem cell lines, to investigate that their fundamental properties enabled them to be embryonic stem cells. This process is often called characterization.

Using specific techniques to determine the existence of specific cell surface markers are characteristically shaped by undifferentiated cells.

Through examining the chromosomes under a microscope is a method to assess whether the chromosomes are damaged or if the number of chromosomes has changed. It does not detect genetic mutations in the cells.

Determining whether the cells can be re-grown, or sub cultured, after freezing, thawing, and re-plating.

Challenging whether the human embryonic stem cells are pluripotent by:

1. Permitting the cells to differentiate instinctively in cell culture;
2. Deploying the cells so they will differentiate to form distinctive cells of the three germ layers; or
3. Inoculating the cells into a mouse with a blocking immune system to examine for the development of a benign tumor called a "teratoma." Since the mouse's immune system is blocked, the inoculated human stem cells are not rejected by the mouse immune system and researchers can detect development and differentiation of the human stem cells. Teratomas characteristically encompass a combination of many differentiated or partly differentiated cell sorts—a sign that the embryonic stem cells is able of differentiating into many cell kinds, Figure 2.6.

STIMULATED EMBRYONIC STEM CELLS TO DIFFERENTIATE

Stem cells can be generally clustered into two classes:

1. Pluripotent stem cells and
2. Multipotent stem cells.

Pluripotent stem cells entail embryonic stem cells (ESCs) and induced pluripotent stem cells (iPSCs), Figure 2.6. The familiar iPSCs are usually found via the nuclear reprogramming of adult somatic cells. Like ESCs, iPSCs are considered by their indefinite development in the culture without a loss of pluripotency in addition to their ability to differentiate into virtually all kinds of somatic cells in the body.

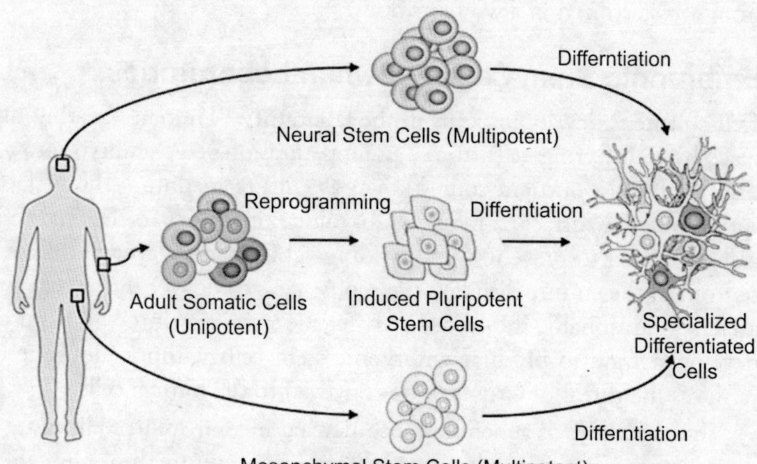

Figure 2.6: Stem cells and their classification.

In difference, adult multipotent stem cells are characteristically originate in a variety of human tissues and organs, such as bone marrow, skin, and central nervous system. Some instances of multipotent stem cells are mesenchymal (MSCs) and neural stem cells (NSCs). MSCs have the possibility to differentiate into diverse tissues of mesenchymal origin, including osteoblasts, adipocytes, and chondrocytes in addition to those of non-mesenchymal origin -like neurons. NSCs, on the other hand, are well recognized for their differentiation into tissues of neuronal lineages, such as neurons and neuroglia.

Researchers differentiated cells of explicit kind of cells such as heart muscle cells, blood cells, or nerve cells, can by monitoring the differentiation of embryonic stem cells. The chemical structure of the culture medium is altered by injecting specific genes. This leads to direct the embryonic stem cells into some explicit cell kind.

Numerous illnesses can be healed if researchers can consistently influence the differentiation of embryonic stem cells into specific cell kinds. The differentiated cells can then heal certain diseases in the future. Parkinson's disease, diabetes, traumatic spinal cord injury, Duchenne's muscular dystrophy, heart disease, and vision and hearing damage might be healed by transplanting cells creäted from human embryonic stem cells.

ADULT STEM CELLS

The fundamental roles of adult stem cells in a living organism are to preserve and repair the tissue in which they are resident. An adult stem cell is considered to be an undifferentiated cell, discovered among differentiated cells in a tissue or organ. The adult stem cell can rejuvenate itself and can differentiate to yield some or all of the main specialized cell types of the tissue or organ.

Adult stem cells ar found in many more tissues than they once thought possible. This finding has led scientists to use adult stem cells for transplants. Adult hematopoietic (bone marrow), or blood-forming, stem cells from bone marrow have been used in transplants for more than 4 decades. Stem cells exist in the brain and the heart, were not at first expected to reside. If the differentiation of adult stem cells can successfully be controlled in the laboratory, they will become the basis of transplantation-based therapies. The implementation of 3D bioprinting for organ manufacturing and implementations would then become a common place.

Adult Stem Cells Functions

Adult stem cells have been identified in many organs and tissues, including brain, bone marrow, peripheral blood, blood vessels, skeletal muscle, skin, teeth, heart, gut, liver, ovarian epithelium, and testis. Adult stem cells are believed to exist in a certain area of each tissue, which is called a "stem cell." Current evidence proposes that some types of stem cells compose the outermost layer of small blood vessels. Stem cells may remain dormant (non-dividing) for long periods of time until they are activated by a normal need for more cells to maintain tissues, or in case of emergency caused by disease or tissue injury, Figure 2.7.

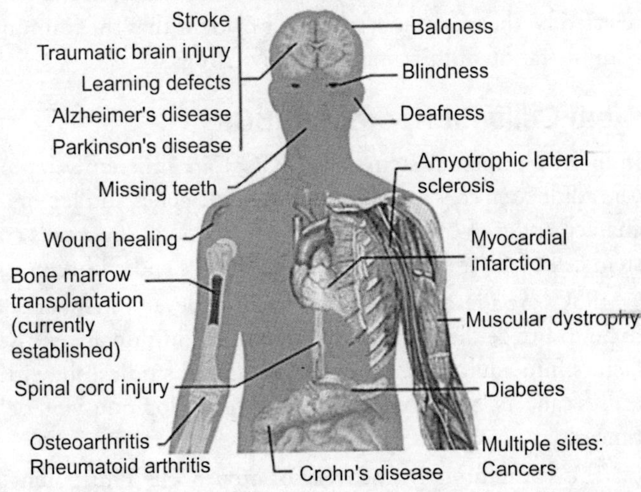

Figure 2.7: 3D Bioprinting Potential Treating Many Diseases

It is now believed that there are a very small number of stem cells in each tissue. Once the stem cells are removed from the body, their ability to divide is inhibited to generate large quantities of stem cells. Scientists around the world are trying to find better ways to grow large quantities of stem cells in many laboratories, and then manipulate them to generate specific cell types to be used in treating injury or disease.

The Similarities and Differences between Embryonic and Adult Stem Cells

There are advantages and disadvantages of Human embryonic and adult stem cells with respect to prospective utilization for cell-based regenerative treatments.

Stem Cells	Type of Differentiated Cell	Environment
Adult stem cells	Limited to tissue of origin	Failed in Culture
Embryonic stem cells	All cell types of the body	Grown in Culture

One main variance between adult and embryonic stem cells is their different capabilities yielding the number and type of differentiated cell kinds they can transform. Embryonic stem cells can yield all cell types of the body because they are pluripotent. Adult stem cells are supposed to be restricted to differentiating into different cell types of their tissue of origin.

Embryonic stem cells can effortlessly be developed in culture. Adult stem cells are rare in mature tissues. Therefore, isolating these Adult stem cells from an adult tissue is cumbersome. Additionally, the methods to expand their numbers in cell culture have not successful. However, the need for large numbers of cells for stem cell replacement is needed.

Scientists fear that tissues derived from embryonic and adult stem cells may differ in the likelihood of being rejected after transplantation. Since relatively few clinical trials it is not certain if tissues derived from embryonic stem cells would cause transplant rejection, compared to transplanted cells derived from human embryonic stem cells.

However, it is not likely that adult stem cells and tissues derived from them, will initiate rejection after transplantation. This is mainly contributed to the ability of patient's own cells to expand in culture, manipulated into assuming a specific cell type differentiation, and then introduce again into the patient. The use of adult stem cells and tissues derived from the patient's own adult stem cells are less likely to be rejected by the immune system. It is evident that the immune rejection can be circumvented by continuous administration of immunosuppressive drugs.

Stem Cells Pluripotent (iPSCs)

Induced pluripotent stem cells (iPSCs) are inherently reprogrammed adult cells to embryonic stem cells. The adult stem cells are obligated to guide genes and factors, which are essential for upholding the defining characteristics of embryonic stem cells. Though these cells content the crucial fundamentals for pluripotent stem cells, still, it is not recognized if iPSCs and embryonic stem cells deviate biologically.

iPSCs are already prized tools for drug development and modeling of diseases. Reseachers are attempting hope to utilize them in transplantation. Additionally, viruses were used to familiarize the reprogramming factors into adult cells. However, in animal studies, the virus used to introduce the stem cell factors often causes cancers. Scientists have begun to develop non-viral delivery strategies that have less chance of causing cancers.

This breakthrough discovery of Stem Cells Pluripotent (iPSCs) has fashioned a powerful new means to "de-differentiate" cells whose developmental destinies had been previously assumed to be determined.

Nonetheless, tissues derived from iPSCs are nearly identical match to the cell donor, which prevents rejection by the immune system. The iPSC strategy produces pluripotent stem cells will help scientists to reprogram cells to repair damaged tissues in the human body. This function is of great importance in 3D organ bioprinting and remedy application.

THE POTENTIAL USES OF HUMAN STEM CELLS

Overcome Obstacles - 3D Bioprinting Technology

Several means were developed in human stem cells used in medical research of human embryonic stem cells that yielded data about the intricate events that happen during human development. The main purpose is to learn the steps of how the undifferentiated stem cells become the differentiated cells that generate the tissues and organs. Understanding turning genes on and off is fundamental to this process. Cancer and birth defects are the most devastating medical conditions happened to abnormal cell division and differentiation. Once the genetic and molecular controls of these processes are clearly understood will result in treasure trophy of information of how cancer and organ defects arise. Such valuable discovery will lead to strategies for therapy and treatment.

Human stem cells are currently being used in 3D bioprinting to test new drugs. New drugs are tested for safety utilizing differentiated cells formed from human pluripotent cell lines. Other types of cell lines have been used in similar ways. Cancer cell lines are used to screen anti-tumor drugs. The availability and the presence of pluripotent stem cells enable drug testing applied in a broader variety of cell types. Nonetheless, to verify drugs effectively, the environment must be identical when comparing different drugs. Researchers must accurately control the differentiation of stem cells into the specific cell type, where drugs are tested.

The formation of cells and tissues as cell-based therapies are perhaps the most important discovery in our 21st century. They are considered vital potential application of human stem cells. Additionally, the cell-based treatment and repair enabled the 3D bioprinting application to be considered as a revolution in medicine. Donated organs and tissues are replacing destroyed tissue and ailing organs. Stem cells, geared to differentiate into specific cell types, provided realistic hoped to millions of patients worldwide. Diseases such as macular degeneration, spinal cord injury, stroke, burns, heart disease, diabetes, osteoarthritis, and rheumatoid arthritis are potential treatment samples as a renewable source of replacement cells and tissues.

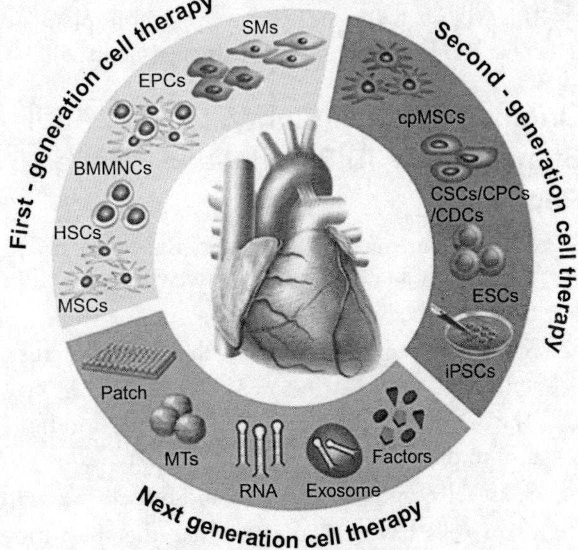

Figure 2.8: Strategies Repairing Heart Muscle - Adult Stem Cells

Healthy heart muscle cells can be generated in the laboratory and then transplanted into patients with chronic heart disease. Additionally, bone marrow stromal cells can be transplanted into a damaged heart to repopulate the heart tissue. Alternatively, injected cells can be injected to cause the repair through secreting growth factors, instead of incorporating cells into the heart, Figure 2.8.

Stem Cells for the Future Treatment of Heart Disease

Cardiovascular disease deprives the heart of oxygen, resulting in the death of cardiac muscle cells (cardiomyocytes). This loss of oxygen triggers a flow of damaging events leading to the formation of scar tissue, heart failure and even death. A potentially new strategy to treat heart failure is through repairing or regenerating damaged heart muscle tissue.

There numerous patients suffer from type 1 diabetes. If the pancreas cells that normally produce insulin are destroyed by the patient's own immune system it is possible to direct the differentiation of human embryonic stem cells in cell culture to form insulin-producing cells. Transplantation therapy ultimately could be used for persons with diabetes.

Stem cells must be easily manipulated to acquire the necessary characteristics for successful differentiation, transplantation, and engraftment. . The stated list below provides a path for successful control of cell-based treatments to bring such treatments to patients. Stem cells must be made to:

1. Proliferate sufficient quantities of tissue.
2. Differentiate desired cell type(s).
3. Survive after transplant.
4. After transplant integrate into surrounding tissue.
5. Function for duration of recipient's life.
6. Avoid harmful consequence to recipient.
7. Avoid immune rejection.

Intensive research is underway to 3D Bioprint (manufacture) tissues that will not be rejected due to immune rejection.

This will open very positive avenues in implementing revolutionary therapies. It is predicted sooner or later that 3D bioprinting stem cells treatments will become a standard treatment.

Origin and Use of Embryonic and Adult Stem Cells in 3D Bioprint and Tissue Repair

1. Stem cells are naturally -renewed as unspecialized cells that able give rise to many cell types of all tissues of the body.
2. Stem cells can be result from the embryo, fetus and adult.
3. The Stem cells have the ability to divide and but also to differentiate to specialized cell types
4. Stem cells can generate nerve and muscle
5. Stem cells have become the base therapies for diseases and disorders.

Defective or damaged cells can be replaced in defective organs and tissues as a cure. Cell differentiation and tissue repair can be directed to distinguish embryonic versus adult sources of stem cells and data for clinical application, Figure 2.9.

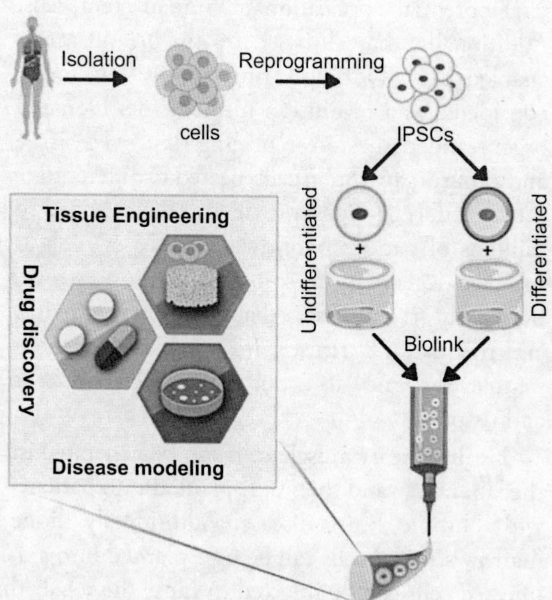

Figure 2.9: Preparing Stem Cells as 3D Printing Bioink

Formation of Organs Utilizing 3D Bioprinting

The understanding of stem cells functions has widely open a new world of therapies stated as follows:

1. Stem cells are primitive cells present in all organisms
2. Stem cells can divide and give rise to more stem cells
3. Stem cells switch to become more specialized cells
4. Stem cells specialize to brain, heart, muscle and kidney
5. Stem cells in early embryos give rise to cells of all tissues of the adult body
6. Stem cells are 'pluripotent'
7. Stem cells involved in tissue replacement and repair
8. Stem cells give rise to cell types already present in the surrounding tissue from which they are derived
9. Stem cells of the bone marrow give rise to hematopoietic cells
10. Adult stem cells are generally regarded as "multipotent."

Isolation Human Stem Cells from Surplus IVF Embryos

The isolation of human stem cells from surplus *in vitro* fertilization (IVF) embryos, and their derivation and use has been a controversial subject in many parts of the world. The controversial is deepened due to the shortage of donor organs and tissues for regenerative medicine to be used as disease therapies for Diabetes, Parkinson, Rheumatoid Arthritis and Myocardial Infarction and other diseases caused by the loss of function of specific cell types.

The recent evidence that adult stem cells have a greater capacity for differentiation than had previously been thought has made adult stem cells to be a positive alternative to organ transplant. It is obvious embryonic and adult stem cells can be equivalent in their capacity to produce large numbers of specific cell types for transplantation.

Embryonic Stem Cells and Teratocarcinomas

More than 2 decades ago, the first embryonic stem cells were isolated. It was discovered that teratocarcinoma, a spontaneous tumor of the testis in humans. The tumors consist of tissues as varied as hair, muscle, bone and even complete teeth. The tumor cells look like a disorganized fetus. Biologists discovered that teratocarcinomas contain undifferentiated stem cells and could be induced by transferring embryos to extra-uterine sites. After stem cells differentiation, the embryonic carcinoma cells are no longer malignant, as therapy for teratocarcinoma, Figure 2.10.

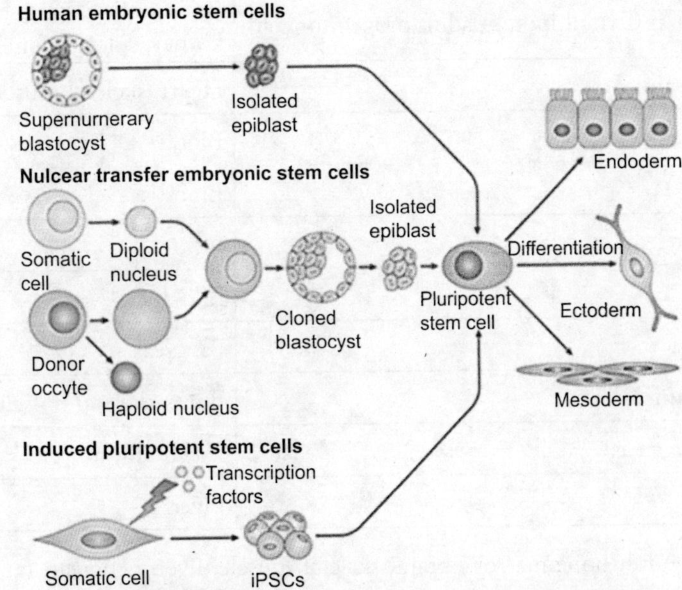

Figure 2.10: From Blastocyst, to Embryonic Stem Cell Line, to Somatic cells - hES: Human Embryonic Stem Cell.

Transdifferentiation of Adult Stem Cells

Achievement implementations conveyed for embryonic stem cells have lately expanded to adult stem cells. Adult stem cell function was presumed beforehand to be limited to cell lineages existing in the organ from which they were derived and to be related solely in their repair. Though, in current years a number of lines of evidence have recommended that these adult stem cells are multipotent and can transdifferentiate into diverse cell lineages, Table 2.1.

Table 2.1: Transdifferentiation of Adult Stem Cells

Tissue of origin	Newly formed tissue
Bone marrow	
Unfractionated	Brain
	Kidney
	Skeletal muscle
Mesenchymal/	Brain
stromal	Bone
	Fat
	Heart
HSC (hematopoietic stem cells)	Liver, lung, skin, gastro-intestinal tract
MAPC (multipotent adult progenitor cell)	Brain, retina, lung, heart, skeletal muscle, liver, intestine, kidney, spleen, bone marrow, blood and skin
Brain	Heart, skeletal muscle, kidney, stomach, intestine, liver
	Blood
	Blood
Skeletal muscle	Heart
Liver	Bile duct
Fat	Pancreas
Skin	Bone, skeletal muscle
Pancreas	Fat, brain, muscle
	Liver

Adult bone marrow, brain, skeletal muscle, liver, pancreas, fat, skin and skeletal muscle have all been shown to possess stem or progenitor cells with the capacity to differentiate or transdifferentiate into cell types other than their tissue of origin. Among all presently known adult stem or progenitor cells, cell populations from bone marrow have shown the highest potential with respect to multilineage differentiation and functional engraftment into host animals.

Origin and Use of embryonic and Adult Stem Cells in Differentiation and Tissue Repair

Where:

- HSC, hematopoietic stem cells;
- MAPC, multipotent adult progenitor cell.

Bone Marrow Stem Cells

Bone marrow stromal sometimes is referred to mesenchymal stem cells, which is a subcategory of cells that is able to be separated by plastic adherence, and have revealed differentiation into various cell types. This bone marrow has differentiated to bone, tendon, cartilage and fat. A cardiomyogenic cell line was recognized by culturing mesenchymal stem cells in the presence of the DNA demethylation agent. Upon initiation, the mesenchymal stem cells showed a cardiomyocyte phenotype, expressed cardiac differentiation markers and commenced spontaneous beating.

Injection of human marrow stromal cells into brains showed cells migrated to successive layers of the brain. Injection of murine marrow stromal cells into the lateral ventricle of neo-natal caused in their migration all through the forebrain and cerebellum and their differentiation into astrocytes and neurons.

Bone marrow comprises Hematopoietic stem cells (HSC) at very low frequency. Hematopoietic stem cells have the capability to re-inhabit the hematopoietic surroundings. Transplanted bone marrow cells, augmented for Hematopoietic stem cells have the capability to differentiate into hepatocytes in liver. Sole Hematopoietic stem cells have the capability to re-inhabit the hematopoietic system, and differentiate into epithelium of lung, skin, liver as well as the gastro-intestinal tract.

Brain Stem Cells

The populations of bone marrow cells have shown that adult neural stem cells have a wider differentiation possibility as predicted. Embryonic and adult stem cells, clonally derived from the forebrain expressing β-gal ubiquitously, were systemically injected. The neural stem cells were able to engraft into the hematopoietic system of hosts to produce a range of blood cell types following the bone marrow transplantation.

The contribution of neural stem cells to the hematopoietic system was confirmed by transplantation of human brain derived cells. Plasticity of neural stem cells was also observed when injecting neuro-spheres derived from single-cell cultures, into the amniotic cavity of chick embryos. Neural stem cell offspring were found, in ectodermal tissues, mesodermal, and endodermal tissues, e.g., heart, kidney, skeletal muscle, lung stomach, intestine and liver. However, no neural stem cell offspring were detected in the hematopoietic system.

Skeletal Muscle Stem Cells

A population of myogenic precursors found in adult skeletal muscle. The self-renewal, often called satellite cells exhibited differentiation in response to physiological and pathophysiological stimuli. The satellite cells have been shown to differentiate into slow-twitch muscle fibers when injected into the heart and improve cardiac function.

A sub-population of skeletal muscle exhibited regeneration of muscle and blood cells after 4-8 weeks after transplantation derived from donor cells.

These findings concluded that muscle-derived human stem cells (HSCs) are derivative of the hematopoietic system rather than the adult muscle ancestor cell population.

Liver Stem Cells

Liver cells often called the oval cells are capable of proliferation. Liver cells are considered the stem-cell partition in the liver. Liver cells provide regeneration of hepatocytes, hepatic stem cells, which give rise to pancreatic cells, producing insulin and glucagon.

Cells from a clonal stem cell line derived from adult liver cells were injected into the left ventricle. Cells were detected in the heart and a cardiomyocyte phenotype was observed. Also, electron microscopy indicated the occurrence of well-organized sarcomeres, intercalated discs and gap junctions, co-localized suggesting their multi-potency.

Pancreatic Stem Cells

The hepatic stem cells have the ability to differentiate into pancreatic cells. Cell fractions from the pancreas were observed to differentiate into hepatocytes after transplantation to the liver. Immediately after transplantation donor-derived cells showed liver-specific proteins and displayed phenotypical resemblance with hepatocytes.

Fat Stem Cells

Liposuction procedures provided human adipose tissue, which is used to isolate a fibroblast-like cell population. These types of cells are called processed lipo-aspirate (LPA) cells. The LPA cells established differentiation into adipogenic, chrondogenic, myogenic and osteogenic cells.

Skin Stem Cells

Cells isolated from the skin of juvenile or adult culture and studied for their ability to differentiate into cell-types of different lineages. *In vitro* experiments demonstrated differentiation of skin-derived cells and clones of individual cells into neurones adipocytes, glial and smooth muscle cells, based on phenotype and cell-specific markers. Furthermore, cultured skin cells from human scalp were also able to produce neural proteins, when induced to differentiate. These skin stem cells were different from mesenchymal or neural stem cells, as determined by their phenotype and the production of nestin and vimentin.

Adult Stem Cells and Tissue Repair

A sub-population of bone marrow cells was acquired stating green fluorescent protein (GFP) collectively. Following intra-ventricular injection of these cells that had experienced experimental myocardial infarction, regeneration of cardiomyocytes was established in 40% of the cells. Besides regeneration of cardiomyocytes, anew fashioned endothelial and smooth muscle cells, prearranged in coronary vessels, were also established. Nine days after transplantation, the percentage of newly fashioned tissue, resulting from the injected bone marrow cells, was 68% of the dead tissues region of the ventricle. Most prominently, cardiac function enhanced following transplantation.

A non-invasive technique to upsurge the number of circulating stem cells by treatment with stem cell factor (SCF) and granulocyte-colony-stimulating factor (G-CSF). Cytokine-treatment showed newly fashioned cardiac tissue in the dead region, and as a result reduced mortality and augmented ejection dead, parameters reflecting cardiac function. This trial approach evaded the high mortality of their preceding discovery and established the positive homing of donor-derived bone marrow cells into the dead region of the recipient. Though, it is not clear which sub-population of bone marrow cells is accountable for regeneration of cardiac tissue.

Another sub-population of bone marrow cells, also called side-population or (SP), expressing β-gal were selected by fluorescent activated cell sorting (FACS) and injected into irradiated subject. SP cells also expressed the adhesion molecule and angiopoietin, all features of endothelial precursors.

After 10–12 weeks subjects with high contributions of engrafted cells were exposed to an investigational ischemia–reperfusion regime to mimic cardiac death. Homing and engraftment of donor-derived side-population (SP) cells were instituted in the dead region. These engrafted cells had differentiated into both cardiac muscle and had fashioned vessel-like constructions. Cardiac cells positive for β-gal were primarily positioned at the border zone of the dead region

Lately, two exploration groups achieved co-culture of adult neural stem cells or bone marrow cells with embryonic stem cells. Both groups reached to the same astonishing deduction: adult stem cells can fuse with embryonic stem cells in culture. Co-culture of bone marrow cells with embryonic stem cells using selection markers that would eliminate the growth of embryonic stem cells develop to colonies with an embryonic stem cell morphology, which they named bone marrow-derived embryonic stem-like (BMESL) cells. *In vitro* differentiation of these BMESL cells into endoderm, ectoderm and mesoderm was observed.

The outcomes from the investigations propose that undifferentiated stem cells derived from adult tissues are not defined to be progenitor cells with restricted differentiation possibility. Somewhat, these cells appear to hold a much broader capability for cellular differentiation that is reliant on and receptive to explicit cues existing in the surroundings of the engrafted site. Though, with the exception of three studies with bone marrow cells, this has not been thoroughly established for any other stem cell of adult origin. Questions of scale-up and directed differentiation for transplantation and tissue repair continue to be answered, just as for embryonic stem cells, but their possibility autologous origin provide them exceptionally favorable prospective in the growth of cell replacement therapies.

Although considerable progress has been made towards understanding the control of differentiation of adult and embryonic stem cells, both have potential but must go rigorous clinical trials.

It is inaccurate to state that adult stem cells are equivalent to embryonic stem cells for treating diabetes, Parkinson and probably other diseases. Adult stem cells have tissue compatibility benefits in transplantation but in definite genetic or autoimmune illnesses, (matched) tissue from ES cells and immunosuppressive drugs may be preferable.

3D Bioprinting can be implemented utilizing the present human embryonic stem cell lines as they are adequate enough to for organ printing applications. Though, when the requirements for transplantation and histocompatibility are elevated isolation of novel lines become vital.

3D Bioprinting Promises

The practical side of printing small-volume organs "may be accomplished within the next few years." Though, the capability to improve vascularization of such organs is at least few years away. The challenge is to figure out what cells to utilize to evade immune rejection.

Tissues can be engineered to enable its structure to create suitable forms applicable for many biomedical applications, and in particularly in 3D Bioprinting applications. Developing tissue engineered products for 3D printing may require a simple approach such as developing a single flat sheet of tissue cells as tissue engineered product. This flat single sheet, filled with one type of cells may be used in patching burns and wounds, as well as to wrap around organs. . Multilayer of skin tissues populated with different types of cells and vascular can be printed with a 3D bioprinter that prints living embryonic cells.

Printing Organs Using Bio Ink Materials

Printing out dental materials and bones were the pioneering products used in 3D biomaterial printers. Complex models of cells, proteins and organs are utilizing bio-ink biomaterials in 3D bioprinters to create various 3D constructs. Additionally, 3D bioprinters are using bio-ink to print in-vitro assays for clinical diagnostics and drug discovery.

The ideal environment is to perform biochemical tests on 3D printed tissues extracted from a patient's own cells. At present using animals as test models may not predict drug toxicity in humans.

3D Bioprinting Utilizing Vitamin B2

Vitamin B2, often referred to as riboflavin, has the ability to create nontoxic effects. It is used particularly in 3D printing of medical grafts.

3D Bioengineered Tissues Reduce Danger of Tissue Rejection

Transplanted organs are often faced with the challenge of organ rejections. 54-60% of worldwide amputated limbs have lost their limbs due to vascular conditions including peripheral artery disease (PAD) and diabetes. PAD and diabetes causing Amputations are projected to reach nearly 3.0 million by 2020. Organ rejection of is the foremost problem for patients experiencing this procedure.

A higher rate of remedial can be achieved through the advancements of bioengineered tissues to reduce or eliminate the consequence of tissue rejection succeeding surgery.

Military and Defense departments worldwide have devoted substantial medical financial resources each year to employ tissue engineering for treatment of traumatic injuries. Undoubtedly 3D bioprinting techniques using tissue-engineering will have a positive impact on the healing methods of traumatic injuries. Parallel to this effort is the healing of tissue loss correspondent with cancer and tissue degeneration.

Life is in the Blood

The major obstacle in 3D Bioprinted organs is printing a network of blood vessels to solve the major obstacle of vascularization.

Challenging Issues with Bioprinting

Avoiding immune rejection is the first order of importance in selecting the appropriate cells to satisfy this vital issue. Subsequent to solving the issue off immune rejection, a bank of several donors' stem cells must be accumulated and categorized to the appropriate cells for the specific organ being printed, in particularly liver cells. A greater problem to face is bioprinting micro-nerves, using a reservoir of nerve generation. Nerve generation is a formidable challenge in issue engineering technology.

Biomechanical Cell Functions

Parts of the body which require human cells to perform biomechanical functions, such as the liver or kidney, are still several years away from reaching human patients. Biomechanical cell functions are still in the feasibility stage. It is not known yet how to keep cells alive at high cell density. Also, adequate cell size needed to match human organs.

A 3D bioprint structure may require in excess of 1 billion functioning cells to reach acceptable function of a liver or kidney. There are several cell types in these organs. 3D bioprinter may not contain sufficient cells types as they only contain one or two cell types to create a 3D printed structure.

A biopsy takes long time to grow enough cells to build satisfactory cell size of a human organ. Scientists are considering placing 3D bioprinter into the operating theatres to directly print a patient's cells and tissue into a wound in the operating room. a large stock of thses cells must be at hand in a frozen state, used when the needs are risen. However, the prospect of treating acute injuries with 3D printing is currently slim.

Regrow Limbs in Human - Undiscovered

Human genome does not have the ability to send signal and stimulate genes to help a new regrow of a human limb. Accordingly, human limb replacements, a reconstruction rather than stimulation is a viable alternative of regrowth.

The American Transplant Foundation stated that about 200,000 patients are awaiting life-saving organ transplant in the United States. Nonetheless, there are many patients are in need to register to be a potential organ donor. However, donor organs must undergo immunosuppressive therapies. Unfortunately, such therapy may limit the quality of the recipient life. 3D bioprinting organoid may provide healthy organs after transplant.

Eventually, 3D-printed organs will go through many advancements and validations before qualifying the system for successful organ manufacturing and transplant.

3D Bioprinting Organs – Affordability

3D bioprinting products may become affordable depending on the ease of printing as well as the supply and demands. However, the more complex products of 3D printed functional organs, which is exceptionally difficult to print would be extremely expensive, and may be limited to the rich willing patients. Patients who may not have the financial means will be forced to register for donor organ waiting list. If the new technology is not covered by the healthcare insurance, it likely will be available only to the rich and elite.

Biology vs. Bioprinting Technology

It is predicted that the cost of healthcare will lead to a race between biology and Bioprinting technology. An associated problem will be related to triage -- deciding, in essence, who deserves to receive the new technological treatments. The situation may be upon us soon where deciding whether, for example, giving an artificial kidney to an 80-year-old, who only has a few years to live, and also need an artificial heart. These are thorny issues if the Bioprinting technology is not apportioned to everyone needs it.

However, this hard economic calculus already is being practiced for organ replacement therapy at present. Perhaps, the current best practices can be extended when the new Bioprinting technology finally is ready to use.

Being First – Serves as Guinea Pigs

Additional subject of apprehension is that the initial patients to experience transplants with 3D-Bioprinted organs will of inevitability be desperate -- and because their doctors are breaking novel ground, they will oblige as guinea pigs. Most patients in clinical trials trust that the involvement has a chance to help them, albeit that characteristically is not the case. The 3D-printed technologies, will bump into failures before scientists determine the right approaches, cells, techniques. Patients may have irrational expectations and may in fact be inferior after trying a novel Bioprinting technology.

Rules and Regulations

There already are procedures to handle patients being unprotected to novel medical technologies. Hospital oversight panels would control donor issues, cells and tissue for informed consent.

Body Enhancement

There may be fields in which guidelines and ethics may be ignored: illegal clinics supplying to athletes and people looking for body reform for personal gratification or to advance an edge for them. It does not seem irrational to assume that people may try to find methods to augment themselves with 3D printing if they can get an advantage. Some athletes already assume substantial jeopardies to increase their performance on the field utilizing illegal ingredients with grave side effects; others might seek augmentations to help achieve their ideal of beauty.

The centralized government might not become tangled in regulating beautifying efforts. For athletes, organizing bodies such as the NCAA and NFL likely would be involved.

Adapted Desktop Printer

Bio-printing process was greatly enabled by the innovative modifications of traditional desktop printers. As reverse engineering methodology was successfully carried out a bio-printer was developed. One side of the bio-printer is used as a printing stage, while the other side of feeding paper through was eliminated. Consequently, the major modification made to desktop printers is the addition of a Z-axis. The Z axis is required to lower the printing area with each successive layer. A trigger switch was incorporated to permit overriding the normal operation of the printer, compelling it to print without paper presence.

A minor detail on the printer was adapted. It can accurately generate human tissues cultures to aid in-vitro drug development or toxicity testing.

Tissue Engineering

The lack of organ transplantations and long waiting lists has given rise to 3D bioprinting of tissue engineering. Tissue Engineering is the practical implementation of the engineering technique dealing with understanding the structure of living tissues in living organs, and the technique to engineer tissues to assist or replace ailing organs.

Tissue engineering has the potential of widespread use for in vitro applications.

Tissue Engineering applications are summarized in the following Table

Table 2.2: Tissue Engineering Applications

Disease	Application
• End-Stage Renal Disease	Use of stem cells Generation of histocompatible tissues
• Liver Disease	Cell-based Therapy
• Cardiovascular Disease	Endothelial cell seeding of vascular grafts Tissue-engineered vascular-conduits Generation of heart valve leaflets Cardiomyoplasty Genetic manipulation *In vitro* conditions

Bioprinting Creating Living Tissue

A particular bio-inkjet printer is developed to cultivate new biological organ transplants or living tissues for making regenerative medicine that repairs ailing organs. Therefore, traditional inkjet cartridge was modified to accommodate biological ink containing the necessary living tissues.

The aim is to build organs layer by layer using stem cells and solve the severe shortage of donor organs to save lives of multitudes of patients. Making or replacing organs is the wave of the future.

Stem Cells Bioprinting Process

Available stem cells taken from the patient, who requires replacement, will be used to manufacture the ailing organ instead of being placed on organ donor lists. This tissue engineering emerging-technology and regenerative medicine is known as Bio-printing process. Primarily, , it is the process that requires the simultaneous alterations of cells, and growth factors through a micro-scale orifice. Bio-Printing applications are presented in Table 2.3.

Table 2.3: Bio-Printing Applications

Bio-Printing	Applications
Printing Organs	• Grow replacement organs
Printing Medication	• Print own medicine at home
Printing Skin	• Laser-printed skin cells
Printing Cartilages and Bones	• Bone-like material that acts as a scaffold • Stem Cells
Printing Replacement Tissue	• Heart Patch • Sheets of cardiac tissue
Printing Human Embryonic Cells	• Test drugs
Studying Cancer with Printed Cells	• Test treatments

Bioprinted Muscles

The print cartridge reservoir represents the greatest challenges the bioprinting technology. One of the biggest challenges is cell aggregation and The sedimentation in cartridge reservoir may contaminate and ruin the cell growth and inhibit development. Also, ejected cells from the nozzle may clog the nozzle. FDA approval is an additional hurdle to overcome. This hurdle is contributed to the unique DNA codes for every patient; therefore, additional regulatory approvals may be required as each patient possesses unique bone, tissue and organs..

Interestingly enough, athletes may be motivated to enhance their physical capacity using additional bioprinted muscles. Beauty pageants may motivate young women to replace body parts, or if determine to stay continually young through replacement instead of aging normally. These are social factors that must be taken into considerations.

CLONED - 3D BIOPRINTING IN THE MEDICAL FIELD

Human embryonic stem cells successfully developed from skin cells has enabled the dawn created of developing new medical treatments as human stem cells were cloned trough transferring the nucleus of another cell.

Skin cells taken from an infant carrying a genetic disease, and then fused them with donated human eggs creating human embryos identical to the infant genetical identity to the child. Stem cells were extracted from those specific embryos. The embryos were used the technique similar to the one used to create Dolly the sheep, in 1996.

Adult cell were taken from the patient's body. The central portion of the DNA was sifted out. This process has immediately followed by injected the substance into a donated human egg, which was also stripped out of its genetic material. Remarkably, as expected the genetic substance extracted from the adult cell instructs the empty egg cell for further development.

Several different cell types were converted by the stem cells just as normal embryonic stem cells derived through this technique of sifting out the central portion of the DNA. Most importantly, not only variety of cells was generated but also the nerve cells, liver cells and heart cells were among the converted cells.

Furthermore, the concern of transplant rejection was eliminated because these re- programmed cells can be created with nuclear genetic material from the same patient. This is a significant step forward in developing effective and safe and stem cell treatments, tailored also to be used in regenerative medicine.

The cloned method does not require the use of embryos in creating the type of stem cell capable of transforming into any other type of cell in the body. The breakthroughs in nuclear transfer have advanced the possibility of human reproductive cloning.

3D Bioprinting Embryonic Stem Cells

The technology of 3D Bioprinting has extended into to stem cells, human body organs and prosthetic human body limbs. All efforts are created for the hope to eliminate the need for organ transplants.

3D bioprinting software was successfully developed to generate 3D human tissue aimed to create a bank of 3D-printed organs tailored to assist thousands of patients who are in the waiting list for organ transplant.

While the embryonic stem cells are resulting from the blastocyst a few days after fertilization, and have the ability to give rise to various types of tissue in the human body, the new advancement in cloning method of 3D-Bioprinting technology have a potential to breakthrough in medical field saving multitudes of patients.

Therapeutic Cloning

Process of Therapeutic Cloning - Somatic cell nuclear transfer can create clones for both reproductive and therapeutic purposes. Figure 2.11, depicts the removal of the donor nucleus for schematic purposes; in practice usually the whole donor cell is transferred.

Somatic Cell Nuclear Transfer (SCNT): is a technique where an egg absent of its nucleus, fused with a donor cell nucleus to create an ovum in a laboratory setting. It is usually used in embryonic stem cell development, or in regenerative medicine. This process is often referred to as "therapeutic cloning."

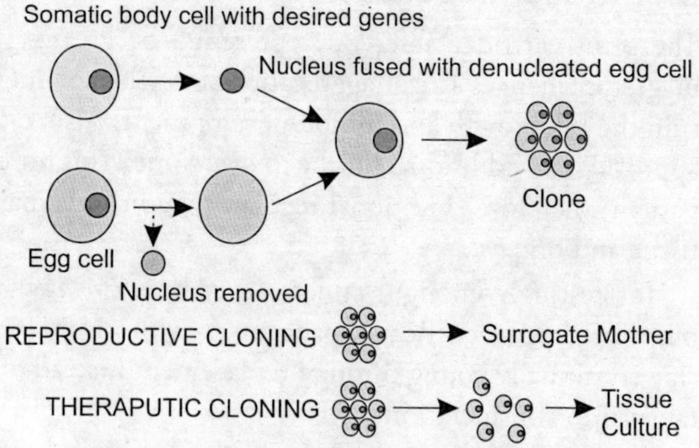

Figure 2.11: Process of Reproductive and Therapeutic Cloning

The technique is also used in reproductive cloning, Figure 2.12.

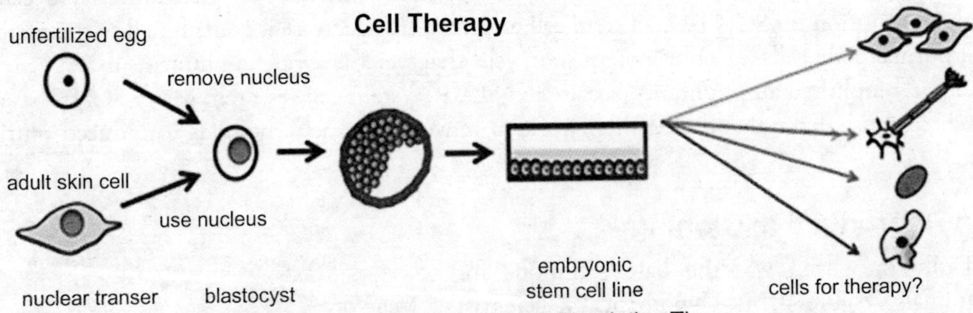

Figure 2.12: Cell Utilized in 3D Bioprinting Therapy

In SCNT the nucleus it contains the organism's DNA of a somatic cell is removed and the rest of the cell disposed. . The nucleus of the somatic cell is inserted into the enucleated egg cell, after the nucleus of an egg cell is removed.

The host cell reprograms the somatic cell nucleus after being inserted into the egg. Currently, the egg containing the nucleus of a somatic cell is electrically stimulated with a shock to begin dividing. This single cell forms a blastocyst many divisions in culture. The DNA of the blastocyst is identical the DNA of the original organism.

SCNT IN STEM CELL RESEARCH

SCNT are often used in stem cell research. Obtaining stem cells from SCNT procedure genetically matched DNA to the donor organism.

Unspecialized cells are able to be produced into a specialized cell. Specialized cells can replace another cell that has been lost in the body. In order to create cell lines that have genes linked to particular disease genetically-customized stem cells may be used. . Parkinson's disease patients may donate their somatic cells. The stem cells resulting from SCNT would have genes that correlate to Parkinson's disease. The disease specific stem cell lines would be used to obtain a greater understanding of the disease.

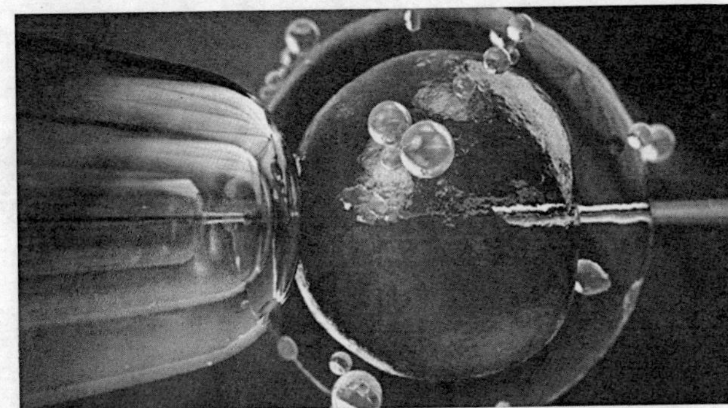

Figure 2.13: Human Embryonic Stem Cell Colony in Embryonic Fibroblast Feeder Layer.

Additionally, cell-based therapies can be genetically-customized ass stem cell lines to transplant to the patient, Figure 2.13. The subsequent cells would be genetically identical to the somatic cell donor. Accordingly, the rejection complications from immune system are avoided.

Presently, there are a few worldwide Laboratories for using somatic cell nuclear (SCNT) transfer technique to produce embryonic stem cells. In the United States, Harvard Stem Cell Institute, the University of California San Francisco, Stemagen (La Jolla, CA), and Advanced Cell Technology. In the United Kingdom, the Human Fertilization and Embryology Authority, Roslyn Institute, and the Newcastle Centre for Life, Newcastle Upon-Tyne.

However, the development and improvement of alternative methods of generating stem cells have impeded the motivation for SCNT-based stem cell research. This decrease is contributed to the methods to reprogram normal body cells into pluripotent stem cells that were developed in humans in 2007. In 2008, this method accomplished an important part in SCNT-based stem cell research. Many SCNT-based stem cell research scientists have abandoned their work to move to the new methods of induced pluripotent stem cells.

SCNT in Reproductive Cloning

Famous Dolly the sheep was the base technique for cloning animals, Figure 2.14. Hypothetically, scientist claim clone humans are theoretically possible using the SCNT technique. However, most scientists including the author of this book strongly believe ***it will not be possible*** to use this technique or any other techniques to develop a human clone. The author believes that the certain inherent pair of proteins in humans namely "Spindle Protein." These proteins could be enucleated (segregated) from the original cell to conduct successful cloning. The author believes an absolute intelligence designer had created such mechanism to prevent human generating chaos out of the orderly matter of creation. It is likely to end up with dysfunctional immovable human cells without a soul.

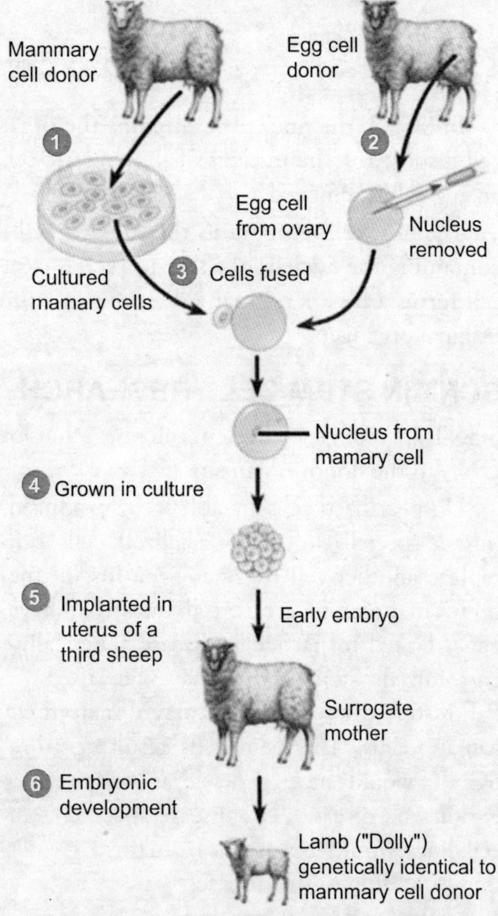

Figure 2.14: Cloning Dolly the Sheep Using SCNT in Reproductive Cloning

Sources of Human Eggs

Proposals to use nucleus transfer techniques in human stem cell research raise a set of concerns beyond the moral status of any created embryo. These have led to some individuals and organizations who are not opposed to human embryonic stem cell research to be concerned about, or opposed to, SCNT research.

Regarding SCNT technique it affects healthy young women to donate large number of eggs, Figure 2.15. Additionally, women need to be persuaded to sell eggs to assist in the development of custom stem cell lines. Women's eggs could then be purchased by the scientist community and then sold again to recipient patients. Young women will be subjected to heavy doses of hormones in order to induce hyper ovulation, to ovulate several eggs at once. The long term consequences and the effect of the unknown public health off this issue may create substantial ambiguity in the mind of young women the long-term effects have not been declared safe to use on a large scale. It has been established that women who receive longer-term treatments with much lower hormones doses are subject to an increase rate of cancer decades later. However, in general, selling body parts is reflecting unethical behavior. Nonetheless, human eggs have been a considered exception to this rule.

A viable alternative to human eggs is to develop a human egg using some stem cell to investigate the possibility of making artificial eggs. Once an artificial human egg is developed successfully, the need for actual human egg is eliminated, to create custom stem cell lines.

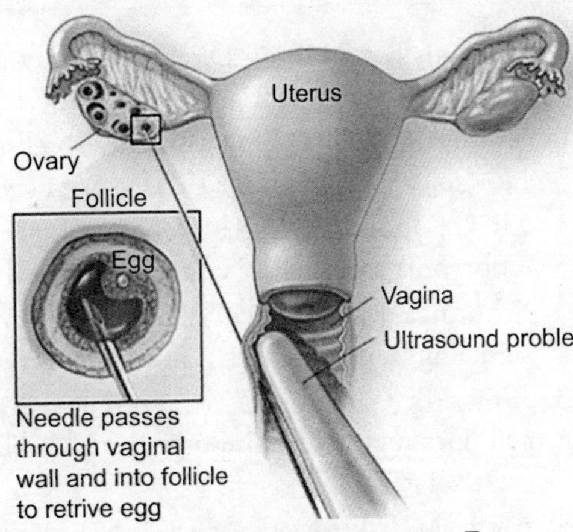

Figure 2.15: Donating a Human Egg - *Courtesy: Mayo Foundation for Medical Education and Research (MFMER)*

Legality of Cloning Treatments

The practice of "Somatic cell nuclear transfer (SCNT)" in the United States is still legal, as it was addressed by federal law in 2005. However, in the United Sates of American it is illegal to clone a human being according to: *H.R.3498 — 114th Congress (2015-2016)*. Several nations adopted a proposal submitted by Costa Rica, calling on member states to "prohibit all forms of human cloning in as much as they are incompatible with human dignity and the protection of human life." However, SCNT is permissible and is still used for research purposes and medical treatments.

In the United Kingdom, **Somatic Cell Nuclear Transfer** (SCNT), is currently legal only for research purposes. SCNT was incorporated into the 1990 Human Fertilization and Embryology Act in 2001. However, to perform or attempt SCNT, permission must be granted from the Human Fertilization and Embryology Authority. The Council of Europe's Convention prohibits Cloning Human Being and appears to ban SCNT.

REFERENCES

1. Mummery C.L., van den Eijnden-van Raaij A.J. (1999) Developmental tumours, early differentiation and the transforming growth factor beta superfamily. Int J Dev Biol 43:693–709. MedlineWeb of ScienceGoogle Scholar
2. Andrews P.W., Przyborski S.A., Thomson J.A. (2001) Embryonal carcinoma cells as embryonic stem cells, Stem cell biology, eds Marshak H.A., Gardner R.L., Gottlieb D. (Cold Spring Harbor Laboratory Press, Cold Spring Harbor, NY), pp 231–265.
3. Thomson J.A., Itskovitz-Eldor J., Shapiro S.S., et al. (1998) Embryonic stem cell lines derived from human blastocysts. Science 282:1145–1147. Abstract/FREE Full Text
4. Reubinoff B.E., Pera M.F., Fong C.Y., Trounson A., Bongso A. (2000) Embryonic stem cell lines from human blastocysts: somatic differentiation *in vitro*. Nature iotechnol 18:399–404.
5. Evans M.J., Kaufman M.H. (1981) Establishment in culture of pluripotential cells from mouse embryos. Nature 292:154–156.
6. Martin G.R. (1981) Isolation of a pluripotent cell line from early mouse embryos cultured in medium conditioned by teratocarcinoma stem cells. Proc Natl Acad Sci USA 78:7634–7638.
7. Smith A.G., Heath J.K., Donaldson D.D., et al. (1988) Inhibition of pluripotential embryonic stem cell differentiation by purified polypeptides. Nature 336:688–690.
8. Moreau J.F., Donaldson D.D., Bennett F., et al. (1988) Leukaemia inhibitory factor is identical to the myeloid growth factor human interleukin for DA cells. Nature 336:690–692.

9. Williams R.L., Hilton D.J., Pease S., et al. (1988) Myeloid leukaemia inhibitory factor maintains the developmental potential of embryonic stem cells. Nature 336:684–687.
10. Smith A.G. (2001) Embryonic stem cells, Stem cell biology, eds Marshak H.A., Gardner R.L., Gottlieb D. (Cold Spring Harbor Laboratory Press, Cold Spring Harbor, NY), pp 205–230.
11. Xu C., Inokuma M.S., Denham J., et al. (2001) Feeder-free growth of undifferentiated human embryonic stem cells. Nature Biotechnol 19:971–974.
12. Schuringa J.J., van der Schaaf S., Vellenga E., Eggen B.J., Kruijer W. (2002) LIF-induced STAT3 signaling in murine versus human embryonal carcinoma (EC) cells. Exp Cell Res 274:119–129.
13. Amit M., Carpenter M.K., Inokuma M.S., et al. (2000) Clonally derived human embryonic stem cell lines maintain pluripotency and proliferative potential for prolonged periods of culture. Dev Biol 227:271–278.
14. Amit M., Itskovitz-Eldor J. (2002) Derivation and spontaneous differentiation of human embryonic stem cells. J Anat 200:225–232.
15. Klug M.G., Soonpaa M.H., Koh G.Y., Field L.J. (1996) Genetically selected cardiomyocytes from differentiating embryonic stem cells form stable intracardiac grafts. J Clin Invest 98:216–224.
16. Li M., Pevny L., Lovell-Badge R., Smith A. (1998) Generation of purified neural precursors from embryonic stem cells by lineage selection. Curr Biol 8:971–974.
17. Mountford P., Nichols J., Zevnik B., O'Brien C., Smith A. (1998) Maintenance of pluripotential embryonic stem cells by stem cell selection. Reprod Fertil Dev 10:527–533.
18. Fujikura J., Yamato E., Yonemura S., et al. (2002) Differentiation of embryonic stem cells is induced by GATA factors. Genes Dev 16:784–789.
19. Reubinoff B.E., Pera M.F., Vajta G., Trounson A.O. (2001) Effective cryopreservation of human embryonic stem cells by the open pulled straw vitrification method. Hum Reprod 16:2187–2194.
20. McWhir J., Schnieke A.E., Ansell R., et al. (1996) Selective ablation of differentiated cells permits isolation of embryonic stem cell lines from murine embryos with a non-permissive genetic background. Nature Genet 14:223–226.
21. Brook F.A., Gardner R.L. (1997) The origin and efficient derivation of embryonic stem cells in the mouse. Proc Natl Acad Sci USA 94:5709–5712.
22. Holden C., Vogel G. (2002) Stem cell lines. 'Show us the cells,' US researchers say. Science 297:923–925.
23. Richards M., Fong C.Y., Chan W.K., Wong P.C., Bongso A. (2002) Human feeders support prolonged undifferentiated growth of human inner cell masses and embryonic stem cells. Nature Biotechnol 20:933–936.
24. Martin G.R. (1975) The formation of embryoid bodies *in vitro* by homogeneous embryonal carcinoma cell cultures derived from isolated single cells. Dev Biol 59:169–187.
25. Martin G.R., Wiley L.M., Damjanov I. (1977) The development of cystic embryoid bodies *in vitro* from clonal teratocarcinoma stem cells. Dev Biol 61:230–244. CrossRefMedlineWeb of ScienceGoogle Scholar
26. Weiss M.J., Orkin S.H. (1996) *In vitro* differentiation of murine embryonic stem cells. New approaches to old problems. J Clin Invest 97:591–595.
27. Rohwedel J., Guan K., Wobus A.M. (1999) Induction of cellular differentiation by retinoic acid *in vitro*. Cells Tissues Organs 165:190–202.

28. Lee S.H., Lumelsky N., Studer L., Auerbach J.M., McKay R.D. (2000) Efficient generation of midbrain and hindbrain neurons from mouse embryonic stem cells. Nature Biotechnol 18:675–679.
29. Liu S., Qu Y., Stewart T.J., et al. (2000) Embryonic stem cells differentiate into oligodendrocytes and myelinate in culture and after spinal cord transplantation. Proc Natl Acad Sci USA 97:6126–6131.
30. Buttery L.D., Bourne S., Xynos J.D., et al. (2001) Differentiation of osteoblasts and *in vitro* bone formation from murine embryonic stem cells. Tissue Eng 7:89–99.
31. Lumelsky N., Blondel O., Laeng P., et al. (2001) Differentiation of embryonic stem cells to insulin-secreting structures similar to pancreatic islets. Science 292:1389–1394.
32. Gratsch T.E., O'Shea K.S. (2002) Noggin and chordin have distinct activities in promoting lineage commitment of mouse embryonic stem (ES) cells. Dev Biol 245:83–94.
33. Tropepe V., Hitoshi S., Sirard C., et al. (2001) Direct neural fate specification from embryonic stem cells: a primitive mammalian neural stem cell stage acquired through a default mechanism. Neuron 30:65–78.
34. Kawasaki H., Mizuseki K., Nishikawa S., et al. (2000) Induction of midbrain dopaminergic neurons from ES cells by stromal cell-derived inducing activity. Neuron 28:31–40.
35. Wichterle H., Lieberam I., Porter J.A., Jessell T.M. (2002) Directed differentiation of embryonic stem cells into motor neurons. Cell 110:385–397.
36. Mummery C., Ward D., van den Brink C.E., et al. (2002) Cardiomyocyte differentiation of mouse and human embryonic stem cells. J Anat 200:233–242.
37. Rathjen J., Lake J.A., Bettess M.D., et al. (1999) Formation of a primitive ectoderm like cell population, EPL cells, from ES cells in response to biologically derived factors. J Cell Sci 112:601–612.
38. Jones E.A., Tosh D., Wilson D.I., Lindsay S., Forrester L.M. (2002) Hepatic differentiation of murine embryonic stem cells. Exp Cell Res 272:15–22.
39. Wiles M.V., Keller G. (1991) Multiple hematopoietic lineages develop from embryonic stem (ES) cells in culture. Development 111:259–267.
40. Slager H.G., Freund E., Buiting A.M., Feijen A., Mummery C.L. (1993) Secretion of transforming growth factor-beta isoforms by embryonic stem cells: isoform and latency are dependent on direction of differentiation. J Cell Physiol 156:247–256.
41. Nishikawa S.I., Nishikawa S., Hirashima M., Matsuyoshi N., Kodama H. (1998) Progressive lineage analysis by cell sorting and culture identifies FLK1+VE-cadherin+ cells at a diverging point of endothelial and hemopoietic lineages. Development 125:1747–1757.
42. Yamashita J., Itoh H., Hirashima M., et al. (2000) Flk1-positive cells derived from embryonic stem cells serve as vascular progenitors. Nature 408:92–96.
43. Weima S.M., van Rooijen M.A., Feijen A., et al. (1989) Transforming growth factor-beta and its receptor are differentially regulated in human embryonal carcinoma cells. Differentiation 41:245–253.
44. Odorico J.S., Kaufman D.S., Thomson J.A. (2001) Multilineage differentiation from human embryonic stem cell lines. Stem Cells 19:193–204.
45. Itskovitz-Eldor J., Schuldiner M., Karsenti D., et al. (2000) Differentiation of human embryonic stem cells into embryoid bodies compromising the three embryonic germ layers. Mol Med 6:88–95.
46. Reubinoff B.E., Itsykson P., Turetsky T., et al. (2001) Neural progenitors from human embryonic stem cells. Nature Biotechnol 19:1134–1140.
47. Schuldiner M., Eiges R., Eden A., et al. (2001) Induced neuronal differentiation of human embryonic stem cells. Brain Res 913:201–205.

48. Assady S., Maor G., Amit M., et al. (2001) Insulin production by human embryonic stem cells. Diabetes 50:1691–1697.
49. Kehat I., Kenyagin-Karsenti D., Snir M., et al. (2001) Human embryonic stem cells can differentiate into myocytes with structural and functional properties of cardiomyocytes. J Clin Invest 108:407–414.
50. Zhang S.C., Wernig M., Duncan I.D., Brustle O., Thomson J.A. (2001) *In vitro* differentiation of transplantable neural precursors from human embryonic stem cells. Nature Biotechnol 19:1129–1133.
51. Levenberg S., Golub J.S., Amit M., Itskovitz-Eldor J., Langer R. (2002) Endothelial cells derived from human embryonic stem cells. Proc Natl Acad Sci USA 99:4391–4396.
52. Kehat I., Gepstein A., Spira A., Itskovitz-Eldor J., Gepstein L. (2002) High-resolution electrophysiological assessment of human embryonic stem cell-derived cardiomyocytes: a novel *in vitro* model for the study of conduction. Circ Res 91:659–661.
53. Xu C., Police S., Rao N., Carpenter M.K. (2002) Characterization and enrichment of cardiomyocytes derived from human embryonic stem cells. Circ Res 91:501–508.
54. Pera M.F. (2001) Human pluripotent stem cells: a progress report. Curr Opin Genet Dev 11:595–599.
55. McDonald J.W., Liu X.Z., Qu Y., et al. (1999) Transplanted embryonic stem cells survive, differentiate and promote recovery in injured rat spinal cord. Nature Med 5:1410–1412.
56. Bjorklund L.M., Sanchez-Pernaute R., Chung S., et al. (2002) Embryonic stem cells develop into functional dopaminergic neurons after transplantation in a Parkinson rat model. Proc Natl Acad Sci USA 99:2344–2349.
57. Kim J.H., Auerbach J.M., Rodriguez-Gomez J.A., et al. (2002) Dopamine neurons derived from embryonic stem cells function in an animal model of Parkinson's disease. Nature 418:50–56.
58. Brustle O., Jones K.N., Learish R.D., et al. (1999) Embryonic stem cell-derived glial precursors: a source of myelinating transplants. Science 285:754–756.
59. Goldstein R.S., Drukker M., Reubinoff B.E., Benvenisty N. (2002) Integration and differentiation of human embryonic stem cells transplanted to the chick embryo. Dev Dyn 225:80–86.
60. Kondziolka D., Wechsler L., Goldstein S., et al. (2000) Transplantation of cultured human neuronal cells for patients with stroke. Neurology 55:565–569.
61. Drukker M., Katz G.,Urbach A., et al. (2002) Characterization of the expression of MHC proteins in human embryonic stem cells. Proc Natl Acad Sci USA 99:9864–9869.
62. Rideout W.M., Hochedlinger K., Kyba M., Daley G.Q., Jaenisch R. (2002) Correction of a genetic defect by nuclear transplantation and combined cell and gene therapy. Cell 109:17–27.
63. Rickard D.J., Sullivan T.A., Shenker B.J., Leboy P.S., Kazhdan I. (1994) Induction of rapid osteoblast differentiation in rat bone marrow stromal cell cultures by dexamethasone and BMP-2. Dev Biol 161:218–228.
64. Friedenstein A.J., Chailakhyan R.K., Gerasimov U.V. (1987) Bone marrow osteogenic stem cells: *in vitro* cultivation and transplantation in diffusion chambers. Cell Tissue Kinet 20:263–272.
65. Ashton B.A., Allen T.D., Howlett C.R., et al. (1980) Formation of bone and cartilage by marrow stromal cells in diffusion chambers *in vivo*. Clin Orthop 1:294–307.
66. Umezawa A., Maruyama T., Segawa K., et al. (1992) Multipotent marrow stromal cell line is able to induce hematopoiesis *in vivo*. J Cell Physiol 151:197–205.
67. Makino S., Fukuda K., Miyoshi S., et al. (1999) Cardiomyocytes can be generated from marrow stromal cells *in vitro*. J Clin Invest 103:697–705.

68. Hakuno D., Fukuda K., Makino S., et al. (2002) Bone marrow-derived regenerated cardiomyocytes (CMG Cells) express functional adrenergic and muscarinic receptors. Circulation 105:380–386.
69. Toma C., Pittenger M.F., Cahill K.S., Byrne B.J., Kessler P.D. (2002) Human mesenchymal stem cells differentiate to a cardiomyocyte phenotype in the adult murine heart. Circulation 105:93–98.
70. Azizi S.A., Stokes D., Augelli B.J., DiGirolamo C., Prockop D.J. (1998) Engraftment and migration of human bone marrow stromal cells implanted in the brains of albino rats—similarities to astrocyte grafts. Proc Natl Acad Sci USA 95:3908–3913.
71. Kopen G.C., Prockop D.J., Phinney D.G. (1999) Marrow stromal cells migrate throughout forebrain and cerebellum, and they differentiate into astrocytes after injection into neonatal mouse brains. Proc Natl Acad Sci USA 96:10711–10716.
72. Mezey E., Chandross K.J., Harta G., Maki R.A., McKercher S.R. (2000) Turning blood into brain: cells bearing neuronal antigens generated *in vivo* from bone marrow. Science 290:1779–1782.
73. Brazelton T.R., Rossi F.M., Keshet G.I., Blau H.M. (2000) From marrow to brain: expression of neuronal phenotypes in adult mice. Science 290:1775–1779.
74. Castro R.F., Jackson K.A., Goodell M.A., et al. (2002) Failure of bone marrow cells to transdifferentiate into neural cells *in vivo*. Science 297:1299.
75. Petersen B.E., Bowen W.C., Patrene K.D., et al. (1999) Bone marrow as a potential source of hepatic oval cells. Science 284:1168–1170.
76. Theise N.D., Badve S., Saxena R., et al. (2000) Derivation of hepatocytes from bone marrow cells in mice after radiation-induced myeloablation. Hepatology 31:235–240.
77. Krause D.S., Theise N.D., Collector M.I., et al. (2001) Multi-organ, multi-lineage engraftment by a single bone marrow-derived stem cell. Cell 105:369–377.
78. Wagers A.J., Sherwood R.I., Christensen J.L., Weissman I.L. (2002) Little evidence for developmental plasticity of adult hematopoietic stem cells. Science 297:2256–2259.
79. Poulsom R., Forbes S.J., Hodivala-Dilke K., et al. (2001) Bone marrow contributes to renal parenchymal turnover and regeneration. J Pathol 195:229–235.
80. Imasawa T., Utsunomiya Y., Kawamura T., et al. (2001) The potential of bone marrow-derived cells to differentiate to glomerular mesangial cells. J Am Soc Nephrol 12:1401–1409.
81. Reyes M., Lund T., Lenvik T., et al. (2001) Purification and ex vivo expansion of postnatal human marrow mesodermal progenitor cells. Blood 98:2615–2625.
82. Soloman, Sabrie "Sensors Handbook, 2nd Edition, McGraw-Hill Publishing Co., NY – ISBN 978-0-07-160570-0
83. Soloman, Sabrie "Sensors and Control Systems in Manufacturing 2nd Edition," McGraw-Hill Publishing Co., NY – ISBN 0-07-059626-3
84. Soloman, Sabrie "Affordable Automation," McGraw-Hill Publishing Co., NY – ISBN 0-07-059633-6
85. Soloman, Sabrie "Sensors Handbook, 1st Edition," McGraw-Hill Publishing Co., NY – ISBN 0-07-059630-1
86. Soloman, Sabrie "Sensors and Control Systems in Manufacturing 1st Edition," McGraw-Hill Publishing Co., NY – ISBN 978-07-10572-4

3 Additive Bio-Manufacturing

THE EVOLUTION OF REGENERATIVE MEDICINE

This chapter provides an exhaustive review introducing and updating tissue engineering used in regenerative medicine for biologists, engineers and medical doctors. It enhances the new technical opportunities available today to reduce comprehension barriers to demonstrate the potential of humanitarian organ Bioprinting.

Definitions:

Curli protein: is a type of amyloid fiber produced by certain strains of enterobacteria

Amyloid: is associated with human neurodegenerative diseases such as Alzheimer's disease.

CsgA and CsgB: The csgBA operon encodes the major structural subunit of curli, *CsgA*, as well as the *CsgB* nucleator protein.

(LGDW = Laser-Guided Direct Writing;
LIFT = Laser Induced Forward Transfer;
BAT = BioAssembly Tool;
hMSCs = human mesenchymal stem cells;
BMSCs = bone marrow stromal cells;
PEG = poly(ethylene glycol);
DA = diacrylate.
BioLP = Biological Laser Printing;
2PP = Two-Photon Polymerization;
ECs = endothelial cells;
SMCs = smooth muscle cells;
ADSCs = adipose-derived stromal cells;
DM = dimethacrylate;

Chitosan Scaffold: A copolymer derived from the alkaline deacetylation of chitin, used in consisting natural polymers, which have recently been developed more quickly and have gained more popularity.

Fabricating 3D Bioprint - Living and Non-living Substrates

Prominent scientists have persuaded bacterial cells to produce biofilms that can incorporate nonliving materials stimulated by natural materials such as bone, a medium of minerals and other substances, including living cells, such as "gold nanoparticles" and quantum dots, and can to be used as viable substances in 3D printing.

3D bioprinting allows the creation of exactly monitored 3D cell models and tissue builds, by engineering anatomically-formed substrates with tissue-like intricacy. Attributable to the high degree of manipulation on structure and arrangement, 3D bioprinting has the outstanding prospective to solve many acute unmet requirements in medical research, counting applications in cosmetics testing, drug research, regenerative medicine, and functional organ replacement. Tailored models of disease can be generated utilizing patient-derived stem cells, such as induced pluripotent stem cells (iPS cells) or mesenchymal stem cells. Contingent upon the implementation, a variety of materials, approaches, and cells can be utilized to harvest the anticipated tissue construct, Figure 3.1.

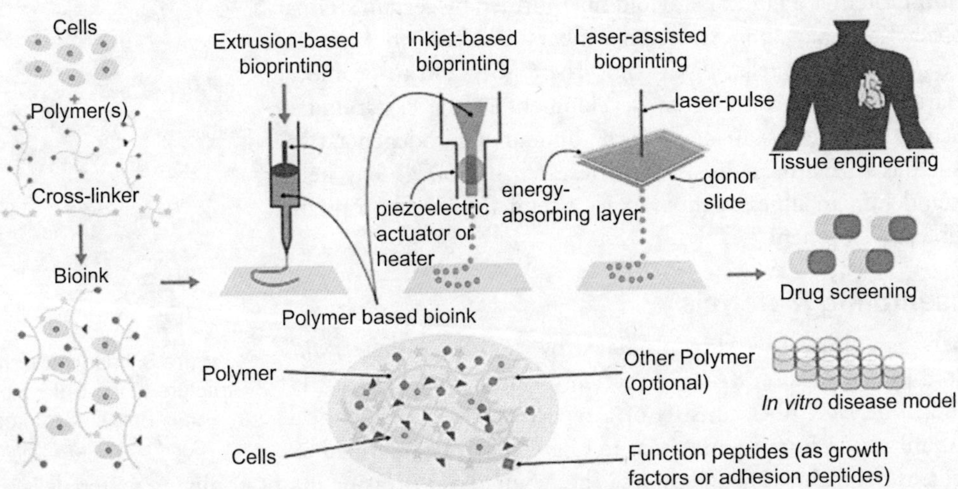

Figure 3.1: 3D Bioprinting Allows the Creation of Precisely Monitored 3D Cell Models and Tissue Constructs
Courtesy: Sigma-Aldrich

These "living materials" syndicate the advantages of live cells, which respond to their environment, yielding complex biological molecules, and yield multiple length scales, with the aids of nonliving materials, which add properties such as conducting electricity or discharging light. These complex biological materials can be used in "*Bioprinting Materials*."

Several biofilms act as natural biological systems, such as biofilms, shells and skeletal tissues, are able to combine multifunctional and environmentally responsive multiscale assemblies of living and non-living components. The genetic circuits and cellular communication circuits may be utilized to induce to regulate Escherichia coli "*curli*" amyloid production. The "*e-coli*" cells can be prepared to perform self-assembling amyloid fibrils across multiple length scales. These results in producing amyloid-based materials that either undergo autonomous patterning, or be externally controlled, Figure 3.2.

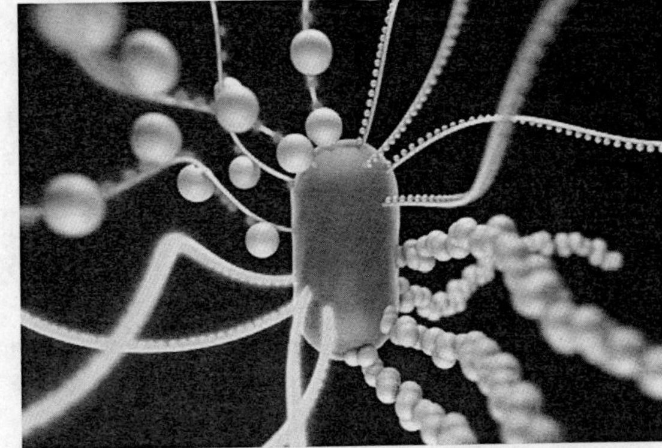

Figure 3.2: Enhancing Biological and Non Biological Substances Creating 3D Bioprinting Materials-
Courtesy: Yan Liang

Bioprinting ink with engineered cells can be produced through Gold nanoparticles (AuNPs) and quantum dots (QDs), interfaced with "curli" fibrils with inorganic materials to create an environmentally responsive biofilm-based electrical switch. This property can produce gold nanowires and nano-rods, co-localize AuNPs with CdTe/CdS QDs to modulate QD fluorescence lifetimes. Also, it can nucleate the formation of fluorescent ZnS QDs. This is the fundamental innovation to synthesize, patter, and control functional composite materials producing Bio-ink for 3D printing.

The new 3D printing materials represent a simple demonstration of the validity of thi innovation generate complex devices such as healing materials, solar cells, and self-diagnostic devices.

3D printing ink is constructed to permit the living and the nonliving worlds together to make hybrid materials that have living functional cells.

The *Curli* protein is a kind of amyloid fiber formed by certain strains of enterobacteria. They are extracellular fibers situated on bacteria such as *e.Coli* and Salmonella, Figure 3.3. These fibers aid to promote cell population behavior via biofilm development in the extracellular environment. Amyloid is related with human neurodegenerative illnesses such as Alzheimer's disease. The research in "*curli*" may help to understand human illnesses thought to ascend from inappropriate amyloid fiber development.

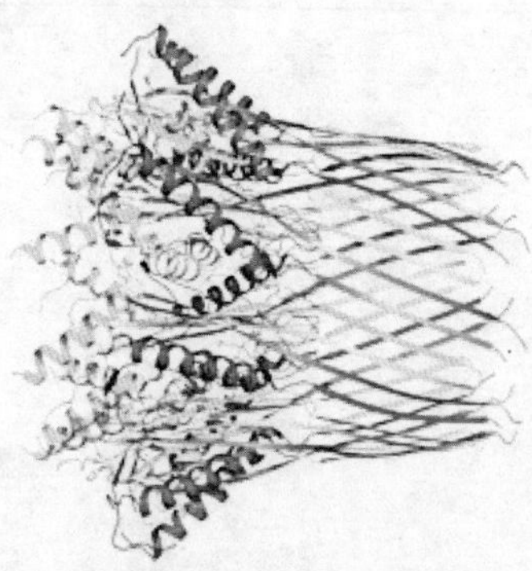

Figure 3.3: Crystallographic Structure of the Entero-bacteria *Curli* Protein Structures
Courtesy: NCBI Education

Self-Assembling Materials

"*Curli* fibers" is a natural biofilms produced by the bacterium"*e. coli.*" The amyloid proteins assist "*e.Coli*" to attach to surfaces, if used as 3D printing ink. The *curli* fiber consists of a repeating chain of identical protein subunits, which is often referred to CsgA. It can be modified by adding protein fragments called peptides. The required 3D printing ink of nonliving materials is produced by these peptides. The presence of nanoparticles of gold enables the use the produced substance as biofilms for 3D printing ink.

Different types of "*curli*" fibers can be developed to produce significant mechanical properties under certain conditions by controlling the biofilms' properties to develop gold nanowires with considerable dots. Additionally, cells to communicate with other cells to change the composition of the biofilm over time.

Initially, scientists disabled the bacterial cells' natural ability to produce CsgA, and then replaced it with an engineered genetic circuit that produces CsgA but only under certain conditions – specifically, when a molecule called AHL is present. This puts control of curli fiber production in the hands of the scientists, who can adjust the amount of AHL in the cells' environment. When AHL is present, the cells secrete CsgA, which forms curli fibers that coalesce into a biofilm, coating the surface where the bacteria are growing.

CsgA was produce and tagged peptides with after the scientists engineered *e.Coli* cells to. The peptides are composed of many clusters of the amino acid histidine, in the presence of a Tc molecule. Scientists continued to manipulate the two types of cells combined to grow in single colony. This allowed engineers to adapt the material composition of the biofilm by manipulating the quantity of AHL and a Tc in the environment within the mixture of tagged and untagged fibers if AHL and Tc are present. Row of gold nanowires that can create electricity once gold nanoparticles were added to the

Cell-Cell Communication

Also, scientists developed abilities to control the composition of the biofilm by allowing cells to coordinate with each other. They coordinated cells produced untagged CsgA and also AHL, which gave rise to other cells to begin producing histidine-tagged CsgA.

Gold particles allowed active "*curli*" to increase. Therefore, cells demonstrated the ability to communicate with each other. Consequently, cells could change the composition of the material over time. This demonstrated the ability to emulate the natural abilities to form bone. The environmental signals are aimed to force bone to create a material as a response, Figure 3.4.

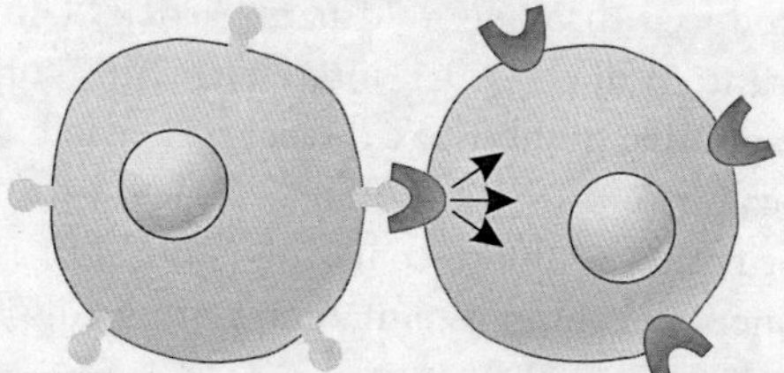

Figure 3.4: Cells Communicating by Interaction between Molecules Protruding from their Surfaces
Courtesy: Source: 2011 Pearson Education, Inc.

Scientists engineered cells to increase quantum dots to the curli fibers by producing curli fibers. Various peptide tag SpyTag, are added to bind them with quantum dots coated with SpyCatcher protein. Quantum dots and gold nanoparticles t, are generated from grown SpyCatcher protein cells along with the bacteria that produce histidine-tagged fibers. .

Batteries and solar cells can be developed through these hybrid materials. Engineers and scientists are also successful in coating the biofilms with enzymes that catalyze the breakdown of cellulose. This could be useful for converting agricultural waste to biofuels. Other important applications include developing diagnostic devices, and building scaffolds for 3D Bioprinting tissue engineering.

Bioprinting Cells to Produce Non-Living, Structural Biomaterials

NASA scientists successfully developed bioprinting engineered cells, which will be able to 3D print different materials, such as wood, bone, and enamel, Figure 3.5.

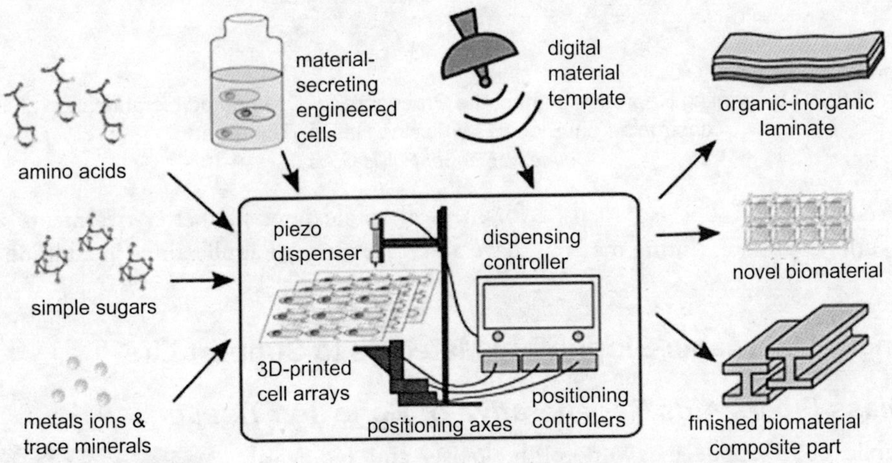

Figure 3.5: 3D Bioprinted Living and Non Living Substances

Courtesy: Image of NASA

This may be considered as an evolutionary biological work to produce Bioprinting substances. Bioprinting holds vast potential. It is already being used to create human tissue, and may soon be used to create matter in the form of non-living biomaterials as well, such as wood or bone. Also, scientists are working on a technique that could allow objects such as spacesuits, and even the body, to be repaired with 3D Bioprinted cells. With this type of technology, the 3D Bioprinter could be the one-stop place for all kinds of replacement parts.

It is believed that Bioprinting 3D arrays of cells engineered will discharge the necessary materials, the abundant *in situ* resources of atmosphere and regolith become organic, inorganic, or organic-inorganic composite materials. Novel, biologically derived materials have become possible to form to be used in 3D bioprinting applications.

Bioprinting cell mass in a gel media, which contains chemical signal, permits the technology to work. A piezoelectric bioprint head prints the gel media. The media produces clusters of cells that interact together and produce the desired biomaterials in a 3D pattern, Figure 3.6.

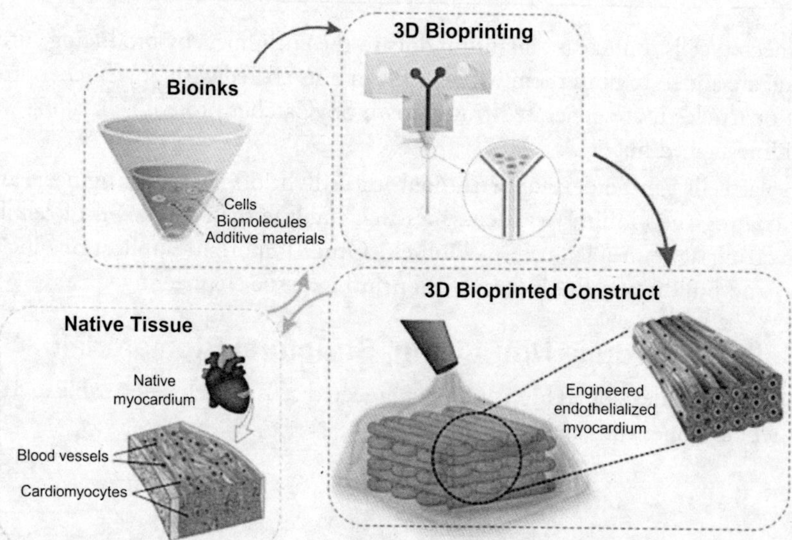

Figure 3.6: 3D Bioprinting Representation of the Procedure for Design and biofabrication of tissue constructs from Bionics Mimicking Native Tissue Parts
Courtesy: Elsevier Publications

The 3D cellular bioprinting is best used for medical applications for better treatments and tailored treatment solutions. 3D bioprinting materials have a massive range of applications in addition to the field of medicine.

Developing 3D Inorganic-Bioprinting Materials to Support Life

Woven Glass Fibers Aids Regenerative of Bone and Tissue

While a public belief about the work of biologists and biological engineers is that their main material concerns is living tissue, the ever more increasing interest in the relations between living and nonliving materials, have led the scientists to cultivate nonliving substances to generate Bioprinting materials. These scientists work, which is related to the life sciences to focus on organic substances with link to non-living materials, Figure 3.7.

Figure 3.7: 3D Matrix of Flexible Glass Fibers Utilized to Help Human Body Self-Repair
Courtesy: Blaine Brownell, AIA,

One example is bioactive silica. It is a type of ceramic biomaterial developed to support the growth of bone and similar organs tissues in the human body. Scientists have developed a 3D scaffold made of bio-active silica.

Electrospinning Weave Matrix of Flexible Silica Fibers

Electrospinning is the technology to knit a matrix of flexible silica fibers, where scientists developed a spongy matrix that matches the fibrous cellular matrix of bone tissue. Accordingly, such bio-friendly material propagated rapid bone growth without adversarial effects.

A Solvent-dissolved or melted polymer is fed through a syringe to the spinneret, Figure 3.8. As polymer reaches the spinneret tip, the initiated voltage creates an electrostatic force that pulls polymer from spinneret to electrode deposition surface. The fiber path, which is initiated with a stable region for approximately 2–3 mm before the polymer jet becomes unstable and causes lateral perturbations prior to fiber placement. 3D Bioprinted a matrix of fiber silicon implanted in a hernia sufferer patient, figures 3.8 and 3.9.

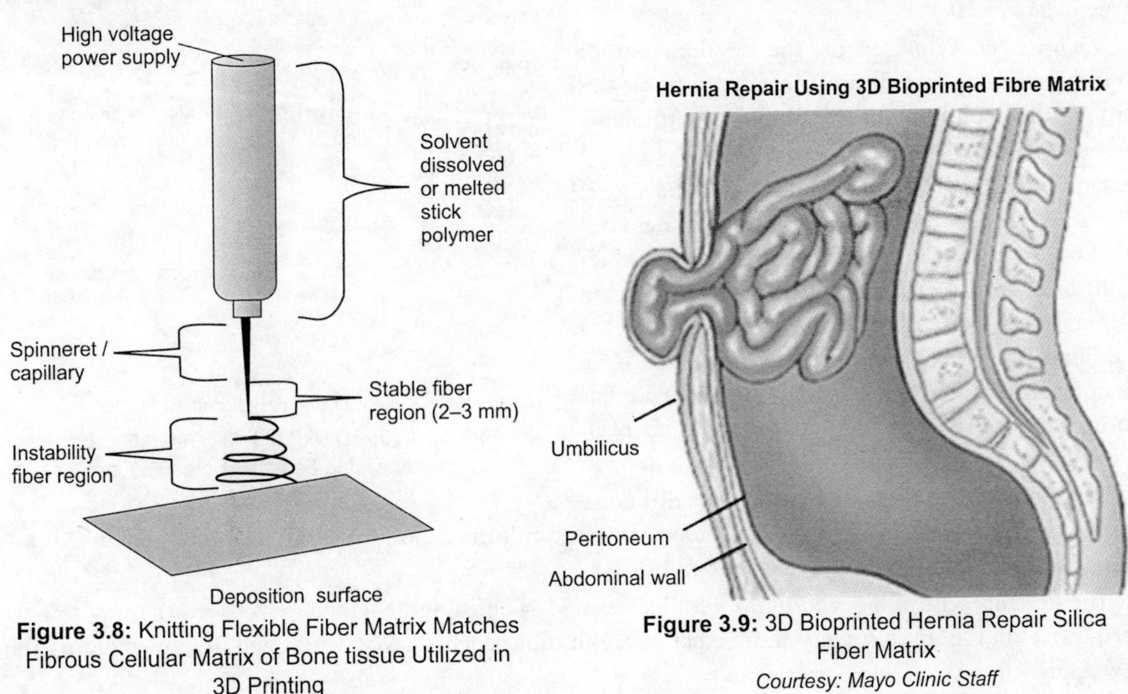

Figure 3.8: Knitting Flexible Fiber Matrix Matches Fibrous Cellular Matrix of Bone tissue Utilized in 3D Printing

Courtesy: Nanotechnology - IOP Publishing Ltd

Figure 3.9: 3D Bioprinted Hernia Repair Silica Fiber Matrix

Courtesy: Mayo Clinic Staff

Bioactive Printing Glass Stimulate Bone Growth

Bioactive printing glass can stimulate bone growth and bond with bone and soft tissue, by generating flexible cotton wool–like bioactive glass fibers, Figure 3.9.

In addition to helping bones repair, the scientists also think the material could be used in dental implants. Additionally, they are developing a bioactive glass with antibacterial properties for use in wound-closure systems.

Laser-Assisted Bioprinting

The development and advancement of Bioprinting is the breed of several combined technologies, including tissue engineering, computer-aided design, machining, fluid mechanics, miniaturization, laser scanning, and automation, , The use of computer-aided transfer processes for patterning and assembling of living and non-living substances assisted also in the creation of bio-engineered structures to serve regenerative medicine, pharmacokinetic and basic cell biology.

Bioprinting represents a fundamental shift in Tissue Engineering. The principle of advanced tissue engineering became more than planning cells onto a biodegradable scaffold, but rather to organize the individual substances, which are biologically relevant components of the tissue depositing them before maturation depositing layer-by-layer. Additionally, giving a better control on the cell distribution into 3D constructs, as well as automating the processes to minimize contamination during tissue engineering process. This will undoubtedly reduce the duration for manufacturing a fully functional engineered-tissue, Figure 3.10.

Scientists concentrated on the development of bone bioprinting to evaluate the new technologies through the application of the Bioprinting principle to bone tissue engineering.

Understanding the influence of the local microenvironment on human mesenchymal stem cell (hMSC) is one of the main goals of the 3D tissue-engineered constructs. The 3D tissue engineering assisted in the study of the effects of hMSC patterns and biomaterials composition, and stiffness on tissue function. The design of novel collagen-based materials through innovative cross-linking procedures has been also studied.

Figure 3.10: 3D Bioprinting Fabricating Bone Construct Using Laser Assisted Bioprinting
Courtesy: Alexandros G. Sfakianakis

Experimental developments incorporated the use of a second generation of Laser-assisted Bioprinter to permit the deposit of living cells and biomaterials in-vitro, in-vivo in safe, sterile and mechanized process.

The development of bioprinting *in vivo*, has assured scientist and engineers has paved the way to new perspectives in robotized surgery and medicine. Significant benefits have been gained in implementing the new approach in tissue engineering.

Bone Tissue Reconstruction through 3D Laser Printing

New innovative technologies have targeted the field of tissue engineering and biomaterials aiming to advance the field of bone tissue reconstruction through 3D laser bioprinting technology. The key element of success is contributed to Bio-imaging as it played a fundamental role forming tissue in an experimental model of defected bone. The development included mechanizing the bioprinting process to produce and control the reproducible regenerated tissue, Figure 3.11.

The employment of a second generation Laser-Assisted Bioprinter has enabled new perspectives applications in regenerative medicine.

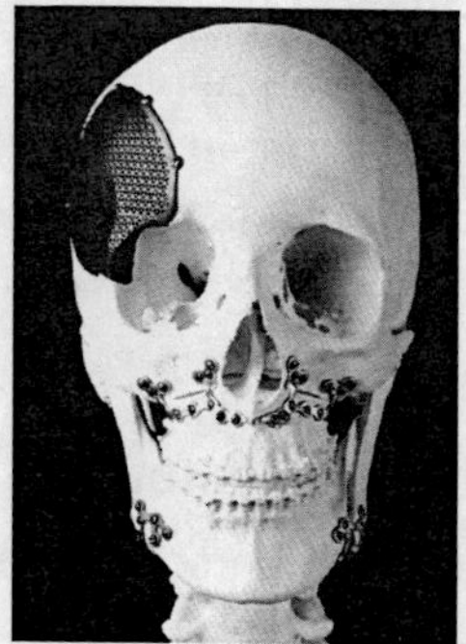

Figure 3.11: 3D Printing Revolutionizing Facial Reconstruction Creating Replacement Pieces Match the Injury
Courtesy: GE Additive

3D Bioprinting Drugs Revolutionize

The application of Bioprint drug is one of the applications that can be tailored specifically to each patient. The challenges are not on the actual 3D bioprinting machine, but on the abilities to generate the drug biodegradable ingredients including the active ingredient in a deliverable form to the bioprinting head in a form of disposable liquid. Using 3D Bioprinter to create a capsule or tablet allows doctors to alter a dosage according to the specific requirements, Figure 3.12.

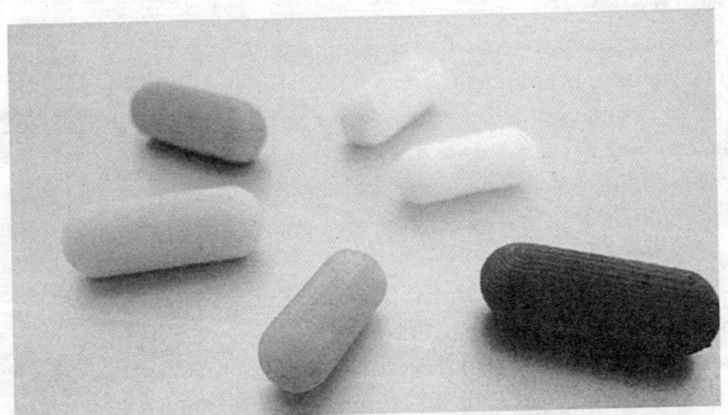

Figure 3.12: 3D Bioprinted Pharmaceutical-Grade Excipients
Courtesy: Drug Development and Delivery

A biodegradable material is the substance that is dissolvable when the drug is consumed by a patient. Additionally, scientists have developed substance to deliver chemotherapeutic compounds for cancer patients. The capsule can be printed in 3D Bioprinter allowing only the needed medicine, which is mainly the active ingredients, to be inserted before sealing it.

The 3D printers combined which the advancement of nano technology were recognized as opportunity for rapid prototyping using this fabrication method in medicine, and pharmacology. The use of nanoparticles with additive manufacturing, 3D Bioprinting became a viable technology to print biocompatible materials. The material can be mixed with antibiotics and other pharmaceutical compounds. Also, the compound can even be a time-release to be naturally broken down in the body over certain time.

It is evident that one of great advantages of the new Bioprinting technology will be its ability to tailor the contents of a drug for particular needs. This could mean a dose of antibiotics could be made stronger or weaker, depending on the requirements of the patient. It would also mean hospitals or pharmacies would not have to wait for deliveries from pharmaceutical companies. As long as they have the drug in question, they can create the dosage in the medical facility, or drug store.

Use of Consumer Printer

3D bioprinting technology is available for use to individual consumer as well as to industry alike. The difference is the scale of the printer, and the scale of the object to print. 3D printers are equipped with several software, which are either resident within the system or available as an open source software to the public. The use of 3 D bioprinter is not proprietary to one community over another community. There are numerous uses of 3D Bioprinters in the field of pharmaceuticals and medical industry. The vast majority of antibiotic implants, which are placed inside a patient undergoing an operation to ensure there is no risk of infection are made out of bone cements. Bone cements, are normally used to anchor joints, such as a hip or knee, have to be mixed by the experienced surgeon and are non-biodegradable. The implant must to be removed once the operation has been completed. These antibiotic implants can be made out of bio-polymers, which can be broken down by the body, thus no additional surgery is required.

Currently, embedding of additives in polymer requires industrial-scale printers to ensure proper dispersion throughout the extruded polymer. The tabletop scale Bioprinting method enables substance dispersion, allowing individuals to customize additives to the desired levels.

3D Bioprinting Hearing Aids

Thus far 3D Bioprinters have been used to create the outer shells for devices such as hearing aids. There are conservatively about 10 million hearing aids in circulation. The great advantage of using a 3D-printed hearing aid is that it gives the user much greater comfort, as it can be adjusted to the exact measurements required. This would simply not be possible if it was mass produced in a factory, Figure 3.13.

Figure 3.13: 3D Bioprinting Hearing Aids
Courtesy: ASIGA PRO 2 3D Printer. Photo via 3DP4ME.

Meanwhile, unit was reported that a woman in the Netherlands was given a replacement jaw, made out of titanium powder, which had been created by a 3D printer. Once the 3D digital design was uploaded, the jaw was split up automatically into 2D layers and then the cross sections were sent to the Bioprinting machine.

Creating 3D Bioprinting for Facial Self-healing

A new breakthrough in "self-healing" polymers encouraged scientists and engineers to further develop limbs and organs with broken veins, and utility pipes, can remedy themselves.

The discovery of the new advanced in self-repairing plastics, which is a hundred times stronger than previously developed in self-repairing plastics, making it capable of fixing scars as wide as 30 mm.

Accordingly, engineers and scientist were enthused by the veins and arteries within the human body, and set positive plans to recreate a synthetic vascular system capable of delivering the a healing substance as an agent needed to repair scars, Figure 3.14.

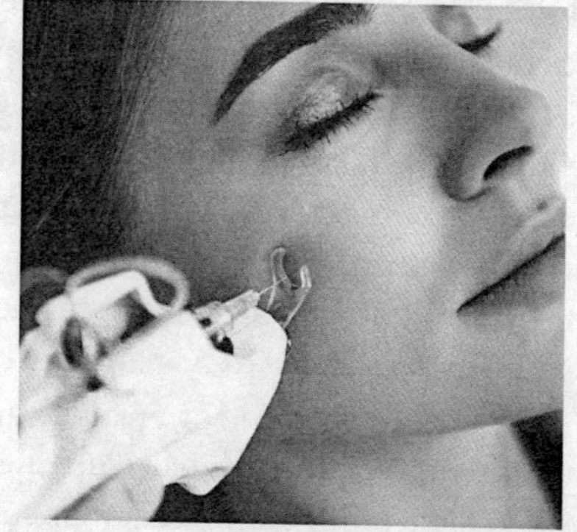

Figure 3.14: Bioprinting Plastic - Human Vascular system
Courtesy: CAN STOCK PHOTO INC

It is demonstrated that microscopic defects are self-healed; however, the re-growth of material wasted through disastrous damage requires a regenerative-like approach. Previously, filled polymers with a capsule containing a specific chemical capable of making repairs were used. As the polymer was penetrated, the healing liquid was released and formed a "bridge" the newly created hole.

Applying the same discovery by reenacting a network similar to the human vascular system, a much faster technique was developed to distribute the healing substance, which delivered in two different parallel paths to spread a the viscous substance over the affected area. When substance is hardened, the plastic regains about 65% of its original strength.

The scars greater than 35mm were filled for a period of 20-30 minutes. The mechanical functions are fully restored within a maximum of three hours. Potentially, the vascular systems are able to eliminate fracture in limbs by preventing small, noncritical cracks from expanding to critical sizes.

While the self-healing polymer innovation is remarkable, , scientists and engineers stated that further and extensive development are needed, in particularly in the area of propagating and in mechanisms to distribute the required substance before the plastic is be able to repair itself, specifically after more devastating ruptures. However, bullet strikes or bomb explosion consequences require exceptional investigations of the consequential scars to remedy.

When damage is unpredictable and uncontrolled, more complex and interconnected vascular networks will be necessary to provide sufficient vascular coverage and redundancy to circumvent channel blockage.

3D Bioprinting Valve Tissue

The seriousness of disease inflicting the Aortic valve (AV) is a serious health consequence that plague citizens of all ages. The most dangerous heart valve disease is the congenital heart valve defects. The newborns are especially vulnerable and it may lead to be fatal if unattended. Surgical replacement is the commonly recommended treatment for the defective AV. The adult standard procedure is fitting a prosthetic valve. However, the younger adults and children may not be able withstand the severity of the surgical procedure. Nonetheless, tissue engineering has proven worthy potential to overcome limitations of nonliving prosthetics, Figure 3.15. Additionally, the shortage of human donor supplies can be avoided through providing living tissues that can integrate to the recipient tissues to grow, and integrate.

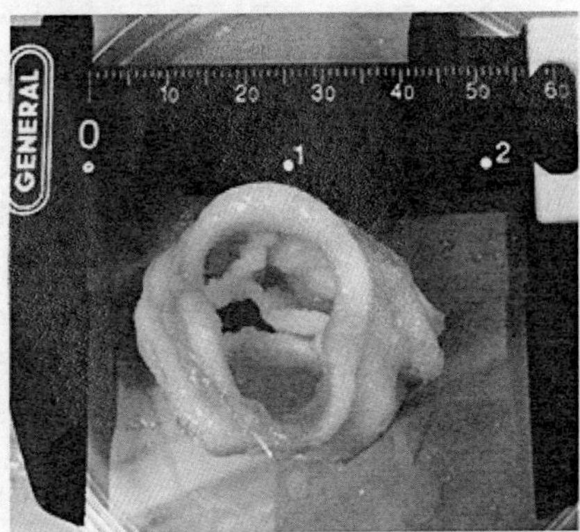

Figure 3.15: 3D Bioprinting Aortic Valve
Courtesy: Elsevier

The critical requirements for tissue-engineered heart valves:

1. The engineered valve must be able to mimic the physiological function of the native valve
2. The natural geometry and performance of the valve root must be maintained as essential factor for healthy coronary blood flow
3. The cusps, and sinus wall must be maintained as essential factor for healthy coronary blood flow
4. Tissue-engineered heart valves must have the same intrinsic asymmetry as the root, which prevents cusp deterioration.

The 3-D Bioprinting is a popular technology in the advanced additive manufacturing field. The 3D bioprinter has been developed to create adequate precision of the 3D structures from human living tissues. This 3D innovation in "Bioprinting" technology has been advanced to manufacture living heart valves, which possess identical anatomical architecture as the original valve.

3D Construction of Aortic Valve

3D Bioprinting can competently produce 3D biological members that are exact in configurations and mechanical intricacy. Employing a computer-assisted design (CAD) and computer aided as well as computer-aided manufacturing (CAM) are fundamental technologies to use in the 3D Bioprinting technology. The bioprinter deposits a thin film of living cells on top of each other. Depositing, accurate configurations, which precisely matches the heart valve dimensions is accomplished through a software routine called "Slicing." Slicing dissects the entire heart configurations into deposited layers of living tissues. The part is built vertically as the layers accumulate.

The Design of biological tissues was successfully facilitated through 3D bioprinting technology. The physical characteristics of the printed object can main their integrity. These characteristics include the natural geometry, stiffness, and biological cues that are required for successful functions. 3D Bioprinting of tissues generates both the biology and the structure of printed organ at the same time. Additionally, the specifics of custom patient's need have become imminent reality. Also, leaning disease pathogenesis made it possible to study drug effectivity an side effects. Making patient-specific tissue models has become a common practice to make living tissue replacements a specific matter directly related to patient's configurations.

Accordingly, bio-materials have been modified for 3D bioprinting technique, which includes special polymers substances, such as anionic polymer alginate, which has beneficial features for biomedical applications. The anionic polymer alginate enjoys low cost, and superior biocompatibility, s well as low toxicity. Alginate-based hydrogels are beneficial for 3D application due to their broad range of viscosities at room temperature for accurate printing.

The smooth muscle cells and valve interstitial cells were successfully 3D bioprinted thanks to utilizing alginate/gelatin hydrogel substances. The living aortic valve conduits were successfully 3D bioprinted with full anatomical resemblance to the native valve.

Moving Forward

Tissue engineering (TE) is defined as *"the application of the principles and methods of engineering and life sciences towards the fundamental understanding of structure–function relationships in normal and pathological mammalian tissues and the development of biological substitutes to restore, maintain or improve functions. (Google Scholar)"*

The advancement of tissue engineering paved the way to successful implementation of 3D tissue Bioprinting technology. In a like manner, the advancements of CAD/CAM combined with laser scanning, slicing techniques, and quantitative imaging analysis paved the innovative implementation of 3D bioprinting technology. United scientists, tissue engineers, manufacturers and industrial entrepreneurs have lead the world in innovative implementations of 3D bioprinting systems that reaped the success and the confidence in saving lives and improving the quality of life of countless recipients worldwide. 3D bioprinting of tissue has become the world standard in complex tissue fabrication. The challenges to advance the 3D bioprinting technology to solve the problems of the higher level of tissue complexity, glandular, and vascularized and innervated functional tissues are still lying ahead of enthusiastic scientists and engineers.

TRANSLATING BIOPRINTING TISSUE ENGINEERING

Platforms to Cancer Research

Technology platforms originally developed for tissue engineering applications produce valuable models that mimic three-dimensional (3D) Bioprinting tissue organization and function to enhance the understanding of cell/tissue function under normal and pathological situations. These models show that when replicating physiological and pathological conditions as closely as possible investigators are allowed to probe the basic mechanisms of morphogenesis, differentiation and cancer, Figure 3.16.

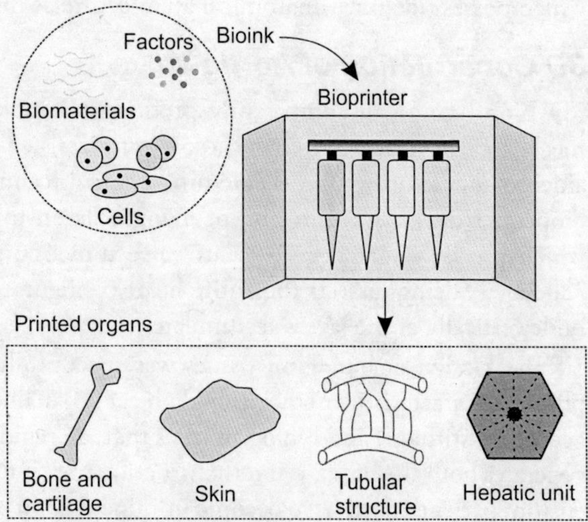

Figure 3.16: Implementations of 3D Bioprinting in Tissue Engineering

Significant efforts investigating angiogenetic processes and factors in tumorigenesis are currently undertaken to establish ways of targeting angiogenesis in tumors. Anti-angiogenic agents have been accepted for clinical application as attractive targeted therapeutics for the treatment of cancer.

Combining the areas of tumor angiogenesis, combination therapies and drug delivery systems is therefore closely related to the understanding of the basic principles that are applied in tissue engineering models.

3D Bioprinting Identified Complex Interacting Matrix Stiffness

The studies of the development and signaling in cancer through the implementation of 3D Bioprinting model systems have led to the identifications the effect of the complex interacting roles of matrix stiffness and composition on growth factor receptors. The therapeutic targets for future cancer therapies are understood through the plasticity, regulation, and suppression, including inhibitors of angiogenesis in tumors.

3D bioprinting is the new delivery systems expected to greatly improve the skill to distribute drugs locally and in therapeutic concentrations to intended location in living organisms.

Development of Scaffold - Matrix-Based Culture Systems

The design and development of scaffold is the fundamental factor to ensure successful and accurate creation of an object through 3D print constructed layer by layer. - Printing culture tissue using 3D Bioprinted culture has given rise to expand the use of the 3D bioprinter in the innovation of tissue engineering, and sparked the design and development of scaffold. Manipulating cell culture is in contrast to the plastic conventional tissue culture. Cell culture delivers more physiological geometries as well as microenvironments that is more closely a representation of the natural extracellular matrix (ECM) cells found *in vivo*. Accordingly, tissue engineering techniques has influenced other biomedical research areas. For instance, tissue engineering paradigms provide physiological models of human tissue that enabled for studying disease pathogenesis as well as screening the effect and toxicity of drugs *in vitro*, Figure 3.15. Therefore, the use of these tissue counterparts may meaningfully subsidize to the development of new therapeutics.

The practical properties of cells can be revealed and manipulated to an extent that is not likely in either 2D cultures or animal experiments. In particular, early signs of tumor growth before effective vascularization appear to be closely replicated. 3D cultures of tumor cells within a short timeframe, progress empty centers that resemble the necrotic areas of *in vivo* cancers. These areas are observed not far from nutrient and oxygen supplies, Figure 3.17. It has become apparent that through the development of a vascular supply, those 3D cultures are better suited than 2D culture techniques to study characteristics relevant to angiogenesis itself.

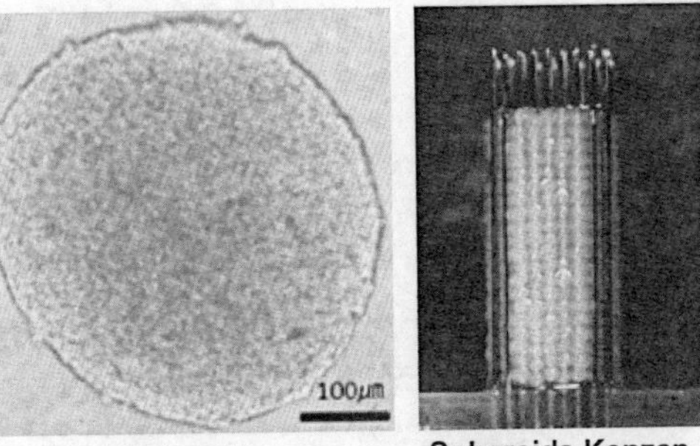

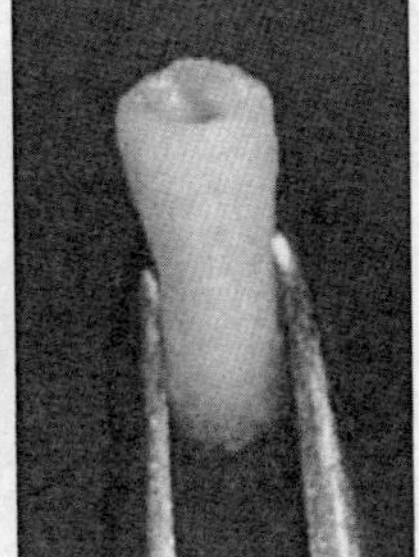

Figure 3.17: New Scaffold-Free 3D Bioprinting Method Available in USA
Courtesy: 3ders.org

Although *in vitro* analysis of angiogenesis provide narrow promises, scientists nd engineers established how characteristics of a 3D fibrin matrix were used as conduit towards growing of suspended endothelial progenitor cells in a way that lumen-containing blood vessel-resembling formation developed, a characteristic not attainable in 2D culture. As for the tumor physiology, the propagation of tumor cells cultured in 3D is slower, thus more physiological than that of monolayer cultures.

3D Bioprinted Cultures Exploring Different Cell-Types

The important benefit of 3D Bioprinted cultures is the communication of different cell-types can be studied, e.g., penetration of tumor spheroids by endothelial cells (ECs)has been verified and it depends not only on the production of pro-angiogenic nature by tumor cells but also on the expression of endothelial cells (ECs).

It was demonstrated that the control of tumor-genesis was considerably influenced by the 3D micro-environment. Therefore, employing 3D cell culture approach would encourage synthetic and biomimetic matrices in tissue engineering to improve *in vitro* and *in vivo* tumor simulation. In addition, developing means to tumor growth. The tissue engineering applications in tumor angiogenesis can be used to support the formation of a vascular network. This may enable the function of endothelial lineage cells, and the pro-angiogenic to grow. The most important goal is to development of anti-angiogenic therapies and new drug delivery systems including growth factor or cell therapy, Figure 3.18. The study of blood vessel development from a tissue engineering using 3D bioprinting technology as well as a tumor angiogenesis background would expedite the development of the novel therapeutic drug.

Figure 3.18: 3D Bioprinting Approaches for Tissue Engineering Applications
Courtesy: Biotecnika

3D Bioprinting Tissue Engineering

Tissue engineering is a term often associated with the term regenerative medicine. Not long ago, scaffold- and matrix-based tissue engineering involved the combination of a scaffold with cells or biomolecules and rise of the repair and regeneration of tissues, Figure 3.17.

The three-dimensionality of cell culture vividly advanced the field of tissue engineering enabling progress of constructs *in vitro*, resulting in the development of biomaterials that enabled cell culture in this new dimension, Figure 3.19. The advances in tissue engineering have led to the design of reliable scaffold to represent the geometry, chemical composition and signaling environment of the natural extra-cellular matrix.

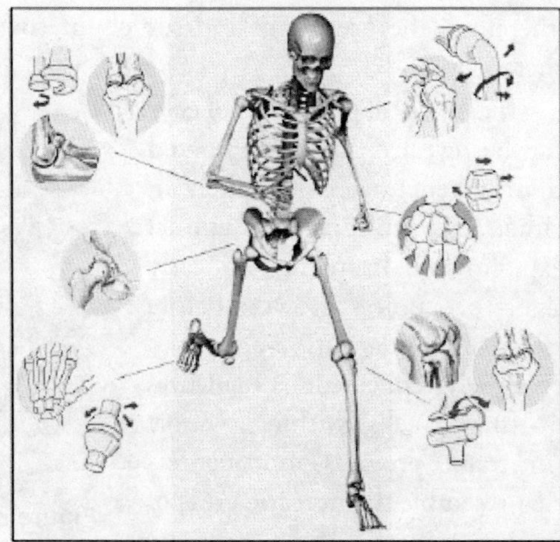

Figure 3.19: Schematic View 3D Bioprinted Joints in Human Skeletal System
Courtesy: Pulsus Group

Tissue engineering additional benefits:

1. It can be a tool biomedical research area.
2. It can be useful for *in vitro* models of human tissue
3. It can be explore disease pathogenesis in cancer
4. It can screen drug effects for the development of therapeutics.

Enhancing Future Cancer Research

The 3D culture systems built on tissue engineering constructs to heal prostate and ovarian cancer. Scientists have successfully shown how tissue engineering technology enabled *in vitro* and *in vivo* tumor modelling enhances future cancer research.

Ovarian Cancer

Scientists and engineers have developed 3D Bioprinting technique to study bone metastasis related to prostate cancer *in vitro* and *in vivo*. Additionally, scientists and engineers have shown how a synthetic biomimetic hydrogel system is used to investigate biological queries related to ovarian cancer.

Stereolithography

Stereolithography is a unique 3D bioprinting system created to construct polymers such as poly-lactide resin or polypropylene according to their favorable mechanical properties. . Cells are often directly encapsulated when using stereolithography 3D Bioprinting technique, hydrogels, Figure 3.20.

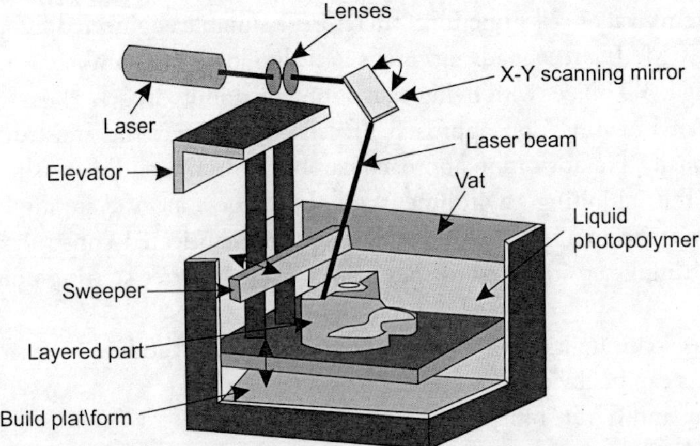

Figure 3.20: Illustration Stereolithography Process Using Layers of Liquid Photopolymerizable Polymers
Courtesy: MilliporeSigma

Bioprinting Very High Resolution 3D Scaffolds

A two-photon polymerization (2PP) technique was implemented for printing 3D scaffolds to generate a very high resolution down to the nano-meter (nm) scale. The photosensitive hydrogel is mixed to the photo-initiator together with the photo-polymerized through absorption enabling a laser beam projected on the sample. The resulting spectrum due to photo-initiator absorption is half-wavelength of the laser radiation for the two-photon polymerization. 3D arrangements can be formed by locating the laser spot as predetermined geometry. The remaining un-polymerized material can be washed away. The scaffolds can be precisely fabricated with resolutions down to 100 nm using the 2PP approach, Figure 3.21.

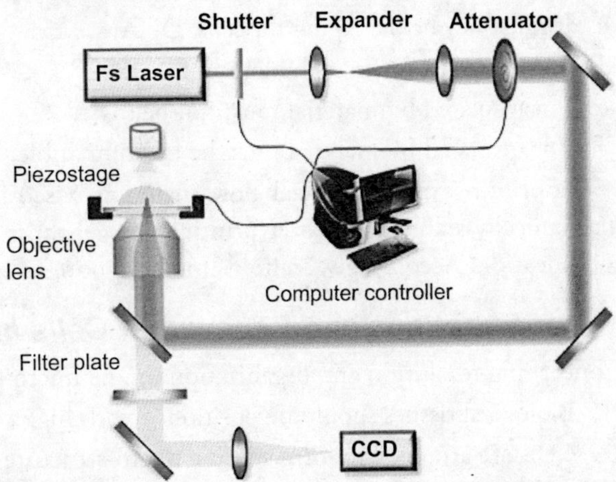

Figure 3.21: Two-Photon Polymerization Microfabrication of Scaffolds: Advanced 3D Bioprinting - Tissue Engineering and Drug Delivery
Courtesy: Royal Society of Chemistry RSC 2019, 9, 34472-34480

3D Tissue Bioprinting

3D tissue printing is referred to bio-fabrication, Bioprinting biological constructs, which extends several cell types and biomaterials simultaneously in various sizes. Scaffolding of tissue engineering traditionally is manufactured through a technique known as salt leaching, gas foaming, or freeze drying. Cellular functionality can best be controlled through 3D milieu. During 2D patterning, 3D cellular micro-environments provide substantial advantage for investigating natural cell behavior, extending to proliferation, migration, differentiation and drug response.

Stereolithography is used to print and bio-plot 3D cellular constructs. 3D tissue bioprinting does not create a local 3D environment for cells; however, 3D cell patterning does. A support structure development is fundamental for the survival of cells functions; therefore, a suitable engineered ECM is required until the cells create their own ECM. The recent advancement in technology is to combine scaffold printing and cell patterning constructing a 3D object with living cells submerged in hydrogels.

Precise placement of biofabrication approach of cells throughout the construct can be successfully accomplished. Additionally, biofabrication approach can be also achieved for the simultaneous placements of several cell types. The inhibiting environments of the process are the sterility during and after the fabrication, e.g., appropriate temperature and pH range. Bio-fabricated constructs, can be printed cell aggregate directly on stimuli-sensitive gels as they will provide a better starting point encouraging faster functional tissue.

3D bioprinting in cell culture is being interchanged with biofabrication .

Basically, the process can be divided into three phases:

1. Bioink preparation, and CAD "blueprint";
2. Printing of the biological solution
3. Cultivation of the printed construct

The post-processing step is stated as follows:

a. Intended to induce maturation and transform the printed construct into a functional tissue.
b. The printing method must not be toxic.
c. It must not be damaging for cells' DNA.
d. The cell must retain its biological function.
e. Encapsulated biomaterial must not be toxic,
f. Encapsulated biomaterial must be biocompatible, and solidify in response to specific stimuli.

Bioprinting require special flow including viscosity and surface tension. In order to accomplish a mechanically stable tissue after printing the gel must provide the capacity to undergo fast and nontoxic solidification. Accordingly, scaffolds must support cell growth and tissue formation *in vitro.*

Mimic Functions and Behavior of Cells in 3D Environments

Concerning resolution and organization on the micro-scale there are some contradictions:

1. Biological tissues should be positioned with high resolution.
2. Cells are arranged to form specific micro-structures.
3. Essential to mimic functions of cells in 3D environments *in vitro* construction of 3D architecture.
4. Cells have organizational capacities as to organize themselves to form tissues.

Considerable amounts of cells must be handled properly in order to create a tissue and processed it as quickly. In most tissues, vasculature is crucial as blood is the conduit of oxygen and nutrients for living cells.

Also, blood eliminates waste. Accordingly, proper perfusion system is vital for cell survival and activity. The success of 3D bioprinting is dependent s on the gelation state of its hydrogel layers.

3D Bioplotting and Inkjet Bioprinting

3D biofabrication technologies used to deposit biological material layer-by-layer to to create a 3D construct. Stereolithography technology employed in , 3D bioprinting, 3D bio-plotting, as well as inkjet bioprinting. Table 3.1, which explains the individual results and details of the studies accomplished using of 3D bioprinting in cell-hydrogel, constructs Table 3.1.

Table 3.1: highlights the main differences of various 3D tissue engineering techniques

Technique	Stereolithography	LIFT/LGDW	Inkjet	Bioplotting
Resolution *Load volume*	1.3 µm nl–mL range	10–100 µm > 500 nL	85–300 µm mL range	Some 100 µm mL range
Scale-up	Depending on the stereolithographic	Not very suitable due to stacking of 2D layers	Multiple print heads	Very suitable, no geometrical limits
Element size	80–125 µm	< 100 µm	30–200 µm	Some 100 µm
Pros	Very high printing resolution possible	High cell densities possible	Best known and cheapest technique	Printing of large constructs in cm range possible
Cons	Only photosensitive hydrogels possible	Not suitable for 3D constructs in the mm range	High shear forces; clogging	Lowest printing resolution of several 100 µm compared to other methods

3D Biofabrication Methods

Stereolithography - Liquid Photosensitive Polymer

A liquid photosensitive polymer is used in the Stereolithography technology, to solidify the liquid through exposure to UV or laser light. Either UV light or Laser will cure the pattern traced and adheres on the resin to the layer below. Upon solidification of a single layer, the following layer of liquid polymer is solidified after being patterned by the laser. A cleaning process is applied as soon as a 3D part is constructed. Cells can then be encapsulated in photo sensitive hydrogel again through induced laser or UV light. It is worthy to note that the laser has the capacity to crosslink (join polymer chain) a specific pattern within a single layer. As a result, layer-by-layer the hydrogel-cell substance is applied to crosslink a special designed pattern. The 3D bioprinting tissue engineering technique has been established to be especially useful for the development of compound structures which could not be traditionally built. Also, the 3D bioprinting of tissue engineering was adapted to make the procedure more applicable to augment cell growth and cell propagation.

Cell-Encapsulation

The most attractive method to bond hydrogel-forming polymers is the use of Photo-polymerization. The Photo-polymerization technique results in a stable matrix for cell encapsulation. Encapsulating cells in hydrogels shields the embedded cells microbial harms that may cause cell-damage.

Laser-Guided Direct Writing (LGDW) - Cell Patterning

Laser-Guided Direct Writing (LGDW) was presented as a 2D cell patterning technique. The major task of the LGDW was to:

1. to successfully pattern cells onto a substrate.
2. Provides a robust possibility to expand the technique into 3D.
3. Facilitate 3D fabrication.

In order to facilitate the fabrication of 3D hydrogel must be layered on top of each deposited cell layer. Each single deposited layer is followed by a process of subjecting cells to the laser beam, and then forces it onto the substrate. Three layers of cells measuring integrated with hydrogel layers of 500 μm thick were interchangeably patterned on top of each other. This method resulted in a accurate 3D pattern of cells. Endothelial cells ECs reached a viable to 89% after utilizing LGDW technology.

Laser Light Inducing Biomaterial Deposition

Biological materials become active after being deposited, usually by utilizing induced laser light techniques, such as:

a. Matrix Assisted Pulsed Laser Evaporation Direct Write (MAPLE DW) – combined with laser induced forward transfer (LIFT).
b. Biological Laser Printing (BioLP)- combined laser induced forward transfer (LIFT).

Skin regeneration and stem cell therapy are the aim to most advancement is 3D tissue engineering. As a result, Skin cell lines, and human mesenchymal stem cells (hMSCs) have been arranged in well-defined patterns. The purpose is to prove the of the influence of the laser induced forward transfer (LIFT) on cell survival, propagation, cell death (apoptosis) , and DNA harm after 3D bioprinting. A cell survival rate of 98% was maintained. Also, all cell types maintained their ability to multiply after deposition. Skin cells and hMSCs did not show an increase of death in living cell, (apoptosis) or DNA damage. The 3D biofabrication methods, Laser-Guided Direct Writing (LGDW), and the laser induced forward transfer (LIFT) techniques have proven their indispensable values as cells are still patterned enabling the whole construct only to become 3D by adding a hydrogel layer after every patterned cell layer.

Inkjet Bioprinting

One of the most promising 3D bioprinting tissue technologies is the Inkjet Bioprinting application. This is contributed according to its unique features of high-output efficiency, lost cost of effectiveness mechanization. Inkjet printing is a unique non-contact technique which receives digital data and slices it layer-by-layer by depositing ink drops on printed layer successively. Figure 3.22, illustrates inkjet printing diagram.

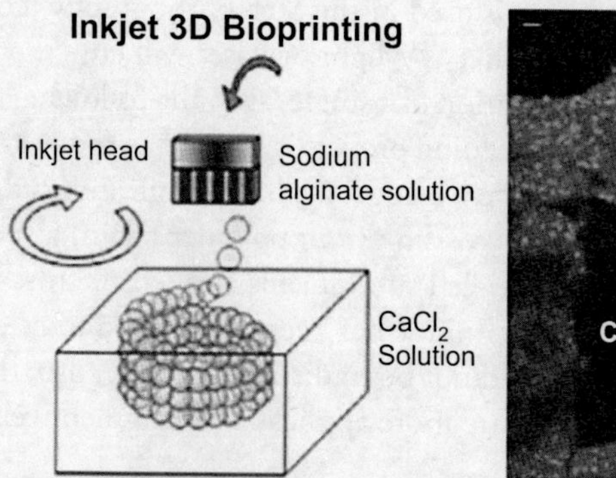

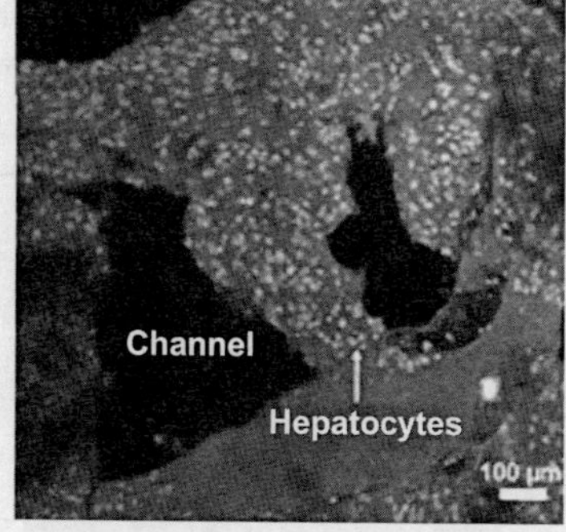

Figure 3.22: 3D Biofabrication - Tubular Structure of 200 μm Using Inkjet Technology

Courtesy: MDPI

Group of cells are dispensing though printing, using a modified inkjet printer pf specific volumes of liquid substance in well designated and defined 3D locations. Thermosensitive gels were used to create arranged layers sequenced on top of each other. An "Aortal" cells systematically reached 3D cellular assemblies. During culture construct, cells have been revealed to fuse within the hydrogel. Several layers of fibrinogen, thrombin and neural cells were printed layer-by-layer. The result is creating a 3D sheet that illustrated healthy functions of the neurons when maintained by using the thermal inkjet printer. Further, cells differentiation capabilities were illustrated. The cell neurons and mesodermal functionality after printing were successfully achieved cells.

The homogeneously sized alginate precursor beads are deposited in circular patterns layer-by-layer into CaCl2 cross-linker solution forming a tubular structure due to gelation; (B) LSCM observation of a 3D bioprinted construct consisting of hepatocytes encapsulated in gelatin/chitosan hydrogel after 6 days of culture.

Bioprinting Ruptures Cells – The Down Fall of Bioprinting

Inkjet printing is affordable, versatile and enjoys high speed printing. However, but due to the small dispensing orifice, high shear forces cells are subjected to high impact forces during operation, which may cause failure or malfunction to the cells. It is also possible that the small orifice size may not dispose ink when printing high cell densities. The industrial print heads used have a nozzle diameter of less about 300 μm. It can print small drop volumes of about 0.015 μL per orifice. Modifications of inkjet technology to safe biological applications are underway, to prevent of droplet ejection and ink clogging.

To improve printing speed utilizing small volumes of about 0.015 μL, inkjet printers may be fitted with multiple nozzles to print several types of cells simultaneous printing of one material. The possible number of different inks is dependent on the amount of ink cartridges in the 3D bioprinter. Each cartridge is filled independently with cells, different hydrogel precursors. Also, substance of some growth factors can fill a printing cartridge. Inkjet printing demonstrated its successfully deposition of different cells including neurons in 3D. Fast dispensing nozzle size of about 30 μm to 300 μm could be accomplished.

3D Bio-plotting - Extrusion of Continuous Filaments

3D bioplotting is a newer techniques used in 3D biofabrication technology. Bioplotting is a technique developed to extrude a continuous filament of substances. it is aa more advantageous system compared to drop-based deposition in inkjet technology. The bioplotting system, the ink is stored in a reservoir equipped with orifices controlled an X-Y-Z Cartesian robotic mechanisms. One

The bio-plotter was originally used to deposit hydrogel scaffolds without cells growing bone tissue engineering. Bone tissue mixed with different hydrogel precursor's solutions and printed with the bio-plotter system. Different fiber diameters and scaffold constructions were created Cell survival was successfully detected for the printed substances. Differences cell survival could be observed depending on thehydrogel used.

A more advanced bio-plotter was used to print biomaterials, biomolecules and cells with a wide range of viscosities. Validation study was performed sustaining survival rate between 46% and 86% depending on the nozzle diameter. The diameter varies between 200-500 μm. Also, the results indicated that the smaller the nozzle tips the less rate of cell survival. The printed path width was twice the nozzle diameter.

Robotic Printing

A multi-axis printing robotic equipped with printing platform was used to provide for 3D biofabrication. Tissue engineering was printed where the printed was encapsulated substance pre-crosslinking with calcium sulfate. The printed result was validated to show the successful application of the multi axis robot in tissue engineering with a cell survival of 94% and an impressive homogenous cell distribution.

Additionally, scientists and engineers erected 3D bio-plotter equipped with fine needles to investigated the feasibility of employing a number of different cell types in several hydrogels.

Bioprinting Artificial Liver

3D Bioprinting hepatocytes and gelatin hydrogel precursors encapsulated in gelatin/chitosan, an artificial liver was made. An advanced method was implemented by printing adipose-derived stromal cells (ADSCs) embedded in gelatin/alginate/fibrinogen hydrogel precursor to form a vascular-like network next to hepatocytes in gelatin/alginate/chitosan. The 3D bioprinted construct consisting of hepatocytes and gelatin/chitosan, illustrated in Figure 3.23.

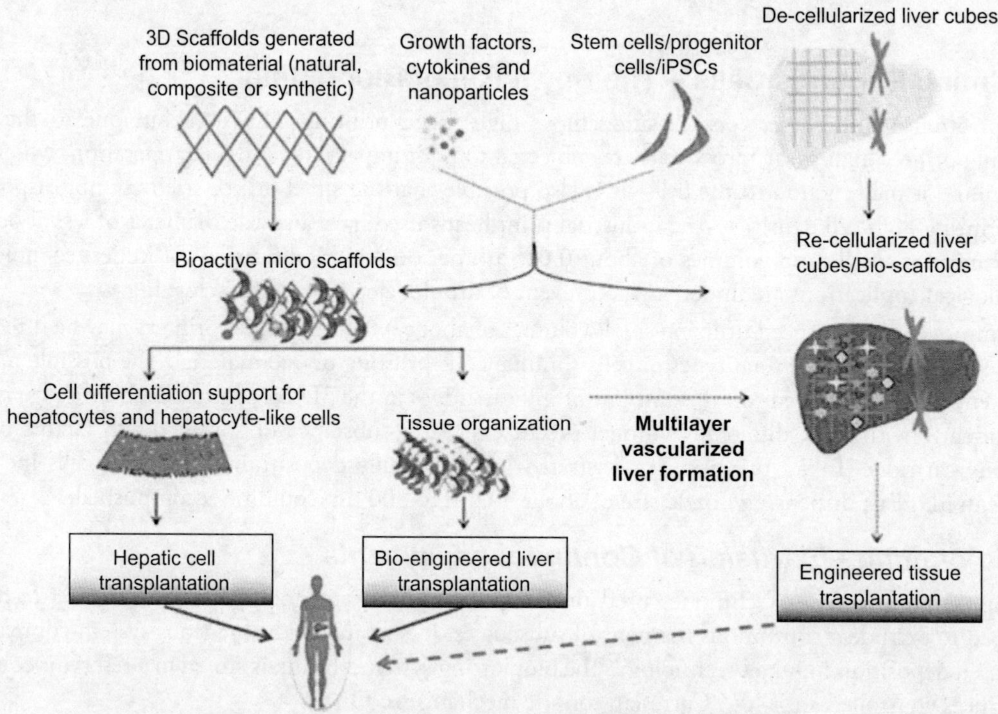

Figure 3.23: 3D Bioprinted Construct of liver Tissue Engineering of Current Approach – Dotted Lines - Proposed Approach
Courtesy: BoD Publisher: Norderstedt, Germany

A 3D bio-fabrication system with an accuracy 10 μm was developed to incorporate a multi-orifice system. The implemented orifice in this system varies from 30-to 500 μm in diameter. Each orifice was activated by an electromechanical mechanism. The 3D bio-fabrication had several possibilities enabling to simultaneously deposit cells, scaffold materials, growth factors, and bioactive compounds.

The cells survived in printed culture more 3 weeks post-printing, and the cells proliferated 6 times more than the original cells. The system data were used to validate the expected specifications that bio-plotter ink that the flow rate is accurate according to the among deposited and the amount proliferated.

Nano-Bioprinting Technique

Scientists and engineers explored a new hybrid nano-bioprinting technique, which syndicates the initial patterning abilities of syringe-based cell droplets with the active patterning aptitudes of super paramagnetic nanoparticles. The data showed after printing, there was no influence on cell capability detected directly, if paramagnetic nanoparticles were present in the alginate. The data showed that in a low viscous alginate gel that cells in paramagnetic nanoparticles could be moved with a magnet depending on the magnet intensity.

In order to construct functional living structures, a depositing mechanism was developed to enable the self-organizing capacity of cells and tissue. The purpose of the living structure is to copy early structure-forming process in the embryo. The process is represents the liquid-like behavior of tissues for bio-ink fusion. Scientists and engineers concentrated their activities on using 3D bioprinter in several phases of the fabrication process:

1. Extruded a cellular of embryonic cardiac cells.
2. Cut embryonic cardiac cells into equal-size multicellular cylinders.
3. Cylinders were incubated and turned into spheroids.
4. The spheroids used as bio-ink cartridge.
5. Bioprinting polymerized hydrogel including vascular endothelial growth factor.

The spheroids fused after and detected that ECs organized into vessel-like structures. Also, this method is also often referred to as scaffold-free tissue engineering. The spheroids were utilized to print the 3D constructs layer-by-layer.

Comparison between the Different Biofabrication Techniques

Inkjet:

Advantages: When various 3D bioprinting techniques were compared to each other it became evident that that inkjet printing is the most simple applied technology reflecting its least expensive method of all cell printers. Also, it has the benefit of a non-contact printing technology. The printing speed is fast. The orifice size varies between 30 μm to 200 μm. It inkjet high resolution vary between 85 μm to 300 μm.

Inkjet:

Disadvantages: High shear forces is generated during cell printing due to the small orifice appear during cell printing. The shear force may damage the cells. Printer high stress and generated heat create harsh printing environments. Delivering viable cells may not be assured due to several technical barriers in printer development. Accordingly, the cell viability decreases significantly, in particularly for a relative larger objects to print, which need longer time to print.

Laser Induced Forward Transfer - LIFT

Printing living cells successfully accomplished by the only available technology LIFT The orifice in LIFT and modified LIFT were efficiently developed to provide successful printing of living cells, Figure 3.24. The orifice development and setup have avoided the risk of printer clogging. Thus, high density cells of 108 cells/ml as well as viscous fluids can be successfully printed with resolutions of 10 μm to 100 μm. A cell viability of 95–100% is assured.

With respect to the influence of cells on substrates during the printing procedure, both inkjet and the modified LIFT techniques displayed comparable droplet velocities of 4–20 m/s and similar impact forces.

Laser-Guided Direct Writing – LGDW

High resolution cell deposition is considered a synonymous with LGDW. Single cells can be subjected to the laser and deposited onto a surface, which permits an accuracy of ±50μ. A small amounts of biological liquid is required, a small 3D construct consisting of a single cell type may be built up, Figure 3.24.

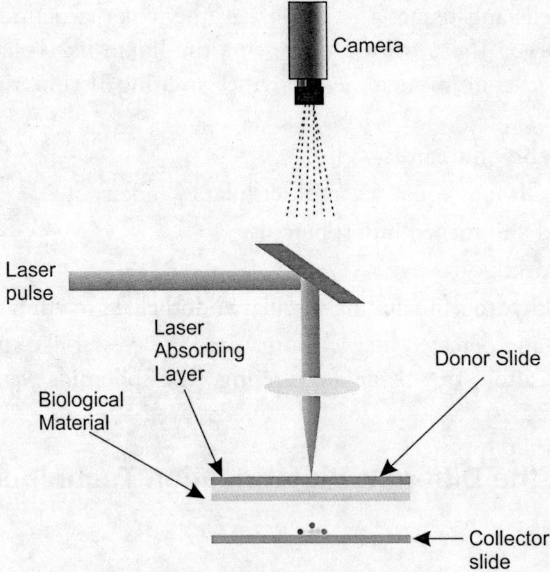

Figure 3.24: A Schematic of Modified "Laser-Induced Forward Transfer LIFT" - Technique for Single Cells
Courtesy: (Gruene et al., 2010; Koch et al., 2012; Ringeisen et al., 2004).

Stereolithography

Stereolithography was the first techniques in 3D Bioprinting layer-by-layer manufacturing providing a layer resolution of up to 1.3 μm, and a laser spot size of 80–125 μm. This technique permits only photosensitive materials. This results in a narrow choice of materials since these hydrogels have to be biocompatible for cell encapsulation.

The user of 3D tissue engineering bioprinter must consider that each of the various bioprinting methods as a competing technology.

The following factors are the main determinant of the type of bioprinter to be used:

1. Resolution of the different techniques,
2. High resolution print required,
3. Object size,
4. Layer widths requirement,

Laser-Guided Direct Writing - LGDW

High resolution prints can be successfully achieved by Laser-Guided Direct Writing - LGDW. Laser-

Guided Direct Writing LGDW is an acceptable technique to investigate cell-cell interactions. The Factors that determine the acceptable technique may be defined as follows:

1. The load volume of the particular printers.
2. Syringe-based deposition.
3. Inkjet based systems to enable large amounts of cell-hydrogel mixture.

The disadvantage of this technology is that a certain amount of the substance remains in the cartridge and/or syringe.

Modified Laser Induced Forward Transfer - LIFT techniques, better used in small amounts of substance as small as 500 nL load volume. This allows a very small deposit of cells in a very small quantities. This is an excellent technique when small amounts of cells are available. However, it is a hindrance when larger cellular constructs are desired.

3D bioplotting

If a larger tissue constructs is required with a few millimeter lengths, 3D bioplotting technology would be the technique of choice, considering that sufficient amounts of biomaterial are available to deposit in a reasonable time. 3D bioplotting can be scaled - up can be achieved if the following factors are needed:

1. If one is interested in building 3D constructs.
2. The scale-up is needed in terms of geometrical size,
3. Required certain number of different cell types within one setup,
4. Increase in printing speed with parallel printing to reduce the process time.

LIFT and LGDW techniques may not be appropriate for scale-up as cells are patterned in a 2D manner onto the surface and the whole construct only becomes 3D through adding a hydrogel layer after every patterned cell layer. The hardware configuration for each layer must be modified.

Modular Tissue Engineering

Building blocks for generating larger amount of tissues can be produced through addressing the difficulties and means in modular tissue engineering reconstructing biomimetic structures through designing microscale tissues.

To generate Modular tissue components, Figure 3.2, multiple approaches described as follows:

1. Cell aggregation,
2. Microfabrication of cell-loaded hydrogels or
3. Cell printing can be utilized.

The above alternatives are ultimately assembled into tissues with specific structures through loading of layers or direct assembly. Many tissues are created by reiterating functional units e.g., the liver lobule. The main purpose of modular tissue engineering is to generate more biomimetic engineered tissues by mimicking the native active units.

Table 3.2: Biofabrication of 3D cell-hydrogel constructs

Technology	Hydrogel	Cells	Fabrication specifications
Sterelithography			
	PEG	Fibroblasts	Laser beam θ ~250 μm; layer thickness: ~250 μm
	PEO, PEGDM	Ovary cells	Ring scaffolds: θ : 5.3 mm; thickness: 1.5 mm; UV laser spot: 250 μm: resolution per layer: 150 μm; x–y resolution: 250 μm
Laser-based biofabrication			
LGDW *BioLP*	Collagen, Matrigel Matrigel	ECs, hepatocytes Osteosacroma cells	Single cell resolution 2 layers of cells separated by a 75 μm layer of hydrogel
LIFT	PEGDA, Alginate, EDTA, blood plasma, Matrigel; Collector slide: agarose	Fibroblasts/ keratino cytes hMSCs, ECs	θ Droplets: 80–140 μm; speed: 1200 cell droplets/ min; scaffold height including 6 layers: 300 μm; focal spot: 45 μm; distance between spots: 75 μm; accuracy: 5 μm
LIFT-2PP	PEGDA	SMCs, ECs	θ laser spot: 45 μm; laser tranferred droplet size: 80–140 μm
Inkjet printing			
Inkjet bioprinter	Fibrin gel (fibrinogen + thrombin) Alginate, fibrinogen, thrombin	Neural cells HeLa cells, ECs	25 orifices with θ 50 μm, resolution: 85 μm; 250,000 drops/s; 5 layers Speed: 20 mm/g; ejection time: 800 Hz; pattern θ : 1 mm; size 5 × 7 mm
	Collagen	SMCs	Size of construct 5 × 5 mm, 5 layers; thickness per layer: 16.2 μm;
	Fibrin gel (fibrinogen + thrombin)	ECs	Size: 10 × 5 × 2 mm; 2 layers
3D bioplotting			
Bioplotter *Envisiontec*	Alginate, Lutrol F127, Matrigel, agarose, methylcellulose	BMSCs	Nozzle: 100–400 μm; 4–10 layers; thickness per layer; 150 μm; spacing 300 μm; speed: 1–30 mm/s; pressure: 0.5–3 bar, size: 20 × 20 mm Nozzle: 200–500 μm; resolution < 5 μm; accuracy < 5 μm; deposition rate: 12 nL/s–1 mL/s; speed: 10 μm/s – 50 mm/s; size: 2 × 2 × 1.5 mm; layer height: 50 – 100 μμm; pressure 1.2 bar
Fab@Home	Alginate	Chondrocytes	θ Nozzle: 840 μm, nozzle precision: 250 μm, width: 1200 μm; height: 800 μm; flow rate: 0.6 mL/s; size: θ: 6 mm × 2 mm height
	Methacrylated hyaluronic acid, Methacrylated ethanolamide, PEGs	Hepatoma cells, epithelial cells, fibroblasts	

Technology	Hydrogel	Cells	Fabrication specifications
Bioplotter	Gelatin/chitosan	Hepatocytes	θ Nozzle: 300 μm: drop volume: 20 nL; lateral resolution: 10 μm: X–Y velocity: 10 mm/s; extruding velocity 30 mm/min; pressure: 0.3 bar; layer height: 180 μm
	Gelatin/alginate elatin/alginate/ fimogen elatin/alginate/ chosan	Neuron cells; br Schwann cells, ADSC, hepatocyte it	θ Nozzle: 250 μm, X–Y velocity: 5 mm/s; extruding velocity 15 s mm/min; layer height; 150 μm; width: 380 μm
Multi-nozzle SFF deposition system	Alginate	ECs, fibroblasts, hepatocytes	Droplet-based or continuous depostion; Nozzle: 30–500 μm; velocity: 10 mm/s; pressure: 0.3 – 2.8 bar; 40 layers
Cell writing system	Alginate with iron oxide nanoparticles	ECs	Size of construct 5 × 5 × 2 mm Printing pressure: 0.3 bar
Dispensing-based deposition system	Mebiol (N-isopropylamid and poly oxyethylene)	Insect cells	Feed speed: 0.5–0.03 mm/s; pressure: 0.3–0.4 bar; line width: 114–300 μm; size 1 × 1 mm
nScrypt bioprinter	Collagen I, agarose	Embryonic cardiac cells, ECs, ovary cells, SMCs, fibroblasts	Tubes: 900–25,000 μm, wall thickness 300 μm

Organ Printing

Self-assembly of cellular aggregates is known as "Organ Printing," which are constructed by several techniques of the bioprinting technologies Figure 3.2. The use of biofabrication or 3D bioprinting of tissue engineering leads to organ Bioprinting. Organ Printing utilizes a bio-fabrication system for placing the whole prefabricated cellular aggregates. However, in segmental tissue engineering, the minor structure units are Bioprinted by a biofabrication device, nonetheless, additional assembly can be achieved through any technique. Cellular aggregates can be one method for constructing the minor units in modular tissue engineering. Organ Bioprinting usually denotes to the implementations of cellular aggregates.

When cell aggregates is matured to tissue spheroids can be utilized as building blocks by letting them to fuse after deposition. The commencement is the fundamental step in organ Bioprinting. Tissue engineering of inherit self-assembly nature uses the structural capacity of cells to build tissues leading to ultimately organs by mimicking natural morphogenesis. For instance, endothelial cells (Ecs) are genetically destined to form blood vessels and construct natural tubular constructions by themselves when the right environments are facilitated. Fusion and structure creation of the multicellular aggregates can be achieved by making a gel with optimum properties.

Cellular aggregates consist of collected several thousands of cells. The processing time span is substantially decreased as a great number of cells that can be printed simultaneously. This technique improved cell survival, as less individual cells are subjected the harsh printing environments within cellular aggregates. Also, through bypassing the nozzle and due to shorter overall printing time.

Organ Bioprinting has a unique advantage, which is the tight control level of cell and ECM placement for numerous biological materials that can be placed simultaneously.

Additionally, organ printing provides the opportunity to deal with the problem of vascularization in thick tissue constructs by controlling the release of growth factors.

A disadvantage of this technique is that the structural support is dependent on the self-assembly process. Consequently, it is less controllable. The 3D bioprinting through self-assembly tissue engineering without scaffolds has been accomplished. Recent advancements concentrate on the vital role of the hydrogels for cell organization and the manufacturing of vascular systems.

Biofabrication Facing the Vascularization Problem

The simplest way to develop perfusion is to create channels into 3D scaffolds or cellular constructs, as illustrated previously in Figure 3.23. Nonetheless, during culture the build may take a different form due to cells proliferating, differentiating or even yielding extra-cellular matrix ECM, and due to the breakdown of the cell supportive matrix. Therefore, the channel structures worsen and limit the medium supply. Utilizing biofabrication techniques, vessel inducing cells, such as endothelial cells ECs were implemented on well-defined locations imitating real vessels and their function. Endothelial cells ECs have been illustrated to form tube-like structures and integrity after 21 days of culture. Recently, the engineered vessels are mostly utilized to perfuse the build with cell culture substance, compared to *in vivo*, where the source of nutrients and the elimination of waste products occur via blood-flow.

The possibility to attach the Bioprinted vascular tree after *in vitro* development in a bioreactor to an *in vivo* host vasculature is significant. It has been revealed that by utilzing bioprinting technology to address this challenge, the first successful results for engineering vessels were acquired.

Organ Bioprinting is at Hand

The technology of 3D Bioprinting tissue engineering is at hand for the following reasons, sated as follows:

1. 3D biofabrication is a energetic growth for which promising results have been gained in the last 2 decades, and which has the possibility to continue to grow considerably.
2. 3D biofabrication extended applications in organ printing for *in vivo* substitutes, tissue reconstructions for *in vitro* biotechnological and pharmacological needs
3. Drug-testing where animal testing may be partially substituted by use of 3D bio manufactured tissue.
4. Drugs—either by themselves or embedded in microspheres or nanospheres—could be deposited within the construct enhancing the tissue formation
5. Scientists and engineers have used inkjet printing technology for cell-based gene therapies
6. Scientists and engineers presented the possibility to transfect genes into cells by inkjet printing besides the accurate delivery of the modified cells to a given target.
7. Recently, many new roles for regenerative medicine are rising in order to meet the needs of the specific applications.

Digital Bioprinting

Digital bioprinting is one illustration where the biomaterial is labeled with magnetic nanoparticles and can be placed in a controlled means as a result of magnetic communications. It is foreseen the cell Bioprinters will be as common for biological, academic, clinical and industrial enterprises as microscopes are today.

Current work in biomedical and medical laboratories exhibited that cell deposition encapsulated in numerous hydrogels is possible.

Cell Recovery after Bioprinting

Scientists and engineers examined cell capability and cell recovery after Bioprinting as a function of dispensing pressure and orifice diameter. Cell capability and cell recovery amplified with decreasing dispensing pressure and/or enlarged orifice diameter; with the most of cell recovery happening within the first 24 hours. Exclusively for reproducibility it is significant to describe a suitable working range keeping variants to a minimum.

To utilize the biofabrication technology for a wide-ranging of implementations counting on diverse materials and feature sizes such as the diameter of the dropped filament or droplet, a substantial validation work on quality and reproducibility still ahead.

Tissue Bio-assembly

Stem cells played a vital role for 3D tissue building since they deliver high possibility for the formation of intricate constructs. Stem cell biology is a actually significant constituent for bio manufacture since stem cells are the fundamental raw resources for tissue bio-assembly.

Cellular environment hydrogel is the encounter in utilizing stem cells for bio manufacture implementations that can be enhanced .Combining the benefits of cell attachment, cell excitation and mechanical stability enabled mimicking the *in vivo* environment to the highest degree. Amendments of mechanical and inherent characteristics and protein in addition to development factor distribution have to be examined for further advancements.

Stem cell guideline and the conservation of a precise phenotype rely on numerous biological influences such as growing factors but also matrix elasticity, which has been revealed in various researches.

Mathematical representations and computer recreations are significant tools in modern Bio manufacturing arrangements to enhance procedure design and for implementations. 3D biomaterial droplets have been extensively studied. Meanwhile it has been revealed that inapt printing limits like the droplets speed can intensely affect the result.

Computer-Aided Tissue Engineering

Experimentations to print chitosan scaffolds with a droplets-based rapid prototyping bio technique were achieved to qualify the pattern. The data presented a harmonious arrangement with the pattern forecasts and also additional research with alginate-hydroxyapatite hydrogels verified the calculated pattern. The agreements between experimental results and simulation were established. Scientists and engineers accomplished noteworthy work on computer-aided tissue engineering, the utilization of computer-aided technology together with biology in tissue engineering. CAD, image handling,

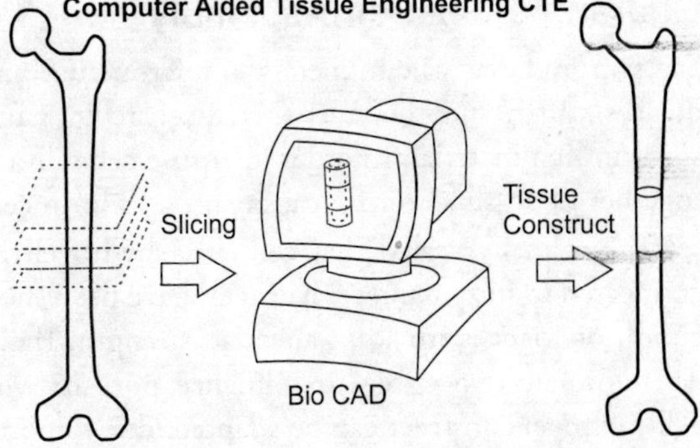

Figure 3.25: Computer Aided Tissue Engineering

engineering and small form factor for modeling, designing, iterations and fabricating of biological tissue and organ replacements has been achieved.

In silico observation of cell fusion, which has lately become the emphasis of much attention, likewise to augment procedure circumstances and the concluding result. Researchers and engineers studied mathematical representations for forecasting post-Bioprinting tissue remodeling; backing up the indication that tissue fluidity may deliver a mechanism for *in vitro* organ construction. utlizing computer reiterations, the builds can be foretold and hence enhanced before Bioprinting. However, this method will necessitate more consideration to launch it as common tool for designing novel biological procedures and 3D Bio engineered tissues.

Successful Implementations

In recent years, a growing number of scientists and engineers have been studying novel techniques for bio manufacturing 3D cellular builds utilizing intricate configurations. The quantity of effort and the formation of novel interests and applications worldwide show the robust growth possibility of this novel arena. These methods are growing progressively. 3D bio manufacturing has been revealed to be appropriate for the fabrication of anticipated scaffold configurations with diverse materials.

Physiological Printing Conditions

The impact of printing conditions on cell capability and scaffold materials, a key aspect for new tissue development, have already been explored thoroughly. If attention was paid to physiological printing conditions such as temperatures below cell denaturation and the application of nontoxic gelling methods, several devices showed no impact on cell capability. This is important since biofabrication methods will only have a future when no impairment of cellular function during processing can be guaranteed. Cell capability, phenotype maintenance, differentiation capabilities as well as cell functionality were found to be stable for those printing various techniques when appropriate conditions were applied.

Biofabrication techniques should be able to repeat living structures down to very small scales. It has been shown that it is not possible, at least for the moment, to achieve such detailed results. Recently, scientists looking into cell fusion and self-assembly have come up with a new concept which might lessen the need to reconstruct tissues down to a cellular level. They found that it was not necessary to yield precise replicas of biological organs, if cells were delivered with the adequate surroundings and signals and that if those requirements are met, the cells will assemble according to their natural function.

Guarantee Cell Self-Assembly

Scales from large scale in meters to micrometer range could be designed and fabricated by 3D bioprinting devices. Also, and the micro-architecture in nanoscale range could then be formed by cells and the surrounding material. In order to ensure cell self-assembly, the devices must be able to print cells as closely together as possible so they can merge and form cellular aggregates.

The results revealed that cells in a hydrogel-rich and cell-poor construct remained viable but rarely connected to form tissues. Hydrogels have been shown to be promising materials for generating appropriate condition for cells providing adequate strength. They can be designed according to a predefined configuration. Therefore, it is possible to influence porosity within the construct. Mechanical characteristics of the cell-hydrogel construct can be adapted, due to porosity adaptation without changing the basic material.

Biofabrication Methods Suitable for Cell Placement

Generally, it may be stated that the biofabrication approaches are appropriate for cell assignment and scaffold manufacture as stated by a pre-described design; but the numbers of technologies are still in their beginning stage and have to be enhanced, particularly in terms of resolution, repeatability and precision of cell placement. One of the most critical fact is to consider is the need of down-scaling to enable single cell placement, as well as avoiding clogging or drying of the cells at the same time, especially for the building of larger constructs.

REFERENCES

1. Langer R, Vacanti JP. Tissue engineering. Science. 1993; 260: 920–6. CrossRef, PubMed, CAS, Web of Science® Times Cited: 4078, ADS
2. Bleiziffer O, Horch RE, Hammon M, et al . T17b murine embryonal endothelial progenitor cells can be induced towards both proliferation and differentiation in a fibrin matrix. J Cell Mol Med. 2009; 13: 926–35.
3. Wu X, Rabkin-Aikawa E, Guleserian KJ, et al. Tissue-engineered microvessels on three-dimensional biodegradable scaffolds using human endothelial progenitor cells. Am J Physiol Heart Circ Physiol. 2004; 287: H480–7.
4. Feder-Mengus C, Ghosh S, Reschner A, Martin I, et al. New dimensions in tumor immunology: what does 3D culture reveal? Trends Mol Med. 2008; 14: 333–40.
5. Paszek MJ, Zahir N, Johnson KR, et al. Tensional homeostasis and the malignant phenotype. Cancer Cell. 2005; 8: 241–54.
6. Weaver VM, Petersen OW, Wang F, et al. Reversion of the malignant phenotype of human breast cells in three-dimensional culture and *in vivo* by integrin blocking antibodies. J Cell Biol. 1997; 137: 231–45.
7. Griffith LG, Swartz MA. Capturing complex 3D tissue physiology *in vitro*. Nat Rev Mol Cell Biol. 2006; 7: 211–24.
8. Langer R, Tirrell DA. Designing materials for biology and medicine. Nature. 2004; 428: 487–92.
9. Lutolf MP, Hubbell JA. Synthetic biomaterials as instructive extracellular microenvironments for morphogenesis in tissue engineering, Nature. Biotechnology. 2005; 23: 47–55.
10. Arkudas A, Tjiawi J, Saumweber A, et al. Evaluation of blood vessel ingrowth in fibrin gel subject to type and concentration of growth factors. J Cell Mol Med. 2008; DOI:DOI: 10;1111/j.1582--4934.2008.00410.
11. Arkudas A, Pryymachuk G, Hoereth T et al. Dose-finding study of fibrin gelimmobilized vascular endothelial growth factor 165 and basic fibroblast growth factor in the arteriovenous loop rat model. Tissue Eng Part A. 2009; DOI:DOI: 10.1089/ten.tea.2008.0477.
12. Hutmacher, DW, Cool S. Concepts of scaffold-based tissue engineering–the rationale to use solid free-form fabrication techniques. J Cell Mol Med. 2007; 11: 654–69.
13. Fischbach C, Chen R, Matsumoto T, et al. Engineering tumors with 3D scaffolds. Nat Methods. 2007; 4: 855–60.
14. Yamada KM, Cukierman E. Modeling tissue morphogenesis and cancer in 3D. Cell. 2007; 130: 601–10.
15. Kim JB. Three-dimensional tissue culture models in cancer biology. Semin Cancer Biol. 2005; 15: 365–77.

16. Zhang H, Wang ZZ. Mechanisms that mediate stem cell self-renewal and differentiation. J Cell Biochem. 2008; 103: 709–18.
17. Sutherland RM, Eddy H, Bareham B, et al. Resistance to adriamycin in multicellular spheroids. Int J Radiat Oncol Biol Phys. 1979; 5: 1225–30.
18. Miller BE, Miller FR, Heppner GH. Factors affecting growth and drug sensitivity of mouse mammary tumor lines in collagen gel cultures. Cancer Res. 1985; 45: 4200–5.
19. Sutherland, RM. Cell and environment interactions in tumor microregions: the multicell spheroid model. Science. 1988; 240: 177–84.
20. Kunz-Schughart LA, Freyer JP, Hofstaedter F, et al. The use of 3-D cultures for high-throughput screening: the multicellular spheroid model. J Biomol Screen. 2004; 9: 273–85.
21. Friedrich J, Ebner R, Kunz-Schughart LA. Experimental anti-tumor therapy in 3-D: spheroids–old hat or new challenge? Int J Radiat Biol. 2007; 83: 849–71.
22. Desoize B, Jardillier J. Multicellular resistance: a paradigm for clinical resistance. Crit Rev Oncol Hematol. 2000; 36: 193–207.
23. Helmlinger G, Netti PA, Lichtenfeld HC et al. Solid stress inhibits the growth of multicellular tumor spheroids. Nat Biotechnol. 1997; 15: 778–83.
24. Santini MT, Rainaldi G, Indovina PL. Apoptosis, cell adhesion and the extracellular matrix in the three-dimensional growth of multicellular tumor spheroids. Crit Rev Oncol Hematol. 2000; 36:75–87.
25. Kleinman HK, McGarvey ML, Liotta LA et al. Isolation and characterization of type IV procollagen, laminin, and heparan sulfate proteoglycan from the EHS sarcoma. Biochemistry. 1982; 21: 6188–93.
26. Lee GY, Kenny PA, Lee EH, et al. Three-dimensional culture models of normal and malignant breast epithelial cells. Nat Methods. 2007; 4: 359–65.
27. Wozniak MA, Keely PJ. Use of three-dimensional collagen gels to study mechanotransduction in T47D breast epithelial cells. Biol Proced Online. 2005; 7: 144–61.
28. Bissell MJ. Modelling molecular mechanisms of breast cancer and invasion: lessons from the normal gland. Biochem Soc Trans. 2007; 35: 18–22.
29. Debnath J, Brugge JS. Modelling glandular epithelial cancers in three-dimensional cultures. Nat Rev Cancer. 2005; 5: 675–88.
30. Abbott A. Cell culture: biology's new dimension. Nature. 2003; 424: 870–2.
31. Bissell MJ, Radisky D. Putting tumors in context. Nature Rev. Cancer. 2001; 1: 46–54.
32. Bissell MJ, Weaver VM, Lelièvre SA et al. Tissue structure, nuclear organization, and gene expression in normal and malignant breast. Cancer Res. 1999; 59: 1757–63.
33. Petersen OW, Ronnov-Jessen L, Howlett AR, et al. Interaction with basement membrane serves to rapidly distinguish growth and differentiation pattern of normal and malignant human breast epithelial cells. Proc Natl Acad Sci USA. 1992; 89: 9064–8.
34. Wang F, Hansen RK, Radisky D, et al. Phenotypic reversion or death of cancer cells by altering signaling pathways in three-dimensional contexts. J Natl Cancer Inst. 2002; 94: 1494–503.
35. Park CC, Zhang H, Pallavicini M, et al. Beta1 integrin inhibitory antibody induces apoptosis of breast cancer cells, inhibits growth, and distinguishes malignant from normal phenotype in three dimensional cultures and *in vivo*. Cancer Res. 2006; 66: 1526–35.
36. Pampaloni F, Reynaud EG, Stelzer EH. The third dimension bridges the gap between cell culture and live tissue. Nat Rev Mol Cell Biol. 2007; 8: 839–45.

37. Weigelt B, Bissell MJ. Unraveling the microenvironmental influences on the normal mammary gland and breast cancer. Semin Cancer Biol. 2008; 18: 311–21.
38. Fischbach C, Chen R, Matsumoto T, et al. Engineering tumors with 3D scaffolds. Nat Methods. 2007; 4: 855–60.
39. Cushing MC, Anseth KS. Materials science. Hydrogel cell cultures. Science. 2007; 316: 1133–4.
40. Soloman, Sabrie "Sensors Handbook, 2nd Edition, McGraw-Hill Publishing Co., NY – ISBN 978-0-07-160570-0
41. Soloman, Sabrie "Sensors and Control Systems in Manufacturing 2nd Edition," McGraw-Hill Publishing Co., NY – ISBN 0-07-059626-3
42. Soloman, Sabrie "Affordable Automation," McGraw-Hill Publishing Co., NY – ISBN 0-07-059633-6
43. Soloman, Sabrie "Sensors Handbook, 1st Edition," McGraw-Hill Publishing Co., NY – ISBN 0-07-059630-1
44. Soloman, Sabrie "Sensors and Control Systems in Manufacturing 1st Edition," McGraw-Hill Publishing Co., NY – ISBN 978-07-10572-4

4 Organ Printing

3D Bioprinting the Aortic Valve

The author deeply desires the readers to recognize the values of maintaining a healthy heart to maintain a high quality of life, and longevity.

"The major worldwide health care issue is cardiovascular diseases. As stated by the World Health Organization, cardiovascular diseases killed thus far 19.9 million people in 2019, which represents 34 % of mortalities. Heart valve failures represent the major cause of these diseases. The number of patients requiring heart valve replacement is forecasted to triple from approximately 290,000 in 2003 to over 850,000 by 2020."

This chapter is devoted to remedy Heart-Valves through the revolutionary 3D Bioprinting technology.

Nomenclatures

To the readers who may not possess general knowledge of biological or medical aspects, the stated abbreviations below will be useful to acquire understanding of organ 3D Bioprinting - the "Aortic Hear-valve."

- 5-HT, serotonin;
- 5-HTT, serotonin transporter;
- αSMA, Alpha smooth muscle actin;
- ACE, angiotensin converting enzymes;
- ADHD, attention-deficit hyperactivity disorder;
- ANF, atrial natriuretic factor;
- ANGII, angiotensin II;
- APOE, Apolipoprotein E;
- ARB, angiotensin receptor blocker;
- AT1, angiotensin II Type 1;
- AVD, Aortic valve disorders;
- B1 and B2, bradykinin receptors;
- BAV, bicuspid aortic valve;
- BK, bradykinin;
- BMP, bone morphogenic protein;
- BMSC, bone marrow derived mesenchymal stem cells;
- BPV, bioprosthetic valve;
- CAVD, calcific aortic valve disease;
- CHD, congenital heart defect;
- LDL, low density lipoprotein;
- LDLr, low density lipoprotein receptor;
- LPS, lipopolysaccharide;
- MDA, *N*-demethylated metabolite 3,4-methylenedioxyamphetamine;
- MDMA, 3,4-methylenedioxymetham-phetamine;
- MGP, matrix gla protein;
- MMP, matrix metalloproteases;
- MPV, mechanical prosthetic valves;
- NADPH, nicotinamide adenine dinucleotide phosphate-oxidase;
- NFκB, nuclear factor kappa B;
- OPG, osteoprotegerin;
- OPN, osteopontin;
- P4HB, poly-4-hydroxybutyrate;
- PEG-DA, poly (ethylene glycol) diacrylate;
- PGA, polyglycolic acid;
- PHO, polyhydroxyalkanoate;
- PLGA, poly (lactic-co-glycolic acid);

- CVD, cardiovascular disease;
- COL, collagen;
- CX40, connexin 40;
- DGEA, collagen-derived peptide motif;
- EC, endothelial cells;
- ECM, extracellular matrix;
- EDMP, Endocardial derived mesenchymal progenitors;
- EDS, Ehlers–Danlos syndrome;
- EDTA, Ethylenediaminetetraacetic acid;
- ELN, elastin;
- EMT, endothelial to mesenchymal transformation;
- eNOS, endothelial nitric oxide synthase;
- FBN1, fibrillin;
- FDA, Federal Drug Administration;
- FZD, frizzled receptors;
- GAG, glycosaminoglycans;
- HMG-CoA, 3-hydroxy-3-methylglutaryl-coenzyme A;
- HVD, heart valve disease;
- KKS, kallikrein-kinin system;
- PLLA, poly (L-lactic acid);
- RANK, receptor activator of nuclear factor kappa B;
- RANKL, receptor activator of nuclear factor kappa B ligand;
- RAS, renin-angiotensin system;
- RGDS, fibronectin-derived peptide motif;
- ROS, reactive oxygen species;
- RVD, rheumatic valve disease;
- SDS, sodium dodecyl sulfate;
- SMM, smooth muscle myosin;
- SPARC, secreted protein acidic and rich in cysteine;
- TEHV, tissue engineered aortic heart valves;
- TGFβ, transforming growth factor beta;
- TIMP, tissueinhibitorofmetalloproteinases;
- TNX, tenascin-x;
- TLR, toll-like receptors;
- VEC, valve endothelial cells;
- VEGF, vascular endothelial growth factor;
- VIC, valve interstitial cells;
- YIGSR, laminin-derived peptide motif

HEART VALVES

Unidirectional flow of oxygenated blood is regulated through the aortic valve to the myocardium and arterial mechanism. Efficient functions of biomechanically and hemodynamically ensured by the normal anatomical configurations and microstructural complexity. Unique cell phenotypes are populated by the yielding cusps that continually remodel tissue for long-term resilience under particularly demanding of mechanical condition. Normal healthy heart valve is sown in Figure 4.1. Unhealthy heart valve is shown in Figure 4.2.

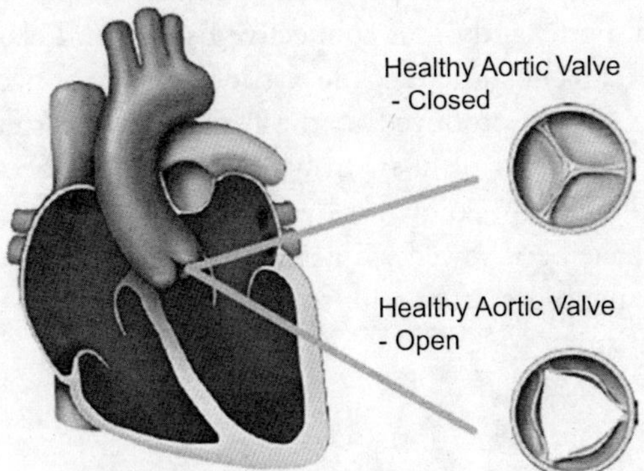

Figure 4.1: Healthy Aortic Heart Valve
Courtesy: UCI Health

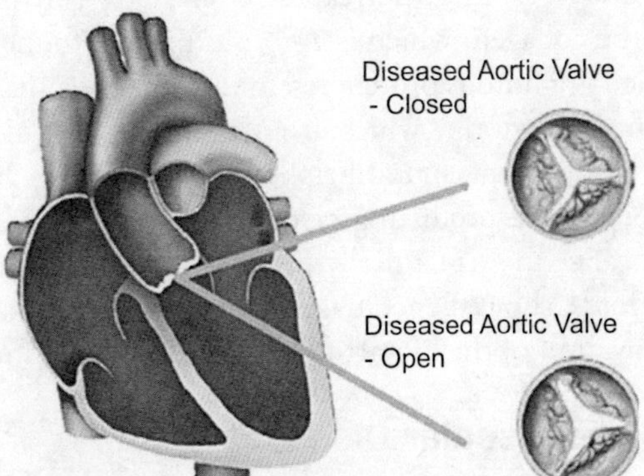

Figure 4.2: Diseased Aortic Heart Valve
Courtesy: UCI Health

Genetic and micro environmental sources cause modification from regular valve homeostasis, which is a state of steadiness or a inclination to reach steadiness, either metabolically within a cell or organism or socially and psychologically within an individual. This alteration or modification leads to hereditary and/or premature structural degeneration Features of atherosclerosis shared with aortic valve stenosis pathobiology. However, its final calcification endpoint is distinct. Although, little is known about the mechanisms of normal valve mechano-biology and disease, it is vital and to have comprehensive medical significance and mechanisms of disease.

This is echoed the rareness of prognostic analytical approaches, initial stage interventional plans, and inactivity in regenerative medicine improvement. Therapy of aortic valve disease benefits greatly by bioprinting tissue engineering exclusive possibility.

Bio-fabricating Heart Valves

Presently-utilized artificial and biological heart valve implants have some drawbacks. Mechanical prostheses, made mainly of carbon, metallic and polymeric mechanisms, need perpetual anticoagulation treatment, which is the treatment utilized in preventing the accumulation of new blood clots, and to treat present clots by averting them from increasing in size. The utilization of anticoagulant frequently causes adversative responses, e.g., thromboembolic difficulties and endocarditis, which is inflammation of the thin membranous lining endocardium of the heart's cavities. It causes Xenogenous, which is foreign substance that has been introduced into the organism. Also, it causes allogenous, which is additional complications in biological prostheses related with immune response, thrombosis and deterioration, and therefore they have a high rate of re-surgery.

Biological Prostheses of Autologous Origin

Complicated surgery usually arises from pulmonary autografts biological prostheses, which is an artificial body part of autologous origin. It frequently weighs the patient with the danger of re-surgery. Consequently, bio-artificial heart valves are reserved. A much better alternative is to utilize "Tissue Engineering." It is utilized to ensure an autologous biological component. The autologous approach is derived from a patient's own stem cells of the body. The artificial heart valve must be biologically well-matched, robust, capable to withstand normal mechanical wear and tear, and capable to grow over time in particularly with children. Accordingly, 3D Bioprinting scaffolds made of synthetic materials are recommended for this purpose, such as poly-lactic acid, polycaprolactone, poly4-hydroxybutyrate, or natural polymers. The natural polymers can be collagen, which is a fibrous protein found in skin, bone, and other connective tissues, and elastin, which is a fibrous protein resembling collagen that is the main constituent of the elastic fibers of connective tissue, and fibrin, which is an insoluble fibrous protein that is produced in the liver from the soluble protein fibrinogen and helps in blood clotting. It forms a network of fibers in which blood cells become trapped, thus producing a clot, or hyaluronic acid, which is acid, also known as hyaluronan, is a clear, gooey substance that is naturally produced by the body. These have been sowed with stem cells. Nanostructured scaffolds, showed encouraging results, in addition to utilizing bioreactors prior to implantation as dynamic cultivation of the bio-artificial grafts into an animal organism.

Cardiovascular Diseases

The major worldwide health care issue is cardiovascular diseases. As stated by the World Health Organization, cardiovascular diseases killed 19.9 million people in 2019, which signifies 34 % of mortalities. Heart valve failures represent the major cause of these diseases. The number of patients demanding heart valve replacement is forecasted to triple from approximately 290,000 in 2003 to over 850,000 by 2020.

Heart Valve Replacement - Conservative Surgery

The aortic, pulmonary, mitral and tricuspid are linked to heart valves . The currently preferred approach is heart valve replacement in the operating surgery treatment of the aortic heart valve diseases. . Conservative surgery is preferred treatment in the situation of mitral and tricuspid valves, of the patient's own native valve, which is, valvuloplasty is preferred in adulthood, Figure 4.2. Heart's mitral valve doesn't close firmly, letting blood to stream retrograde in the heart. Tricuspid valve is made of three triangular parts, mainly that between the right atrium and ventricle of the heart. Rheumatic fever is an impediment of neglected strep throat that can impair heart valves, as well as the tricuspid valve, casing the tricuspid valve regurgitation later in life, Figure 4.3.

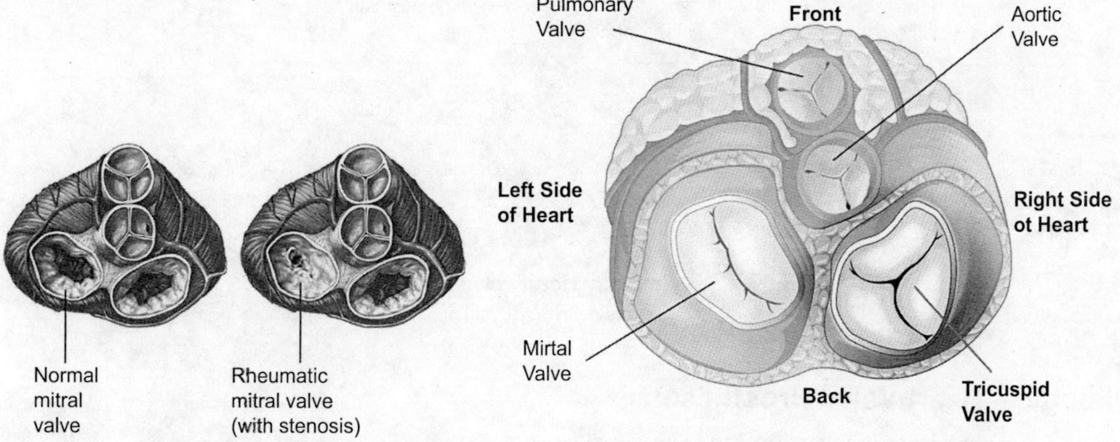

Figure 4.3: Rheumatic Mitral Valve Disease - *Courtesy: A.D.A.M., Inc.*

Figure 4.4: Tricuspid Valve Disease *Courtesy: The Society of Thoracic Surgeon*

Inherited heart defects: Some heart deficiencies present at birth (congenital) may disturb the tricuspid valve.

While tricuspid and pulmonary valve surgeries are rare among adults, they are achieved in circumstances of substantial regurgitation. Nonetheless, circumstances have been conveyed in which all four kinds of heart valves were substituted with prostheses, Figure 4.4.

Mechanical and Biological Heart Valve Prostheses

Heart valve prostheses can be classified into two elementary groups, Sated as follows:

1. Mechanical prostheses
2. Biological prostheses.

Mechanical prostheses are prepared of pyrolytic carbon filled with metallic and polymeric mechanisms. These devices pose risks associated with valve mechanism. The risks include thrombus formation; thrombosis is the formation of a blood clot, known as a thrombus, within a blood vessel. It prevents blood from flowing normally. Also the risk extends to thromboembolic complications, which causes formation in a blood vessel of a clot (thrombus) that breaks loose and is carried by the blood stream to plug another vessel. Also, the risk includes endocarditis, which is an infection of the endocardium; it is the inner lining of the heart chambers and heart valves that happens primarily on the surface of foreign materials. There are other types of risks associated with permanent and long-lasting anticoagulation therapy. This includes hemorrhage bleeding impediments happening at 1 % patients per year.

Long-term endurance of patients, which exceeds 20 years, with mechanical heart valves, and without anticoagulation therapy and without complications has been also maintained, Figure 4.5.

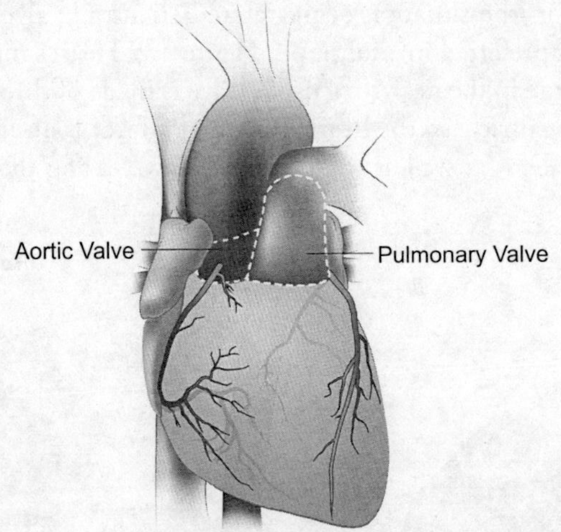

Figure 4.5: Pulmonary Heart Valve Replacement
Courtesy: CryoLife, Inc.,

Biological Heart Valve Prostheses

Patient's organism immune reaction can be invoked thrombosis and undergo degeneration through Xenogenous and allogenous biological heart valve prostheses; complicated surgery with an increased risk of mortality is a consequence of collecting and implantation of autologous prostheses.

Consequently, construct bio-artificial valves by tissue engineering methods provides a substitute method to advance the quality of heart valve prostheses, Figure 4.6.

1. The perfect valve prosthesis should be easy implantable,
2. should develop a physiological hemodynamic enactment deprived of structural worsening
3. should be non-thrombogenic, nonimunogenic and silent in its performance.

Tissue-engineered cardiac valve with the suitable biomechanical characteristics are exceptionally

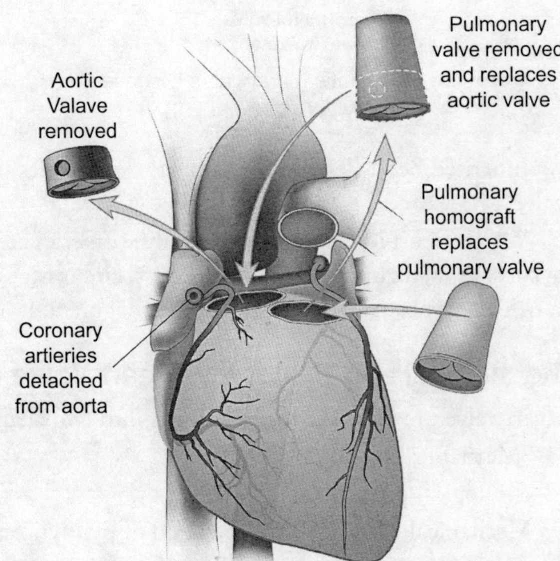

Figure 4.6: Pulmonary Valve Removed – Aortic Valve Removed – Coronary Arteries Detached
Courtesy: CryoLife, Inc.,

significant. They compensate their biological characteristics. Without this type of tissue-engineered cardiac valve high pressure and stresses would consequence in valve catastrophe, with possibly life-threatening significances.

Mechanical Valve Prostheses

There are three main kinds of mechanical valves stated as follows:

1. caged-ball valves,
2. tilting-disk valves and
3. bileaflet valves,

The caged ball type was the first artificial heart valve in 1952, which utilized a metal cage to house a silicone elastomer ball. There were several alterations conducted on each type of heart valves.

Caged-Ball Valves

As the blood pressure in the heart chamber, during contraction of the heart, surpassed the value in the external of the chamber, the ball was pressed alongside the cage and permitted the blood to stream. At the end of the heart's contraction, the pressure inside the chamber fell, so the ball progressed back against the base of the valve, creating a seal.

A similar valve comprised of a silicone ball walled in a cage shaped by metallic wires formed from the valve casing. Caged ball valves have a high inclination to formulate blood clots; consequently, the patient must have a high amount of anti-coagulation. Furthermore, the movement of the ball harms the blood cells. Consequently, while caged-ball valves attained the first long-term achievement in prosthetic heart valve technology, the device was discontinued in 2007.

Tilting-Disc Valves

Another more progressive kind of artificial heart valves is characterized by tilting-disc valves. Tilting-disk valves had a individual round device, directed by a metal support. These valves were fabricated of a metal ring sheltered by a textile material of extended polytetrafluoroethylene (ePTFE). In this textile piece, the suture threads were stitched to grip the valve in place. Two metal backings aided a metal ring to grip a disc, which opens and closes as the heart thrusts blood via the valve.

The disc is typically manufactured of a very solid carbon material, pyrolytic carbon, allowing the valve to operate properly for many years deprived of deterioration. Mechanical valves models have the disc classified into two portions, which open and close similar to a door.

St. Jude Medical Bi-leaflet Valve

St. Jude Medical Corporation introduced a new bi-leaflet valve mechanical design valve. It contains two hemispherical leaflets that rotate about struts, which is held to the valve casing. The two leaflets of the valve are the casing and struts. They are fabricated of pyrolytic carbon and contain no stainless steel compound. Earlier valves contained stainless steel compounds. Only titanium is still an element of the casing body in some designs. Backflow is a disadvantage of the bi-leaflet valves; they provide more natural blood flow. They are still not considered ideal. However, they are better than the caged-ball or tilting-disc grafts. These bi-leaflet valves have a better opening area, and improved tolerance acceptance by the body. Also, they are the least thrombogenic of the synthetic valves types. They are often used on moderate anticoagulation treatment.

Biological Valve Prostheses

Biological prostheses suffer lack of development as an adverse reaction imbedded artificial mechanical body part as prostheses, which is an artificial body part. Biological prostheses became more reliable and commercially available when stented glutaraldehyde-fixed valves, taken from animal origin, are implemented, Figure 4.7. The most reliable biological prostheses at present are stated as follows:

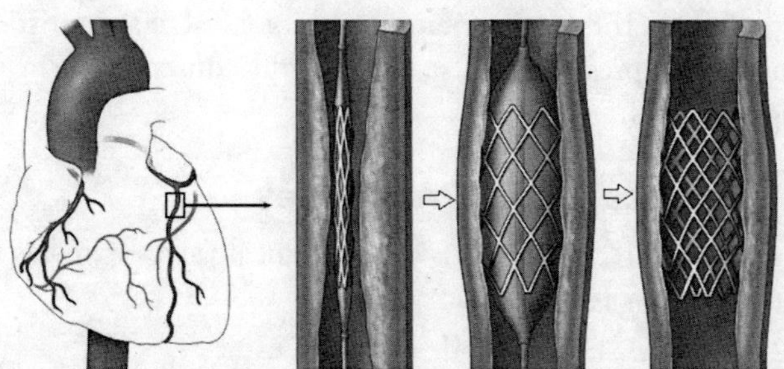

Figure 4.7: 3D Printed Heart Stent - Installing Heart Stent
Courtesy: Image: Central Georgia Heart Cente

a. porcine xenograft valves and bovine pericardial valves, which are implants derived from other species than human. Porcine xenograft has been used as a barrier to the skin for over 30 years, and due to its wound healing promoting factors. Pericardium is a tough, double-layered membrane which covers the heart.
b. allograft valves (also called homo-grafts, and taken from human givers other than the patient).
c. autografts, which are taken from autologous patient's stem cells.

Valve Dysfunction

Fibrotic tissue production is evoked by xenografts and allografts. Xenografts are subject to cellular rejection more or less like what is observed in organ allografts. This can cause valve dysfunction with occurrence of 2-4 % patients/year. The occurrence of disruptive thrombosis is about 0.2-1.8 % per year. 65 % of patients under age of 60 receive re-surgery 15 years after their implantation of these kinds of biological valve implants. Both xenografts and allografts are increasing age of the allograft donor. Also, younger people and teenagers recipients have been related with allograft failure. Recipient in this case requires for allograft valve re-surgery owing to faster valve deterioration. The allograft in right ventricular outflow tract reconstruction has 5 year survival and 25 % of the recipients have less than 1 year of age, 61 % for those between 1 and 10 years, and 81 % for those older than 10 years. Degeneration due to age-dependency can be triggered by several reasons.

a. immunological reaction
b. advanced overgrowth of the embedded allograft by patient's fibrotic cells.

Young patients display enlarged hemodynamic burdens on the allograft, as they are more lively and have higher metabolic requirements.

c. Valve procurement,
d. duration of the allograft ischemia,
e. time between valve harvest and valve implantation, and
f. surgical method for implantation influence the long-term durability of the implant allograft.
g. Dehiscence of the prosthesis is also a serious complication that affects both bio-prostheses and mechanical prostheses.

Percutaneous biopsy is a biopsy that is obtained by putting a needle through the skin in order to obtain tissue within the body for examination. The primary positive percutaneous valve substitute in a human patient was shown on pericardial and other valves primed from non-valve tissues. The Stem cells were shaped from bovine jugular vein stitched into a platinum stent. It was located in the pulmonary valve location, Figure 4.8.

Aortic valve was achieved effectively in another positive human percutaneous imbedding.

PULMONARY AUTOGRAFTS IN CHILDREN

Pulmonary autografts have been revealed to be unaffected to long-term deterioration. Pulmonary autografts were deep-rooted to be capable to cultivate in children. Also, pulmonary autografts can be utilized to substitute the aortic origin and all diseased tissue in aortic valve endocarditis. The graft can last longer than the patient as a sign for utilizing an autograft in mturer age of the patients. However, a pulmonary autograft embedded into the aortal location is an intricate and needs surgery. Mortality increased by 20 % of patients need re-surgery 10 years after operation.

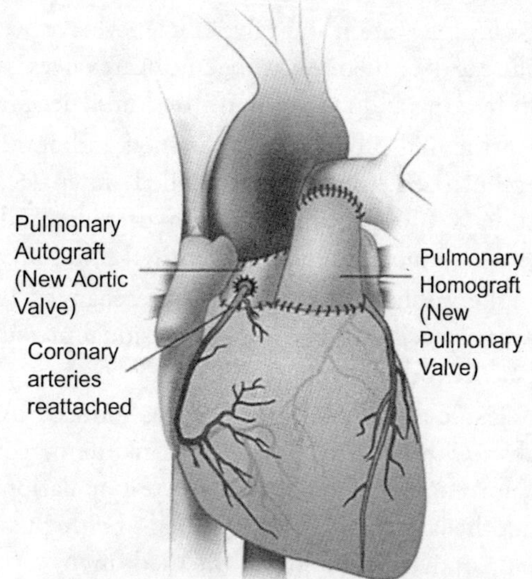

Figure 4.8: New Pulmonary Valve - New Aortic Valve – Coronary Arteries Reattached
Courtesy: CryoLife, Inc.,

This kind of surgery encompasses substituting the aortic valve by the autologous pulmonary valve, and the pulmonary valve is then substituted by a pulmonary homograft.

Constructing Valves Using Bioprinting Autologous Cells

New method to the growth of autologous heart valves is accessible by means of tissue engineering. It includes building valves utilizing Bioprinting autologous cells and extracellular matrix (ECM) molecules, or a bioactive synthetic material mimicking the normal ECM. This procedure permits the histological construction of the ordinary valve to be entirely valued. Profound understanding of this structure, and also of the characteristic features of the cell kinds existing in heart valves, is essential for effective heart valve tissue engineering.

Tissue-Engineered Heart Valves

Currently, the following approaches are used for developing tissue-engineered valves:

Cell debris, assumed to be the cause of the following effects:

1. The antigenicity of xenogenic tissues, such as bovine or porcine pericardium, porcine small intestinal submucosa, and porcine or ovine heart valves
2. In addition, porcine aortic valves need to be cross-linked by glutaraldehyde before implantation, which can cause cytotoxic and other adverse reactions.
3. Glutaraldehyde-fixed pericardium was resistant to collagenase degradation and cell infiltration, but significantly prone to increased calcification.

Homo-grafts, which do not need glutaraldehyde cross-linking, can last 15-20 years and have better mechanical classifications than acellular glutaraldehyde-fixed xenografts. This promising conduct probable outcome from the existence of interrelated sheets of collagen, layers and tubes of elastin, highly non-linear mechanisms, anisotropy and viscoelasticity of native valve tissue.

Catastrophic Failure of Decellularized Xenogenic Heart Valves

Tissue preservation and valve efficiency have been impacted by the decellularization technique. Catastrophic failure of decellularized xenogenic heart valves is the result of violent decellularization by enzymes, detergents, spit freezing and radiation, utilized in SynerGraft technology.

Structural proteins were retained and appeared to be intact, by deoxycholic acid which is completely decellularized heart valves. distilled water decellularized the carotid arteries Also 0.025 % trypsin, 1 % Triton X-100 and ammonium hydroxide produced many structural deviations of the ECM, such as less dense and more extended adventitial collagen, as well as enlarged spacing between fibrils.

The mechanical properties were changed by the alterations, which tangled a higher ductile modulus, lower extensibility, and reduced residual pressure of the decellularized arteries. These suggested substantial fiber network disturbance.

Tissue regeneration *in vivo* was induced by the implantation of decellularized allograft and xenograft valves with effective propagation of the matrix by interstitial cells. However, implant endothelialization is not enough. Comprehensive re-population of detergent-decellularized ovine pulmonary valves with endothelial cells was accomplished *in vitro* in a dynamic bioreactor. EC were adapted in the bioreactor, to a comparatively high flow of the medium by a progressive upsurge of flow rate from 0.1 l/min to 2.0 l/min.

If differentiated EC are less available for seeding on the valve, mesenchymal stem cells (MSC) can be extracted from the bone marrow and phenotypically transformed into EC *in vitro*. MSC also deliver a basis for fibroblasts/myofibroblasts and normal muscle cells.

When MSC were sown onto a decellularized porcine pulmonary valve pre-coated with fetal bovine serum and fibronectin, they populated the scaffold, and endothelium liner, and the existence of fibroblasts and myofibroblasts was showed in the valve implant.

Mechanical and Bioprinted Biological Heart Valve Prostheses

At present, mechanical and biological heart valve prostheses have many disadvantages. Consequently, substantial innovations were accomplished to make bio-synthetic heart valves with the advantage of autologous biological element utilizing tissue engineering techniques. Though, methods for the expansion of the tissue-engineered heart valves still exhibit some restrictions. Encouraging methods appear to be the utilization of altered decellularized xenogenic or allogenic scaffolds

The progress of materials comprising polymerized extracellular matrix and captured cells allows an analysis of molecular mechanisms of cell-toextracellular matrix connections, cell response to mechanical stimulus, as well as cell-cell communication in co-cultures of diverse cells kinds.

Some forward-thinking technologies grip out high confidence, e.g. rapid prototyping and nano-spider technology for generating a nano-structural design to stimulate the propagation of scaffolds with cells.

Bioprinting Cell - Organs 3D Gels

Cell Bioprinter was industrialized to allow researchers, biologists and engineers to position cells *in situ*ations that mimic their individual places in organs. Though, this technology was restricted to the printing of two-dimensional (2D) tissue builds. New method was advanced telling the utilization of thermosensitive gels to produce consecutive layers for cell printing. The capability to drop cells on previously printed consecutive layers delivers a real chance for the understanding of three-dimensional (3D) organ Bioprinting.

Printing Complex 3D Organs with Computer-Controlled

Organ printing permits researchers to print intricate 3D organs with computer-controlled, precise insertion of different cell kinds, by a procedure that can be accomplished in predetermined time. Scientists demonstrated that cell accumulation can be positioned in the consecutive layers of 3D gels stimulating them for fusion to happen. Scientists projected the slight thickness of the gel that can be produced by depositing the liquefied substance at room temperature onto a heated substrate. The generated cell aggregates with the corresponding size to ensure a direct contact between printed cell layers during consecutive printing cycles.

Scientists sowed that the closely-positioned cell aggregates could fuse in two kinds of thermo-sensitive 3D gels. The data strongly supported the viability of the organ-printing technology.

Shortage of Donor Organs

The severe shortage of donor organs has led to scientists to comprise synthetic mechanical organs, xenotransplantation, which is utilizing animal organs, tissue Bioprinting, and regenerative medicine. Synthetic organs have been developed, but they can considerably lessen a patient's quality of life and may yield unwanted side effects. Animal organs are a likely methodology, particularly when it includes the utilization of organs from transgenic animals with a reduced capacity to encourage acute immune response after transplantation.

The potential spreading of animal viruses is still great concern. Regenerative medicine, which is the reparation of incapacitated diseased organs *in vivo* by gene therapy, as well as by cell transplantation, is also a likely cultured methodology. In the early stages of a disease gene therapy and cell transplantation, as well as drug therapy, are most effective. However, organ replacement at terminal phases is still needed for. Tissue Bioprinting, which is the developing of organs from a patient's own cells, is an excellent methodology to solve the shortage of donor organs.

Rigid Scaffolds from Polyglycolic

The 3D bioprinting tissue encompasses the utilization of inflexible scaffolds from isolated cells and polyglycolic acid, which is known as "PGA," It is generating organs based on the evidence that seeding cells in a bioreactor on porous biodegradable scaffolds will be adequate. Though, there are at the minimum several issues with this technique:

1. Cell diffusion and seeding is not very effective. It is not uniform throughout the scaffold.
2. Organs comprise of many types of cells, and the necessity to place diverse cell kinds in precise locations is difficult in solid scaffold design.
3. The rigid scaffolds made from PLA are not suitable for engineering contractile tissue, such as heart and vascular tubes.
4. The absence of vascularization is causing difficulties in utilizing solid scaffold seeding technology.

The absence of vascularization has established to be very real for constructing blood vessels, but it cannot be modified for intricate 3D organs. Implanting technology can only be implemented to multilayer tubular organs. It also suffers from the lack of a direct positioning mechanism.

Fabricating physical models of hard tissues, tissue scaffolds, and custom-made tissue implant prostheses can be employ rapid prototyping in3D Bioprinting technology. 3D bioprinting technology has been utilized to yield innovative scaffolds with manageable porosity and channel sizes. It can possibly permit for vascularization. The technique foremost restriction is the absence of polymers with suitable mechanical characteristics for soft-tissue builds. This method also has the disadvantage that cells must be positioned in the exact locations in the 3D printed scaffold.

Bio Cell Bioprinter Place Cells in Positions Mimicking their Positions in Organ

Bio cell printer is developed to empower placing cells in positions that mimic their positions in an organ. The 3D Bioprinter can deposit up to nine different solutions of cells into a precise position using distinct software. A novel prospect for extending the Bioprinting technology to 3D is formed by the utilization of thermo-reversible gels. Biodegradable nontoxic, thermo-reversible gels are utilized as a sort of "paper" on which tissue constructions can be Bioprinted, and the cells are the "ink," Where, consecutive layers could be Bioprinted by depositing another layer of gel onto a previously Bioprinted surface, Figure 4.9.

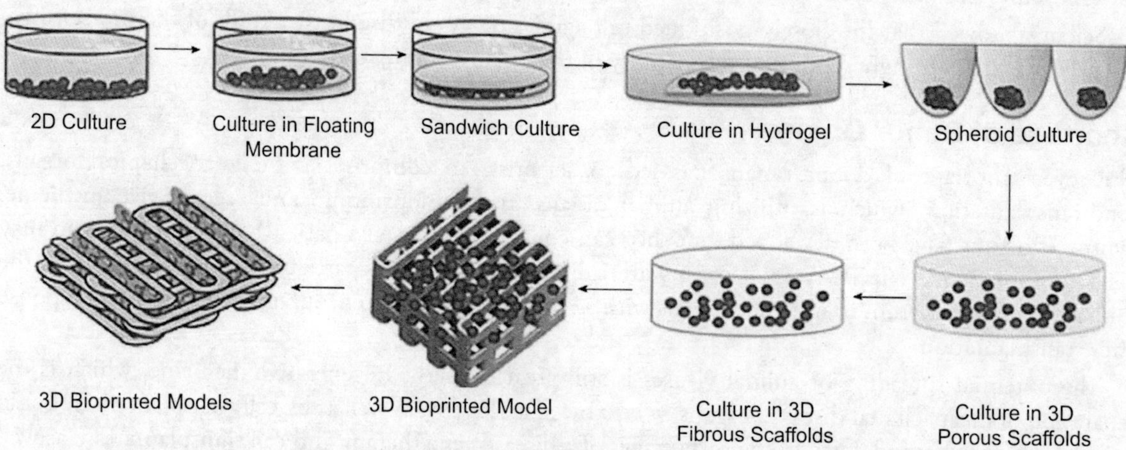

Figure 4.9: Preparing 3D Bioprinting Tissue Culture
Courtesy: Baishideng Publishing Group Inc.

Bioprinting organ allows Bioprinting intricate 3D organs, with positioning diverse cell kinds in 3D manufactured organs, in a limited time. Cell collections can be positioned in a 3D gel for fusion to happen. It is projected the slight thickness of gel that can be produced by depositing the liquefied substance at room temperature onto heated substrate. Founded on this projected constraint, cell collections were designed with suitable sizes to guarantee a direct interaction between Bioprinted cell collections during successive Bioprinting cycles. Placing cell collections close enables fusion of 3D gels, Figure 4.9.

Cell Bioprinting Procedures

The following procedure have been developed to maintain accurate 3D Bioprinting

1. Germ-free stainless steel needles with luer plastic pivots were reserved on ice pending they were full of the polymer gel of interest.
2. They were then attached into the ethanol-sterilized luer fixtures of the print head.
3. Upkeep was undertaken to uphold a needle temperature below 4°C.
4. It is significant to uphold a low temperature when at work with collagen as it will gel permanently and clog the needles at room temperature.
5. The Bioprint head and the Bioprinter were positioned onto heated petri dishes.
6. The dishes were positioned into particularly designed aluminum blocks that were reserved at 36°C by a heater.
7. The heater, blocks, dishes, and Bioprinter were placed into a germ-free laminar flow hood.

Preparation of Collagen Gels

The following procedure was equipped for effective gel preparation:

1. Collagen gels were prepared using lyophilized calf skin collagen.
2. A 3.33 mg/ml collagen solution was prepared by slowly dissolving the collagen in acetic acid (1%) at 0°C.
3. The gel was prepared by adding 156.5 μl of sterile, filtered 10× DMEM and 30 μl of sterile 0.34N NaOH to each milliliter of collagen solution.
4. The subsequent solutions were reserved on ice until they were utilized.

Estimation of Gel Thickness

The following technique was also established for calculating gel thickness:

1. The thickness of the gel layers was assessed by depositing a sequence of drops onto the heated petri dishes.
2. Each drop was permissible to gel before a new, smaller drop was positioned on top of the gel.
3. The sequence of staggered gel layers thus formed was positioned under a microscope and observed.
4. The micro-movement of the microscope objective was measured as the focal plane shifted from one layer to the next.

Toward Organ Bioprinting

The adhesiveness of gels for cells can be modified. The data show that cell aggregate fusion can be achieved in the thermo-reversible gel. Also, the data showed that sequential layers of thermo-reversible gels can be reproduced generating a thickness corresponding to the cell aggregate diameter. This thickness allows cell aggregates to be laced in closely enough to fuse. Both 3D collagen and thermo-reversible gels were shown to allow fusion of closely-placed cell aggregates, Figure 4.10.

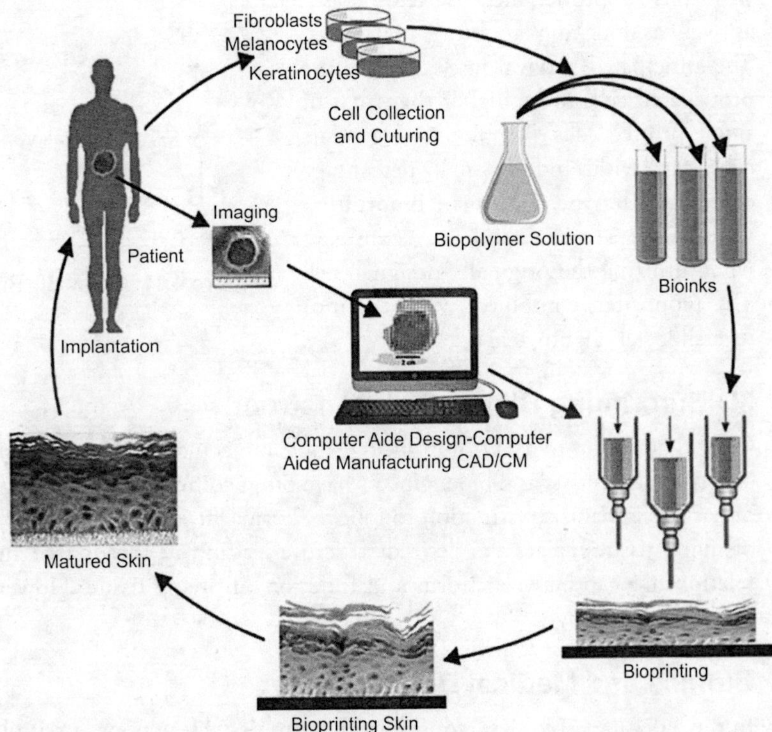

Figure 4.10: Skin Tissue Bioprinting - Creating Organ from Synthetic and Natural building Blocks
Courtesy: Springer Nature Switzerland AG

Cumulative cells fusion affects the 3D structures to shrink. This factor must be reflected in the design of organ blueprints. Though, it is advisable to observe that further extracellular matrix ECM will be shaped by the cells. Therefore, as long as extra glue for cell cumulative fusion to serve as a binding device for helping the self-assembly of Bioprinted tissue builds.

Also, this extra extracellular matrix can behave as a safeguard between neighboring structures, enabling allowable tolerances of a blueprint.

Adhesion can be enhanced through the utilization of gel with RGD ligands. On the other hand, aggregates confined inside gel deposits can be Bioprinted. Fusion takes place when deposits are placed close to each other. Cell attachment time determines when to print the successive layer. Soft-tissue adhesives could dramatically stop the builds from being washed off during consecutive Bioprinting cycles.

A corresponding growing factor could be added in slow-release microspheres to hasten and direct the growth. Cell adhesion could be promoted in optimized thermo-reversible gel. It results in better cell endurance as well as more operative cell aggregate fusion.

Avoiding Cell Death

The possible disadvantage of cell demise in 3D Bio-printed is cell death during cells aggregates. Scientists devised various means to overcome apoptosis and the death of cells in tissues inside the aggregates through incorporating survival factors, such as the basic fibroblast growth factor, utilizing temporary genetic alterations of cells with anti-apoptotic, such as "telomerase" as well as blocking apoptotic pathways. The anticipated survival rate is enable the propagated cells to be higher than that of single printed cells. 3D bioprinting tissue cells can withstand a small percentage of cell death. The 3D organ bioprinting technology is successful to construct when utilizing the originally-designed cell 3D bioprinter combined with thermo-reversible gel, Figure 4.11.

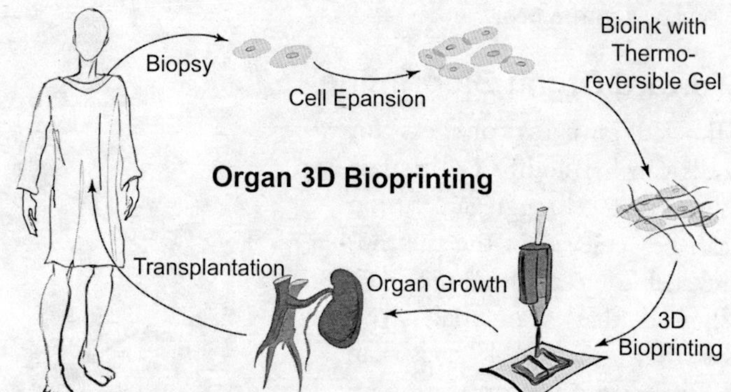

Figure 4.11: Organ 3D Bioprinting with Thermo-Reversible Gel
Courtesy: The Conversation

3D Bioprinting Billion Cell Construct

Careful attention to a 3D Bioprinting single molecule leads to successful aggregate to the whole organism construct. Scientists and bioengineers have often contemplated about the ideology of engineering principle of part design and construction to follow "Form, Fit, and Function." Would the experimental architecture of living tissues create the desired structure leading to the desired function? 3D bioprinting closed the relationship gap between "form and function" in living tissue. However, 3D bioprinting technology has several disadvantages.

Bionics and Medical Devices

In the 3 decades "Bionics," considered human tissue functions are similar to basic mechanical and electrical systems, close loop servomechanisms. Scientists and engineers tried to mimic components of body systems to successfully interface with the actual body anatomy and be completely unified. This conceptual framework has proven useful in practice, with contemporary work applied to human patients through surgical implants such as knee, hip, and limb prostheses, pacemakers; and cochlear as well as retinal devices are successful example to accomplish seamless integration between machine and body anatomy. However, many living tissues could be substituted with electromechanical components to improve the quality of patients' life

No matter how innovative materials are used in constructing artificial organs, they cannot substitute tissues in the body whose purposes are closely tied to their biochemistry. Liver function, for example, can metabolize acetaminophen and alcohol, while inventive polymers cannot.

Cells Function Responsible for Body Biochemistry

Cells biochemistry in the body is the major practical power in the human body. The determinations prior to 3D bioprinting to separate cells from their native environment *in vivo*, will guarantee successful application for therapeutic devices. Live cells can be loaded into bioreactor chambers outside the body in extracorporeal liver-assist devices. The cells can then be linked in a closed loop with host blood circulation. The blood circulation creates the intended biochemical benefit from cells in the device to affect the patient. This method includes dialysis of blood during kidney failure, which has caused illnesses that are not appropriate in long-term therapies.

Figure 4.12A: Complex Anatomy - Leonardo da Vinci Understood the Interpenetrating Networks of Lung Vasculature and Branched Airways (c. 1500) *Courtesy: European Union Leonardo Digitale.*

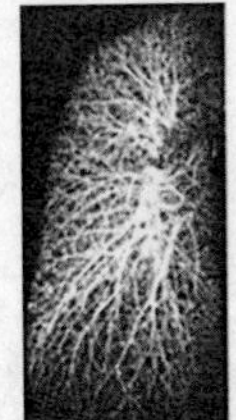

Figure 4.12B: Complex Anatomy - Lung Vasculature - Reconstructed - Computed Tomography *(CTS scans).*

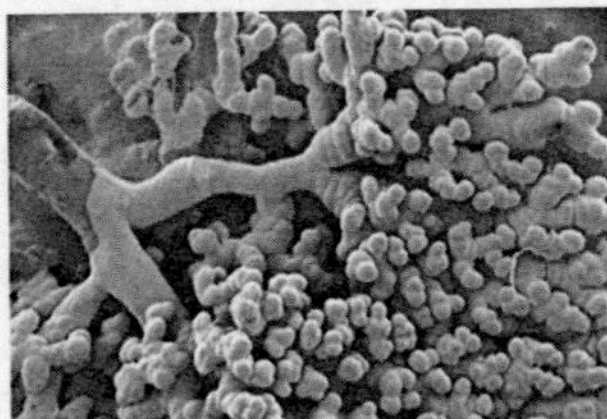

Figure 4.12C: Complex Anatomy - Air Sac - Adult Rat Lung *(electron micrograph).*

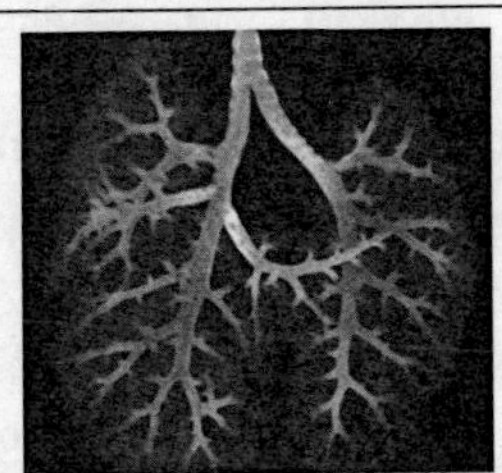

Figure 4.12D: Complex Anatomy - Embryonic Mouse Lung - Branching Morphogenesis *– (Optical Projection Tomography Image)*

Unfortunately, Cells placed into devices outside the human body are increasing at the bottom of a Petri dish bear little similarity to the delicate anatomical difficulty found in the human body. Lung organs , heart organ, brain organs, kidney organs, and liver organs are saturated by vascular systems to carrying air, lymph, blood, urine, and bile. These experiments still do not lead us to a clear path toward physical comprehension of these tissues in the laboratory, Figure 4.12. Nonetheless, a full understanding of tissue organization is not fundamental developmental biology, before improving the quality of life for patients suffering from damaged or diseased organs.

Transplanting Organs from Human Donor

Immunosuppressive therapy is the life line in transplanting human organs from a willing donor into a needy recipient. Similarly, segregated cells have been proven to provide biochemical benefits to the host, even when injected or placed at unusual sites inside the recipient. The prospect of using a patient's own stem cells to develop living models of their functional biochemistry. Developing cellular implants offers revolutionary changes to scientists, engineers, researchers and healthcare.

Stem cells inducing pluripotency cells are the most effective discovery to develop new ways to construct new organs. The following issues are in the mind of developers, stated as follows:

a. cell number and cell types,
b. 3D arrangement of cells,
c. local extracellular
d. and micro-environmental considerations

These issues remain largely intractable because of difficulties in placing and culturing cells in three-dimensional - 3D. The embryonic organ propagates containing considerable count of cells change differentiation as a function of cell population and micro-environmental characteristics. The, greater count of cell populations rapidly die due to the lack of adequate oxygen and nutrient transport.

The revolutionary advances in 3D printing are being adapted to operate within the soft, wet environments where cells function best. As a result of 3D printing successfully accelerate producing heterogeneous physical objects of high complexity, many biologists and bioengineers have gained unprecedented access to great advances in tissue architecture.

Scar Tissue Formation

Scar tissue formation in skin wounds can be blocked by the use of a bio-compatible, nontoxic insert made from a distinct construction of collagen and glyco-samino-glycans. The nontoxic implant assists tissue regeneration and ingrowth of functional nerves and blood vessels. This type of regeneration lacks the ability to regenerate hair follicles, and sweat glands.

3D Bioprinting of Tissue Engineering to treat scars occurred in heart cardiac tissues. Tissues are consisted of several types of cells accumulated into hierarchal constructions, Figure 4.13. 3D Bioprinting can be utilized to construct efficient tissues from cells and scaffold-forming materials.

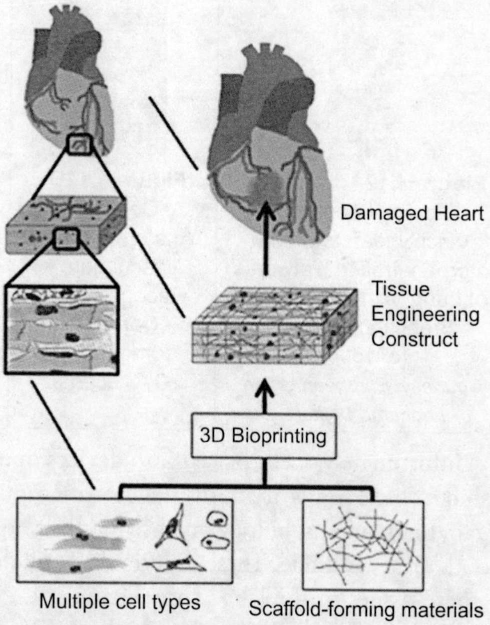

Figure 4.13: Schematic Representation of 3D Bioprinting for cardiac tissue engineering treating Cardiac Tissues scars Using cells and scaffold-forming materials
Courtesy: Springer Nature Switzerland-AG

Cells, Scaffold, and Bioactive Factors

3D printing of tissue engineering has evolved to combine cells, scaffolds, and bioactive factors to build a model grafting. Starting with steady progress in the utilization of 3D printing of simple thin tissues such as cornea and bladder, successful functional restorations were accomplished in human patients. The success was also contributed to the use of only one or two cell kinds. The body was capable to sufficiently create up the change to achieve anticipated function.

A thin construct requires thin tissues, and thin tissues also require fewer cells. A thin construct was implanted of hepatocytes subcutaneously. The implanted cells were noted to regenerate and function normally. However, the thin constructs cannot substitute for a whole liver. As tissue builds were scaled up. Growing too much thin tissues to be cellularized have formed solid organ as therapies for humans.

The slightest therapeutic inception for solid organ substitution is projected at 5–10 billion functioning cells. Also, the cell density represents a vital to succeed in organ replacement. .

It is known fact that cells must come close to have direct contact with neighboring cells in order to function. These criteria may take place during embryonic morphogenesis and growth, homeostasis, or during wound remedy. Accordingly, several Petri dishes to regenerate 10 billion functioning cells will not resolve the cells organization function to form functioning constructions, where their suitable characteristic is strengthened.

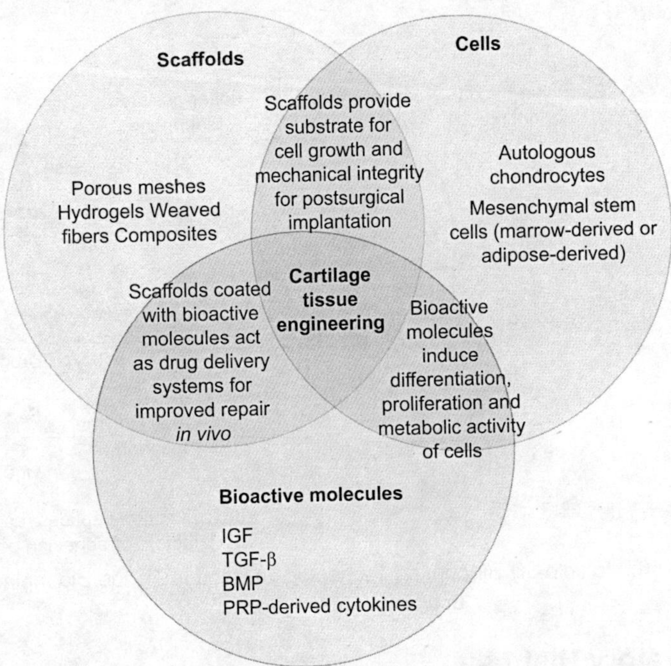

Figure 4.14: Interlink among Cells, Scaffolds, and Bioactive Molecules
Courtesy: Nature Publishing Group

3D bioprinting of tissue engineering solved the raised concerns of the following issues:

1. Which anatomical mechanisms are essential and adequate for tissue function?
2. Which are extra trappings?
3. Do micro-tissues are required with multiscale vasculature.
4. Are organ on-a-chip systems satisfactory?

Decellularized organs, which are then recellularized, have established their great usefulness for tissue engineering implementations.

To answer this question, the 3D bioprinting of tissue engineering are providing precisely the competences, automation, and reproducibility that are intrinsically built in.

3D Bioprinting Layer by-Layer

The methodology to re-organize living cells has been adopted technologies from the microprocessor industry.

The beginning of 3D Bioprinting for the bio-manufacture of biological objects of subjective difficulty provides a considerable change in tissue engineering. 3D Bioprinting is an additive technology. Rather than starting with a block of material and removing what is undesirable in a subtractive process, additive bio-manufacturing starts from nothing and selectively builds, one layer at a time, a bio-object of interest according to computer instructions.

A substantial display of technologies is currently in utilization, and all are in actuality adaptable to bio-engineering living tissues, Figure 4.16. An additive approach is exclusive among manufacturing technologies because it gives the user independent access to every (X,Y,Z) coordinate Access to each coordinate can make 3D printing slow. Each time a print resolution is doubled, the number of required coordinates scales by a factor of eight. The capabilities of 3D printing are best demonstrated in the fabrication of structures that cannot be made in any other way.

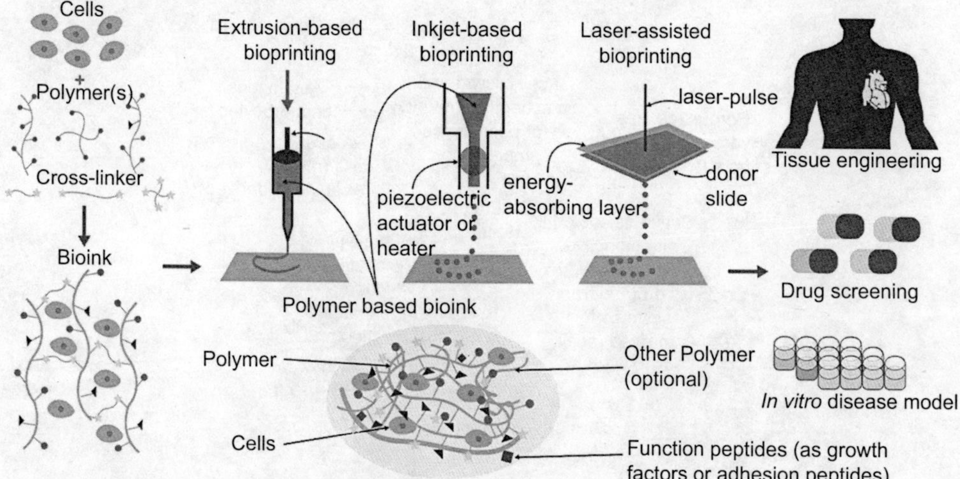

Figure 4.16: Multiple 3D Bioprinting Technologies Used in Tissue Bioengineering
Courtesy: Millipore-Sigma

Treating in Respiratory Distress

3D Bioprinting was utilized in biology and medicine as a substitution of the commonly utilized resins and polymers with biocompatible, nontoxic materials. This approach was utilized to successfully treat respiratory distress.

3D Bioprinted Tracheal Splint

Noninvasive body scanning was utilized to chart the tracheal flaw. A design of a tracheal construct was made, and then 3D Bioprinted the tracheal implant, which is manufactured completely from Poly-epsilon- caprolactone - PCL, Figure 4.16.

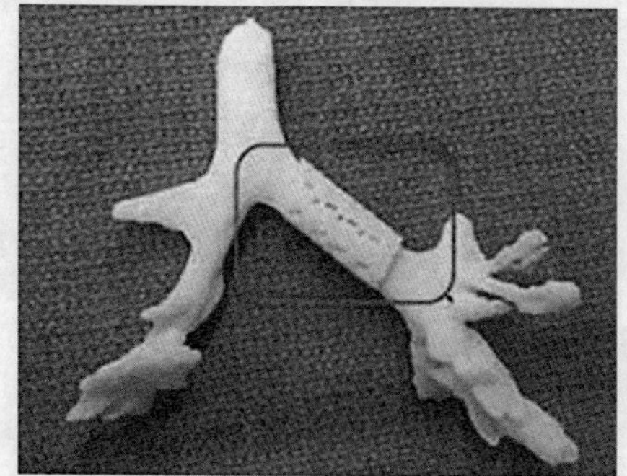

Figure 4.16: 3D Bioprinted Tracheal Implant
Courtesy: Joseph Stromberg

The splint is often fully resorbed within few years, while the patient respiration has substantially improved.

Similarly, dental prostheses can be manufactured with 3D Bioprinters in a secluded dental practice. Patient-precise custom made denture can be printed at the comfort of the doctor and the patient like, as an advantage of the comprehensive clinical adoption of the 3D Bioprinting technology. To adapt 3D

Bioprinting for the manufacture of living tissues, cells and extracellular matrix (ECM) are combined as ink in a process known as bioprinting. Inkjet printing, light-projection photolithography, and syringe-based extrusion allow the selective deposition of tissues in reproducible and heterogeneous patterns.

Microcomputed Tomography

Microvascular cells are currently inkjet Bioprinted in fibrin. The cellular aggregates are made from fibroblasts, which can be extruded among supporting tissues of complex carbohydrate, and primary aortic cells can be printed in the shape of their parent valve based on digitized microcomputed tomography. In addition to ink attentions, viscosity and crosslinking chemistry, and cell handling are the major challenges. Some of the most desired cell types for Bioprinting are hepatocytes, which are fragile cells in culture. They may not survive the 3D printing process itself.

Dissimilarity to these additive cellular inks, brief Bioinks can be bioprinted, and then selectively removed later. A process similar to wax casting, this sacrificial molding strategy substitutes most of the precision of specific cell placement for accurately structuring the negative space in tissues. Smooth channels and tubes can be patterned and perfused to keep resident cells alive at densities not currently possible with bioprinting. The combination of sacrificial Bioinks with bioprinting methods may yield a hybrid strategy giving the best characteristics from each technique, e,g., heterogeneous cell patterning and perusable vasculature. The open source software development, made designs and code are legally available to anyone. Similarly, hardware design, and software tool chain for 3D bioprinters are public knowledge.

The STL File Only Describes 3D Surface Volume

The result is an explosion of more than 575,000 3D Bioprinters in operation worldwide by researchers, scientists, engineers, and hobbyists enthusiastic to benefit with focused scientific exploration. However, the most universal digital file format used for 3D Bioprinting—the stereolithography file (STL)—lacks any hierarchy to represent the structure of living tissue. The STL file only describes the surface of a 3D volume and contains no information about its internal space. The National Institutes of Health (NIH) has launched a new data bank for 3D printing which may help to address this standardization problem.

Popularization of the hardware and software tools required for 3D Bioprinting indicate the technology is becoming accessible even to non-experts. The diversity of benefits in 3D Bioprinting, coupled with the wide distribution of printers themselves, mean an overabundance of openings exist for applying 3D Bioprinting to biology and medicine.

An Abundance of Opportunities for 3D Bioprinting Cascading Signals

3D bioprinting applications have enabled specific cell types to be positioned 3D arrangements. Additionally, 3D Bioprinting dictated the crosstalk between cells in all layers providing the experimental platform on which to test advanced multiscale computational models. Biological abilities to reconstruct models *in vitro*-control became simplified through 3D Bioprinting.

Cooperativity and Morphogenesis

3D bioprinting highlighted a greater understanding of the techniques which can pattern cell populations needed to explain their behavior in all three dimensions, morphogenesis and cooperativity in large-scale tissues. Angiogenesis models, and the emergent of new blood vessels from pre-existing blood vessels, are progressing into 3D multiscale models. Cellular activities and instantaneous data acquisition enable valuable monitoring, direct cellular motions, tissue deformations and stresses.

Disease Progression

3D Bioprinting models will bring a clearer understanding of constructing living organism applicable representations of disease progression. *In vitro* 3D Bioprinted models of tumor biology have provided valuable opportunities in cancer. Transportation models have had complications making accurate calculations of chemotherapeutic possible because of the difficulties of measuring, authenticating, and correlating mass transport directly in patients. Most tumors are heterogeneous and unique to each patient. Understanding cancer forces continued to improve which will benefit from 3D printing technology, correlating to human clinical data.

Pharmaceutical Applications –Organ –on-Chip

The potential of organ-on-a-chip systems can be physically modeled to reflect human cells behavior. This organ-on-chip may bring high-efficiency drug testing to cultured, vascularized human tissues manufactured with 3D printing. Since patient specific reactions to drugs are hard to predict, 3D bioprinting can make it possible to test drugs on cells cultured taken directly from specific patients. Accordingly, 3D bioprinting helps to predict the most effective therapeutic remedy and tolerable certain dose.

A lung on a chip equipped with air-filled channel represented by yellow color, and blood-like solution-filled channel represented by blue color, both channels are lined with human cells to mimic Bio-organ function, Figure 4.17.

An eye-on-a-chip, equipped with microfluidic channels represented by yellow color, which provide nutrients to cells positioned at the center within the scaffold of circular configurations. The Scientific personnel have linked a micro-engineered simulated functional eyelid that mimics regular intermittent human blinking, Figure 4.18.

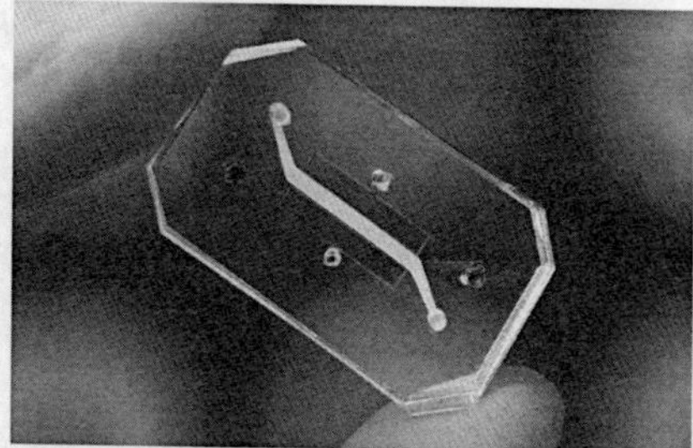

Figure 4.17: 3D Bioprinted Lung On-Chip
Courtesy: Wyss Institute – Harvard University

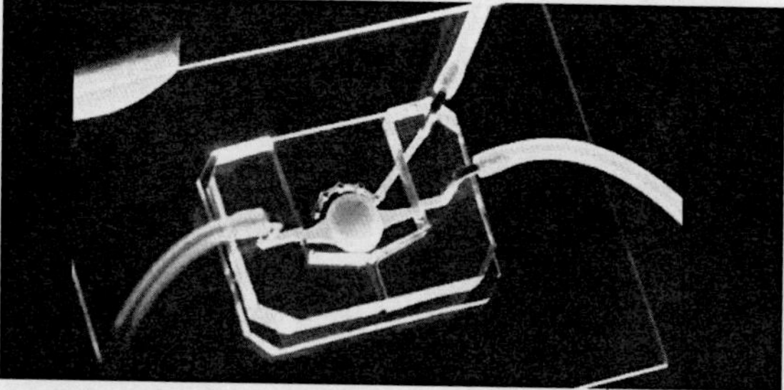

Figure 4.18: 3D Bioprinting Emulating Eye-On-Chip – *Laboratory of Dan Huh – University of Pennsylvania*

Figure 4.19: 3D Bioprinting Mini Female Reproductive System On-Chip
Courtesy: Northwestern University

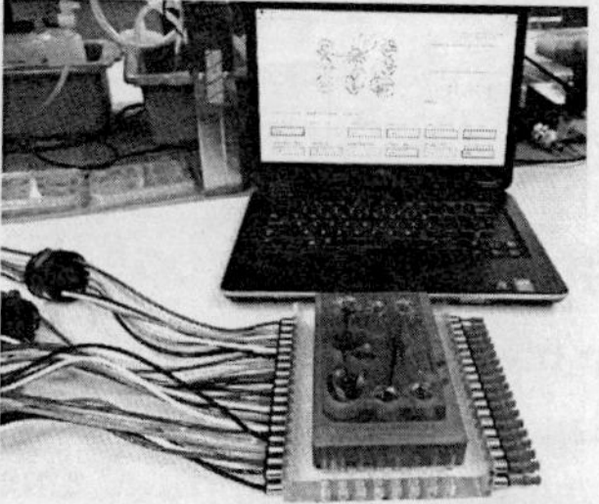

Figure 4.20: 3D Bioprinted Multi-Organ Platform On-Chip
Courtesy: Physiomimetics, MIT

A mini-model of "Female Reproductive System," where the blood is represented by the blue color fluid, which flows through pits encompassing mini organs, Figure 4.19.

A number of scientists have undertaken the concept of developing organ-on-chip to connect diverse organ chips together to regenerate organ simulation or perhaps the whole human body.

A multi-organ platform linked to correlated software, where each cavity grips a tiny physiological system, such as the lung, the gut, or the central nervous system. Channels are developed to permit fluids to flow through them mimicking physiological cardiac response, Figure 4.20.

Organ-On-Chips is a development similar to the illustrated Brain-On-Chip, Figure 4.21. The Brain-ON-Chip holds miniature hollow channels filled with considerable number of living human cells and tissues, Figure 4.21.

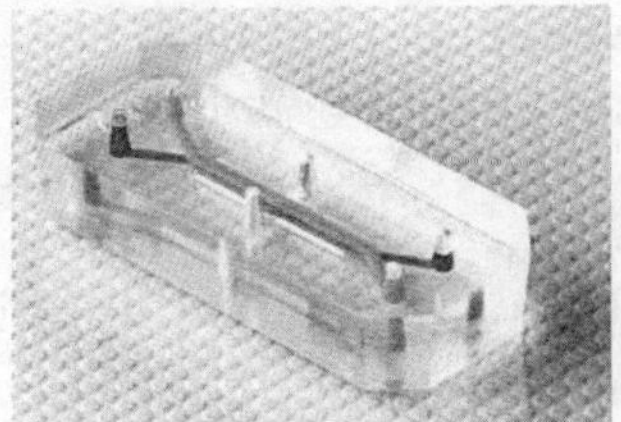

Figure 4.21: 3D Bioprinted Brain-On-Chip
Courtesy: Emulate, Inc.,

Oxygen and Nutrients Viabilities for Living Cells

The main concern of tissues engineering containing living cells is the inevitable need for a continuous supply of oxygen and nutrients to maintain cell viability. The growth of human cell aggregates can survive by diffusion alone, as long as the living cell thickness does not exceed several hundred micrometers thick. However, large cell aggregates suffer death of living cells, due to the difficulties in transporting oxygen and nutrients to living cells, and discarding waste products out of tissue constructs.

Bioprinting New Vascular Networks

In order to grow living tissue grafts that can survive beyond the dissemination limits of oxygen in the interstitial fluids, It may require to construct new vascular networks to be connected into the host vasculature blood flow. This allows the prevascularized implanted tissues to be integrated into the host vasculature. This may take up to several weeks. The process has to exceptionally fast to prevent necrosis of billion-cell constructs. Necrosis can happen within hours, either in engineered tissues or in donor organs for transplantation.

It is noted that the biological structure of matching organs harvested from diverse people have exclusive vascular constructions. However, these organs can work similarly for each person. Although the main arteries and veins are inherently encoded and formulate during embryogenesis, the microvasculature is fashioned according to local forces and needs.

It is of interest to recognize that the vessel architecture of the retina is more distinct among people than their fingerprints. The overall transport of blood components that results from vessel architecture is a principal factor defining healthy and diseased tissue. Accordingly, to solve transport dilemma in engineered tissues, it is probably that there is more than one architectural solution is possible, Figure 4.22.

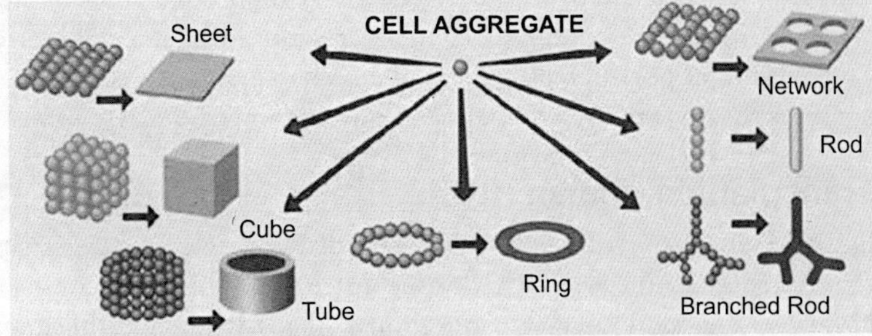

Figure 4.22: Step 1, Bioprinting Cell Aggregate
Courtesy: 3D Printing Industry - Scott J Grunewald

Scientists for the first time were able to successfully utilize the advancements in Bioprinting technology, to 3D Bioprint human tissue incorporating the intricate functional vascular network. Optimistically, bring us closer to replacing impaired organs. The technology of Bioprinting organ is highly sophisticated and exceptionally complex. Bearing in mind that we hardly understand precisely exactly how the human body works, in particularly, how cells are incredibly organized to form a well-functional network of circulatory system, Figure 4.23.

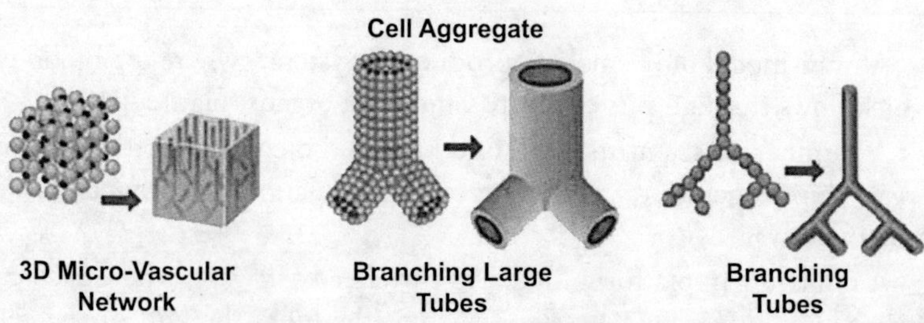

Figure 4.23: Bioprinting Vessel-Like Tubes
Courtesy: 3D Printing Industry - Scott J Grunewald

The experts and researchers utilized an engineering manufacturing procedure to create the vascular network by generating a pattern prior to constructing the organ tissue around it. Utilizing an intricate 3D Bioprinter to fabricate a network of very small connected fibers constructed from a glucose molecule, such as agarose. This glucose pattern is immersed in a protein hydrogel media, which can solidify when subjected to visible light. At that moment, the network of bioprinted fibers was detached and substituted with human endothelial cells. A week later the cells were self-organized and shaped a sequence of stable vessels-like tubes, Figure 4.23.

Miraculously, the cells precisely begin to self-organize and generate circulatory structures, leading to the formation of a pattern of the circulatory system, until the creation of an entire functional organ is completed. More amazingly, is the natural formation of solving the hidden secret of creating bioprinted organs, Figure 4.23.

The central difficulty related with 3D Bioprinting is the sequential processes leading to successful functional human organs to imitate the complexity of the human body, incorporating intricate network of veins, arteries and capillaries. The human circulatory network system is exceptionally cumbersome aimed primarily to oxygenate several nanometers of blood vessels throughout all cells of the human body, Figure 4.24.

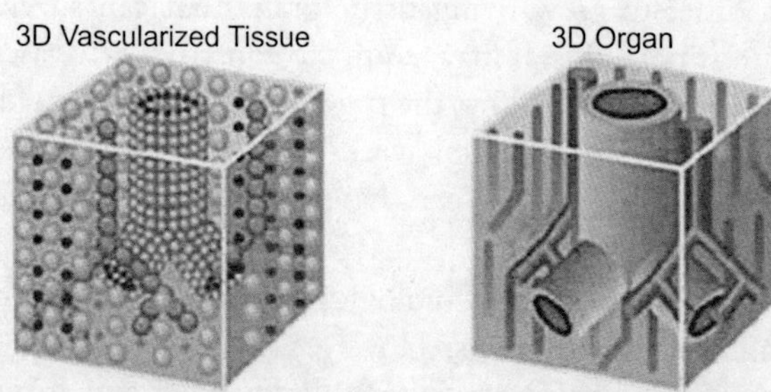

Figure 4.24: Bioprinting 3D Organ
Courtesy: 3D Printing Industry - Scott J Grunewald

A team of scientists from the universities of MIT, Harvard, Stanford and Sydney have effectively Bioprinted efficient and highly organized vascular network. While, this success is a positive step in organ Bioprinting it is still far from true implementations to replicate a whole and purposeful replacement of a human organ. A genuine regeneration of intricate and functional organs is the hope of every patient and physician need.

Transplanted Human Hearts

Distribution of nerves to an organ part of native vasculature is significant for widening of the blood vessels, especially the arteries and vasoconstriction. This is for controlling blood pressure all over the body. However, the neurovascular connection perhaps is not life-threatening for early efforts to create living tissue. Transplanted human hearts, for instance, are not surgically tangled to the nerves of the recipient; they pulse at their individual pace. The heart can be activated by being near to the sensory nerves to be innervated and reinstated feeling in its regenerated skin substitutes.

It is likely that the re-innervation will function in other engineered organ systems. Nonetheless, Individual neuronal procedures extending to few meters in length may not be deposited with 3D bioprinting anytime shortly. However, the application of 3D Bioprinting inside thinning gels delivers unique opportunity for neural printing, because long strands are easily deposited.

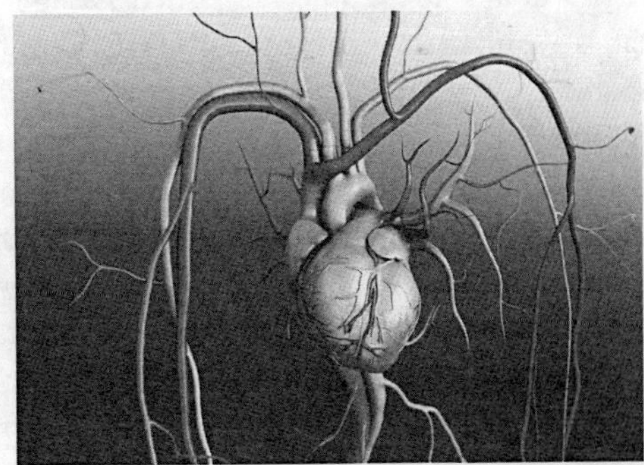

Figure 2.25: Transplanting Human Heart as the Only Viable Alternative for a Patient Remedy
Courtesy: Ugreen/iStockphoto

Heart transplantation is the well-recognized customary cure for patients that have no other means to recover due to their progressive heart failure. Physicians and surgeons are continually seeking novel ways to advance remedies, Figure 4.25.

3D Bioprinting - GMP (Good Manufacturing Practice)

Regulatory pathway is required to bring the experimental trials of living cells conducted at the laboratories out to the clinic for the treatment of human patients. Dealing with the construction for a given tissue construct developed for human treatment will necessitate addressing the challenges of the compliance with the "Good Manufacturing Practice" (GMP). Constructs fabricated by 3D Bioprinting encompassing living cells, are subject to a GMP regulations.

Some of these GMP requirements are stated as follows:

1. How to keep tissue construction sterile
2. Quality assurance
3. The venture funding for human clinical trials.

When incorporating GMP in the design of 3D Bioprinting systems can only profit the translational progress from research to development, notwithstanding the upsurge in cost. In particularly, there are direct chances for designers of biomaterial-inks and 3D Bioprinters to commercialize the work beyond research.

Standardizing and automating tissue assembly, based on open-source, will be carried on to assist in reproducibility across laboratories. New techniques in measuring engineered tissue function are required for better mathematical algorithms to construct the cells. Non-invasive imaging and other associated practices advanced mainly for evaluating animal models can be implemented to tissue engineering research requirement.

Cellular grouping in vascularized tissues is streamlined into four organizations, which are rarely reiterated together in engineered tissue constructs. Soluble blood constituents vary considerably in size, concentration, and biochemistry, and each has discrete objectives and mechanisms for transferring tissue construct.

Bioprinting: Creating Human Organs

The substantial advancements in the field of 3D Bioprinting, it has become possible to print out human organs. The major contributor is the development of 3D Bioprinter technology, relying on the process of additive manufacturing, 4.26.

3D Bioprinting, as discussed previously, is the procedure of successive addition of bio-materials from the printer that ultimately grows into an organ or a biological membrane. The Bioprinter receives a copy of the 3D design in "Computer Aided Design" CAD forms that is required to print out. The 3D Bioprinter is then capable to shape on several layers constructing an actual physical object.

In manufacturing applications additive manufacturing technology has been used for many years. While the additive manufacturing technology is very useful in saving time and costs, the price is often a constraint. Large enterprises have used these 3D printers extensively to fabrication glass, metal, ceramics, and several other elements related to the manufacturing processes.

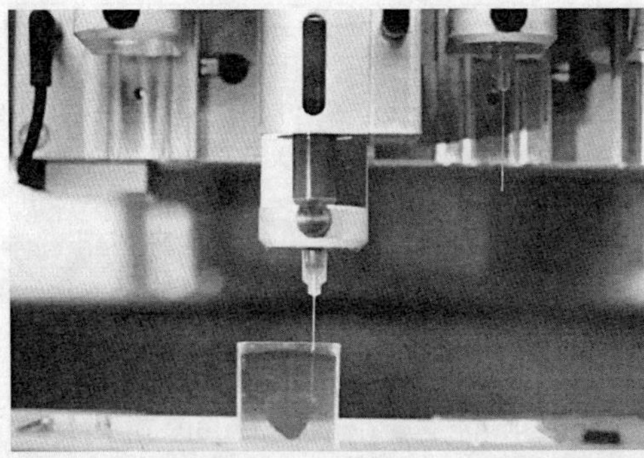

Figure 4.26: 3-D-printed heart engineered by Tel Aviv University scientists
Courtesy: Tomer Applebaum

The promising results of modern 3D Bioprinters inspired scientists, researchers, and engineers to replicate the processes in the medical, biological, and pharmaceuticals fields.

Once the first human tissue bioprinters was successfully produced a flood gate in 3D Bioprinters was opened. The Bioprinters were proficient of Bioprinting any human tissue that may be useful in regenerative therapies or medical research. Medical transplants are utilized by practitioners to create needed human tissue. Undoubtedly, the technology became a beacon of hope in the medical industry.

Bio-ink Made Living Human Cells

The 3D Bioinks in the Bioprinters creates human tissue utilizing additive manufacturing technology. The 3D Bioink is composed of live human cells. The contents of Bioink are successfully mixed up of human cells. The Bioprinter is able to generate human tissues. Several scientists, researchers, and engineers utilized Bioprinters to produce diverse human tissues. The initial development scientists, and engineers produced is a 3D skin printer to aid in wound healing.

Bioprinting Kidney Cells, Heart Valves, Knee Cartilage, and Bone Implants

Scientists, engineers, and researchers also have 3D Bioprinted out kidney cells. Several universities successfully utilized 3D Bioprinters to print out heart valves, knee cartilage, and bone implants.

After these successful achievements, different human organs were printed by several scientists, researchers and engineers. While these impressive well-documented achievements are encouraging, as stated previously, the main challenges are still the ability to 3D Bioprint functional human organs.

In Vitro-Drug Testing and Testing of Toxicity in New Drugs

A 3D Bioprinter was developed to arrange cells from the Bioink in micro spheres. This arrangement allowed the cells to continue flourishing after the manufacture of the tissue. The 3D Bioprinter emphases are on creating human stem cell tissue as contrasting to the tissue created by previous Bioprinters. The new developed 3D Bioprinter creates precise human tissue useful *in vitro*-drug testing and testing of toxicity in new drugs.

These advancements in 3D Bioprinter is effective in producing human tissue. Scientists and researchers were successful to control the amount of tissue freed during the 3D bioprinting process. The cost of the bioink was substantially reduced over time. The Bioprinter, which was able to produce viable stem cells, has led to absorbing the cost of the research and development of novel drugs. Ultimately, these benefits will be passed on to consumers leading to lower drug costs. Also, the capability to harvest stem cells is a constructive step for organ transplants.

Improving Bioprinter in Human Tissue

Many institutions are working to progress on the present state of technology in the Bioprinting of human tissue. The projected purpose of creating a multidisciplinary group is working on creating functional human tissue in the next few years. Scientists are still working on producing reliable Bioink.

Revolutionize the Medical Industry

Extensive experiments may be conducted to assist validating a cure for a specific disease through 3D Bioprinting Technology, which holds potential for the growth of more effective medicines. 3D Bioprinting technology regenerates organs to assist people who may need transplants. The time to receive a donor's organ expose the recipient to uncertainty of life, while the 3D Bioprinting technology holds the possibility of refining such circumstances. More importantly, the illegal trade in human organs soon will be diminished, sustained by the abundance of viable organ tissue.

Human Body on a Chip –Mini Physiology

The knowledge of generating minuscule varieties of human organs has been on sight for two decades. The aim is to Bioprint a miniscule size depiction of an organ on a chip that can seamlessly mimic the tasks and responses of the entire organ, Figure 4.20.

Lungs on Small Chips

Recent advances in the field of 3D Bio-Organ printing made it possible to mimic the real human organs. The miniature organs were propagated in 2D and have limited abilities to mimic the true function of human organ. 3D Bioprinting technology empowered completely functional miniature organs to be produced, such as the lungs, on small chips, Figure 4.17.

Body on a Chip

The effects of drugs on human organs may not be well understood their true reaction inside the body through only 3D Bioprinting models. However, "Body on a chip" may succeed to give a prediction of how those drugs will interact inside a living body, where several organs interrelate with one another. "Body on a Chip" is an electronic chip configured to represent 3D of all correlated channels to convey the true function of various interrelated organs

Alive cells are composed from actual human tissues used in 3D Bioprinting. The living cells are assembled into 3D tissues and small organs. These small organs are then relocated on a chip in the same organization as they are present in a human body, 4.20.

Channels Resembling Blood Vessels

Connecting organs through channels must resemble blood vessels, precisely as they are connected in a real human body. Artificial liquid solution fills the channels representing the blood vessels. . The soluble liquid acts as an alternative to blood. This fluid carries nutrient supplies to the miniature organs, which can be used to provide drugs to microorganisms to study the response of the organs to them.

The electronic chip is configured to contain sensors to continually monitor the changes in temperature and chemistry throughout the chip. This chip represents the exact behavior of the body's physiology. The sensors simulate the responses of the human body in a microchip, Figures 4.22, 4.23, and 4.24.

Body on a Chip – Drug Testing

"Body on a chip" provides several gains over current methods for drug testing. These are stated as follows:

1. **Direct Human Testing:** "Body on a chip" delivers a technique to test the special effects of drugs, disease causing organisms and radiations, directly on human cells. This is superior technique than the animal testing in the initial stages
2. **Testing Through Miniature Organs - Low Fluid Consumption:** The miniature chip warrants that only a small amount of fluid is needed for the operational of the miniature organs. Also, small amount of drugs required to produce desired effects. This saves cost and materials of the fluid and drugs.
3. **Increased Portability:** The "body on a chip" device is portable as does not exceed 5cm in size.
4. **Increased Process Control:** The minute chip equipped with sensors allows the process ease of monitor and control.

Revolution in Bioprinting Medical Science

3D Bioprinting combined with "body on a chip" will transform drug development and boost the comprehension of the standard performance of human body and the effect of diseases in changing body performance. The consequences of "body on chip" are stated as follows:

1. Drug Development
   a. Revolutionize drug development
   b. Eliminate the need of years of cellular and animal testing before testing drugs on human body.
   c. Reduce the cost and time required for drug development.
   d. Faster drug to the market leads to inexpensive drug
   e. Safer and more effective.
   f. Decrease in health care costs
2. Chemical Warfare

"Organ-On-Chip" is funded by USA *The Defense Threat Reduction Agency (DTRA)* - Recognizing that wars scattering throughout the world, USA is actively compelled in taking part in them. Unfortunately, there is an eminent threat of chemical and nuclear weapons may be used against the United States of America. Effective protective measures can be modeled describing radioactive weapons interacting with human tissues.

Models of Human Physiology and Pathology

It is not possible to comprehend human body and the diseases that disturb it through direct testing on human tissues, but on animals, which may or may not applicable to human organs. However, "Body on a chip" will provide clear appreciation of the human body physiology and the effect of drugs on it.

A Cure for Poisonous Gas

When poisonous gas was released in the environment, severe side effects developed due to the poisonous gas. The "body on a chip" was available for testing a certain drug. By administering the novel drug to patients, within few hours the ailment was subsided.

3D Bioprinted Breast Implants for Cancer Patients

3D printed breast was developing as an implant for cancer patients who suffered a surgical operation for breast cancer in which the surgery is limited to the removal of the visible and palpable tumor only (lumpectomies). Surgeons removed the cancerous substances and left the residual breast tissue. Unfortunately, the breasts have become misshaped and required reconstruction, Figure 4.27.

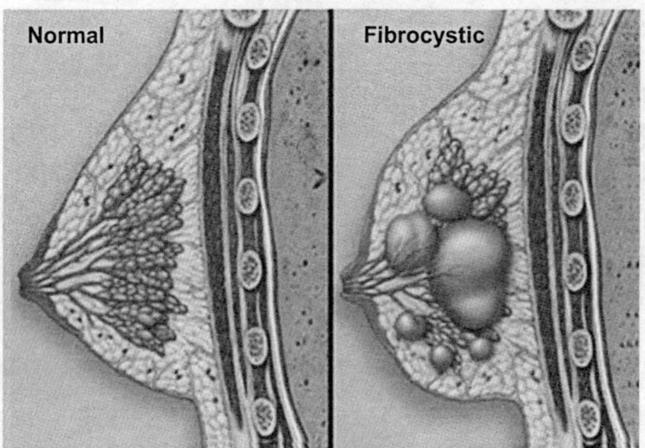

Figure 4.27: Bioprinted Breast Implants
Courtesy: Mayo Foundation for Medical Health

About 200,000 women a year suffer lumpectomies. The amputation of the entire breast to have a complete reconstruction is feasible.

Fortunately, a popular rebuilding method called "fat grafting." Fat grafting contains suctioning fat cells out of the patient's body, and then inserting them into the vacuum the tumors left behind. As the fat may be reabsorbed into the patient's body this method may fail.

3D Bioprinted grafts overcame the issue of fat re-absorbed. Additionally, the fat was created utilizing the women's individual cells, avoiding the risk of rejection.

Scientists and medical doctors have removed fat cells through liposuction from the patients. The fat cells preserved, and then used as Bioink in 3D Bioprinter printing the necessary lump for implant.

3D Bioprinter printed the implants. The Bioprinter lays down a biodegradable gel taking the form of a scaffold identical to the patient's tumor. The Bioprinter filled the scaffold with the patient's individual cells. It is then implanted at the tumor voids.

REFERENCES

1. Nicodemus GD, Bryant SJ. Cell encapsulation in biodegradable hydrogels for tissue engineering applications. Tissue Eng Part B Rev. 2008; 14: 149–65.
2. Fischbach C, Kong HJ, Hsiong SX, et al . Cancer cell angiogenic capability is regulated by 3D culture and integrin engagement. Proc Natl Acad Sci USA. 2009; 106: 399–404.
3. Serban MA, Scott A, Prestwich GD. Use of hyaluronan-derived hydrogels for three-dimensional cell culture and tumor xenografts. Curr Protoc Cell Biol. 2008; 10: Unit 10–14.

4. Prestwich GD. Evaluating drug efficacy and toxicology in three dimensions: using synthetic extracellular matrices in drug discovery. Acc Chem Res. 2008; 41: 139–48.
5. Liu Y, Shu XZ, Prestwich GD. Tumor engineering: orthotopic cancer models in mice using cell-loaded, injectable, cross-linked hyaluronan-derived hydrogels. Tissue Eng. 2007; 13: 1091–101.
6. Scaife CL, Shea JE, Dai Q, et al . Synthetic extracellular matrix enhances tumor growth and metastasis in an orthotopic mouse model of pancreatic adenocarcinoma. J Gastrointest Surg. 2008; 12: 1074–80.
7. Zhang S, Gelain F, Zhao X. Designer self-assembling peptide nanofiber scaffolds for 3D tissue cell cultures. Semin Cancer Biol. 2005; 15: 413–20.
8. Li L, Beniash E, Zubarev ER, et al . Assembling a lasing hybrid material with supramolecular polymers and nanocrystals. Nat Mater. 2003; 2: 689–94.
9. Silva GA, Czeisler C, Niece KL, et al . Selective differentiation of neural progenitor cells by high-epitope density nanofibers. Science. 2004; 303: 1352–5.
10. Gelain F, Bottai D, Vescovi A, et al . Designer self-assembling Peptide nanofiber scaffolds for adult mouse neural stem cell 3-dimensional cultures. PLoS ONE. 2006; 1: e119. 1–12.
11. Sargeant TD, Oppenheimer SM, Dunand DC, et al . Titanium foam-bioactive nanofiber hybrids for bone regeneration. J Tissue Eng Regen Med. 2008; 2: 455–62.
12. Zhang S. Designer self-assembling Peptide nanofiber scaffolds for study of 3-d cell biology and beyond. Adv Cancer Res. 2008; 99: 335–62.
13. Kim MS, Yeon JH, Park JK. A microfluidic platform for 3-dimensional cell culture and cell-based assays. Biomed Microdevices. 2007; 9: 25–34.
14. Birgersdotter A, Baumforth KR, Porwit A, et al . Three-dimensional culturing of the Hodgkin lymphoma cell-line L1236 induces a HL tissue-like gene expression pattern. Leuk Lymphoma. 2007; 48: 2042–53.
15. Raeber GP, Lutolf MP, Hubbell JA. Molecularly engineered PEG hydrogels: a novel model system for proteolytically mediated cell migration. Biophys J. 2005; 89: 1374–88.
16. Ehrbar M, Rizzi SC, Schoenmakers RG et al . Enzymatic formation of modular cell-instructive fibrin analogs for tissue engineering. Biomaterials. 2007; 28: 3856–66.
17. Lutolf MP, Weber FE, Schmoekel HG et al . Repair of bone defects using synthetic mimetics of collagenous extracellular matrices. Nat Biotechnol. 2003; 21: 513–8.
18. Lutolf M P. Integration column: Artificial ECM: expanding the cell biology tool box in 3D. Integr Biol. 2009; 1: 235–41.
19. Ehrbar M, Rizzi SC, Schoenmakers RG et al . Biomolecular hydrogels formed and degraded via site-specific enzymatic reactions. Biomacromolecules. 2007; 8: 3000–7.
20. Kupatt C, Horstkotte J, Vlastos GA, et al . Embryonic endothelial progenitor cells expressing a broad range of proangiogenic and remodeling factors enhance vascularization and tissue recovery in acute and chronic ischemia. FASEB J. 2005; 19: 1576–8.
21. Fukumura D, Jain RK. Tumor microvasculature and microenvironment: targets for anti-angiogenesis and normalization. Microvasc Res. 2007; 74: 72–84.
22. Murasawa S, Asahara T. Gene modified cell transplantation for vascular regeneration. Curr Gene Ther. 2007; 7: 1–6.
23. Wei J, Blum S, Unger M, et al . Embryonic endothelial progenitor cells armed with a suicide gene target hypoxic lung metas-tases after intravenous delivery. Cancer Cell. 2004; 5: 477–88.

24. Rosol TJ, Tannehill-Gregg SH, LeRoy BE, et al . Animal models of bone metastasis. Cancer. 2003; 97: 748–57.
25. Meijer GJ, de Bruijn JD, Koole R, et al . Cell-based bone tissue engineering. PLoS Med. 2007; 4: e119. 0260–4.
26. Soloman, Sabrie "Sensors Handbook, 2nd Edition, McGraw-Hill Publishing Co., NY – ISBN 978-0-07-160570-0
27. Soloman, Sabrie "Sensors and Control Systems in Manufacturing 2nd Edition," McGraw-Hill Publishing Co., NY – ISBN 0-07-059626-3
28. Soloman, Sabrie "Affordable Automation," McGraw-Hill Publishing Co., NY – ISBN 0-07-059633-6
29. Soloman, Sabrie "Sensors Handbook, 1st Edition," McGraw-Hill Publishing Co., NY – ISBN 0-07-059630-1
30. Soloman, Sabrie "Sensors and Control Systems in Manufacturing 1st Edition," McGraw-Hill Publishing Co., NY – ISBN 978-07-10572-4

5 3D Printing Scaffolds

ADVANCED BIOSENSING TECHNOLOGY

The Role of MEMS – NANO-MEMS in 3D Bioprinting

While stem cells are the fundamental ingredient in 3D Bioprinting organs, it could not be accomplished, as it may render useless, without the substance that can hold cells together in coherent manner to allow the cells to take the intended shape. This substance is the "scaffold" upon which all the cells will proliferate to take the intended form of the organ or membrane. Thanks to the science and art of the scaffolding technology 3D Bioprinting came to fruition to represent is vital parameter to advance and implement the natural outcome of the Bioprinting technology.

Equally impressive is the advancements of MEMS (Micro-Electromechanical-Machining-System and 3D Biosensors Technologies dealing with micro-nano sensing devices. The recent objective is to utilize the use of 3D Bioprinting combined with Nano-MEMS to miniaturize the development of bio/medical sensors devices.

Our study in this chapter is focused on several selected techniques of Bio-fabricating scaffolds chosen out of a host of scaffolding techniques geared to support of healthy and robust methodology to 3D Bioprint organs a human membranes.

The Nano-MEMS - 3D Bioprinted Biosensors will substitute the current cumbersome methodologies necessitating intricate expensive tools operated by highly skillful professionals. 3D Bioprinted Biosensors technology employs intelligent scale-down biomedical sensors for *in vivo* diagnostics of "Gastro-Intestinal" Tract, GI tract. 3D Biosensors will provide real-time monitoring of biological response, revealing molecular communications happening at the gut-microbiome-brain-axis, Figure 5.1.

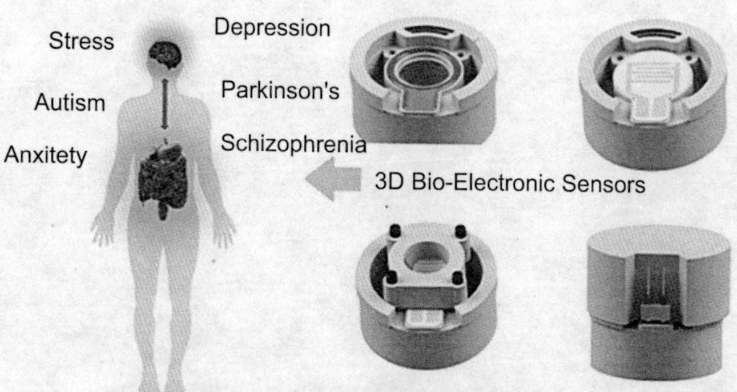

Figure 5.1: Gut-Microbiome-Brain and Associated Pathologies – 3D Bioprinted Biosensors
Courtesy: University of Maryland

3D Electrochemical Bioprinted Biosensors detect molecular and cellular change in aqueous environments. The entire 3D Bioprinted sensor can be swallowed by encapsulating it along with it own miniature power supply in form of miniscule battery to pass through GI tract. The Bioprinted sensors are equipped with wireless communication circuitries to transmit molecule Biosensing data for external analysis.

Real-time Monitoring of Physiological Events

The lifesaving diagnostic data will convey accurate status of the PA "better pancreatic adenocarcinoma." The 3D Bioprinted electrochemical Biosensors is an indispensable tool for evaluating the patient's gut-microbiome. Also, it utilized to identify the metabolic state of the gut microbiome in patients showing both GI and "neurological disorders," solving the missing-links in molecular communication at the Gut-microbiome-brain-axis.

Future Biosensing applications will need high reliability to accomplish real-time observation and monitoring of physiological actions, integration of biosensors into feedback-based devices, revealing of toxins and advanced diagnostics.

Such performance will necessitate biosensors with excellent resolution, specificity, and output, as well as the capability to simultaneously detect several analyses. While these burdens have yet to be totally comprehended, new advances in bio-fabrication may permit sensors to achieve the high spatial resolution essential, and bring us closer to understanding devices with these capabilities. To this end, new advances in bio-fabrication techniques are presented. Precisely, giving important rank on bioprinting techniques, e.g., micro-contact 3D Bioprinting, 3D inkjet Bioprinting, and laser direct-write, LDW may prove crucial to biosensor fabrication and scaling.

Recent biosensors have employed these Bio-manufacture techniques with success, and further improvement may allow higher performance, including multiplexing multiple analytes, or cell types within a single biosensor. Recent developments in "3D bioprinting and their potential to create biosensors with live cells encapsulated in 3D microenvironments play a central role in evolving Biosensors. Such developments in bio-fabrication will increase biosensor usefulness and obtainability, with influence grasped in many interdisciplinary fields, as well as in the clinic.

Sensing Biomarker Differences of Patient-to-Patient

For analytical, environmental, and clinical implementations of biosensors, it is necessary to be capable to sense, measure, and report a multitude of diverse analytes speedily and concurrently. Frequently, revealing of a single analyte, such as a disease marker or protein, is not enough for analysis or disease screening. This has become particularly pertinent in current years as patient-particular treatments are being established as functional of patient-to-patient biomarker differences.

As the abilities of biosensors remain to progress, these new biosensors must have the capability to multiplex large amounts of data, with high dynamic variety and a low signal-to-noise ratio, if possible in a single sensor. Additionally, as scientists progress an extensive variety of discovery methods, the scaling and multiplexing of Biosensing detectors and transducers becomes a bio-fabrication challenge.

Detecting Proteins, DNA, Cells, and Toxins via 3D Bioprinted Biosensors

Present biosensors requirement are essential to detect proteins, DNA, cells, and toxins, between other analytes. Normally, the discovery of multiple analytes is preferred for carrying out diagnostics, observing, and toxicity screening as well as abundant other Biosensing implementations.

These necessities call for refined biosensor manufacture. Biosensors need numerous transduction components in order to process many analytes independently, so high-throughput manufacture of transducers and/or sensors is of boundless advantage to the biosensor field. For instance, current biosensors have utilized high throughput protein modelling for bacteria detection, toxin detection, and amount of protein communication. Extra kinds of high-throughput biosensor devices necessitate refined manufacture technology, such as micromechanical cantilever collections to detect chemicals. Otherwise, many new biosensor-based researches intention only to detect single compound but endeavor to do so with very high sensitivity.

Detecting Multiple Analytes

In order to examine or detect many analytes biological consequences with high output, it is essential to harness manufacture technologies for biosensors claims. Bio-manufacture technologies have been progressing in current years, mainly cell-based manufacture technologies that allow cellular encapsulation and development in 3D micro settings. Although there have been many recent bio-manufacture developments, predominantly concerning new materials or arrangements of materials, concentrating on those developments in 3D printing and deposition technologies will be most appropriate to biosensors.

Miniaturizing Biosensors

It has been distinguished that reducing Biosensors that depend on surface capture of molecules results in inferior achievement, due to conveyance of target molecules to the Biosensor. Nonetheless, modelling, as contrasting to flow-based distribution, permits direct transfer of molecules, cells, or a specimen to be analyzed to an anticipated setting. This direct conveyance considerably progresses binding kinetics over arbitrary communications. Mode based on high-throughput Biosensing characteristically includes depositing a protein, antibody, or other molecule of attention to a substrate, and gauging binding of another molecule or cell.

Lithography-Based Modelling

Modelling is typically achieved by way of "lithography-based approaches," which utilize elastomeric patterns with small release aspects manufactured from silicon wafer masters. For "lithography-based modelling" implementations, a photomask is utilized to allow choosing polymerization of photoresist, on silicon, to generate a master. In the instance of micro contact printing, this master may then be utilized as a negative to manufacture many elastomeric imprint or molds, naturally out of polydimethylsiloxane - PDMS.

Bioprinting - Inkjet and "Laser Direct Writing -LDW"

Self-assembled mono-layers - SAMs of many polymers can be patterned onto surfaces utilizing an implant, though chemical communication between the modeled polymer and the substrate restricts the polymers and substrates that may be utilized. "Soft lithography methods" are contact-based, and include high pressure on an imprint to convey material from a donor substrate. As an alternative, patterning can be accomplished utilizing non-contact deposition techniques, counting on inkjet printing and laser direct writing - LDW. Non-contact techniques permit the direct modelling of material, and cells, wherever the precise binding chemistry is not as essential, Figure 5.2.

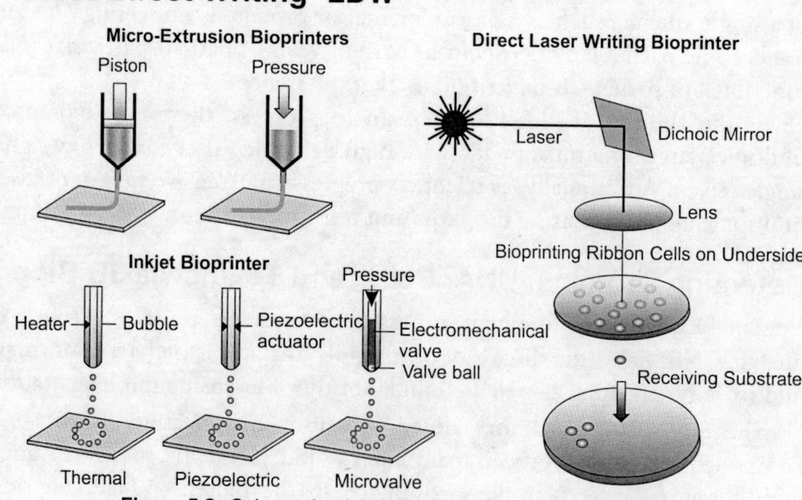

Figure 5.2: *Schematic - Inkjet Bioprinting – Laser Direct Writing*
Courtesy: Biomedical Engineering Societies (BMES)

Non-contact deposition techniques, such as LDW, embrace diverse benefits over other methods for Biosensor implementations. Distinctive contact-based modelling includes a chemical or physical communication with the substrate, and typically needs limiting growth to a precise domain. The unlimited growing of cells is an essential principle for repetitive cell performance. With unusual cell performance, cell analyte manufacture may be affected, thus spreading a mistake in the Biosensing system. To avoid substrate alteration that may impact cell function, non-contact based approaches are utilized. LDW and inkjet Bioprinting are both non-contact bioprinting techniques applicable of high-throughput non-contact deposition. These two systems vary in their achievement, with distinctive trade-offs in throughput, tolerance, and precision. LDW is a more precise and accurate deposition method, whereas inkjet Bioprinting provides superior throughput.

For high-throughput requirements, modelling procedure characteristically restrain proteins via absorption to a surface. Though, numerous Biosensors are live-cell tools that sense biological responses and cell attitude in response to inducements. Merely gauging the reaction of a cell to a static influence may not be adequate for some applications, and data about dynamic reaction and longer-term fluctuations in cell behavior may be anticipated. Although there are some biosensors that can monitor certain molecular altitudes, such as real-time glucose observation *in vivo*, real-time measurement systems normally do not spread outside one analyte because of the precise recognition methods.

Förster Resonance Energy Transfer (FRET)

Some know-how that inspect real-time variations in cell attitude embrace Förster resonance energy transfer -FRET, bioluminescent resonance energy transfer -BRET, or other fluorescent-based exposure systems that react to an adhesion occurrence. Though, these methods are also characteristically restricted to a limited analytes since fluorescence emission overlay. Generally, the amount attainable in real-time Biosensing is rather restricted, Figure 5.3.

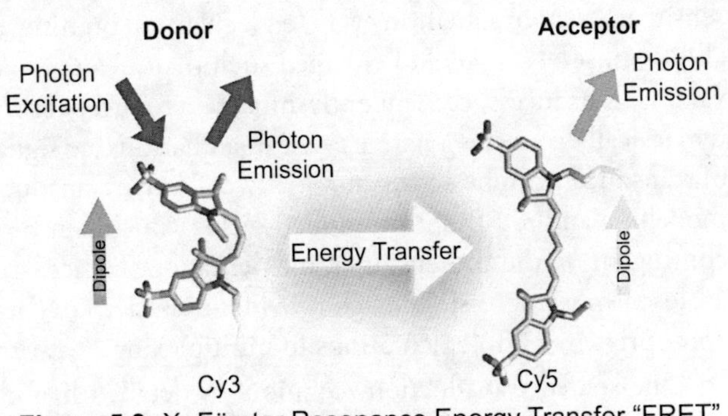

Figure 5.3: X- Förster Resonance Energy Transfer "FRET"
Courtesy: PLUS ONE

A FRET occurrence begins once a molecule considered by a "donor" and "acceptor" couple diffuses into a volume of adequately large laser strength close the focal point (left). The amounts for sensed photon radiations for the acceptor n_A and donor n_D are logged until the molecule diffuses out of the focal volume (top right), Figure 5.3. Throughout the donor stimulation either a photon is radiated or energy is non-radioactively transported to the acceptor and radiated with rates that are contingent upon the molecular conformation (lower right), Figure 5.3.

Modelling Technologies

Biosensing may profit from the application of modelling skills to yield live cell tools, since modelling permits the assigning precise analytes or cells in characterized as spatial locations. Modelling distinct analytes, or even transducers, in parallel, characterized positions can be utilized to examine exact impetuses instantaneously. This concurrent examination not only raises throughput, but does so in a way that synergistically raises the data one can collect from analytes and cellular reactions. Numerous modelling tools are studied, comprising those that can pattern proteins and sustainable cells.

Current bio-manufacture progresses in modelling, mainly with living cells and in 3D, may aid to progress the Biosensors arena. It is understandable that high output and speedy screening of many markers may be essential for point-of-care (POC) diagnostics. Nevertheless, inspecting dynamic cellular reaction via living-cell assays is also a significant property that requires to be studied by novel Biosensors, and this may be possible by means of high-output means.

3D Bioprinted Biosensing Detection and Transduction

Biosensors, even those manufactured with refined Bioprinting procedures, contain three fundamentals:

- the biological signal (analyte),
- the transducer(s), and
- the detector/readout.

The biological indicator is what one is attempting to sense, though transducers permute this indicator into one that can be sensed and calculated for analysis. The biological indicator happens as expected, so most of the manufacturing engineering includes the transducer and detector. Also, detectors mostly can state variations in light, voltage, pH, and/or absorbance, so more engineering choice lies in the transduction fundamentals than in the indicator.

With the purpose of detect analytes at minor scales; refined transducers must be intended that can recognize, calculate, and potentially focus a material, antibody, chemical, or other biologic. To identify a sensitive extent of signal, an event, e.g., chemical binding, a micromechanical reaction, or an alteration in cell ability must be sensed and enlarged such that it can be measured, monitored, and compared to anticipated values. Transducers, consequently, must be a robust part of the Biosensor, as they are the only component that dynamically converts a native event to a quantifiable signal. Contingent on the implementation, a diversity of transducers can be active, and many collective transducers are illustrated in Figure 5.4. Concentrating on modelling and 3D Bioprinting methods is valuable in Biosensing, a transducer must be prudently selected, contingent on the Biosensing implementation. Hence, transducers that may be configured when designing a Biosensor are accessible. Each provides exclusive advantages and difficulties, with some transducers better appropriate for implementations in multiplexing, real-time Biosensing, or manufacture.

One shared transduction technique is detailed ligand adhesion. When a ligand of interest adheres to a cell, antibody, protein, or chemically-adapted surface, the adhering event can be identified by many techniques. Approaches for identifying fluorescent protein-protein adhering comprise enzyme-correlated immunosorbent assays - ELISAs, which can be achieved in high-throughput manner, Figure 5.4.

Technique	Schematic	Example
Immunofluorescence	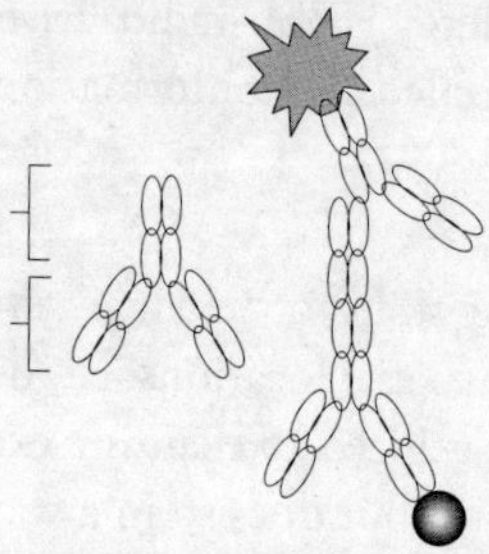	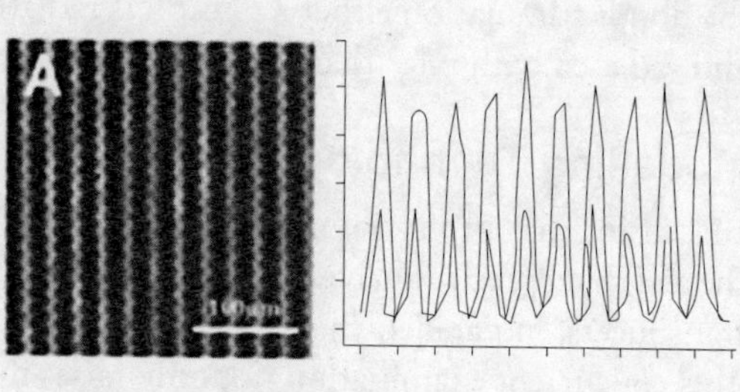

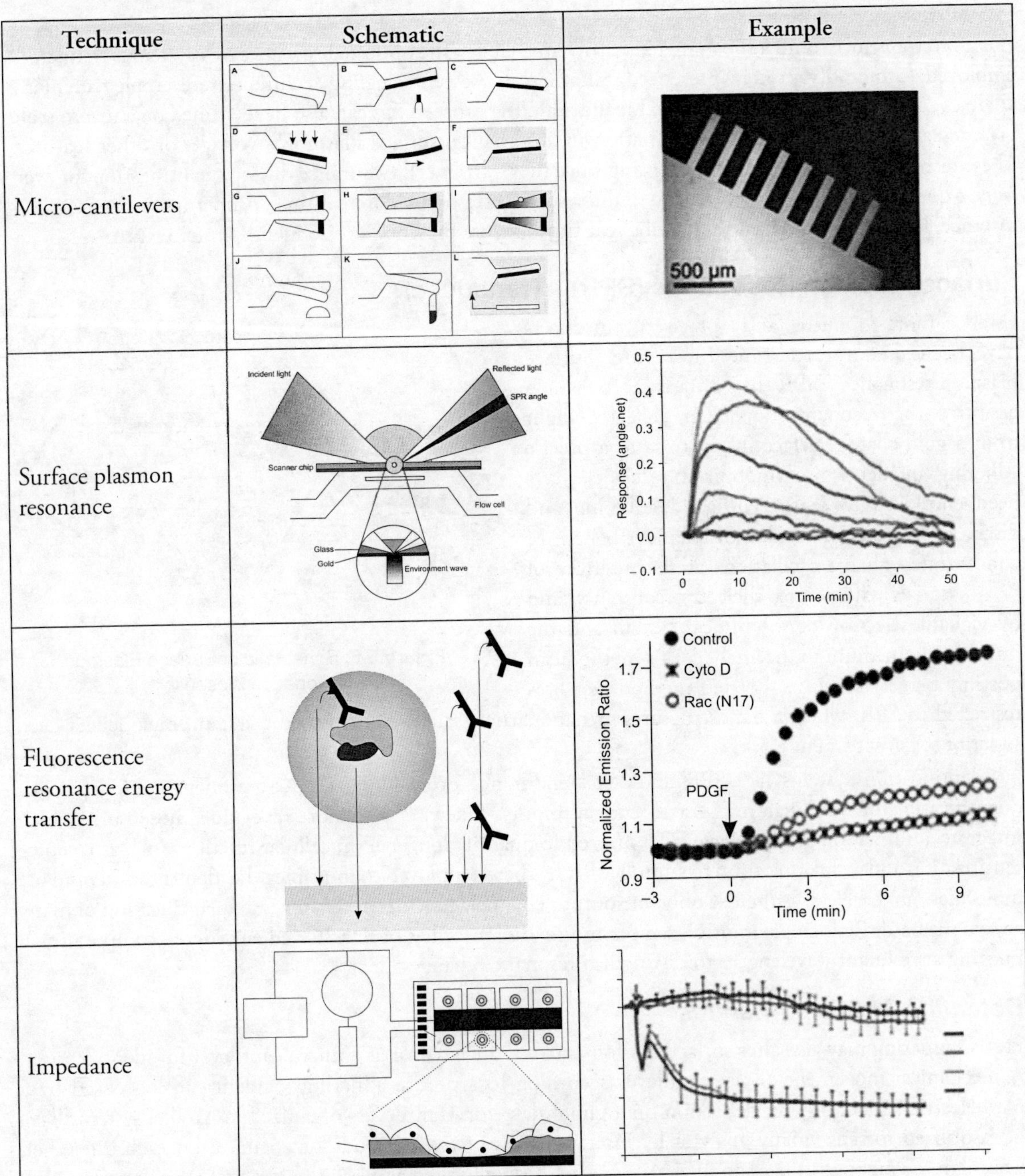

Figure 5.4: Schematics of 3D Bioprinted Immunofluorescence Biosensor Transducers

Courtesy: Published by MDPI AG

Dynamic Biosensing

Though, these data signify a static biological formation, i.e., cells are characteristically lysed, so any time-based data of the reaction is vanished. Other static biosensors contain the modelling of protein, or a chemically energized transducer onto a substrate to identify an analyte. The optical absorbance of a system is frequently employed to quantify protein presence, and when fluorescent labeling techniques are employed, many analytes can be identified concurrently. Ligand adhering proceedings can also be identified on a macro scale via ancillary transducers, such as a variation in absorbance, surface Plasmon's, voltage, or other features. These techniques allow dynamic Biosensing since they permit cell endurance during quantification, or even *in vivo* quantification and any variation in the system are responded in real-time. Amperometric sensors, for instance, have been employed to describe real-time glucose presence and to identify micro RNA.

Surface Plasmon Resonance (SPR) - Transducers

Lately, more refined and precise transducers have been used for real-time Biosensing. Surface plasmon resonance - SPR -based transducers, which quantify the resonance angle of reflected light from a gold-coated surface in response to a precise adhering incident to immobilized receivers, have been employed to sense cell/ligand adhering and discharge or exact molecules. SPR enjoys benefit of concentrated charge oscillations at the interface of two materials with reverse dielectric constants, and by vigilant selection of the optical system and the transducing medium to the analyte, this method can possibly be very sensitive. Related methods, such as restricted to SPR, where a nanostructures or nanoparticles are presented to the system, may deliver even superior sensitivity, Figure 5.5.

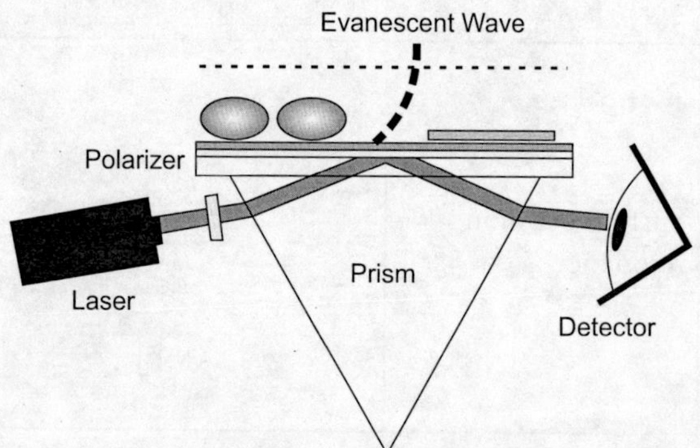

Figure 5.5: Schematic of Surface Plasmon Resonance Biosensor

Identification systems like SPR allow cell endurance over a long-term experiment, which permits quantification of a cellular reaction to convey inputs. Likewise, resonant waveguide Biosensors, which integrate a diffraction grating, have been utilized to quantify longer-term cellular reaction, such as receptor sensitization and cellular reaction to stimuli. SPR has been employed to multiplex the identification of many molecules, nonetheless, as there is only one output channel, concurrent sensing and quantification of many analytes is hard. Reducing a multiplexed plasmonic Biosensor has been achieved employing computational imaging at resonant wavelengths to distinguish resonance changes.

Detecting Multiple Analytes

Precise adhering may also alter surface tension and activate a sensor as a micro cantilever for identification. A mechanical movement of the cantilever according to a precise adhering incident can be sensed by a piezoelectric sensor, or laser deflection on a photo detector. Cantilevers organized in parallel systems have been utilized to sense many analytes by inserting a distinctive antibody or chemical on each cantilever. This high throughput utilization of cantilevers has been allocated as a "synthetic nose." Also, Microfluidics has been utilized to scale ligand adhering methods. Microfluidics can be utilized to flow proteins, explicit antibodies, or analytes to explicit positions. After molecules stream adhere to the substrate they can be utilized to seize analytes and adhere fluorescent antibodies. Both micro-cantilevers and microfluidics permit the position of adhering to be controlled. Position control allows multiplexing many elements onto a biosensor, since each element is distinct, and can be sensed autonomously.

Nonetheless, adhering is not continually essential for transduction and sensing, and ligand adhering may be utilized in conjunction with other transduction approaches. In biological systems, a minor variation in the microenvironment can affect a quantifiable modification in pH or absorbance, signal cells to yield an influence that can be sensed, or modification the intricate impedance of a membrane. pH variations and absorbance variations are not constantly biologically pertinent, predominantly in real-time systems. Strong cells need a distinct pH, and gauging absorbance is most valuable after presenting fluorescent labels to the system.

Electric Cell Impedance Sensing (ECIS)

Sensitive methods for real-time Biosensing that do not need a adhering incident frequently necessitate a refined transduction system. Other sensing approaches for system-wide variations in cell attitude comprise impedance-based transduction, such as Electric Cell-substrate Impedance Sensing (ECIS), in which membrane impedance is restrained, and impedance variations according to cell attitude can be sensed, such as those proficient in reaction to the existence of a toxin. ECIS has been shrunk to a biochip, thus ECIS may be scalable if several such chips can be manufactured. SPR signals have also been employed to transduce a cellular reaction to a stimulus. System-wide variations, like ECIS and SPR, may be valuable for the recognition of single analytes, but when multiplexing sensors or when multiple analytes are essential, more exact manufacture approaches may be essential for progressively intricate sensors. Nonetheless, sensors that can sense variations in impedance or SPR across a membrane permit the detection of reversible reactions on cells, and a return to system balance after a stimulus is eliminated. It is needed to sense variations in cellular attitude over time, as cells may recuperate from a stimulus or have a late reaction. Though these approaches permit cell endurance and can be employed for real-time Biosensing, the foremost difficulties of transduction approaches that do not need ligand adhering are the restricted output and lack of specificity to what is being sensed, Figure 5.6.

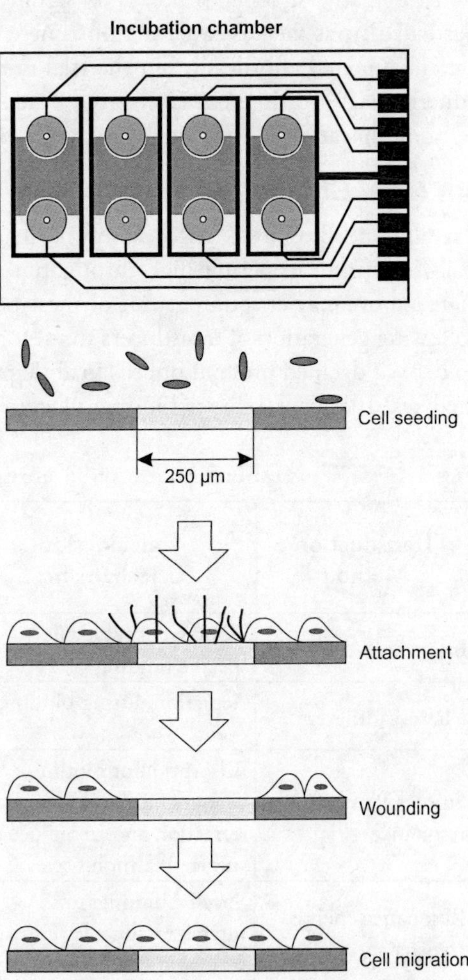

Figure 5.6: Wound healing - Electric Cell-Substrate Impedance Sensing (ECIS)
Courtesy: Elsevier

Electrode Artificial Neural Network

Existing Biosensor requests can apply cells in a diversity of manners to yield cell-based Biosensors. In these methodologies, cells can operate in many parts for the sensing system. Formerly, it was defined that ligand adhering can be implemented directly to cells with the utilization of enzyme-linked fluorescence. In this implementation, the transducing component is the fluorescent antibody destined to a precise protein. Nevertheless, there are numerous other claims of cell-based Biosensors that apply living cells as the transducing component. Cells hold natural mechanisms to sense exact physiological analytes or circumstantial variations, thus present a single benefit for practice as sensing elements.

There are numerous present instances of cell-based transducers in Biosensing. One such instance is the practice of artificial neural networks on an electrode, where the neurons change an circumstantial modification to electrochemical signal. Here, neuroblastoma cells were sensitive to exact composites in insecticides, but comparable Biosensors could be intended with cells sensitive to diverse mixtures. In another instance, circumstantial variations allow changes in cell metabolic action that can be gauged by variations in the extracellular pH. Here, the rate at which cells emit acidic harvests of metabolism was calculated in the attendance of diverse materials. Furthermore, a cell can sense the signal from an analyte or surrounding factor interrelating with cellular receptors to a precise protein or soluble factor through secondary cellular messengers and a physiological reaction. This soluble factor or protein can then be directly sensed by means of approaches such as fluorescently tagged antibodies, or by adhering to a secondary cell that will yield a reaction to the circumstantial modification identified by the main cell. This kind of signaling happens in blood sugar homeostasis, where pancreatic beta cells in the existence of high glucose focus secrete insulin that will signal secondary cells to transform glucose to glycogen.

Particularly at lesser scales, it is significant that Biosensing proficiencies have high sensitivity and particularity, as well as fast reaction times. For Biochemical sensing by exact adhering, these necessities are encountered effortlessly, but the trial in evolving the skill lies in selecting and constructing the precise adhering factor of interest. Conversely, once Biosensing encompasses variations to cell conduct, or a long-term modification in the system, reactions may happen at much lengthier time scales.

An Ideal Biosensor

An ultimate Biosensor is therefore one that is precise, can document reversible reactions, functions in real-time, and can permit high-throughput sensing of numerous analytes. Whereas the Biosensor sensors stated above may encounter some of these necessities, Table 5.1, new manufacture methods may allow the following generation of transducers to meet many more. Particularly, bioprinting methods can be arranged to deposit detailed proteins or cells in anticipated settings, which permits for high throughput and multiple analytes. The adaptability of bio-manufacture and bioprinting will allow robust biosensors to be established.

Table 5.1: Selected Transduction Technique Performance - Biosensing.

Transduction Method	Transduction Mechanism	Simultaneous Multiple Analyte Detection +++	Real-Time Capability +++	Fabrication Speed and Customizability +++
Immunofluorescence	Fluorescent molecule specific binding	++	+	+++
Micro-cantilevers	Actuation during binding event	+++	+	+
Surface Plasmon resonance	Adsorption or binding changes local index of refraction and resonance of surface Plasmon waves	-	+++	+++
Resonance energy transfer	Specific binding or interaction enables emission from and detection of target	++	+++	+
Impedance	Impedance of cell membrane measured	-	+++	++

Biosensing Implementations

Bioprinting encompasses the allocation of material such as cells to a substrate. It often provides the abilities to apply in Biosensing implementations, which encompasses "Rapid Deposition" and modeling of proteins or other Biomolecules. There are several Bioprinting techniques modified for utilization in constructing Biosensors. Several 3D Bioprinting techniques, incorporates "electro deposition," which enables the transmission of thin films of metallic "nanoparticles" to a substrate. Nanoparticles may be called "nanowires." The transfer is achieved through an electric field. 3D Bioprinting thin films of metals can be utilized to develop circuits essential for a Biosensor. Also, a 3D Bioprinting thin film is vital for some immunoassays or microarrays. "Electro deposition" is utilized for 3D Bioprinting thin films of biological substance material as proteins, enzymes, nucleic acids, polysaccharides and bacterial cells.

While Biosensing implementation have utilized thin films, the emphasis is mainly on 3D Bioprinting methodologies, which can be utilized to precisely "deposit" a variety of Biologics and mammalian cells. Direct 3D Bioprinting biologics and cells enable utilization of the several types of sensing.

Bioprinting Modeling Techniques

Table 5.2, illustrates instances of the 3D Bioprinting methods presented below. Besides, 3D Bioprinting precision allows high-throughput analysis to speedily distinguish multiple signals. For instance, fast modeling multiple proteins at diverse focuses can allow recognition of threshold levels to produce a cellular reaction, or to promote cellular adherence for parallel experimentations. Though, "electro deposition" can be shared with masking a microarray of electrodes, to generate a thin film with spatial monitor.

Table 5.2: Selected patterning biofabrication methods, performance, and some materials that have been patterned.

Modeling Technique	Method	Resolution +++	Throughput +++	Printable materials library +++
Micro-contact Bioprinting	Contact-based method, typically micro-stamp.	+++	++	++
Ink-jet Bioprinting	Droplet ejection from nozzle involving thermal piezoelectric or pressure	+	+++	++
Matrix-assisted pulsed laser evaporation direct-write	Non-contact deposition via pulsed laser directly onto gel with cell suspension	+++	+	+++
Laser-induced forward transfer	Non-contact deposition via pulsed laser on sacrificial layer	+++	++	+++

Printing Cells at Various Colony Sizes

With the purpose of multiplying the recognition of diverse analytes and the outcome on diverse kinds of cells, elements of the system must be spatially detached. There are many transduction methods, that diverse analytes can affect each other. Besides, if numerous focusses of a solitary analyte must to be confirmed, multiplexing by 3D isolation of experimental circumstances is essential.

3D Bioprinting cells or protein at various group sizes, at dissimilar concentrations, may have a consequence on cellular performance, and can impact how cells answer to outside signals. Many

3D Bioprinting methods are obtainable, and they can usually be distributed into two major categories:

a. Contact-based printing and
b. Non-contact-based printing.

Contact-based method needs direct contact between an acceptance of substrate and the surface from which a material is donated. It frequently occurs with high pressure to enable the transmission to the receiving substrate.

Non-contact-based Bioprinting, instead, comprises the transmission of cells to a receiving substrate through a discharge incident. Upon discharge, the transported cells are neither touching the donor nor receiving substrate. Both 3D Bioprinting methods have advantages and disadvantages, contingent upon the anticipated application. A diversity of Bioprinting approaches is illustrated in Figure 5.7.

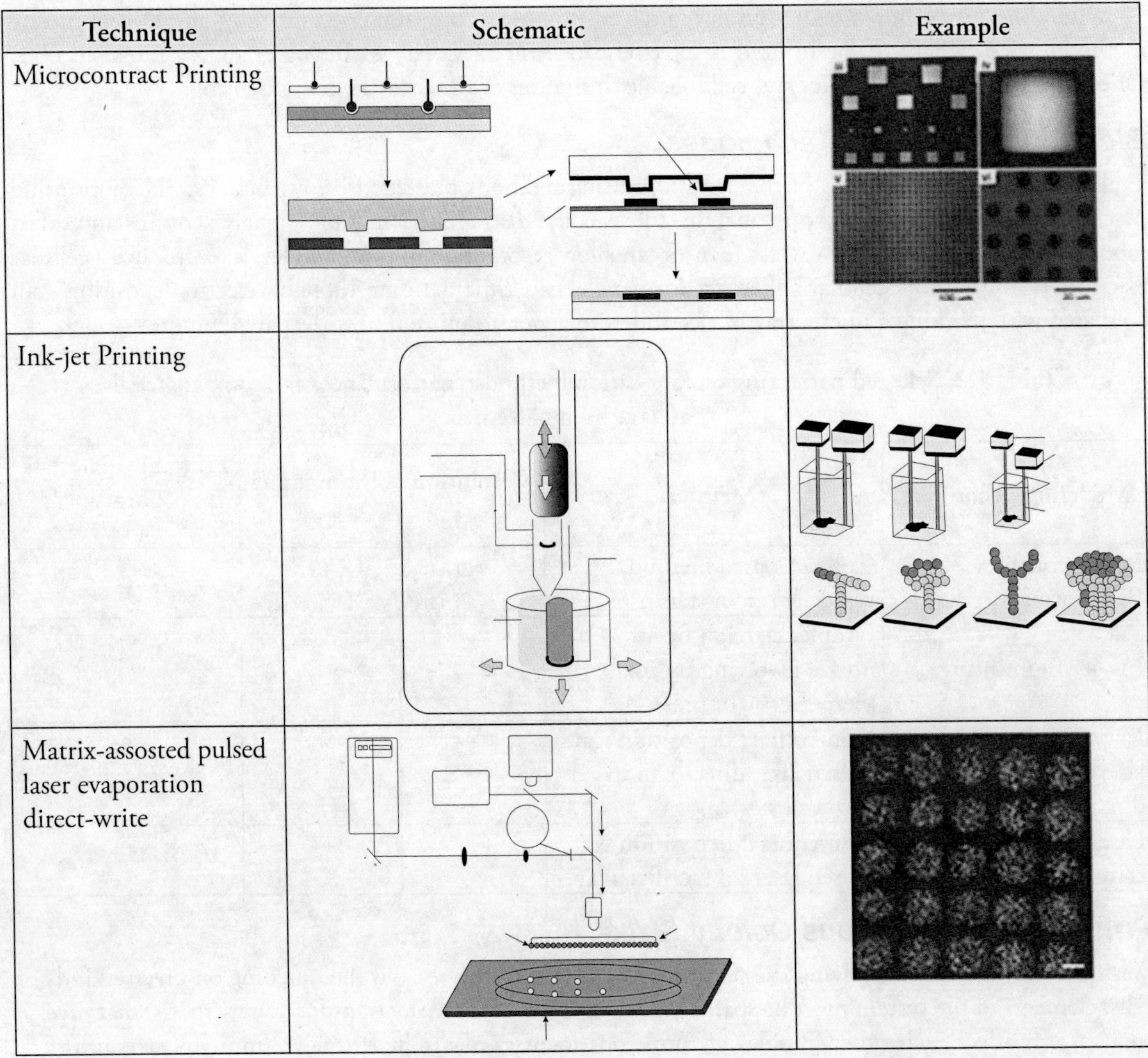

Technique	Schematic	Example
Microcontract Printing		
Ink-jet Printing		
Matrix-assosted pulsed laser evaporation direct-write		

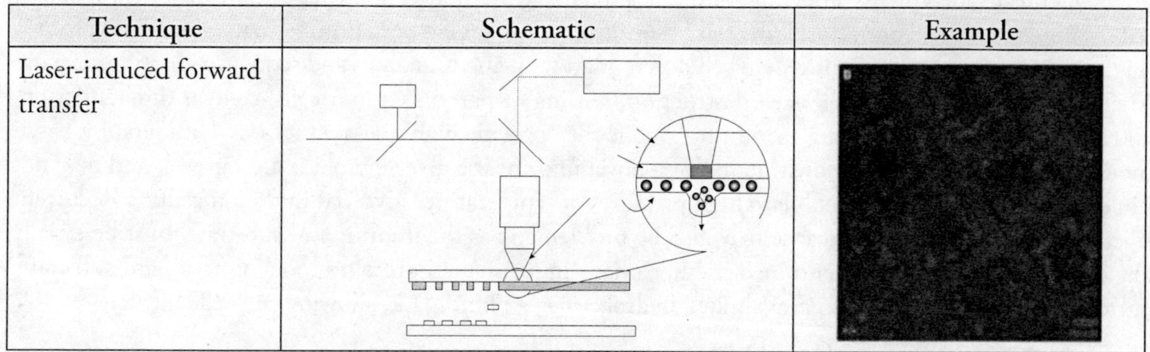

Technique	Schematic	Example
Laser-induced forward transfer		

Figure 5.7: Schematics Selected Biofabrication Modeling - Micro-contact Bioprinting, Inkjet Bioprinting, Laser Direct Write, and Laser Induced Forward Transfer

Courtesy: MDPI AG

Contact-Based Bioprinting

Contact-based 3D Bioprinting approaches comprise a mask or material that touches the substrate to oblige the cells to a precise domain. Varieties of contact-based 3D Bioprinting methods are dependent on soft "photolithography." Detailed micro-modeling of proteins can be achieved by utilizing a "photolithographic" mask and photoresist, joined with chemical handlings to openly generate protein micro-modeling from a silicon master, or to construct patterns or micro-stamps from "polydimethylsiloxane - PDMS."

The micro-modeling are generated once the stamp is covered with the anticipated material, and stamped with high force onto the substrate. Over and above their utilization in contact deposition from a mold, detachable masks have also been made to protect areas of the substrate. When this kind of mask is utilized, the template is implemented to the substrate, immersed with protein, and then detached. Cells or proteins are only placed on the unprotected areas. After the mask is detached, a high-fidelity form is resulted.

Fabricate Hollow Microfluidic Channels

Utilizing these "lithography-based methods," proteins can be precisely formed on flat or even curved surfaces, which is beneficial for regulating cell adherence or other reactions. Micro contact 3D Bioprinting, in particular, can be utilized to place multiple layers of proteins. Also micro contact is used to construct hollow microfluidic channels, by generating a stamp with 3D features.

Customizing microfluidic channels can also be done to permit valves and pumps, both of which are useful in Biosensing implementations due to the possibility to regulate input of an analyte. The micro-fluidic systems allowed by "lithography-based methodologies" enable transfer or exclusion of fluids to precise domain, and generation of grades of soluble factors that can indicate cell conduct. 3D fluid distribution ensures the capability to construct very intricate patterns and to transport proteins, drugs, or cells to distinct areas of a micro atmosphere. These detecting tools amalgamate the high-throughput features of microfluidics with cell imaging to obtain large amounts of information in real-time.

Transferring Protein Surface with Micron-Level Resolution

3D Contact-based Bioprinting is mainly valuable for moving a certain protein to a surface with micron-level precision. Protein modeling can be achieved by direct transmission from a master or reusable imprint, utilizing a detachable mask to avoid cells from conferring to undesired areas of the substrate, or indirectly, by constructing micro-fluidic channels. Channels made by 3D contact-based Bioprinting can have comparable modeling precision as direct modeling, but they also enable 3D non-contact deposition with the capability to flow fluid to assigned positions. Direct 3D Bioprinting proteins has been utilized to manipulate cell forms, migration signals, and to examine stem cell differentiation. Genetic analysis and sequence identification use Micro-fluidic chambers in single-cell destruction, sensing of partying tumor cells, cellular differentiation, and conveyance for *in vitro* drug examining. While they provide high spatial exactness, "lithography-based modeling" methods need the construction of a novel imprint if a diverse mold is anticipated, and may not directly ,mold cells to a surface due to high pressures or temperatures involved in the procedure. Recurrent micro-molding on the same surface may also be problematic as the imprint and substrate must be exactly aligned. Though the above-mentioned methods have implemented protein modeling to a hard substrate, current micro-molding has also cross linked hydrogels via a photomask, allowing 3D cell culture.

Atomic Force Microscopy (AFM)

In addition to this 3D "lithography-based methods," contact-based Bioprinting means also comprise atomic force microscopy AFM-based deposition such as dip-pen "nano-lithography - DPN." DPN utilizes an atomic force microscopy AFM point to 3D deposit an "Bioink" onto a solid surface with similarity for the Bioink, similar to 3D micro contact Bioprinting. Several materials have been established for this method, but the Bioink and substrate must still have a similarity for one another.

Nanometer-Level printing Resolution

AFM-based procedures provide sub-micron level precision and even nanometer-level 3D printing resolution. As cells are configured at the nano level, this type of modeling may show vital, mainly with Biosensors connecting cells. Generally, 3D contact Bioprinting and microfluidics may be utilized to a diversity of cell-based and protein-based biosensors at numerous scales.

Non-Contact Bioprinting

3D Non-contact Bioprinting methods include transmission of a material to a substrate, with only the relocated material approaching into actual contact with the substrate. Consequently, the high forces related with an imprint or elimination of a mask from the surface, are not needed to this approach more suitable for softer materials. 3D Non-contact Bioprinting provides a better adaptability in the substrates onto which materials can be 3D Bioprinted, such as soft hydrogels, and the 3D non-contacting nature permits layer-by-layer deposition to be utilized. There are numerous non-contacting approaches to deposit biological materials, as well as feasible cells, to substrates in controlled patterns.

3D Non-contact Bioprinting is less dependent on surface chemistries than contact-based approaches, letting deposition of a broader variety of biomaterials, to a broader diversity of getting substrates. Though, along with this added adaptability, there can be a loss in Bioprinting resolution with respect to 3D contact-based approaches.

Biomaterials and Bioinks

Biomaterials and bio-inks, for bio-sensing implementation, are acute when sensing a biological signal. Non-contact Bioprinting technique provides a greater variety of biomaterial choices, which, in turn, give superior biosensor customizability. While AM additive manufacturing methods can be well-matched with a extensive choice of biomaterials, only a subset can be utilized for Biosensing as the biomaterial may have to be biocompatible, have a specific attraction, or fall within a specific viscosity range. Viscosity is predominantly significant in 3D nozzle-based Bioprinting approaches because of potential clogging matters.

Cells may also sense and react to biomaterials of dissimilar viscosities and stiffness, from a biological perspective. Other features of biomaterials that may affect bio-sensing comprise the hydrophobicity/hydrophilicity of the chemical substrate, as well as the pore size in the case of 3D geometries. These assets can affect complex ligand and cellular binding, in addition to cellular performance.

Inkjet Bioprinting for Non-Viscous Materials

A specific 3D non-contact Bioprinting method may be designated based on the bio-sensing presentation and anticipated biomaterial characteristics within the system. A collective method for 3D Bioprinting relatively non-viscous biomaterials is 3D "inkjet Bioprinting," which usually includes thermal, piezoelectric, or pneumatic discharge of a droplet from a nozzle or bioprint head. 3D "Inkjet printing" has been modified to Bioprint both proteins and living cells.

Traditional 3D "inkjet 3D Bioprinters" have been adapted for biomaterials and cells, and layer-by-layer Bioprinting permits for the construction of intricate assemblies. While high throughput is likely, and pattern customization, counting numerous cell or biomolecule kinds, is relatively easy, 3D "inkjet" methods have inferior precision than 3D "lithography-based" and "laser-based" patterning methods. The high-throughput features of 3D "inkjet technology" are attractive to biosensor implementations, but specifying positions for exact ligand adhesion or precise molecular boundaries may be hard.

Numerous laser-based methods have also been utilized for non-contact biomaterial modeling. Lasers are frequently employed to openly crosslink photo-active biomaterials into favorite shapes untiring a photomask or two-photon polymerization, but, herein, the emphasis is on procedures utilized for direct deposition of a biomaterial. Moreover to the capacity to transmit soft biomaterials, laser-based deposition approaches permit for an extensive variety of biomaterials to be utilized, far better than those that can be laser crosslinked.

Print Ribbon and Receiving Substrate

The 3D laser-based deposition is predominantly striking for Biosensing implementations. Laser direct writing - LDW is an advancing transmission method employed for 3D Bioprinting feasible cells, proteins, soluble factors, microbeads, and DNA. A characteristic LDW arrangement contains of two coplanar plates, a "Bioprint ribbon" and a delivery substrate. The ribbon comprises biomaterial to be deposited to the substrate, with an original coating containing of two parts:

a. a sacrificial layer and
b. a transfer layer.

The laser is concentrating on the sacrificial layer, and when the laser is pulsed, the sacrificial layer desorbs and expels biomaterial from the Bioprint ribbon onto the receivable substrate, which can be utilized to construct a programmed array or construction of biomaterial or cells. Frequently, the receiving substrate is covered with a hydrogel to pad the expelled biomaterial. There are presently numerous dissimilar differences of LDW precisely adapted for Bioprinting. The fundamental ideologies of biomaterial deposition continue unchanged. The single feature in the method of this difference of LDW is that a single matrix is utilized as the sacrificial layer and the allocate layer. Characteristically, the ribbon coating is a polymer or hydrogel, such as gelatin, that suspends the anticipated transmit biomaterial. Matrix-Assisted Pulsed-Laser Evaporation Direct Writing, MAPLE-DW has beforehand been revealed ability to construct simple biosensors, such as a dopamine electrochemical sensor.

Laser Induced Forward Transfer (LIFT)

Laser induced forward transfer (LIFT) is additional frequently employed type of "LDW." LIFT utilzes a ribbon with a distinguishing sacrificial and transmission layer. The sacrificial layer in a LIFT arrangement is naturally a thin metal film that can be volatilized by the laser used. Underneath the metal film is a layer with the transmitted material, usually some sort of polymer or hydrogel. Like MAPLE-DW, LIFT scientists have investigated with fabrication of biosensors. As a an example of LIFT's biosensor fabrication capability, a DNA microarray was generated with specific DNA strands, which could be sensed by fluorescence microscopy when hybridized with corresponding fluorescent strands. Traditional equipment can be used to detect 3D printed material and/or interacting analytes, and the modeling resolution provided by LIFT permits customization and rapid fabrication of a biosensor. While LIFT and MAPLE-DW trail the same mechanisms, LIFT may provide more caring of the transfer material because of the specific sacrificial layer.

LDW has many unique variations from other biological 3D printing and modeling methods. First, LDW has proven its ability to 3D print with a extensive variety of polymeric material, which is vital for biological materials, particularly feasible cells. Some of the materials currently employed comprise alginate, gelatin, collagen, fibrinogen, chitosan, and hyaluronic acid.

This extensive variety of likely hydrogels or hydrogel mixtures provides users the skill to choose precise materials for their anticipated results. For example, when slight communication between cells and the biomaterial is preferred, gelatin may be chosen because its crosslinking is thermally rescindable, freeing cells to drift unimpeded. Further, the ribbon can be quickly changed during LDW, permitting the user to alter transmission material, as is needed for generating a co-culture cell system. Moreover, LDW permits deposition to any planar receiving substrate material. For modeling connecting cells, there is usually a thin film utilized for softening the influence of the expelled biomaterial. A diversity of dissimilar hydrogels has been utilized as a softening layer or as a 3D post-printing scaffold. Novel approaches have even utilized transplantable surfaces such as cardiac reinforcements that are moved to *in vivo* surroundings after Bioprinting. Other classical modeling and 3D Bioprinting approaches may be restricted by issues comprising biomaterial viscosity and high shear rates, thus significantly limiting their contestant biomaterial selections.

Metrology Biofabricating Sensors

Output, tolerance, accuracy, precision and custom tailor are significant structures to study in technologies employed for manufacturing biosensors. LDW has proven the capability to generate models of custom made spacing with 5 μm resolution sensitivity. The large level of spatial manipulation of the deposited biomaterial permits the manufacture of complex and very precise and sensitive biosensors.

In addition, LDW is a completely mechanized system with CAD/CAM controls. This control permits for not only completely automated fast manufacturing of intricate designs, but for added custom made of model that can differ between receptions of substrates. Though, LDW's high level of spatial control over deposited biomaterial may make large-scale manufacturing unfeasible as of the time necessary to produce a large, high-resolution design. In many instances, it may be more suitable to utilize a substitute modeling method, or a mixture of approaches.

New Generation of 3D Bioprinted Biosensing Devices

Novel development of Biosensing sensors could be comprehended by relating the many 3D biofabrication techniques to implementing biosensors. If cell-based biosensors are used, 3D micro surroundings can demonstrate more physiologically pertinent than 2D substrates. Multiplexing elements of the biosensor, such as examining an analyte against several cell kinds in 3D arrangements, or several analytes versus one cell kind, may yield more pertinent outcomes if cells are in 3D micro-surroundings.

This kind of multiplexing may produce much more data than autonomously examining testing an individual analyte with an individual cell kind in an individual biomaterial. Only by 3D modeling is vigorous, *in vitro*, 3D, high-throughput, cell-based Biosensing likely. There may be extra benefits to manufacturing 3D biosensors. Analytes roaming through a bulk structure made of natural polymeric biomaterial will acquaint more physiologically-relevant diffusive characteristics than those in 2D, which can impact time-based features and the receptiveness of the biosensor. *In situ*ations where a biosensor is made to mimic natural tissue, this may demonstrate to be a significant aspect to reproduce. These constructions have the extra possibility to even be incorporated for real-time detection *in vivo*.

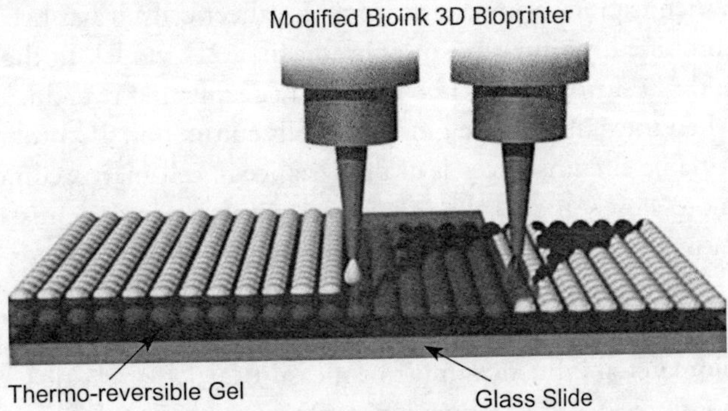

Figure 5.8: 3D Bioprinting Layer by Payer

Additionally, with layer-by-layer 3D Bioprinting technique, Figure 5.8, numerous cell kinds inside the bulk Biomaterial can be spatially organized to yield a signal flow, where one cell kind performs as a sensor and provides reaction to a secondary cell kind to achieve an anticipated behavior. When rapid analyte discovery is necessary, channels can be built to facilitate movement of the analyte to the sensor. This ability could demonstrate useful in a combined biosensor/therapeutic presentation. For instance, a 3D cell-based detector could be manufactured for diabetics, in which sugar amount could be sensed fast by the insulin-production beta-cells embedded in an incorporated construct. These cells could activate an insulin reaction to adjust blood sugar levels consequently. With 3D manufacture technologies, it becomes likely to combine biosensors with a therapeutic remedy.

Advancements in Biosensing Techniques

As the needs and implementations of Biosensing spread, the incorporation of biofabrication techniques into biosensor origins will be dominant. In order to multiplex a diversity of signals and appraise cellular reactions in 2D and 3D, refined sensors must be capable to distinct and quantify analytes of concern. There have been numerous recent advances to Biosensors technology in recent years. By combining these advances with the latest enabling biofabrication methods, mainly contact-based and non-contact-based modeling, even more advances in Biosensing technology can be accomplished. This combination of sensors and fabrication advances can lead to the next generation of Biosensors, with higher degree of sensitivity, throughput, and dynamic array within an individual sensor.

Methods such as micro-contact Bioprinting, inkjet Bioprinting, or LDW enable specific analytes to be located in discrete places, thereby permitting the reactions on a single cell kind to be examined quickly. Comparable methods can be utilized with many cell kinds. Advances in non-contact Bioprinting can deeply affect 3D cell-based Biosensing, which may deliver more physiologically applicable cellular reactions.

Layer-by-layer and micro bead methods will permit the manufacture of 3D biosensors, with all the benefits of multiplexing. As biosensors advance to 3D, there is even the possibility to integrate biosensors into graft therapeutics. The interaction of advances in Biosensing and biofabrication, much more than the developments of these methods individually, have the possibility to be very influential for future Biosensing, research, diagnostic, and therapeutic treatments.

Stem Cell Differentiation - Cancer Drug Screening

2D cell specimen has been the stage for numerous significant scientific findings, fluctuating from studies connecting stem cell differentiation to cancer drug discovery. Though, there are limits on the translation of 2D cellular system replicas, as there are key transformations among the method cells interact and operate in 2D and 3D surroundings, predominantly due to cell union and adhesion-based signaling.

Breast Cancer Cells

When exploratory malignant breast cancer cells, it has been established that the 3D adhesion conduits are combined, yet they are autonomous in a 2D model. In the situation of drug screening, the reaction of 2D and 3D surrogate specimen cells may deeply fluctuate due to this kind of operational change. For instance, when integrated into a biosensor, cells cultured in 2D probably will have dissimilar operational reactions to variable circumstances as of the variance in cell-matrix communications from 2D to 3D specimen. Towards more closely model the *in vivo* cell setting, many scientists have moved to 3D cell culture stages utilizing numerous biomaterials to characterize the cell atmosphere.

Numerous dissimilar biomaterials are presently being employed to generate 3D settings for cell research. With the purpose of imitating the normal tissue atmosphere for the cell kind cultured, numerous biomaterial influences must be measured, such as elastic modulus, hydrophobicity, matrix communication points, e.g., cell combination domains, biomaterial dilapidation rates, and permeability. To regulate these characteristics, logically occurring polymers exhibited in the body are frequently utilized to develop 3D cell scaffolds. Some logically happening polymers presently employed in 3D cell culture comprise collagen, hyaluronic acid, and fibrinogen/fibrin. Extra polymers, not naturally happening in the human body, such as alginate and chitosan, are being utilized for their exceptional and needed features.

Micro Beads and Microcapsules

Small-transmission means, specifically micro beads and microcapsules, are being examined for implementations within biosensor technique. Micro beads are micron-scaled spheroids, made from cross linkable polymers that are employed in hydrogel construction. Presently, these 3D micro-surroundings are being discovered for determinations linking appropriation of small molecules, proteins, drugs, or means to encapsulate cells. As micro beads are made from hydrogel biomaterial, they are almost 98% water by mass, permitting for fast diffusion into and out of the bead. Moreover, by regulating characteristics and/or handing the hydrogel polymer, many structures of the micro bead can be custom tailor, such as pore size, elastic modulus, degradation rate, and permeability. These properties permit for exclusive biosensor construction utilizing micro beads with condensed molecules or cells. As the organic component in the micro bead is expended or freed, the sensor will direct a biological reaction.

So as to a biosensor to work appropriately, it may necessitate the control a biological component on a transducer. Nevertheless, live cells are not easily immobilized, and may travel away from the sensor on a similar substrate. Similar issues ascend when attempting to restrict cells into a specific area though they are booming and be subjected to 2D occurrences.

Cell Encapsulation

Cell encapsulation is one likely answer to this difficulty. Micro beads and micro capsules are prevalent micro-encapsulation techniques, at a scale suitable for utilization in a biosensor. Alike to micro beads, micro capsules are micron-scale spheroids, though, instead of a 3D matrix, a spherical outside shell compels the configuration, within which encapsulated cells may travel. Additionally, these techniques provide the capability to fit the size of the atmosphere and total number of encapsulated cells. Classical micro bead and micro capsule manufacture techniques are incapable to exactly place the manufactured constructions in exact settings, which would be essential to several signals. Nevertheless, a novel technique permitting for the one-step manufacture and modeling of cell-comprising micro beads could resolve this difficulty.

For the manufacture of 3D builds, there are some diverse ways scientists are following; illustrations of such builds are presented in Figure 5.9. One method utilizes injection molding to manufacture large builds. In this process, cells are suspended in a cross linkable hydrogel, or hydrogel mixture, before injection into a mold and following cross-linking of the hydrogel. This method permits for the manufacture of large or small builds, with the overall geometric control being restricted only by the quality/detail of the mold. New image-guided technique is exploiting MRI and micro-CT permitting for the manufacture of highly comprehensive tailored molds.

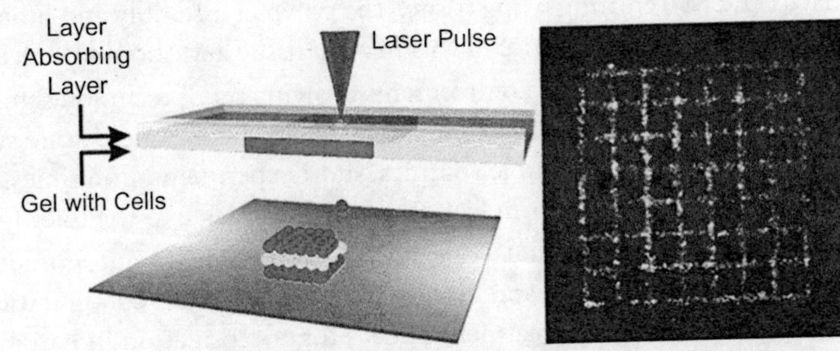

Figure 5.9: 3D Bioprinted Patterned Structures Utilized in Biosensing Implementations

COMMON SCAFFOLD-HYDROGELS

Theoretically, injection molding is well-matched with any cross linkable polymer substance; even hydrogels slow to set are practical due to the biomaterial is inhibited by the mold's configurations. Collective hydrogels and blends thereof, utilized for injection molding comprise collagen, gelatin, alginate, hyaluronic acid, PLGA and agarose. Though injection molding has been made known to be a cost-saving tool for speedily manufacturing 3D tissue builds to an exact total configurations, it provides insignificant to no spatial control of cells or other suspended substances, nor the capability to generate internal construction. In order to better control spatial arrangement, layer-by-layer procedures have been utilized to construct 3D builds.

Scaffold -3D Bioprinting Droplets of Hydrogel

Such 3D constructs are generated utilizing cross linkable hydrogels and predictable modeling techniques, achieved in a layer-by-layer manner. To achieve 3D layer-by-layer Bioprinting, "inkjet and LDW" users have established dissimilar manufacture means. Inkjet users have accomplished layer-by-layer manufacture with unceasing extrusion of hydrogel precursor strand into a crosslinking substance. A diverse method is developed to 3D Bioprint droplet of hydrogel precursor onto a receiving substrate and then crosslinks the layer utilizing a nebulizer. The cross linked layer is then utilized as the novel Bioprinting surface. In the instance of LDW, scientists are reaching layer-by-layer manufacture by coating their substrates in a hydrogel precursor and/or partly crosslinking the hydrogel layer, allowing, then cross-linking the cell suspension, then placing an additional hydrogel layer. Comparable to 3D inkjet Bioprinting, the Bioprinted hydrogel layer aids the novel Bioprinting surface, thereby permitting thick, multiple-layered constructions to be accomplished. Additionally, the dimension between cell layers can be controlled by utilizing a blade coater to regulate the height of hydrogel precursor.

Bioprinting Tissues for Drug Screening, and Organ Transplant

The field of tissue engineering goals is to produce tissues in the laboratory for specimen, drug screening, and eventually for organ transplantation purposes. To accomplish this, cells are predictably cultivated in mono-specimens or co-specimens, and seeded on or encapsulated inside biologically well-matched polymeric builds. These builds are grown in "Petri dishes," stimulated vessels or microfluidic tools. Inside these bioreactors, the polymer is gradually degraded by the cells over time and substituted with the extracellular matrix (ECM), creating living tissue. The polymer assembly and other micro surroundings signals can affect the operational performance of this *in vitro* tissue and should be elected prudently.

Micro scale technique complements evolving set of techniques for manipulating biological Biomaterials in the context of tissue engineering. Major benefits of micro scale systems comprise the necessity for only miniscule reagent and sample volumes, short experimentation times, cost-efficiency, and physical reduction of the experimental platform from the bench top scale to the micrometer- and millimeter-scale.

The small experimental scale also permits for an autonomous control over several experimental parameters, e.g., number and density of cells or size and configuration of the cell-laden polymer structure. This enables controlled treatment of cells for encapsulation in natural or synthetic Biomaterials.

Microfabrication Techniques Creating 3D Cell-Containing Materials

Microfabrication techniques have been employed in a diversity of methods to produce 3D cell-containing Biomaterials. This contains encapsulating cells in gel-based micro droplets, forming cell-containing fibers and micro tubes from gel precursor Bio-solutions, electro-spinning and -spraying polymers to generate Bio-gel droplets and fibers containing encapsulated cells, micro molding viscous cell suspensions into micro scale particles, and Bioprinting biomaterials and cells on a substrate to produce tissue building blocks. The outcome polymeric architectures are porous or permeable to small molecules, permitting nutrients and oxygen to reach the encapsulated cells and metabolic waste products to diffuse away from the cells.

All in all, the implementation of micro scale approaches to create cell-containing polymer structures provides a high level of regulation over the tissue construct procedure. As such, it allows the expansion and study of substituting biological tissues. Common hydrogels utilized in bioengineering and their potential crosslinking procedures are highly effective. Recent advanced micro scale techniques are utilized for producing cell-laden hydrogels. The implementation may be extensive for these micro scale methods in the setting of tissue Bio-engineering and cell culture.

Micro Scale Technologies

Diverse micro scale approaches are utilized for creating micro gel assemblies. These include "soft lithography," "photolithography," 3D bioprinting as well as various microfluidic approaches, "electro-spraying and -spinning." These methods permit for the creation of micro structures comprising only gels, as well as micro structures comprising gels and cells.

Soft Lithography Bioprinting

The term "soft lithography" defines a set of approaches for utilizing elastomeric, rubber and polymer structures based on duplicate molding and self-assembly, and it contains techniques such as micro molding and capillary molding. Cast and replica molding methods commonly utilize chemically cross linked Biomaterials, such as the universally utilized poly, dimethyl siloxane - PDMS. PDMS is a visually see-through two-part elastomer, containing of a bio-liquid base similar to silicone oil and a hardening agent. Upon combining, the two elements start to form chemical links at junctions of interacting polymer strands; this curing procedure proceeds at room temperature but can be thermally accelerated. When dispensed or spin-coated on a solid master, the bio-liquid mixture conforms to all features and hardens within one to a few hours.

Cellular Attachment, by surface patterning with proteins - fibronectin

As expected, hydrophobic, PDMS can be made less lethal for certain cell culture implementations by eliminating uncross linked PDMS molecules. It may also be further operated to permit for cellular adherence, e.g., by surface modeling with proteins such as fibronectin. Nevertheless, PDMS is specially utilized as a master for micro molding of hydrogel particles, such as PEG or GelMA, instead of direct cell encapsulation.

In micro transfer molding a bio-liquid pre-polymer is implemented to a cross linked PDMS master such that the bio-liquid seals out all gaps in the master. The extra bio-liquid is then pressed out with a glass slide, Figure 5.7. The pre-polymer is permitted to harden and continue adhere to the glass slide when the master mold is detached. This method frequently produces structures that are a few μm to ~150 μm thick, although thicker configurations can also be shaped. Alternatively, a PDMS master can be forced onto a bio-liquid prepolymer, compelling the bio-liquid to fill and seal the empty spaces prior to crosslinking.

Encapsulated Human Umbilical Vein

Biological Cells can be encapsulated in the produced gel constructions merely by suspending the gel in the gel prepolymer bio-solution. For instance, employing this technique NIH-3T3 fibroblasts were amalgamated in a hydrogel based on methacrylate hyaluronic acid. In addition, thermo-reactive micro molded hydrogels could be utilized when dynamic micro particles and -molds were necessary. The enlargement and shrinking aspects of hydrogels can be exploited in dynamic implementations, such as when a modification in pore size of the hydrogel is needed. This effect can be utilized to govern the infusion of biomolecules through the particle. Furthermore, any cells and cell aggregates in contact with a dynamic hydrogel experience the variations in scaffold size as a shift in the scaffold topography and reacted to it. To accomplish this outcome, PNIPAAm is utilized in combination with agarose, another thermos-reactive bio-material, and encapsulated human umbilical vein endothelial cells (HUVEC), NIH-3T3 fibroblasts and HepG2 cells in the hydrogels. Individual PNIPAAm master structures are manufactured on a flat substrate. By regulating the adjacent temperature, the structures could be encouraged to shrink or expand. Then, a second hydrogel prepolymer, agarose, is deposited in the places between the PNIPAAm master structures and crosslinked to formulate micro-particles, Figure 5.10.

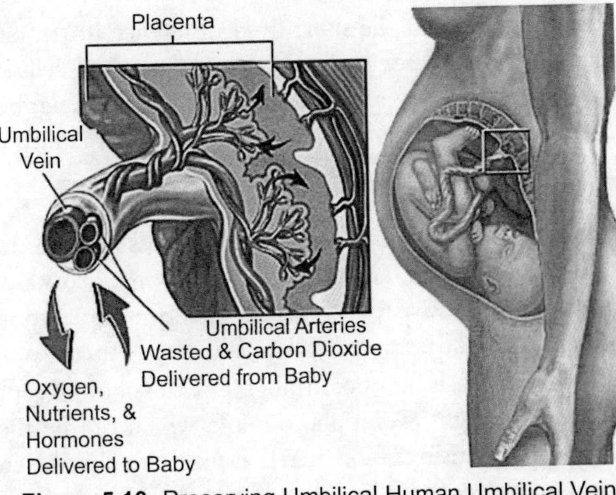

Figure 5.10: Preserving Umbilical Human Umbilical Vein
Courtesy: Anatomy and Physiology

Upon increasing the surrounding temperature, the PNIPAAm particles augmented, thus compelling the agarose molecules until they detached from the substrate. Thus, this technique utilized dynamic micro constructions to generate a secondary set of agarose micro particles.

Comparable to micro-transfer molding, micro-molding in capillaries requires that the liquid prepolymer fills narrow capillary channels formed by a PDMS master and a glass slide, where it hardens. The prepolymer is strained into these micro scale channels by capillary force. This has been observed in micro capillary channels with both open and closed ends. Inside closed channels the confined air diffuses into PDMS, permitting the progressing prepolymer bio-solution to entirely fill the capillary. Micro molding in capillaries has been utilized to pattern UV-sensitive polymers such as polyacrylates as well as macromolecules such as immunoglobulins.

Culture Medium mimics Blood Vessel-like Structures

A characteristic instance of micro molding utilizing a solid 3D design is illustrated in Figure 5.11, where a bio-liquid low-temperature gelling agarose bio-solution was raised in temperature to 70 °C, implemented to a patterned silicon bio-substrate and permitted to harden. Upon gelation at room temperature, the 1 cm thick agarose layer comprising the design bio-imprint was peeled off and closed with a distinctly set as flat agarose slab to produce a hydrogel micro channel. Cells were amalgamated in the gel utilizing the same procedure, subsequent in a high original capability, demonstrating that the short-lived difference in temperature did not have a robust cytotoxic result. This channel was used for perfusion of amalgamated cells with specimen medium to mimic blood vessel-like arrangements. It was revealed that cells exist in neighboring to the bio-fluidic channel had the uppermost capability, by reason of the transmission of media into the porous gel.

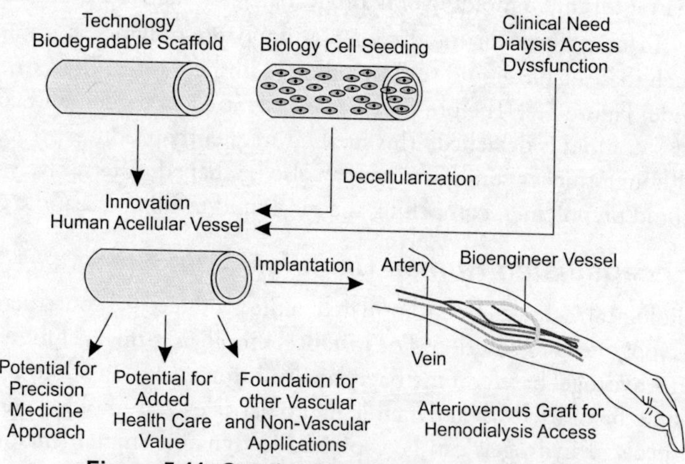

Figure 5.11: Creating Blood Vessel-like Structures
Courtesy: Fresenius Medical Care

Photolithography 3D Bioprinting

Hydrogel precursors can be planned to crosslink consuming light. In this photo cross linking applications, liquid bio-materials, i.e., photoresists, gel precursors are formed into firm constructions by making radical-based photosensitive responses. The energy essential to start the photosensitive chemical response is typically brought by light in the UV range, although other wavelengths may also be utilized.

More precisely, in photolithography the photo receptive bio-material can be placed onto a substrate and subjected to light over a photo mask, which can be Bioprinted on a clear or chrome substrate. The parts of the photosensitive biomaterial available to light be cross linked founded on the design of the mask. The enclosed parts remain liquid and can be detached by washing with a mounting agent. The created polymer designs have the same height and can be considered planar. To formulate more intricate 3D structures, a number of recurrences of the procedure are essential. In each step, a novel layer of the prepolymer is presented and visible to light under a dissimilar mask.

In particular photo sensitive bio-materials, such as photoresists, UV light is absorbed by molecules that create a photo acid. Creating photo acid relieves the protons. Consequently, the shaped photo acids trigger the ring-shaped epoxy groups and purpose as photo originators of the polymerization procedure. Hereby novel protons are freed and can enable another polymerization response. Similar free fundamental polymerization responses take place in a variety of polymers and biomaterials, such as PEG-DA - PEG-diacrylate, poly-oligo, ethylene oxide, monomethyl ether methacrylate), MeHA, methacrylated hyaluronic acid, and GelMA. While free extremists can be destructive to cells, this radical-based photo lithography procedure can be prepared cell-compatible by cautiously selecting the photo initiator and the energy of UV brought to the encapsulated cells.

Microfluidics Used in Drug Discovery, Stem Cell Biology, and Oncology

Cells and atoms can be employed with excellent fidelity inside micrometer-sized micro networks or chambers, permitting trials that necessitate firmly governed cellular micro surroundings. This and other biomedical engineering implementations of microfluidics are employed in drug discovery, stem cell biology, oncology, monitoring cellular behavior, angiogenesis, migration, cellular interactions, and tissue culture. Precisely, subjects regarding 3D cell-tissue developments have been addressed, counting satisfactory spreading of tissues with nutrients and gases as well as their stimulus with chemical, mechanical and electrical signals.

One of the favored prevalent implementations of microfluidics for a group of cell-containing biomaterials depend on hydrodynamic, flow concentrating in glass capillaries and nozzles to condense cells inside droplets or jets of polymer bio-solutions. These assemblies are based on the laminar stream of the sample enclosed by an unmixable phase. The example either contains cells suspended in an un-cross linked hydrogel or two detached streams of cell bio-solution and a hydrogel.

Generation of Droplets Containing hydrogel Prepolymer, with or without Cells

The internal or discrete phase of the subsequent flowing constructions is typically an aqueous bio-solution. The outer, incessant phase is usually non-toxic oil such as corn oil, water-in-oil suspension, or a lower-viscosity aqueous bio-solution, water-in-water suspension. The flow rates of all streams are attuned to allow shearing off of discrete droplets of the inner phase, due to the shear stress produced by the uninterrupted phase. This arrangement permits for the creation of droplets comprising the hydrogel prepolymer, with or deprived of cells, even in water-in-water suspensions.

Such suspensions yield improvement of the modification in viscosity of the two mixed stages, such that for a short-lived period the two stages can be well-thought-out unmixable. Micro suspensions of dextran were created in a bio-aqueous PEG solution, by exciting and by microfluidic flow concentrating. In the first case, methacrylated dextran, dexMA was utilized, which permitted for photo crosslinking of the gel precursor. Furthermore, to flow focusing, hydrogel droplets can be produced in T-junction microfluidic constructions. In this case, the bio-aqueous and oily phases do not co-flow and aqueous droplets are occasionally sheared off by the faster flowing oil stream. In both flow focusing and T-junction devices, the utilization of a fatty phase can perform to reduction the cell capability.

Even when non-toxic bio-liquids like corn oil are implemented to shear off hydrogel droplets, the amalgamated cells close to the droplet and particle surface are in communication with the surfactant and the oil. Then, the quantity of nutrients obtainable inside the micro particles is often only adequate for a short culture period, particularly in compactly packed particles. Therefore, it is significant that the cross linked particles be washed soon after manufacture and moved to a reservoir containing cell medium. Additionally, the washing process should be tender, evading solvents and high centrifugation rates.

Reduced Droplet Size

Several investigations concentrated on the encapsulation of cells in hydrogel droplets by means of bio-microfluidics. For instance, an alginate bio-solution was autonomously presented and a solution comprising Ca2+ ions into a bio-microfluidic channel and recognized a brief co-flow of the two species. Droplets of the bio-aqueous solutions were sheared off spasmodically according to the shear stress produced by the high flow rate of oil. Separate alginate droplets were revealed to gel by adding Ca2+ ions, which substitute Na+

ions as they spread through the droplets. Lessening the cross linker flow rate and upssurge the flow rate of alginate affected in a condensed droplet size.

Moreover, mechanical characteristics of the gel were influenced by the degree of crosslinking, which hinged on the crosslinking time and the cross linker concentration. In the same way, distinct Na-alginate suspensions could be shaped at the tip of a rotating micro nozzle by gravity and cross-linked in CaCl2 solution. The droplet dimensions was governed utilizing the inner nozzle diameter and its speed of rotation.

The flow concentrating on arrangement utilized for simple hydrogel micro-droplet generation also permits the making of so-called Janus droplets, Figure 5.10. These droplets comprise two different halves, which are created by two mixable solutions, co-flowing sample solutions. Such mixed constructions are of interest in cell co-culture implementations. The Janus particle group method is emphasized in Figure 5.12. Here, suspensions of carbon black and titanium oxide molecules in an acrylic monomer bio-solution were co-flowed to harvest Janus microspheres with white and black halves in an aqueous solution. The gel was thermally crosslinked outside the fluidic module. Non-spherical Janus micro particles could also be created constructing dyed micro-dumbbells with different organic and inorganic polymer regions Figure 5.13. Lastly, more intricate particles could be created, comprising three isolated units. The typical two miscible phases labeled with fluorescent green and red dyes) were supplemented by a third hydrogel precursor solution prior to UV-crosslinking.

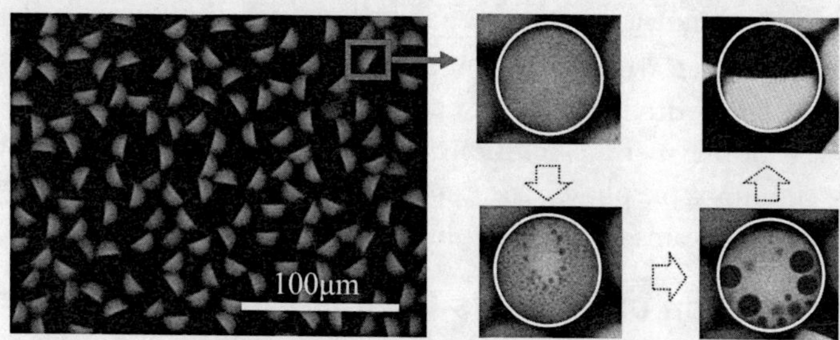

Figure 5.12: Size of Janus Droplets
Courtesy: Advanced Science News

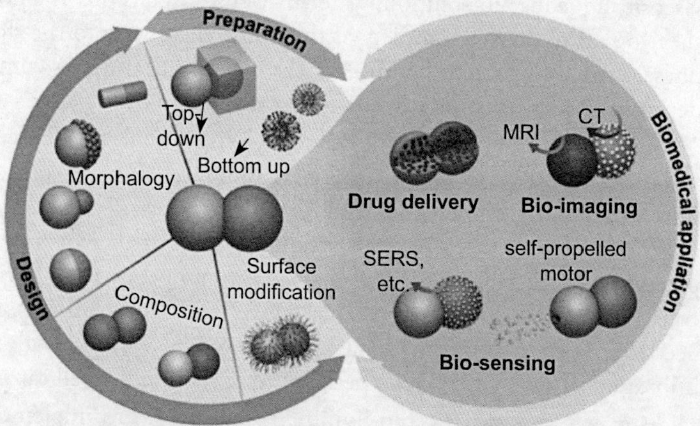

Figure 5.13: Janus Particles: Design, Preparation
Courtesy: Elsevier

Extrusion of Gels for Cell Encapsulation

Electro-spraying and Spinning Bioprinting

As well as microfluidics, electro-spraying and electro-spinning are two approaches that can also be employed for generation of cell-encompassing gel molecules and fibers, Figure 5.14. Nanoscale polymer fibers created complete electrospinning are normally decided in large porous complex. Cells can adhere to and multiply within such a structure. Nevertheless, the emphasis is on microscale fibers, which are adequately large for cell encapsulation.

The physical science of electro-spinning and electro-spraying in hydrogels can be concise stated as follows:

1. Supplying a large electrical power to a prepolymer solution dripping from a syringe needle generates charging of the miniscule droplets. Once surface tension is lesser than the subsequent electrostatic repulsion, the miniscule droplet expands until a stream of bio-liquid is shaped. At low surface tension droplets commence to formulate electro-spraying.

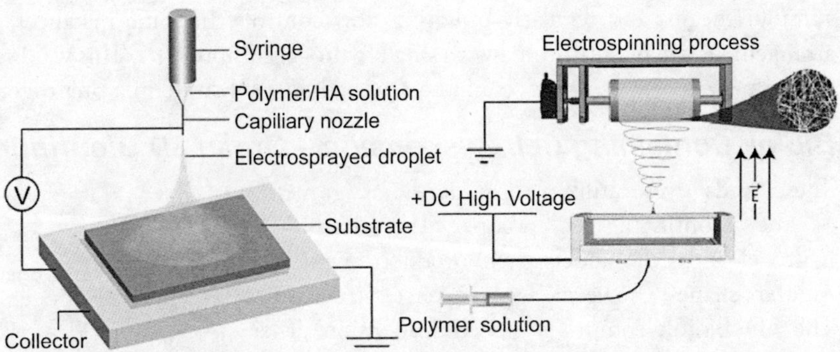

Figure 5.14: Principles of Electro-spraying and Electro-Spinning Bioprinting

2. In contrast, at a sufficiently high surface tension, the charged bioliquid stream rests steady (electro-spinning). In the former situation the jet dry out and charges focus on the jet surface. Electrostatic repulsion once more motivates the expansion of the jet. The produced hydrogel stream is to end with setting down onto a substrate, where it turns gels into fibers. The ultimate form of the electro-spun fiber is governed by a number of parameters stated as follows:
   a. the inner diameter of the needle;
   b. supplied electrical potential and flow rate;
   c. the molecular weight,
   d. viscosity and concentration of the biomaterial; and
   e. the dimension between the needle and the collection substrate.

Usually, both electrospinning and -spraying arrangements depend on an individual syringe pump to motivate the flow. Nonetheless, a formed droplet microfluidic arrangement can be utilized as another substitute. A flow-focusing arrangement could be supplied to produce droplets ahead of their subjection to an electric field. By regulating the droplet configuration and spacing they were capable to govern the number of encapsulated molecules in each formed droplet.

3D Bioprinting Creates Physiological Models of Human Organs

One of the main objectives of tissue engineering is to grow practical 3D Bioprinted models of tissues and organs. These representations would then be employed to investigate the properties of drugs and diseases inside the human body. Creating complete organs in a laboratory is exceptionally intricate. Consequently, regularly only portions of discrete tissues are developed. Though, there are methods to model these tissues in methods that represent their normal corresponding reference.

3D Bioprinting Utilizing Computer-Controlled Deposition of Cells

3D Bioprinting is one modern approach to generating 3D tissues *in vitro*, via cell-laden micro-gels. The 3D bioprinting technology exploits digitally controlled deposition of cells droplets and structural biomaterials such as hydrogels. By introducing cell models utilizing a 3D Bioprinting head mechanism, the cells normally migrate and multiply in the prescribed models. The Bioprinted substance is then polymerized and moved into an incubating compartment for culture. The incubating compartment can be shaped out of a simple "Petri dish," or an intricate microfluidic tool equipped with channels as means of perfusion, electrodes as stimulus means, and regularly pulsating flow controls. In some instances, a large cell-encasing polymer arrangement can be substituted with small hydrogel components. This was established positively illustrating that mm-sized agarose "rods" could be dropped on a substrate in a preprogrammed form.

Bioink Containing Cell Suspension – Organ 3D Bioprinting

These "**rods**" were utilized to guide the assignment of the Bioprinted cell groups. The combined hydrogel and cell models then produced working tubular shapes; however with diverse structures. The 3D Bioink comprises of cell suspension. The biodegradable hydrogel are 3D Bioprinted layer by layer. As a consequence, intricate, preprogrammed 3D cell models inside the subsequent constructions can be created, an aspect that is challenging to accomplish through jetting or electro-spinning. A schematic of the bioprinting procedure and a printed organ structure are illustrated in Figure 5.15.

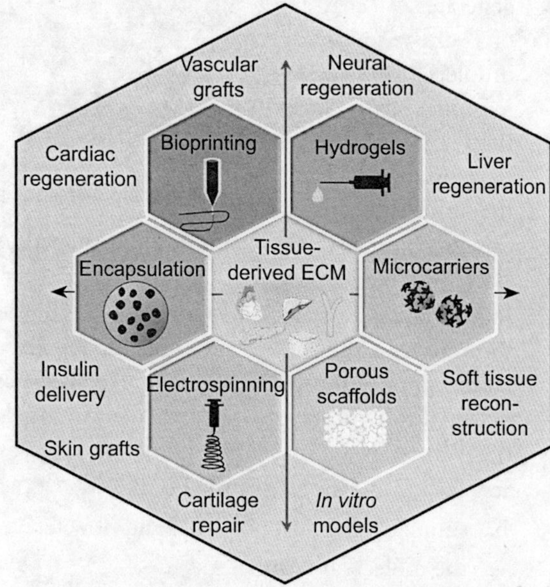

Figure 5.15: Schematics of Tissue Engineering Used in Organ 3D Bioprinting
Courtesy: John Wiley & Sons, Inc.

Affordable Inkjet Bioprinters

Two kinds of 3D bioprinters are usually used:

a. retro-fitted inkjet Bioprinters and
b. extruders Inkjet Bioprinters

Either Inkjet 3D printers are inexpensive and simple to adapt to 3D Bioprinting biological substance, but are frequently clogged with large cell proliferation.

Extrusion Bioprinters with Several Heads

Equally, pressure-driven extrusion devices incline to be more expensive, but are better matched to implementations exploiting high cell concentrations. Industrial bioprinters comprise high precision pressure and temperature monitoring system to maintain cells viability and preserve a low-viscosity polymer bio-solution. These 3D bioprinters also utilize several Bioprinting heads for concurrent printing of several constructions. Alternatively, many Bioprinting heads can be utilized for printing diverse cell kinds and biomaterials.

3D bioprinting has been utilized extensively in tissue engineering implementations. Instances of created tissues contain operational, synchronized beating cardiac tissues grownup from embryonic cardiac and endothelial cells; mammalian bladder tissue; vascular constructions from porcine aortic smooth muscle cells; and human skin models from keratinocyte and fibroblast layers in collagen. Alternatively, bioprinters could also be utilized to govern the positioning of growth factors in cellular microenvironments. This method was implemented to lead the differentiation of stem cells into muscle and bone cells and morphing of osteoblasts into bone cells.

Benefits by means of hydrogels for cell and tissue investigations embrace their cytocompatibility, degradability, and porosity. Furthermore, hydrogels are readily well-matched with microscale manufacture methods, such as "soft and photo-lithography," flow concentrating, electro-spinning, bioprinting and others. These methods permit for exceedingly exact and precise modeling of hydrogel constructions and cell encapsulation.

Some of the approaches are better appropriate for definite implementations than others. For instance, soft and photolithography procedures are acquiescent to creating microscale scaffolds on the order of some hundred μm, while constructions accomplished of encapsulating single cells are more effortlessly created utilizing flow concentrating inside microfluidic tools.

Also, good large-scale spatial arrangement of the scaffolds can be effortlessly attained by means of soft lithography, photolithography and bioprinting, while electro-spinning and -spraying provide random placement of the manufactured constructions. Therefore, the select of the microscale method to create cell-comprising hydrogel molecules mainly hinge on whether single-cell and screening trials are to be directed and whether the purpose is to accumulate tissues *in vitro* from the bottom up or reorganize it from the top down.

Small, spatially governed structures are preferred, though in the greater structures, together with hydrogel fibers can be valuable. Moreover, the kind of cell utilized is also significant: Particular cells, e.g., mainly cardiomyocytes, are extra sensitive to their surroundings, temperature, shear stress, pH than fibroblasts, enabling them to be less likeminded with microfluidic approaches than with soft and photolithography.

Nevertheless, all current methods permit the formation of well-governed tissue builds for drug screening, stem-cell biology, and possibly development of discrete organ tissues. As the decisive goal, the "organ-on-a-chip" perception has previously been exploited for mimicking cardiac and lung tissues, the gastrointestinal villi, and even for progress and successful implanting of a human bladder.

In the foreseeable future, cell-encompassing micro-gels will expectedly endure to be active in tissue engineering implementations. Additionally, they could possibly be valuable for drug screening and therapeutic research. Future investigation activities will also expected to be contingent upon the progression of a high-output platform for creating microscale cell-encapsulating hydrogel constructions. Preferably, such a platform would be well-suited with a variety of hydrogel materials and cells; it would deliver a steady manufacture system, and be able of creating molecules of diverse configurations and forms. This level of flexibility is presently only obtainable by 3D bioprinting methods, as they allow creation of any preprogrammed 3D shape.

Harvesting Tissues and Organs

Tissue 3D Bioprinting is an interdisciplinary and multidisciplinary arena. It has revealed great potential in creating living substitutes for collected tissues and organs for transplantation and reconstructive surgery. Materials and manufacture techniques are critically significant for tissue engineering in planning temporary, artificial extracellular matrices - scaffolds, which sustain 3D Bioprinting, 3D tissue development. The idea of 3D Bioprinting tissues is carefully connected to tissue engineering and materials science.

3D Bioprinting Tissue – Limbs on Demand

The time of walking to a hygienic human 'body shop' is almost at hand. A patient may demand a prescription for a damaged limb, a deteriorating liver, or a dysfunctional heart. Furthermore to the physiological, medical, and genetic data of the patient, the doctor will also gather 3D images of the patient's residual limb with comprehensive anatomical assemblies, bone, cartilage, tendons, ligaments, blood vessels, muscles, nerves, and skin, as well as the outside configurations.

The physician may gather the patient's spittle or other body liquids to excerpt genetic material DNA, and also a micro portion of tissue or bone marrow to acquire planting cells for development.

Once the individual patient departs the hygienic body shop, a digital record is utilized by a tissue expert to scheme the construction a mirror image of missing limb founded on the balancing residual organ, utilizing many biomaterials that mimic the extracellular matrices "ECM" of the tissues of the limb, known as scaffolds or templates.

Then, cells – from a collective cell source, precisely intended for individual patient, or stored cells developed from the patient's personal cells – are planted onto these scaffolds. These properly arranged cell-scaffold mechanisms are then developed distinctly and/or constructed in a special cavity, bioreactor that offers the correct and necessary nutrients, monitoring particles , such as proteins, development factors, and differentiation factors, physical and motivational stimuli, temperature, pressure, and mass transport circumstances for cell propagation, differentiation, and tissue/organ development.

Whereas the tissue/organ is recreating, the scaffolding biomaterials dissolve and vanish, leaving naught foreign in the body. The re-stimulated developed limb precursor will be surgically embedded and readjusted onto the patient during the following visit to the human body shop.

The manufacturing engineered tissues would have the ability to produce, model, and remodel in harmony with the active alterations of the physiological setting of the body. The embedded limb will assimilate into the body. The novel limb will nurture and age as the body's natural limb. This futuristic news is a sample of the arena of 3D Bioprinting tissue engineering is hoping to accomplish.

Bioprinting Tissues and Scaffolds

Bioprinting Tissues has been categorized as an interdisciplinary field that implements the fundamentals of "Advanced 3D Bioprinting Bioengineering" and the life-sciences headed for the progress of biological alternatives that give back, uphold, or recover tissue usefulness. There are three methodologies in tissue engineering:

1. the use of sanctified cells or cell alternatives to substitute those cells that feed the necessary function;
2. the distribution of 3D Bioprinted tissue-promoting constituents, such as development and differentiation factors, to directed locations;
3. rising cells in 3D scaffolds.

The utilization of sanctioned cells or tissue-promoting constituents is taken into account once the flaws are small and well limited. To 3D Bioprint tissues in real configuration scale and prearranged forms, these two methods are totally restricted. Consequently, the third method, i.e. rising cells in 3D scaffolds, has become progressively lively. In this method, scaffolds play an essential part. They direct cells to cultivate, synthesize extracellular matrix and other biological particles, and enable the creation of efficient tissues and organs.

There are a few basic necessities that have been extensively accepted for manipulating polymer scaffolds.

- *First,* a scaffold mist has high porosity and proper pore-size.
- *Second,* a high surface area is required.
- *Third,* biodegradability is usually essential, and a proper dissolution rate is desirable to match the rate of new tissue realization.
- *Fourth,* the scaffold must have the essential mechanical reliability to uphold the predesigned tissue assembly.
- *Fifth,* the scaffold must not be toxic to the cells, i.e. bio compatible.

- Sixth, the scaffold should positively interrelate with cells, as well as improved cell adhesion, growing, relocation, and differentiated formation.

Bioprinting Biomaterials for Scaffolds

Polymers, particles, are the principal materials for scaffolds in numerous tissue engineering implementations, as well as bone and other mineralized tissues. A restricted number of inorganic materials are utilized in bone and mineralized tissue engineering study.

Solid-State Scaffolds

The 3D Bioprinting materials in this classification can shape hard, steady permeable constructions to aid as predesigned 3D scaffolds. They usually do not liquefy or melt *in vitro* tissue culture circumstances, in a water medium, or when set *in vivo*.

Linear Aliphatic Polyesters

Poly (glycolic acid) (PGA), poly (lactic acid) (PLA), and their copolymers poly (lactic acid-co-glycolic acid) (PLGA) are a cross-correlated family of linear aliphatic polyesters, which are most regularly utilized in tissue engineering. These polymers dissolve through hydrolysis of the ester bonds.

PGA is one of the most extensively utilized scaffolding polymers. According to its comparatively hydrophilic tendency, PGA degrades fast in water solutions or *in vivo*, and cannot maintain mechanical coherence between two and four weeks. It has been reformed into nonwoven fibrous materials as one of the most extensively utilized scaffolds in Bioprinting tissues today.

PLA for Scaffold Bio-fabrication

PLA is also extensively utilized for scaffold Bio-fabrication. The additional methyl group in the PLA reiterating unit, associated with PGA, creates more hydrophobic, decreases the molecular affinity to water, and creates a slower hydrolysis rate. It takes several months or even years for a PLA scaffold or implant to weaken its mechanical cohesion *in vitro* or *in vivo*. To attain transitional degradation rates between PGA and PLA, various lactic and glycolic acid ratios are utilized to create PLGAs. These polymers, PLA, PGA, and PLGAs, are amongst the limited artificial polymers approved by the US Food and Drug Administration, FDA- for specific human clinical implementations.

There are other linear aliphatic polyesters, such as poly (ε-caprolactone) (PCL), and poly (hydroxy butyrate) (PHB), which are also utilized in tissue bioengineering studies. PCL disintegrate at a considerably slower rate than PLA, PGA, and PLGA16. The slow deprivation makes PCL less favorable for overall tissue engineering implementations, but more favorable for long-term grafts and controlled release implementations. PCL-based copolymers have lately been manufactured to advance dissolution properties. PHB is manufactured by microorganisms via fermentation. PHB and PHB-based copolymers disintegrate very slowly as of their hydrophobic tendency and, consequently, are less prevalent compared with PGA, PLA, and PLGA for tissue engineering implementations.

Bioprinting Materials for Hydrogel Scaffolds

Hydrogels are cross-linked hydrophilic polymers that encompass great amounts of water deprived of dissolution. Injectable hydrogels are favorable contestants for certain Bioprinting tissue implementations since of the capacity to fill unevenly formed tissue faults, the allowance of minimally invasive processes such as arthritis joints surgeries, and the ease of combination of cells and bioactive agents.

Among the artificial hydrogels, poly (ethylene glycol) (PEG) is commonly investigated. PEG-diacrylamide has been manufactured, and a photo-polymerization step can be achieved in communication with cells. There still some apprehension over the toxicity of cross-linking agents. Nonetheless, the most important restriction of PEG hydrogel for tissue engineering scaffolds is its lack of degradability. Determinations are being made to convey degradability to PEG by either articulating copolymers with PLA, PGA, and PPF, or by presenting enzymatically degradable linkages into the PEG backbone.

Semi-inter-penetrating polymer complexes involving of poly(N-isopropylacrylamide) and linear poly(acrylic acid) chains have been recognized as artificial hydrogel scaffolds. Diblock co-polypeptide amphiphilic encompassing charged and hydrophobic parts have lately been described to formulate hydrogel scaffolds with capable thermal steadiness and mechanical characteristics. More study is required to appraise these novel artificial hydrogels for tissue engineering implementations.

Seaweed Alginates

Alginates, polysaccharides from seaweeds, cross-linked utilizing calcium sulfate (CaSO4) have been utilized as cell distribution means of transporting *in vivo* tissue engineering research. Though, one of the main difficulties of utilizing CaSO4 is that gelation kinetics is problematic to govern, and the subsequent construction is not even. A technique has lately been established to govern the gelation rate so that cross-linking happens consistently to formulate operationally steady and mechanically robust 3D alginate gels.

Collagen, along with porous froth, has also been utilized as a hydrogel for a diversity of tissue restoration and regeneration research, containing blood vessel, skin, cartilage, ligaments, and tendons. Collagen has also been combined with other ordinary polymers, such as chitosan, to produce compound hydrogel scaffolds.

Fibrin, associated with clotting, is another gel-creating fibrillary protein that has been discovered for tissue engineering implementations. Fibrin gels are characteristically arranged by merging fibrinogen and thrombin solutions to capture cells and formulate tissue engineering builds. The fibrin gel has capable characteristics, but its dissolution rate is quick, which could be perplexing for certain implementations.

There are also synthetic proteins, such as elastin-mimicking polypeptides advanced with recombinant DNA technique, which may aid as biomimetic hydrogel scaffolds. This novel synthetic technology may also permit the design of chemical construction for functionalities, possibly for artificial hydrogels or solid-state scaffolds.

Scaffold Bio-fabrication Technologies

In the body, tissues are prearranged into 3D constructions as efficient organs and organ systems. To engineer efficient tissues and organs effectively, the scaffolds have to be planned to enable cell delivery and lead tissue regeneration in 3d.

Textile Technologies

Earlier tissue engineering scaffolds encompassing fibrous environmental polymer fabrics were fashioned utilizing textile technologies. PGA, PLA, and other semi-crystalline polymers can be administered into fibers utilizing textile techniques. One of these scaffolds extensively utilized in tissue engineering study is PGA nonwoven scaffold, Figure 5.16. These PGA scaffolds have been utilized either unaided or shared with other biodegradable polymers for the engineering of cartilage, ureter, intestine, blood vessels, heart valves, and other tissues.

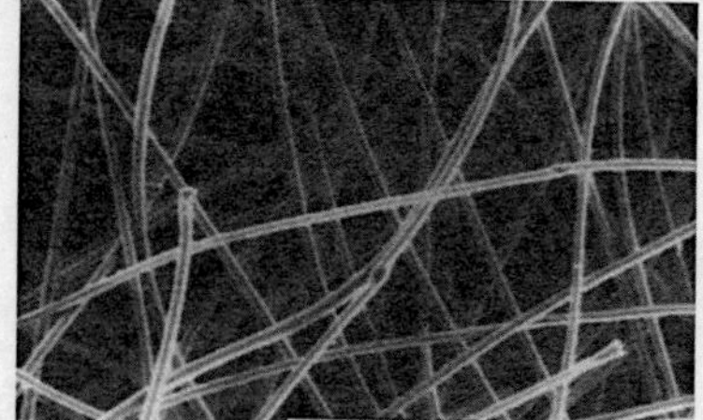

Figure 5.16: Nonwoven Scaffold - 15 μm Fiber Diameter

Though, there are several limits of PGA nonwoven scaffolds, such as low adhesion strength, fast degradation rate, trouble in governing pore outline, and restricted fiber diameter differences.

Polymer Solution Separates in Two Phases

A consistent multicomponent system, under certain circumstances, becomes thermodynamically unbalanced and inclines to isolate into more than one phase in order to decrease the system free energy. A polymer solution separates into two phases.

a. a polymer-rich phase and
b. a polymer-lean phase.

After the solvent is removed, the polymer-rich phase solidifies. Phase-separation methods have been used to manufacture porous membranes for filtration and separation. Though, the pores shaped using such methods typically have diameters on the order of a few to tens of microns and are often not consistently circulated, which is not appropriate for Bioprinting tissue applications. Controlled phase separation procedures, mainly thermally induced phase separation, have lately been investigated for scaffold manufacture.

3D CAD for 3D Bioprinting

One of the mutual inadequacies of the manufacture skills is the absence of precise control of the 3D pore construction of the scaffolds. To solve this problem, digital solution in computer-assisted design and manufacture (CAD/CAM) are being adopted.

One of the compact, free-form manufactures, rapid prototyping methods, 3D Bioprinting, was employed. With CAD/CAM technique, intricate-formed objects were planned utilizing CAD software. Scaffolds were manufactured by 3D ink-jet Bioprinting a binder onto successively placed polymer powder layers. Nevertheless, the minuteness of the powder particles and the binder agent drops, are restricted to a few hundred microns. The precision of locating the 3D Bioprinting nozzle is also restricted. Consequently, the precision of the technique is very restricted. Likewise, other rapid prototyping methods such as "Fused Deposition Modeling, FDM" and "stereo lithography" are also being investigated for scaffold construction.

Rapid prototyping techniques have characteristic inadequacies such as limited material selection and inadequate resolution. The smallest thickness of the scaffold dimensions could not be directly constructed. An inexpensive technique of negative mold manufacturing has sooved this problem of manufacturing resolution. Scaffold construction utilizing rapid prototyped negative replica made through the following procedures:

a. A computer-generated, 3D negative replica of a scaffold with a cubical pore shape, Figure .17a
b. The negative replica of the scaffold made utilizing a rapid prototyping machine, Figure 5.17b
c. A photograph of the polymer scaffold fabricated utilizing the negative replica, Figure 5.17c
d. A scanning electron micrograph of the internal pore structure of the scaffold.

A reverse manufacture method has been utilized to manufacture a negative replica of the scaffold. A polymer solution is cast into such a mold and hardened after draining the solvent, Figure 5.17. The mold is then dissolved away to shape the polymer scaffold with the intended 3D pore network. The scaffold is more consistent, but the feature resolution is not better.

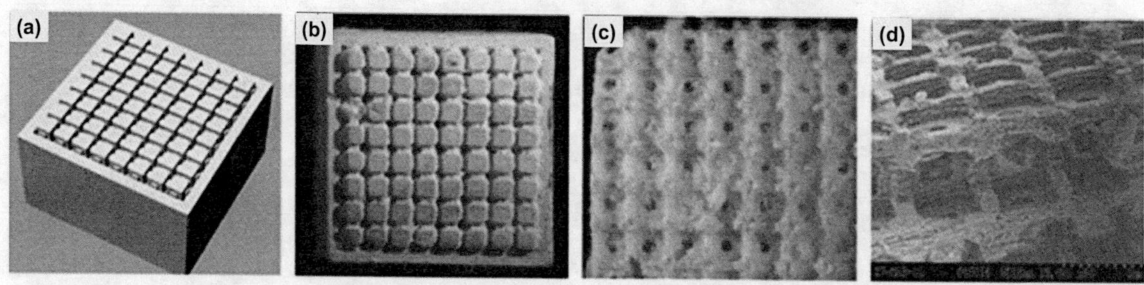

Figure 5.17: Scaffold Fabrication Using Rapid Prototyped Negative Replica

To accomplish higher resolution for scaffolds with well-controlled consistent spherical pores, paraffin spheres are manufactured by a dispersal technique. These paraffin spheres are then conveyed into a 3D mold of the designed configuration. The spheres are bonded together through a heat treatment procedure. A polymer solution is cast into the paraffin assembly in the mold. After draining the solvent, the paraffin sphere assembly is dissolved away. In this manner, a consistent spherical pore construction is formed. Prominently the charrcteristics created have considerably better resolution than those attainable with present rapid prototyping techniques. Furthermore, investment in costly equipment is not essential, which permits the technique to be simply modified an industrial setting. Figure 5.18, (a) Describes PLLA scaffolds, with paraffin spheres, and (b) PLGA (85/15), with paraffin spheres.

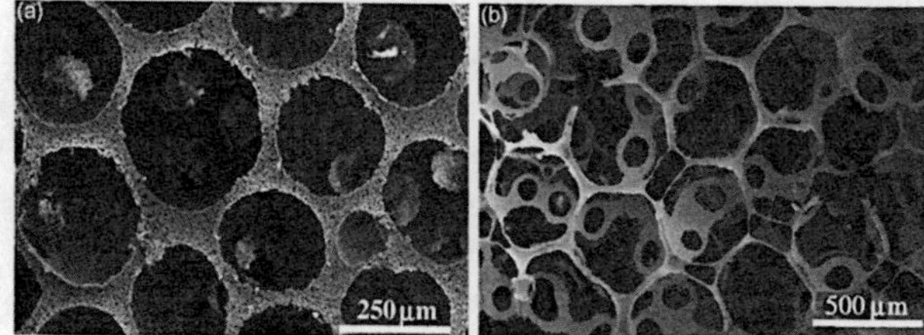

Figure 5.18: Micrographs of Polymer Scaffolds with Interconnected Spherical Pore

Bioprinting Nano Scaffolds

Scaffolds serve as temporary, artificial extracellular matrices "ECM" to facilitate cells and support 3D tissue recreation. Consequently, it is frequently advantageous to mimic certain aspects of a normal extracellular matrix in scaffold design. It is now well recognized that numerous biologically functional particles, extracellular matrix constituents, and cells interrelate at the nanoscale. Henceforth, nano-characteristics synthetic scaffold design is one of the capable novel areas in Bioprinting tissues.

Bioprinting Nano-fibrous scaffolds

Collagen is a main usual extracellular matrix element, and enjoys a fibrous construction with fiber bundles changing in diameter from 50-500 nm124 and 125. To mimic the nano-fibrous building, a few technologies have been advanced to engineer nano-fibrous scaffolds.

Electrospinning

Electrospinning was first presented in early 1930s to fabricate industrial or household nonwoven fabric products. The method has been revitalized over the past decade to make biodegradable and/or biocompatible polymers, macromolecules into fibrous fabrics with an average fiber diameter at micrometer or nanometer scales for tissue engineering scaffolds. To arrange such fibers utilizing electrospinning, a polymer solution is enforced through a capillary, making a drop of polymer solution at the tip, Figure 5.19.

A high voltage is supplied between the tip and a grounded receiving target. When the electric field strength overwhelms the surface tension of the droplet, a polymer solution jet is originated and hastened towards the receiving target. As the jet travels over the air, the solvent vaporizes and a nonwoven polymer fabric is shaped on the target. To create preferential alignment or/and tubular construction, an electrically grounded rotating drum is utilized as the receiving target. A few synthetic polymers (e.g. PGA, PLGA, PCL, and synthetic polypeptides) and natural macromolecules (e.g. collagen and fibrinogen) have been treated into fibrous nonwoven scaffolds for tissue engineering research.

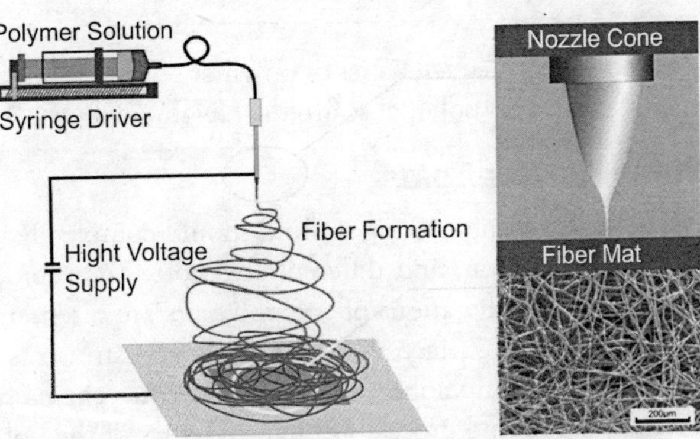

Figure 5.19: Schematics - Electrospinning Process
Courtesy: Oxolutia

Though, there are encounters in utilizing this method to manufacture intricate 3D scaffold forms and inner pore networks. Furthermore, the average fiber diameter is typically on the larger side of the extracellular matrix fibers, occasionally dropping in the micrometer range.

Self-Assembly

Self-assembly is additional stimulating study area, particularly where nano-shape and/or patterned biological constructions are concerned. Self-assembly is chracterized as the independent association of modules into shapes or constructions short of human involvement. Such a plan has been utilized in emerging nano-fibrous materials with possibility as tissue engineering scaffolds.

To compete with the triple helical structure of collagen, peptide-amphiphiles (PAs) have been manufactured. A PA contains of a collagen sequence peptide head group linked to a long-chain ester lipid. The peptide head groups formulate a triple helical construction like collagen, though the lipid tails companion via hydrophobic communications to prompt and/or alleviate the 3D structure. Nonetheless, supramolecular fiber creation has not been established.

To formulate supramolecular nano-fibers, more intricate PAs have been manufactured to encompass some physical domains. These PAs can formulate nano-fibers 5-8 nm in diameter and over 1 μm in length, which are essentially tubular micelle tubes with hydrophobic tails in the center and hydrophilic heads on the outside. The construction is steadied by disulfide bonds. Ionic self-complementary oligopeptides have been manufactured encompassing alternating parts of ionic and hydrophobic amino acids. These peptides have been utilized to formulate fibrous constructions with a fiber diameter of ~10 nm, on the lower side of extracellular matrix fibers. The fibers have a β-sheet structure as an alternative of helical one. Self-assembled nano-fiber systems are, so far, restricted to hydrogel arrangement. To advance such hydrogel scaffolds further, production cost, mechanical characteristics, degradability, and macrospore construction essential to be addressed.

Nanocomposite Scaffolds

Polymer/inorganic compound substances have been established for mineralized tissue engineering implementations. To mimic the size scale of mineral crystals in bone and other mineralized tissues, nano-sized hydroxyapatite (nano-HAP) has been multiplied with synthetic polymers or natural macromolecules to manufacture nano compound scaffolds. Nano-HAP/polymer compound scaffolds have not only enhanced the mechanical characteristics of polymer scaffolds, but also considerably improved protein adsorption over micro-sized HAP/polymer scaffolds. Heightened protein adsorption advances cell adhesion and purpose.

Bioactive Scaffolds

The ideal tissue engineering scaffold should completely interrelate with cells, as well as improved cell bond, growing, migration, and differentiated role. To accomplish these positive cell-scaffold communications, surface or bulk alterations of the polymers are often employed. The surface characteristics can be diverse by either bulk or surface alteration. Bulk adjustment is characteristically understood by copolymerization or functional group adherence to the polymer chains before scaffold manufacture, and typically changes the mechanical and processing characteristics of the polymers. Surface adjustment can be carried out after a porous scaffold has been manufactured. For instance, plasma treatment alone or followed by chemical adjustment has been utilized to adapt polymer thin films and porous scaffolds.

Such methods are most active on 2D film surfaces or very thin 3D constructs. In an intricate porous 3D scaffold, the surface is not just the outside surface, but also the internal 3D surfaces. A replicated body fluid has been utilized to adapt the chemical configuration of the internal 3D pore surfaces of polymer scaffolds. This biomimetic procedure has been revealed to be active at presenting nano-sized, bone-like apatite into the internal pore surfaces *in situ*, and may lead to enhanced scaffolds for bone tissue engineering. More 3D surface adjustment methods are desirable.

To program scaffolds with biological directives, transfer of bioactive molecules and genes has been combined into the scaffold design for tissue engineering. A model of a bioactive scaffold combining nano-fibrous construction, 3D biomimetic surface adjustment, and controlled factor release capability is shown in Figure 5.20.

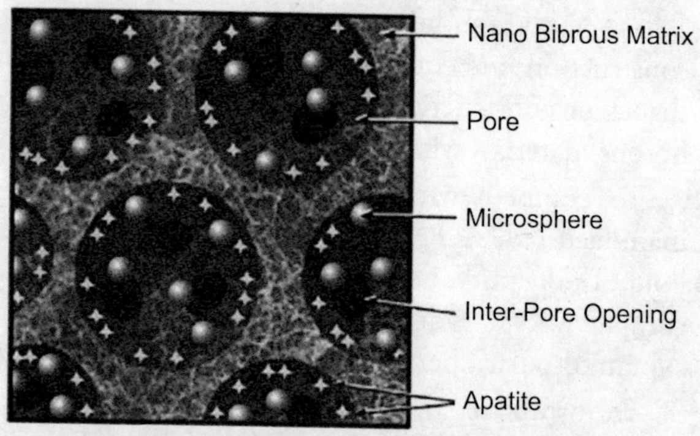

Figure 5.20: Biomimetic Nano Scaffold

The scaffold incorporates the novel nano-fibrous construction of an interconnected pore network with microspheres for controlled release of putative recreating factors. The nano-fibrous scaffolding design utilizes the architectural structures of collagen, provided that a high surface area for cell adherence and new matrix deposition, and an open structure permitting an interactive atmosphere for cell-cell, cell-nutrient, and cell-signal molecule exchanges. The bone mineral mimicking apatite enhances the osteo-conductivity of the scaffold. The biodegradable microspheres release regenerative factors in a controlled fashion in a targeted local atmosphere.

REFERENCES

1. Soper, S.A.; Brown, K.; Ellington, A.; Frazier, B.; Garcia-Manero, G.; Gau, V.; Gutman, S.I.; Hayes, D.F.; Korte, B.; Landers, J.L.; et al. Point-of-care biosensor systems for cancer diagnostics/prognostics. Biosens. Bioelectron. 2006, 21, 1932–1942, doi:10.1016/j.bios.2006.01.006.
2. Tothill, I.E. Biosensors for cancer markers diagnosis. Semin. Cell Dev. Biol. 2009, 20, 55–62, doi:10.1016/j.semcdb.2009.01.015.
3. Taitt, C.R.; Golden, J.P.; Shubin, Y.S.; Shriver-Lake, L.C.; Sapsford, K.E.; Rasooly, A.; Ligler, F.S. A portable array biosensor for detecting multiple analytes in complex samples. Microb. Ecol. 2004, 47, 175–185, doi:10.1007/s00248-003-1011-1.

4. Curtis, T.M.; Widder, M.W.; Brennan, L.M.; Schwager, S.J.; van der Schalie, W.; Fey, J.; Salazar, N. A portable cell-based impedance sensor for toxicity testing of drinking water. Lab Chip 2009, 9, 2176–2183, doi:10.1039/b901314h.

5. Heeres, J.T.; Hergenrother, P.J. High-throughput screening for modulators of protein-protein interactions: Use of photonic crystal biosensors and complementary technologies. Chem. Soc. Rev. 2011, 40, 4398–4410, doi:10.1039/b923660k.

6. Lang, H.P.; Baller, M.K.; Berger, R.; Gerber, C.; Gimzewski, J.K.; Battiston, F.M.; Fornaro, P.; Ramseyer, J.P.; Meyer, E.; Guntherodt, H.J. An artificial nose based on a micromechanical cantilever array. Anal. Chim. Acta 1999, 393, 59–65, doi:10.1016/S0003-2670(99)00283-4.

7. Huang, Y.; Chen, J.; Shi, M.; Zhao, S.; Chen, Z.-F.; Liang, H. A gold nanoparticle-enhanced fluorescence polarization biosensor for amplified detection of T4 polynucleotide kinase activity and inhibition. J. Mater. Chem. B 2013, 1, 2018–2021, doi:10.1039/c3tb00025g.

8. Aristotelous, T.; Ahn, S. Discovery of β2 adrenergic receptor ligands using biosensor fragment screening of tagged wild-type receptor. ACS Med. Chem. Lett. 2013, 4, 1005–1010, doi:10.1021/ml400312j.

9. Hartati, Y.W.; Topkaya, S.N.; Maksum, I.P.; Ozsoz, M. Sensitive detection of mitochondrial DNA A3243G tRNALeu mutation via an electrochemical biosensor using meldola's blue as a hybridization indicator. Adv. Anal. Chem. 2013, 3, 20–27.

10. Lee, S.; Kim, G.-Y.; Moon, J.-H. Detection of 6-benzylaminopurine plant growth regulator in bean sprouts using OFRR biosensor and QuEChERS method. Anal. Methods 2013, 5, 961–966, doi:10.1039/c2ay26136g.

11. Loh, Q.L.; Choong, C. Three-dimensional scaffolds for tissue engineering applications: Role of porosity and pore size. Tissue Eng. B 2013, 19, 485–502, doi:10.1089/ten.teb.2012.0437.

12. Gruene, M.; Pflaum, M.; Hess, C.; Diamantouros, S.; Schlie, S.; Deiwick, A.; Koch, L.; Wilhelmi, M.; Jockenhoevel, S.; Haverich, A.; Chichkov, B. Laser printing of three-dimensional multicellular arrays for studies of cell–cell and cell–environment interactions. Tissue Eng. C Meth. 2011, 17, 973–982, doi:10.1089/ten.tec.2011.0185.

13. Seyama, T.; Suh, E.Y.; Kondo, T. Three-dimensional culture of epidermal cells on ordered cellulose scaffolds. Biofabrication 2013, 5, doi:10.1088/1758-5082/5/2/025010.

14. Terzaki, K.; Kalloudi, E.; Mossou, E.; Mitchell, E.P.; Forsyth, V.T.; Rosseeva, E.; Simon, P.; Vamvakaki, M.; Chatzinikolaidou, M.; Mitraki, A.; et al. Mineralized self-assembled peptides on 3D laser-made scaffolds: A new route toward "scaffold on scaffold" hard tissue engineering. Biofabrication 2013, 5, doi:10.1088/1758-5082/5/4/045002.

15. Morimoto, Y.; Tsuda, Y.; Takeuchi, S. Reconstruction of 3D Hierarchic Micro-Tissues Using Monodisperse Collagen Microbeads. In Proceedings of IEEE 22nd International Conference on Micro Electro Mechanical Systems (MEMS 2009), Sorrento, Italy, 25–29 January 2009; pp. 56–59.

16. Yu, J.; Du, K.T.; Fang, Q.; Gu, Y.; Mihardja, S.S.; Sievers, R.E.; Wu, J.C.; Lee, R.J. The use of human mesenchymal stem cells encapsulated in RGD modified alginate microspheres in the repair of myocardial infarction in the rat. Biomaterials 2010, 31, 7012–7020, doi:10.1016/j.biomaterials.2010.05.078.

17. Solorio, L.D.; Vieregge, E.L.; Dhami, C.D.; Alsberg, E. High-density cell systems incorporating polymer microspheres as microenvironmental regulators in engineered cartilage tissues. Tissue Eng. B 2013, 19, 209–220, doi:10.1089/ten.teb.2012.0252.

18. Lee, J.-Y.; Choi, B.; Wu, B.; Lee, M. Customized biomimetic scaffolds created by indirect three-dimensional printing for tissue engineering. Biofabrication 2013, 5, doi:10.1088/1758-5082/5/4/045003.
19. Payne, G.F.; Kim, E.; Cheng, Y.; Wu, H.-C.; Ghodssi, R.; Rubloff, G.W.; Raghavan, S.R.; Culver, J.N.; Bentley, W.E. Accessing biology's toolbox for the mesoscale biofabrication of soft matter. Soft Matter 2013, 9, 6019–6032, doi:10.1039/c3sm50527h.
20. Song, W.; Yu, X.; Markel, D.C.; Shi, T.; Ren, W. Coaxial PCL/PVA electrospun nanofibers: Osseointegration enhancer and controlled drug release device. Biofabrication 2013, 5, doi:10.1088/1758-5082/5/3/035006.
21. Squires, T.M.; Messinger, R.J.; Manalis, S.R. Making it stick: Convection, reaction and diffusion in surface-based biosensors. Nat. Biotechnol. 2008, 26, 417–426, doi:10.1038/nbt1388.
22. Situ, C.; Buijs, J.; Mooney, M.H.; Elliott, C.T. Advances in surface plasmon resonance biosensor technology towards high-throughput, food-safety analysis. Trends Anal. Chem. 2010, 29, 1305–1315, doi:10.1016/j.trac.2010.09.003.
23. Ligler, F.S.; Taitt, C.R.; Shriver-Lake, L.C.; Sapsford, K.E.; Shubin, Y.; Golden, J.P. Array biosensor for detection of toxins. Anal. Bioanal. Chem. 2003, 377, 469–477, doi:10.1007/s00216-003-1992-0.
24. Volcke, C.; Gandhiraman, R.P.; Basabe-Desmonts, L.; Iacono, M.; Gubala, V.; Cecchet, F.; Cafolla, A.A.; Williams, D.E. Protein pattern transfer for biosensor applications. Biosens. Bioelectron. 2010, 25, 1295–1300, doi:10.1016/j.bios.2009.10.016.
25. Xia, Y.; Whitesides, G.M. Soft Lithography. Annu. Rev. Mater. Sci. 1998, 37, 550–575.
26. Khademhosseini, A.; Langer, R.; Borenstein, J.; Vacanti, J.P. Microscale technologies for tissue engineering and biology. Proc. Natl. Acad. Sci. USA 2006, 103, 2480–2487, doi:10.1073/pnas.0507681102.
27. Wilson, G.S.; Hu, Y. Enzyme-based biosensors for *in vivo* measurements. Chem. Rev. 2000, 100, 2693–2704, doi:10.1021/cr990003y.
28. Plaxco, K.W.; Soh, H.T. Switch-based biosensors: A new approach towards real-time, *in vivo* molecular detection. Trends Biotechnol. 2011, 29, 1–5, doi:10.1016/j.tibtech.2010.10.005.
29. Kellenberger, C.A.; Wilson, S.C.; Sales-Lee, J.; Hammond, M.C. RNA-based fluorescent biosensors for live cell imaging of second messengers cyclic di-GMP and cyclic AMP-GMP. J. Am. Chem. Soc. 2013, 135, 4906–4909, doi:10.1021/ja311960g.
30. Odell, I.D.; Cook, D. Immunofluorescence techniques. J. Invest. Dermatol. 2013, 133, doi:10.1038/jid.2012.455.
31. Hide, M.; Tsutsui, T.; Sato, H.; Nishimura, T.; Morimoto, K.; Yamamoto, S.; Yoshizato, K. Real-time analysis of ligand-induced cell surface and intracellular reactions of living mast cells using a surface plasmon resonance-based biosensor. Anal. Biochem. 2002, 302, 28–37, doi:10.1006/abio.2001.5535.
32. Bastiaens, P.I.; Squire, A. Fluorescence lifetime imaging microscopy: Spatial resolution of biochemical processes in the cell. Trends Cell Biol. 1999, 9, 48–52, doi:10.1016/S0962-8924(98)01410-X.
33. Ouyang, M.; Sun, J.; Chien, S.; Wang, Y. Determination of hierarchical relationship of Src and Rac at subcellular locations with FRET biosensors. Proc. Natl. Acad. Sci. USA 2008, 105, 14353–14358, doi:10.1073/pnas.0807537105.
34. Wegener, J.; Keese, C.R.; Giaever, I. Electric cell-substrate impedance sensing (ECIS) as a noninvasive means to monitor the kinetics of cell spreading to artificial surfaces. Exp. Cell Res. 2000, 259, 158–166, doi:10.1006/excr.2000.4919.

35. Mcauley, E.; Plopper, G.E.; Mohanraj, B.; Phamduy, T.; Corr, D.T.; Chrisey, D.B. Evaluation of electric cell-substrate impedance sensing for the detection of nanomaterial toxicity. Int. J. Biomed. Nanosci. Nanotechnol. 2011, 2, 136–151, doi:10.1504/IJBNN.2011.040998.
36. Baird, C.L.; Myszka, D.G. Current and emerging commercial optical biosensors. J. Mol. Recognit. 2001, 14, 261–268, doi:10.1002/jmr.544.
37. Quinn, J.G.; O'Neill, S.; Doyle, A.; McAtamney, C.; Diamond, D.; MacCraith, B.D.; O'Kennedy, R. Development and application of surface plasmon resonance-based biosensors for the detection of cell-ligand interactions. Anal. Biochem. 2000, 281, 135–143, doi:10.1006/abio.2000.4564.
38. Charych, D.H.; Nagy, J.O.; Spevak, W.; Bednarski, M.D. Direct colorimetric detection of a receptor-ligand interaction by a polymerized bilayer assembly. Science 1993, 261, 585–588, doi:10.1126/science.8342021.
39. Bai, S.; Li, S.; Yao, T.; Hu, Y.; Bao, F.; Zhang, J.; Zhang, Y.; Zhu, S.; He, Y. Rapid detection of eight vegetable oils on optical thin-film biosensor chips. Food Control 2011, 22, 1624–1628, doi:10.1016/j.foodcont.2011.03.019.
40. Wang, W.; Han, J.; Wu, Y.; Yuan, F.; Chen, Y.; Ge, Y. Simultaneous detection of eight food allergens using optical thin-film biosensor chips. J. Agric. Food Chem. 2011, 59, 6889–6894, doi:10.1021/jf200933b.
41. Chen, C.; Xie, Q.; Yang, D.; Xiao, H.; Fu, Y.; Tan, Y.; Yao, S. Recent advances in electrochemical glucose biosensors: A review. RSC Adv. 2013, 3, 4473–4491, doi:10.1039/c2ra22351a.
42. Hamidi-Asl, E.; Palchetti, I.; Hasheminejad, E.; Mascini, M. A review on the electrochemical biosensors for determination of microRNAs. Talanta 2013, 115, 74–83, doi:10.1016/j.talanta.2013.03.061.
43. Homola, J.; Yee, S.S.; Gauglitz, G. Surface plasmon resonance sensors: Review. Sens. Actuator B Chem. 1999, 54, 3–15, doi:10.1016/S0925-4005(98)00321-9.
44. Mayer, K.M.; Hafner, J.H. Localized surface plasmon resonance sensors. Chem. Rev. 2011, 111, 3828–3857, doi:10.1021/cr100313v.
45. Tiefenthaler, K.; Lukosz, W. Sensitivity of grating couplers as integrated-optical chemical sensors. J. Opt. Soc. Am. B 1989, 6, 209–220, doi:10.1364/JOSAB.6.000209.
46. Fang, Y. Non-invasive optical biosensor for probing cell signaling. Sensors 2007, 7, 2316–2329, doi:10.3390/s7102316.
47. Goral, V.; Wu, Q.; Sun, H.; Fang, Y. Label-free optical biosensor with microfluidics for sensing ligand-directed functional selectivity on trafficking of thrombin receptor. FEBS Lett. 2011, 585, 1054–1060, doi:10.1016/j.febslet.2011.03.003.
48. Dong, Y.; Wilkop, T.; Xu, D.; Wang, Z.; Cheng, Q. Microchannel chips for the multiplexed analysis of human immunoglobulin G-antibody interactions by surface plasmon resonance imaging. Anal. Bioanal. Chem. 2008, 390, 1575–1583, doi:10.1007/s00216-008-1849-7.
49. Cetin, A.E.; Coskun, A.F.; Galarreta, B.C.; Huang, M.; Herman, D.; Ozcan, A.; Altug, H. Handheld high-throughput plasmonic biosensor using computational on-chip imaging. Light Sci. Appl. 2014, 3, doi:10.1038/lsa.2014.3.
50. Raiteri, R.; Grattarola, M.; Butt, H.-J.; Skládal, P. Micromechanical cantilever-based biosensors. Sens. Actuator B Chem. 2001, 79, 115–126, doi:10.1016/S0925-4005(01)00856-5. 51.Battiston, F.; Ramseyer, J.-P.; Lang, H.; Baller, M.; Gerber, C.; Gimzewski, J.; Meyer, E.; Güntherodt, H.-J. A chemical sensor based on a microfabricated cantilever array with simultaneous resonance-

frequency and bending readout. Sens. Actuator B Chem. 2001, 77, 122–131, doi:10.1016/S0925-4005(01)00683-9.

52. Wee, K.W.; Kang, G.Y.; Park, J.; Kang, J.Y.; Yoon, D.S.; Park, J.H.; Kim, T.S. Novel electrical detection of label-free disease marker proteins using piezoresistive self-sensing micro-cantilevers. Biosens. Bioelectron. 2005, 20, 1932–1938, doi:10.1016/j.bios.2004.09.023.
53. Carrascosa, L.G.; Moreno, M.; Álvarez, M.; Lechuga, L.M. Nanomechanical biosensors: A new sensing tool. TrAC Trends Anal. Chem. 2006, 25, 196–206, doi:10.1016/j.trac.2005.09.006.
54. Zhang, H.; Shepherd, J.N.H.; Nuzzo, R.G. Microfluidic contact printing: A versatile printing platform for patterning biomolecules on hydrogel substrates. Soft Matter 2010, 6, 2238–2245, doi:10.1039/c000385a.
55. Zheng, C.; Wang, J.; Pang, Y.; Wang, J.; Li, W.; Ge, Z.; Huang, Y. High-throughput immunoassay through in-channel microfluidic patterning. Lab Chip 2012, 12, 2487–2490, doi:10.1039/c2lc40145b.
56. Lo, C.M.; Keese, C.R.; Giaever, I. Impedance analysis of MDCK cells measured by electric cell-substrate impedance sensing. Biophys. J. 1995, 69, 2800–2807, doi:10.1016/S0006-3495(95)80153-0.
57. Xiao, C.; Luong, J.H.T. On-line monitoring of cell growth and cytotoxicity using electric cell-substrate impedance sensing (ECIS). Biotechnol. Prog. 2003, 19, 1000–1005, doi:10.1021/bp025733x.
58. Hynes, A.; Mohanraj, B.; Schiele, N.; Corr, D.T.; Plopper, G.E.; Dinu, C.Z.; Chrisey, D.B. Cell-based detection of bacillus cereus anthrax simulant using cell impedance sensing. Sens. Lett. 2010, 8, 528–533, doi:10.1166/sl.2010.1306.
59. Ferentinos, K.P.; Yialouris, C.P.; Blouchos, P.; Moschopoulou, G.; Tsourou, V.; Kintzios, S. The use of artificial neural networks as a component of a cell-based biosensor device for the detection of pesticides. Procedia Eng. 2012, 47, 989–992, doi:10.1016/j.proeng.2012.09.313.
60. McConnell, H.M.; Owicki, J.C.; Parce, J.W.; Miller, D.L.; Baxter, G.S.; Wada, H.G.; Pitchford, S. The cytosensor microphysiometer: Biological applications of silicon technology. Science 1992, 257, 1906–1912, doi:10.1126/science.1329199.
61. Salimi, A.; Noorbakhsh, A.; Rafiee-Pour, H.-A.; Ghourchian, H. Direct voltammetry of copper, zinc-superoxide dismutase immobilized onto electrodeposited nickel oxide nanoparticles: Fabrication of amperometric superoxide biosensor. Electroanalysis 2011, 23, 683–691.
62. Xiang, C.; Kung, S.; Taggart, D.K.; Yang, F.; Thompson, M.A.; Gu, A.G.; Yang, Y.; Penner, R.M. Lithographically patterned nanowire patterning electrically continuous metal nanowires on dielectrics. ACS Nano 2008, 2, 1939–1949, doi:10.1021/nn800394k.
63. Peng, S.; Zhang, X. Electrodeposition of CdSe quantum dots and its application to an electrochemiluminescence immunoassay for α-fetoprotein. Microchim. Acta 2012, 178, 323–330, doi:10.1007/s00604-012-0844-z.
64. Loget, G.; Wood, J.B.; Cho, K.; Halpern, A.R.; Corn, R.M. Electrodeposition of polydopamine thin films for DNA patterning and microarrays. Anal. Chem. 2013, 85, 9991–9995, doi:10.1021/ac4022743.
65. Novak, S.; Maver, U.; Peternel, Š.; Venturini, P.; Bele, M.; Gaberšček, M. Electrophoretic deposition as a tool for separation of protein inclusion bodies from host bacteria in suspension. Colloid Surf. A Physicochem. Eng. Asp. 2009, 340, 155–160, doi:10.1016/j.colsurfa.2009.03.023.

66. Ammam, M.; Fransaer, J. Two-enzyme lactose biosensor based on β-galactosidase and glucose oxidase deposited by AC-electrophoresis: Characteristics and performance for lactose determination in milk. Sens. Actuator B Chem. 2010, 148, 583–589, doi:10.1016/j.snb.2010.05.027.
67. Poortinga, A.T.; Bos, R.; Busscher, H.J. Controlled electrophoretic deposition of bacteria to surfaces for the design of biofilms. Biotechnol. Bioeng. 2000, 67, 117–120, doi:10.1002/(SICI)1097-0290(20000105)67:1<117::AID-BIT14>3.0.CO;2-6.
68. Suginta, W.; Khunkaewla, P.; Schulte, A. Electrochemical biosensor applications of polysaccharides chitin and chitosan. Chem. Rev. 2013, 113, 5458–5479, doi:10.1021/cr300325r.
69. Moffat, T.P.; Yang, H. Patterned metal electrodeposition using an alkanethiolate mask. J. Electrochem. Soc. 1995, 142, 220–222, doi:10.1149/1.2048452.
70. Fernandes, R.; Wu, L.; Chen, T.; Yi, H.; Rubloff, G.W.; Ghodssi, R.; Bentley, W.E.; Payne, G.F. Electrochemically induced deposition of a polysaccharide hydrogel onto a patterned surface. Langmuir 2003, 19, 4058–4062, doi:10.1021/la027052h.
71. Chen, P.-C.; Chen, R.L. C.; Cheng, T.-J.; Wittstock, G. Localized deposition of chitosan as matrix for enzyme immobilization. Electroanalysis 2009, 21, 804–810.
72. Chen, C.S.; Mrksich, M.; Huang, S.; Whitesides, G.M.; Ingber, D.E. Micropatterned surfaces for control of cell shape, position, and function. Biotechnol. Prog. 1998, 14, 356–363, doi:10.1021/bp980031m.
73. Duarte Campos, D.F.; Blaeser, A.; Weber, M.; Jäkel, J.; Neuss, S.; Jahnen-Dechent, W.; Fischer, H. Three-dimensional printing of stem cell-laden hydrogels submerged in a hydrophobic high-density fluid. Biofabrication 2013, 5, doi:10.1088/1758-5082/5/1/015003.
74. Raof, N.A.; Schiele, N.R.; Xie, Y.; Chrisey, D.B.; Corr, D.T. The maintenance of pluripotency following laser direct-write of mouse embryonic stem cells. Biomaterials 2011, 32, 1802–1808, doi:10.1016/j.biomaterials.2010.11.015.
75. Barron, J.A.; Rosen, R.; Jones-Meehan, J.; Spargo, B.J.; Belkin, S.; Ringeisen, B.R. Biological laser printing of genetically modified Escherichia coli for biosensor applications. Biosens. Bioelectron. 2004, 20, 246–252, doi:10.1016/j.bios.2004.01.011.
76. Gaebel, R.; Ma, N.; Liu, J.; Guan, J.; Koch, L.; Klopsch, C.; Gruene, M.; Toelk, A.; Wang, W.; Mark, P.; et al. Patterning human stem cells and endothelial cells with laser printing for cardiac regeneration. Biomaterials 2011, 32, 9218–9230, doi:10.1016/j.biomaterials.2011.08.071.
77. Britland, S.; Perez-Arnaud, E.; Clark, P.; Mcginn, B.; Connolly, P.; Moores, G. Micropatterning proteins and synthetic peptides on solid supports: A novel application for microelectronics fabrication technology. Biotechnol. Prog. 1992, 8, 155–160, doi:10.1021/bp00014a010.
78. Folch, A.; Jo, B.H.; Hurtado, O.; Beebe, D.J.; Toner, M. Microfabricated elastomeric stencils for micropatterning cell cultures. J. Biomed. Mater. Res. 2000, 52, 346–353, doi:10.1002/1097-4636(200011)52:2<346::AID-JBM14>3.0.CO;2-H.
79. Ostuni, E.; Kane, R.; Chen, C.; Ingber, D.E.; Whitesides, G.M. Patterning mammalian cells using elastomeric membranes. Langmuir 2000, 7811–7819.
80. Selvarasah, S.; Chao, S.H.; Chen, C.-L.; Sridhar, S.; Busnaina, A.; Khademhosseini, A.; Dokmeci, M.R. A reusable high aspect ratio parylene-C shadow mask technology for diverse micropatterning applications. Sens. Actuator A Phys. 2008, 145, 306–315.
81. Jackman, R.J.; Wilbur, J.L.; Whitesides, G.M. Fabrication of submicrometer features on curved substrates by microcontact printing. Science 1995, 269, 664–666, doi:10.1126/science.7624795.

82. Tien, J.; Nelson, C.M.; Chen, C.S. Fabrication of aligned microstructures with a single elastomeric stamp. Proc. Natl. Acad. Sci. USA 2002, 99, 1758–1762, doi:10.1073/pnas.042493399.
83. Beebe, D.J.; Moore, J.S.; Yu, Q.; Liu, R.H.; Kraft, M.L.; Jo, B.H.; Devadoss, C. Microfluidic tectonics: A comprehensive construction platform for microfluidic systems. Proc. Natl. Acad. Sci. USA 2000, 97, 13488–13493, doi:10.1073/pnas.250273097.
84. Gu, W.; Zhu, X.; Futai, N.; Cho, B.S.; Takayama, S. Computerized microfluidic cell culture using elastomeric channels and Braille displays. Proc. Natl. Acad. Sci. USA 2004, 101, 15861–15866, doi:10.1073/pnas.0404353101.
85. Unger, M.A.; Chou, H.P.; Thorsen, T.; Scherer, A.; Quake, S.R. Monolithic microfabricated valves and pumps by multilayer soft lithography. Science 2000, 288, 113–116, doi:10.1126/science.288.5463.113.
86. Jeon, N.L.; Dertinger, S.K. W.; Chiu, D.T.; Choi, I.S.; Stroock, A.D.; Whitesides, G.M. Generation of solution and surface gradients using microfluidic systems. Langmuir 2000, 16, 8311–8316, doi:10.1021/la000600b.
87. Chiu, D.T.; Jeon, N.L.; Huang, S.; Kane, R.S.; Wargo, C.J.; Choi, I.S.; Ingber, D.E.; Whitesides, G.M. Patterned deposition of cells and proteins onto surfaces by using three-dimensional microfluidic systems. Proc. Natl. Acad. Sci. USA 2000, 97, 2408–2413, doi:10.1073/pnas.040562297.
88. King, K.R.; Wang, S.; Irimia, D.; Jayaraman, A.; Toner, M.; Yarmush, M.L. A high-throughput microfluidic real-time gene expression living cell array. Lab Chip 2007, 7, 77–85, doi:10.1039/b612516f.
89. Pitaval, A.; Tseng, Q.; Bornens, M.; Thery, M. Cell shape and contractility regulate ciliogenesis in cell cycle-arrested cells. J. Cell Biol. 2010, 191, 303–312, doi:10.1083/jcb.201004003.
90. Théry, M. Micropatterning as a tool to decipher cell morphogenesis and functions. J. Cell Sci. 2010, 123, 4201–4213, doi:10.1242/jcs.075150.
91. Guilak, F.; Cohen, D.M.; Estes, B.T.; Gimble, J.M.; Liedtke, W.; Chen, C.S. Control of stem cell fate by physical interactions with the extracellular matrix. Cell Stem Cell 2009, 5, 17–26, doi:10.1016/j.stem.2009.06.016.
92. Clark, P.; Britland, S.; Connolly, P. Growth cone guidance and neuron morphology on micropatterned laminin surfaces. J. Cell Sci. 1993, 105, 203–212.
93. Lee, L.H.; Peerani, R.; Ungrin, M.; Joshi, C.; Kumacheva, E.; Zandstra, P.W. Micropatterning of human embryonic stem cells dissects the mesoderm and endoderm lineages. Stem Cell Res. 2009, 2, 155–162, doi:10.1016/j.scr.2008.11.004.
94. Tang, J.; Peng, R.; Ding, J. The regulation of stem cell differentiation by cell-cell contact on micropatterned material surfaces. Biomaterials 2010, 31, 2470–2476, doi:10.1016/j.biomaterials.2009.12.006.
95. Pal, R.; Yang, M.; Lin, R.; Johnson, B.N.; Srivastava, N.; Razzacki, S.Z.; Chomistek, K.J.; Heldsinger, D.C.; Haque, R.M.; Ugaz, V.M.; et al. An integrated microfluidic device for influenza and other genetic analyses. Lab Chip 2005, 5, 1024–1032, doi:10.1039/b505994a.
96. Irimia, D.; Tompkins, R.G.; Toner, M. Single-cell chemical lysis in picoliter-scale closed volumes using a microfabricated device. Anal. Chem. 2004, 76, 6137–6143, doi:10.1021/ac0497508.
97. Stott, S.L.; Hsu, C.; Tsukrov, D.I.; Yu, M.; Miyamoto, D.T.; Waltman, B.A.; Rothenberg, S.M.; Shah, A.M.; Smas, M.E.; Korir, G.K.; et al. Isolation of circulating tumor cells using a microvortex-generating herringbone chip. Proc. Natl. Acad. Sci. USA 2010, 107, 18392–18397, doi:10.1073/pnas.1012539107.

98. Yu, M.; Stott, S.; Toner, M.; Maheswaran, S.; Haber, D. Circulating tumor cells: Approaches to isolation and characterization. J. Cell Biol. 2011, 192, 373–82, doi:10.1083/jcb.201010021.
99. Lai, N.; Sims, J.K.; Jeon, N.L.; Lee, K. Adipocyte induction of preadipocyte differentiation in a gradient chamber. Tissue Eng. C Meth. 2012, 18, 958–967, doi:10.1089/ten.tec.2012.0168.
100. Sung, J.H.; Shuler, M.L. *In vitro* microscale systems for systematic drug toxicity study. Bioprocess Biosyst. Eng. 2010, 33, 5–19, doi:10.1007/s00449-009-0369-y.
101. Soloman, Sabrie "Sensors Handbook, 2nd Edition, McGraw-Hill Publishing Co., NY – ISBN 978-0-07-160570-0
102. Soloman, Sabrie "Sensors and Control Systems in Manufacturing 2nd Edition," McGraw-Hill Publishing Co., NY – ISBN 0-07-059626-3
103. Soloman, Sabrie "Affordable Automation," McGraw-Hill Publishing Co., NY – ISBN 0-07-059633-6
104. Soloman, Sabrie "Sensors Handbook, 1st Edition," McGraw-Hill Publishing Co., NY – ISBN 0-07-059630-1
105. Soloman, Sabrie "Sensors and Control Systems in Manufacturing 1st Edition," McGraw-Hill Publishing Co., NY – ISBN 978-07-10572-4

6 3D Bioprinting Regenerative Medicine

3D BIOPRINTING OF COMPLEX LIVING TISSUES

3D Bioprinting, containing compound living active tissues, is the basis of a new technology in regenerative medicine. The 3D Bioprinting has now authoritatively renowned in the medical world, regenerative healing is conceivable through this new development of the technology. It has unlocked new frontiers in regenerative medicine. Through the 3D Bioprinting, that is already demonstrated infallible in the medical world, made this novel progression in know-how the next revolution in regenerative healing, Figure 6.1.

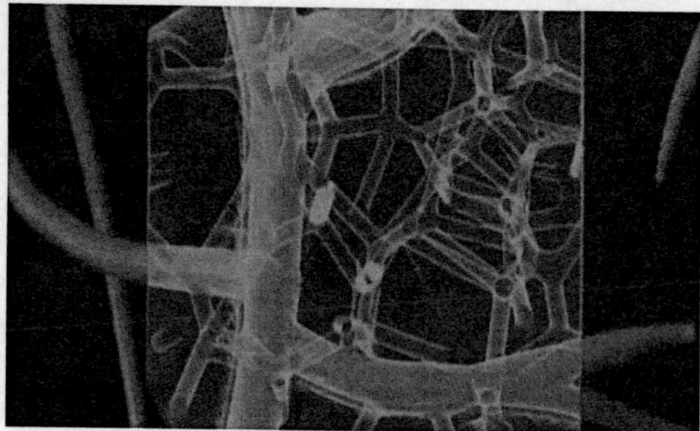

Figure 6.1: 3D Bioprint Collagen Allows Fabricate Fully Functional Components of Human Heart.
Courtesy: Carnegie Mellon University College of Engineering

The purpose of the 3D Bioprinting procedure is enabling Biological objects to be Bioprinted in countless sophisticated practices by means of digital models. 3D Bioprinting has newly extended a secure hold in the medical design deep thinkers, industrialists and creative developer working in a multitude of Healthcare industries. Even in the multifaceted world of medicine, the expertise has been finding complex implementations across a variety of fields, including orthopedics, prosthetics and dentistry. In the meantime, 3D Bioprinting implementation is virtually already in use, with more than forty million printed hearing aids allegedly being utilized everywhere the world.

If additional indications have ensured that the Bioprinting technology is advancing as desired. The recent financial research conducted by "*Transparency Market Research*" forecasts that the medical 3D Bioprinting market is maturing as its financial -worth has advanced from $354 million last year in 2018 to $965 million in 2019. This conducted financial statement is as good a sign as any and that the healthcare assets, with respect to the future of 3D Bioprinting are secure as it is being continual and expanding.

As scientists' accepting of the technology advancements, the progressively refined applications are being forged to work with the body's natural healing inherent mechanisms. Contemporary 3D Bioprinting techniques are capable to generate porous structures, letting formation of "scaffolds," which boost crushed bones to re-knit properly, Figure 6.2.

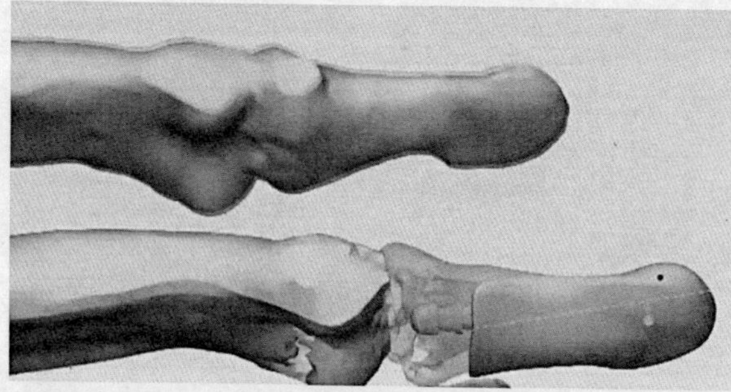

Figure 6.2: A Floridian Patient Receives World First 3D Bioprinted Titanium Finger-Bone Replacement in USA
Courtesy: Winks News – Florida USA

3D Bioprinting is being employed, in orthopedic surgery, to print permeable materials for bone stabilization. The titanium material is the most suitable material for orthopedic grafts. If porous grafts through 3D Bioprinting took place, the patient's bone rapidly changes causing the injured area to stabilize fast to endure the surgery. This is type of implant is a regenerative medicine that is long-lasting.

Bioprinting in Regenerative Medicine

If today's medical 3D Bioprinting practices suggestion at a better prospective for regenerative medicine, the next upsurge of the technology is beholding to fully exploit on that vision. New Bioprinting techniques permit the printing of cell-loaded imitations of human tissues, typically seized together in liquid or hydrogel method, with which inventors are discovering a new boundary of regenerative restorative.

Print-on-Demand Organs

The attention-seeking goal of bioprinting could be print-on-demand organs to expel recipients transplant waiting-lists forever. Of course, years will fade away before Bioprinters are intelligent enough to repeat the multifaceted functions of human organs for safe implantation in patients, but today's researchers are starting the voyage towards that endpoint by presenting cells to the printing process, and safeguarding they can endure it. The field's progress is progressively linked to stem cell study, as the pluripotency of embryonic stem cells could generate the diverse stock desired for both study and ultimate usage in hospitals.

The most likely application is "the joints," where one may transfer from bone to cartilage. The body contains flawless variations of tissue categories to provide the body the inherit abilities to form bones and other articulated organs. Researchers fashioned a 3D-printed bladder through 3D printing. The stem cell may be structured and guided to become bone in one part and cartilage in another part.

3D Bioprinting Developments

Laboratories around the world are employed to thrust onward the discipline of bioprinting. Tendon and Cartilage have been recognized as a good preliminary topic, specified the relatively small nutritional supplies of the tissue's chondrocyte cells. The survivability of these cells aids avoid one of the compound tasks in the bioprinting procedure, which is to preserve printed cells active and nurtured, usually achieved by their close vicinity to the body's network of capillaries. 3D Bioprinting permits the formation of "scaffolds" that boost shattered bones to re-knit.

The final objective for Bioprinted cartilage is receiving adequate attention. Researchers are pointing to generate a 3D Bioprinting system - perhaps in combination with a robotic surgical arm - that can print cartilage directly in the body to mend or substitute injured joint tissue.

Further than cartilage, progress is also being completed on the bioprinting of more compound alive tissues. California-based Organovo, USA Figure 6.3, lately offered the world's first 3D bioprinted human liver tissue employing its NovoGen bioprinting platform. The printed tissue integrates numerous cell types, the patterns of which bear close similarity to the natural construction of liver tissue.

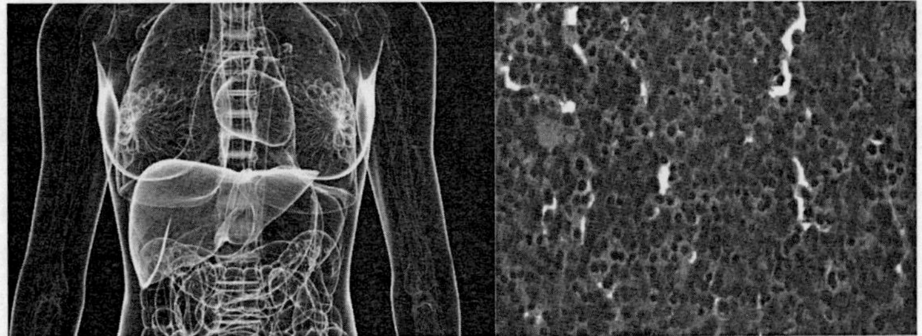

Figure 6.3: 3D Bioprinting Functional Liver Tissue

Courtesy: Oreganovo's 3D Liver Tissue

Testing toxicity of Experimental Drugs and Treatments

While impressive advancements are accomplished in Bioprinting technology of organ printing similar to the Bioprinted of liver tissue, there are no distinctive sign yet that "made-to-order" organs are immediately imminent. However, a more instant application for these living tissues of the liver tissue to examine the effectiveness or toxicity of investigational drugs and cures, Figure 6.4.

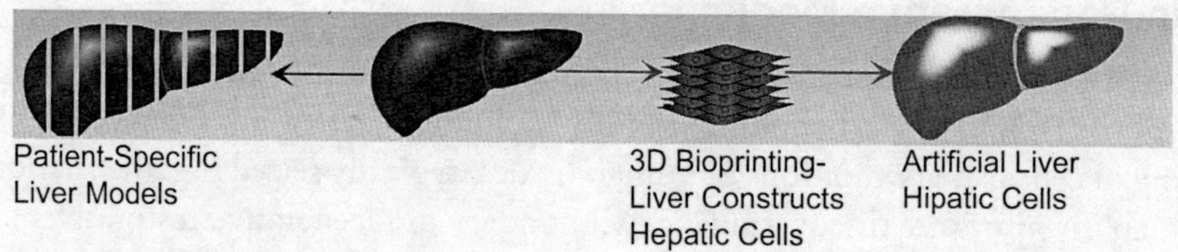

Figure 6.4: 3D Bioprinting Liver Constructs with Hepatic Cells

Body on a Chip

Equally, the so-entitled "body on a chip" for Regenerative Medicine is utilizing Bioprinting approaches to generate distinct cell build that imitate core organ performance, which are then incorporated with artificial blood circulation network. This generates a micro type of the body's vital systems, sufficiently enough to fit on to a microchip, and the flow can be utilized to distribute inoculations or additional agents to test responses.

Economic and Medical Regeneration

Experts ultimately would like to print cartilage straight in the body. As 3D printing and bioprinting remain novel issue to medical claims, there is also the opportunity to generate new areas of high-tech industry - economic as well as medical recreation.

While still in its initial phases of progress, Bioprinting embraces the prospective to open up a courageous new world of therapeutic novelty. Once the supplementary progressive Bioprinting pattern is set and accepted within the medical institutions, safety and monitoring procedure are clearly difficulties yet to be overwhelmed in many areas, the potentials are virtually limitless in the years ahead.

The Revolutionary Power of Bioelectronic Medicine

Interestingly enough, the healing power of medicines that express the electrical semantic of the body could develop realism sooner rather than later, thanks to a motivated strategy propelled by a few pharmaceutical corporations employing Nano technology in drug and treatment discovery. Nano-scale devices could connect the electrical semantic of the human body, possibly reforming personalized medicine.

Though small particles or biologic drugs touch all body tissues - those that may be unhealthy, and those that may not – bio-electronic medicine, which will utilize the prevailing and marginal nervous systems to modulate the drive of diseased organ systems. This modulation has the benefit of binding a "hard-wired" anatomic pathway, which has been evolutionarily designed to govern the purpose of an organ system in an exact means.

To make these medicines an authentic remedy many experts who strongly believe bioelectronic medicine combined with 3D Bioprinting technology could have a massive influence through numerous therapeutic areas - from diabetes to heart failure, rheumatoid arthritis to gut motility. Prominent pharmaceutical and medical research institution have invested heavily in designating creative experts along with formidable capitals in place to continue exploring this novel technology with the purpose of advancing medicine to convey and speak the electronic language of the body as a revolutionary diagnostic in the foreseeable future.

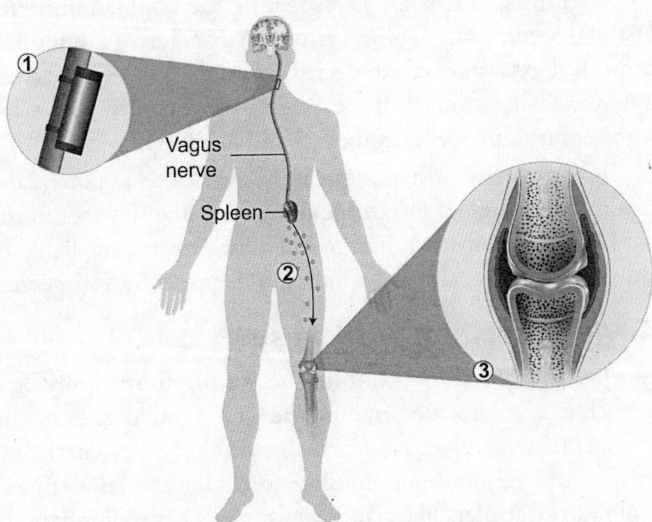

Figure 6.5: Neuromodulation for rheumatoid arthritis
Courtesy: Royal Pharmaceutical Society Publication

The crucial Bioelectronic mechanisms are accessible. Nonetheless, exceedingly dedicated specialists are vital to take part in the diverse disciplines - engineering, biology, and informatics to deliver the technical community the assurance that there are a long term strategies and promises to carry out such revolutionary medicines to patients, Figure 6.5.

The electrical stimulus applied to the vagus nerve through an implantable bioelectronic device, Figure 6.5, impacts splenic T-cells and macrophages to decrease internal manufacture of tumor necrosis factor, which may cause the death of cells in the tissue. The electrical stimulation decreasing the TNF, is an inflammatory mediator, leading to reduced rheumatoid arthritis symptoms in some patients.

The micro-regulator stimulates the vagus nerve, figure 1. It sends signals through the vagus and splenic nerve to trigger a reduction in the activation of T cells and macrophages in the spleen as indicated in (1), Figure 6.5. Also, there is a reduced production of systemic inflammation mediators and reduced activation of circulating immune cells indicated in (2), Figure 6.5. Additionally, this electrical stimulation leads to decreased inflammation, joint damage and pain.

Approaches to Bioelectronic

The Bioelectronics research and development organization in USA propose 3-phase approach to accomplish Bioelectronic medicine.

- Initiation of a new capital investment program to support 20 projects will rapidly enthuse studies in areas that can strengthen Bioelectronic medicines around the world.
- Forming global conferences for the Bioelectronic studies on a regular basis, which aims to encourage scientists to meet, exchange information and review initial results, also provide a \$1milion invention prize.
- Generating "Possible Venture Capital," - a new \$50million venture capital endowment to capitalize the development of bioelectronic medicines and technologies.

The initial investment is utilized in developing an exceptionally novel technique to cure inflammatory illnesses, such as rheumatoid arthritis and inflammatory bowel disease, utilizing implantable devices that arouse the body's vagus nerve.

The investigational study was completed in patients with rheumatoid arthritis. Prior to biologic therapy, the patients involved in the study had *unsuccessful* treatment to contain cancerous cells, a small molecule first-line agent utilized prior to the electronic biologic therapy. Six of considered eight patients had clinical trials, and of those six, two had full disease remission.

Presently, scientists are investigating the population of an advanced phase that has unsuccessful both first and second line biologic drugs. A conclusive randomized will be monitored this investigation with controlled experimental, to decisively validate the effectiveness of this method. While the timeline is hard to forecast with accuracy, if the designated treatment channel is successful it is likely to find an endorsement within four to five years in the US and in Europe within two to three years.

The entire new cluster of diseases was released as potentials of bioelectronic medicine research. Also, the expected outcome of electronic device treating illnesses must have the guarantees to deliver according to the set expectations required by the bioelectronic medicine treatment device. The expected device would require an invasive surgery, which can be preset by a surgeon.

A Wide Spectrum of Diseases

Not just inflammatory conditions that will, futuristically be qualified to be preserved with bio-electronic medicine. Undoubtedly, one of the important details in this field of electronic medicine is the aspect therapeutic areas. Electronic medicine that have the neural element may detect and cure metabolic situations such as diabetes and gut motility to cardiovascular settings, such as hypertension and heart failure, to respiratory disorders like asthma and COPD as well inflammatory and marginal pa*in situ*ations, all have a neural element.

The neural element is capable to disturb disease development. Though, scientists are self-assured that bio-electronic medicine is a novel means of influencing biology, which tolerates significant variety of diseases. On top of establishing which disease category can be best preserved through this method, there are numerous other contests and non-entities that will look to make bioelectronic medicine a certainty.

Refining the accuracy of nerve stimulus is important to ensure eminent success. There are already devices that rouse these nerves today, such as pace-makers and defibrillators, but they are very macro in their result. The electronic medicine will influence the whole network of nerve fibers.

Bioelectronic Medicines

Researches in bioelectronic medicine have generated numerous crucial boundaries that can be presented to a precise nerve without destroying it. The technology may listen in on a specific nerve without being unsettling to the objective of bioelectronic nerve devices. It is clear that the cost of generating such practical solution is carefully measured. Accomplishment is important if a microscopic object is generated and presented on numerous marginal nerves in a minimally invasive manner by most reasonable means. This will allow droves of patients around the world at a cost that works for the entire healthcare system. Generating exact, bioelectronic medicines with advanced resolution, and closed loop arrangements will be provided to patients internationally, hopefully as the predicted and in the time expected.

Development in this field will endure the tests of time, the cost involved as well as the challenges of both the older medical device companies, and innovative participants in pharma and biotech fields. Experts, an specialists trust that the method to Bioelectronic growth is very convincing to patients to receive innovative cure for painful diseases.

Integrating 3D Printing into Product Development Lifecycle

Additive Manufacturing, Bioprinting, is rapidly reconfiguring the corporations' need in numerous design and development products. Although there's an important amount of hyped publicity regarding the technology there are also numerous amount of genuine details to incorporate 3D Bioprinting into product development lifecycle as stated below:

1. Timelines and finances are the two major contests to any product expansion team. The Once the *request* is made, Figure 6.6, there is no technology is well suited than 3D Bioprinting to meet the requestor demand. The digital model is created using CAD the necessary scaffolding technique together with the slicing layers the 3D Bioprinter can provide real life object considering the Bioink is available.
2. Touch and feel is significant constituent to evaluating a prototype, to *discover* any flaws to improve model development. With a multi-material 3D Bioprinted prototype, allow a product to be felt and measure its weight and regulate if it will fit with previously generated constituents. With a 3D printed prototype any pattern or tooling path can be generated from a CAD model to deliver Bio-fabricated object as expected.
3. Reiterate on conceivable design concepts before the final *composition* of the Bioprinted product goes to full production. This earns resourcefulness and an exclusive aptitude to connect the cost of development. From project drafts to project execution, an entirely practical CAD model is created. The 3D Bioprinting can provide the visual illustration of the perception into real entity that may aid sell the novel Bioprinted product.

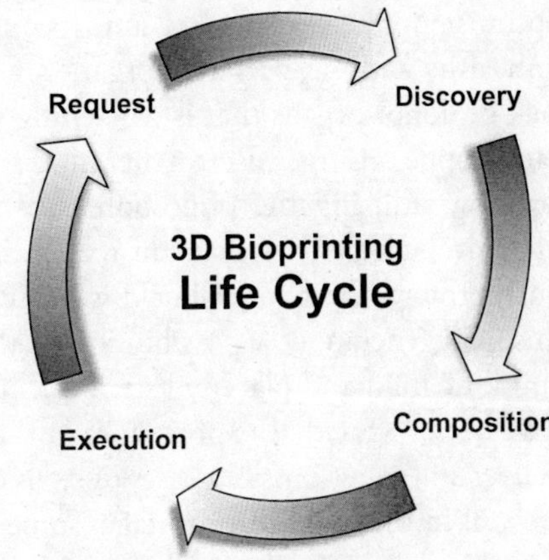

Figure 6.6: Bioprinting Life Cycle

Pioneering the Bioprinting Technology

Ever since the beginning of 3D Bioprinting in 1896, heart surgery has advanced into a very successful arena with the foundation of novel treatment choices for "formerly" terminal diseases. The formation of these new cures stimulated the discovery and development of novel technology (e.g. the heart–lung machine), new surgical materials and devices (e.g. suture material, grafts, mechanical and biological valves), in addition to the development of new medications (e.g. heparin and cyclosporine), and the expansion of new surgical procedures by groundbreaking surgeons (e.g. cross-circulation, aortic surgery and heart transplantation).

Beginning with the first productive suture of a laceration of the heart in 1896, added developments trailed, surrounding numerous fields and proposing managing solutions for formerly uncontrollable cardiovascular diseases. Inventions contain the following land marks:

- Suture for a stab wound of the heart, Germany, 1896.
- Ligature of the patent ductus, USA, 1938.
- Resection of a coarctation of the aorta, Sweden, 1944.
- Palliative oxygenation in 'blue babies,' USA, 1944
- First utilization of the heart–lung machine, USA, 1953
- Cross-circulation, USA, 1954
- Valve replacement USA, 1960
- Coronary artery bypass grafting, Russia, 1964
- Coronary artery bypass grafting, USA, 1967
- Left heart assist device, USA, 1966
- Heart transplantation South, Africa, 1967
- Heart transplantation, USA, 1968
- Total artificial heart, USA, 1969

Bioprinting - Novel Therapies

New treatments resulting from re-creative medicine and tissue engineering technology may deliver new optimism for patients with wounds, end-stage organ failure, or other medical issues. Presently, patients with unhealthy and incapacitated organs are frequently cured with transplanted organs. Nonetheless, there is a lack of donor organs that is worsening yearly as the people ages and as the number of new cases of organ catastrophe upsurges. Researchers and scientists in the field of regenerative medicine and tissue engineering are now utilizing the philosophies of cell replacement, material science, and bioengineering to build biological alternatives that can reestablish and uphold usual function in unhealthy and wounded tissues. Furthermore, the stem cell field is a fast progressing part of regenerative medicine, and new innovations in this field generate new possibilities for this kind of treatment. For example, new sorts of stem cells, such as amniotic fluid and placental stem cells that can avoid the ethical issues related with embryonic stem cells, have been revealed. The procedure of therapeutic cloning and the generation of prompted pluripotent cells deliver still other possible bases of stem cells for cell-based tissue engineering applications. While stem cells are still in the investigation stage, some treatments ascending from tissue engineering activities that make use of autologous adult cells have already arrived the clinical set, representing that recreating medicine grasps much potential for the year to come.

The Advent of Organ Transplant

Patients' pain from unhealthy and injured organs is often cured with transplanted organs, and this action has been in practice for over 50 years. In 1955, the kidney development was the first entire organ to be substituted in a human, when transplanted organ between identical twins. Several years later, surgeon performed genetically different tissues of kidney transplant from a non-genetically identical patient into another. This transplant, conquered the immunologic obstacle manifest a novel era in medicine, and unlocked the door for utilizing transplantation, as a resource of treatment for dissimilar organ systems.

As current medicine upsurges the human lifetime, the elderly population propagates, and the need for donor organs raise with it, because elderly organs are usually more disposed to failure. Nevertheless, there is now a grave lack of donor organs, and numerous patients in desperate needs of organs will die though waiting for transplants. In addition, even if an organ turn out to be obtainable, rejection of organs is quiet a main difficult in transplant patients notwithstanding developments in the approaches utilized for the inhibition of the immune response subsequent the transplant process. Even if rejection does not happen, the need for lifelong use of the inhibition of the immune response treatments leads to several complications in these patients.

The Advent of Tissue Engineering

These complications have led doctors and scientists to find novel grounds for substitutes to organ transplantation. In the 1960s, a normal development happened in which scientists started to associate new instruments and materials sciences with cell biology, and a novel field that is now labeled "tissue engineering" turn out to be as the break of a dawn after long night. As further researchers from diverse fields emanated together with the shared objective of tissue replacement the field of tissue engineering grew into more formally recognized.

Tissue engineering is today a distinct as it may inter related fields which implement the ideologies of engineering and life sciences to the advance of biological alternates that desire to uphold, reinstate or progress tissue function. At that moment, afterward the detection of human stem cells in the early 1980s, the field of stem cell biology acquired form and proposed that it may one day be conceivable to gain and utilize donor stem cells in tissue engineering plans, or maybe even revitalize endogenous stem cells and utilize them to rejuvenate deteriorating organs in adult patients, Figure 6.7.

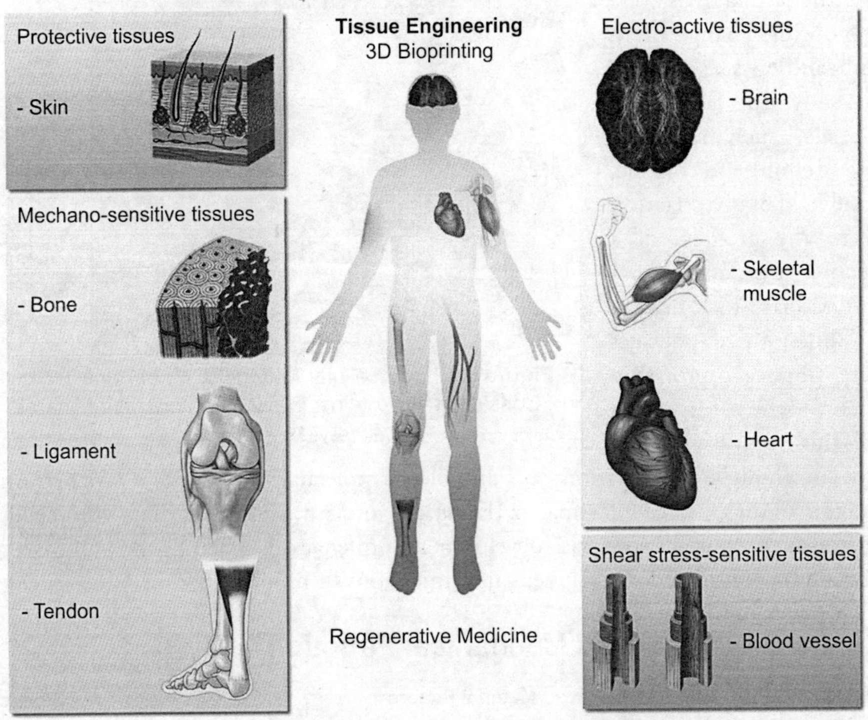

Figure 6.7: Tissue Engineering – Regenerative Medicine
Courtesy: Elsevier

The arenas of stem cells, cell transplantation, and tissue engineering all have one amalgamating concept- the rejuvenation of living tissues and organs. Thus, in 1999, the term regenerative medicine was coined. Researchers and scientists have endeavored to engineer effectively all tissue of the human body.

Regenerative Medicine Strategies

The field of regenerative medicine comprises numerous parts of technology, such as tissue engineering, stem cells, Bioprinting, and cloning. Tissue engineering, one of the leading areas of regenerative medicine, trails the values of cell transplantation, Bioprinting, materials science, and engineering in the course of the development of biological alternates that can reinstate and preserve regular function, Figure 6.7.

Tissue engineering approaches frequently classified into two classes:

a. the utilization of acellular scaffolds, which is contingent on the body's normal capacity to restore for good orientation and way of novel tissue development, and
b. the utilization of scaffolds sowed with cells. Scaffolds are characteristically initiated by manufacturing artificial scaffolds or by eliminating cellular constituents from tissues through mechanical and chemical influence to yield "acellular," collagen-rich matrices. These matrices incline to gradually worsen on implantation and are normally replaced by the extra-cellular-matrix (ECM) proteins that are secreted by the in-growing cells, Figure 6.8. Cells can also be utilized for treatment through inoculation, also with carriers such as hydrogels or alone.

Biomaterials in Regenerative Medicine

In the past, synthetic materials were presented to Bioprint, change or to reconstruct unhealthy tissues or parts in the human body. The fabrication of new materials, such as tetra-fluoro-ethylene (Teflon) and silicone, unlocked a novel field of investigation that directed to the progress of a extensive selection of devices that could be functional for human utilization. While these devices could deliver physical provision or replacement, the functional constituent of the innovative tissue was not reinstated. Though, investigations in cell biology, molecular biology, and biochemistry permitted a better considerate of the extra-cellular-matrix (ECM), Figure 6.8, and its interface with cells in the tissues of the body, as well as communications with development influences and their ligands, and as a consequence, new biomaterials were considered with these communications in mind.

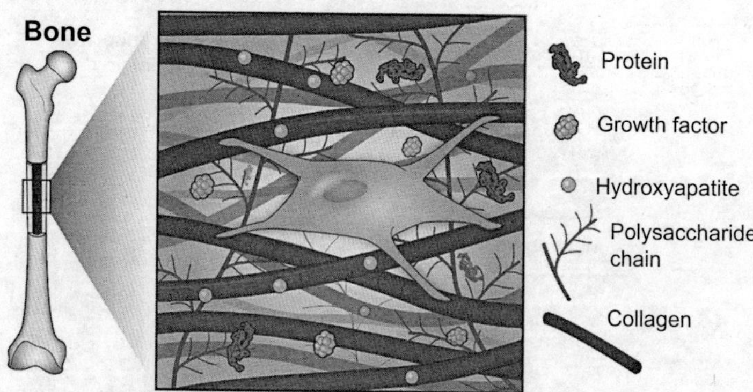

Figure 6.8: Extracellular Matrix (ECM) of Bone Tissue - Consisting of Proteins, Growth Factors, and Minerals
Courtesy: Semantic Scholar

Replicating the Biological and Mechanical Function

In tissue engineering, biomaterials reproduce the biological and mechanical function of the natural ECM found in tissues in the body. Biomaterials deliver a 3D space in which cells can confer, breed, and procedure novel tissues with suitable assembly and function. They also permit for the distribution of cells and proper bioactive factors (e.g., cell adhesion peptides, development factors) to anticipated locations in the body. Because most mammalian cell kinds are anchorage-dependent and will decease if no cell-adhesion substrate is obtainable, biomaterials deliver this substrate whereas permitting transfer of cells with high loading competence. Biomaterials can also deliver mechanical provision contrary to *in vivo* forces so that the predefined 3D assembly of a tissue-engineered organ is upheld throughout tissue expansion.

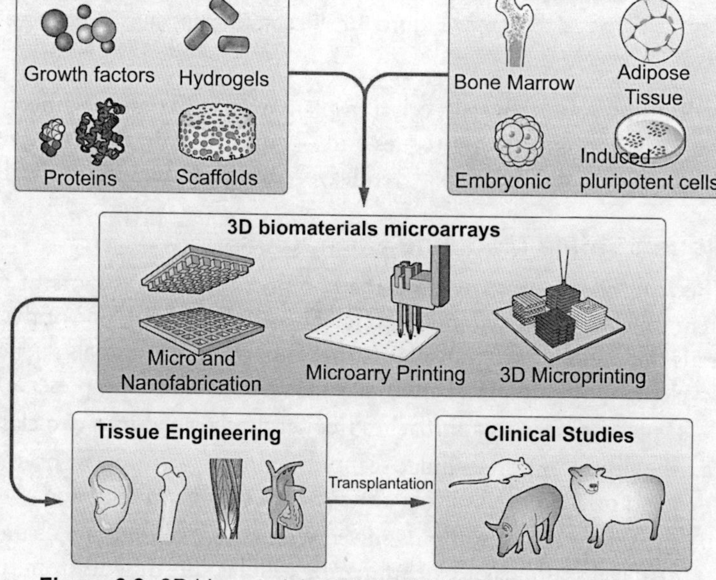

Figure 6.9: 3D biomaterials Utilizing Material Factors Technology, Stem Cell Technology, 3D Bioprinting Materials, Tissue Engineering, and Clinical Studies
Courtesy: Semantic Scholar

The model biomaterial should be environmentally safe and bioresorbable to upkeep the replacement of ordinary tissue without persuading swelling. Mismatched materials are destined for an inflammatory or foreign-body reaction that ultimately leads to rejection or cell death. Because biomaterials provide temporary mechanical support though the cells experience 3-D rearrangement into tissue, a correctly selected biomaterial should permit the engineered tissue to uphold satisfactory mechanical reliability to sustain itself in early growth, whereas in late expansion, it should have begun deprivation such that it does not delay further tissue development. The degraded products, if formed, should be removed from the body through metabolic paths at a passable rate to safeguard that the concentration of these deprived products in the tissues rests at an acceptable range.

Normally, three categories of biomaterials have been exploited for engineering tissues:

- normally subsequent materials (e.g., collagen and alginate),
- acellular tissue matrices (e.g., bladder submucosa and small intestinal submucosa), and
- Artificial polymers such as polyglycolic acid (PGA), polylactic acid (PLA), and poly (lactic-co-glycolic acid) (PLGA).

These classes of biomaterials have been verified regarding their biocompatibility. Unsurprisingly resulting materials and acellular tissue matrices have the likely gain of biological acknowledgment. Nonetheless, synthetic polymers can be shaped reproducibly on a large scale with controlled characteristics such as strength, degradation rate, and microstructure.

Cell Therapy - Native Cells

Once inherent cells are utilized for tissue engineering, a minor portion of donor tissue is detached into discrete cells. These cells are extended in culture and either injected directly back into the host or attached to a support matrix and then re-implanted.

The basis of donor tissue can be heterologous (such as bovine), allogeneic (similar kind, diverse individual), or autologous. The favored cells to use are autologous cells, wherever a biopsy of tissue is found from the host, the cells are detached and extended in culture, and the extended cells are implanted into the same host. The use of autologous cells, while it may result an inflammatory reaction, evades rejection, and thus the deleterious side effects of immunosuppressive medications can be circumvented.

Preferably, both physical and functional tissue substitution will happen with negligible difficulties when autologous inherent cells are utilized. Though, one of the restrictions of applying cell-based regenerative medicine methods to organ replacement has been the inherent strain of rising precise cell kinds in great amounts. Even once some organs, such as the liver, have a high regenerative volume *in vivo*, cell development and development *in vitro* may be hard. By investigating the advantaged sites for dedicated precursor cells in exact organs, as well as discovering the circumstances that encourage differentiation, one may be capable to overcome the complications that restrict cell development *in vitro*. For example, urothelial cells could be grown-up in the laboratory set in the past, but only with restricted growth.

Numerous procedures were advanced over the past two decades that recognized the undifferentiated cells and reserved them undifferentiated throughout their development stage. With the use of these approaches of cell culture, it is now likely to grow a urothelial strain from a single sample that originally covered a surface area of 1 cm^2 to one covering a surface area of 4,202 m^2- the corresponding of one football field - within 8 weeks. These studies designated that it should be likely to gather autologous bladder cells from human patients, grow them in culture, and return them to the donor in adequate amounts for reconstructive objectives.

Main improvements have been accomplished within the past decade on the likely growth of a diversity of prime human cells, with exact methods that style the use of autologous cells for clinical application possible.

Most existing approaches for tissue engineering hinge on a sample of autologous cells from the unhealthy organ of the host. Though, for various patients with widespread end-stage organ failure, a tissue culture may not harvest sufficient regular cells for growth and transplantation. In other occurrences, chief autologous human cells cannot be prolonged from a specific organ, such as the pancreas. In these circumstances, stem cells are intended as being an substitute foundation of cells from which the anticipated tissue can be resulting. Stem cells can be resulting from unwanted human embryos, human embryonic stem cells, from fetal tissue, or from adult sources (bone marrow, fat, skin), Figure 6.10.

Human Embryonic Stem Cells

As for any treatment of organ disorder of clinical medicine, stem cell usage requires an extensive consideration of the underlying pathophysiology of the illness. It is significant that medical professionals who embrace these new approaches into their professional duties to hold a reasonable comprehension of the normal course of the targeted illness. It is also vital for allowing medical professionals to have a heightened understanding of diverse cell sources, with investigation of the stated population of native stem and progenitor cells, Figure 6.10.

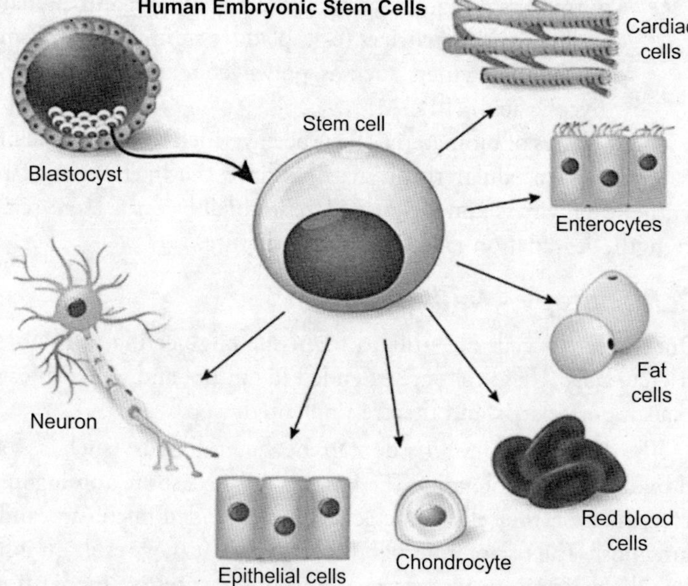

Figure 6.10: Human Embryonic Stem Cells

Human embryonic stem (hES) cells display two extraordinary characteristics:

- the capability to multiply in an undifferentiated but pluripotent state (self-renewal), and
- the capability to differentiate into numerous specialized cell kinds.

They can be quarantined by pronouncing the inner cell mass from the embryo during the blastocyst stage (5 days post-fertilization) and are typically developed on feeder layers containing of mouse embryonic fibroblasts or human feeder cells, Figure 6.10. Additional current accounts have revealed that these cells can be full-grown deprived of the utilization of a feeder layer and thus evade the disclosure of these human cells to mouse viruses and proteins.

These cells have established endurance in culture by preserving their undifferentiated state for at least 80 passages once developed by use of present available procedures. In addition, hES cells are capable to differentiate into cells from all three embryonic germ layers *in vitro*. Skin and neurons have been shaped, representing ectodermal differentiation. Blood, cardiac cells, cartilage, endothelial cells, and muscle have been shaped, demonstrating mesodermal differentiation. Pancreatic cells have been shaped, representing endodermal differentiation. Furthermore, as additional proof of their pluripotency, embryonic stem cells can shape embryoid bodies, which are cell accumulations that encompass all three embryonic germ layers whereas in culture and can form teratomas *in vivo*. Though, there are many ethical and religious alarms related with hES cells since embryos are destroyed in order to get them. Thus, the use of these cells is presently forbidden in various countries.

Stem Cells from Somatic Cell Nuclear Transfer

Stem cells for tissue engineering might also be created through cloning processes. There has been substantial attention in the field of nuclear cloning since the birth of the cloned sheep Dolly in 1997, but essentially, Dolly was not the initial animal shaped by using nuclear transfer, Figure 6.11.

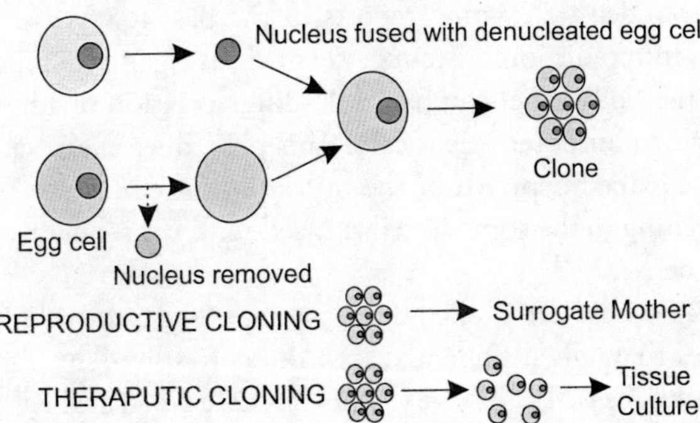

Figure 6.11: Stem Cells from Somatic Cell Nuclear Transfer
Courtesy: Royal Society of Biology

In fact, frogs were the original positively cloned vertebrates resulting from nuclear transfer. Though, in the frog trial, the nuclei utilized for cloning were resulting from non-adult bases. In fact, live lambs were resulted in 1996 by utilizing nuclear transfer as well, but they were manufactured from differentiated epithelial cells resulting from embryonic discs. The implication of Dolly was that she was the original mammal to be resulting from an adult somatic cell by use of nuclear transfer. Since then, animals from numerous types have been fledged by utilizing nuclear transfer technology, as well as cattle, goats, mice, and pigs.

Two kinds of nuclear cloning, reproductive cloning and therapeutic cloning, have been designated, and a better appreciation of the alterations between the two kinds may aid to ease some of the dispute that surroundings these technologies, Figure 6.11.

Barred in most countries for human solicitations, reproductive cloning is utilzed to harvest an embryo that has the equal genetic material as its cell source. This embryo can then be implanted into the uterus of a female to deliver augmentation to an infant that is a clone of the donor. Alternatively, therapeutic cloning is utilized to harvest early stage embryos that are explanted in culture to yield embryonic stem cell lines whose genetic material is matching to that of its source.

These autologous stem cells have the likelihood to develop virtually any kind of cell in the adult body, and therefore would be valuable in tissue and organ substitution practices. Consequently, therapeutic cloning, which has also been named somatic cell nuclear transfer, can bring another basis of transplantable cells. As stated by data from the "Centers for Disease Control CDC" and Prevention, a projected 3,000 Americans die every day of illnesses that might have been cured with stem cell-derived tissues. With present allogeneic tissue transplantation procedures, elimination is a recurrent impediment due to immunologic mismatch, and immunosuppressive drugs are characteristically obligatory. The utilization of transplantable tissue and organs subsequent from therapeutic cloning could lead to the evasion of immune reactions that characteristically are interrelated with transplantation of non-autologous tissues, Figure 6.11.

Though hopeful, somatic cell nuclear transfer technique has definite restrictions that necessitate supplementary investigation before this method can be used extensively in tissue or organ replacement treatment. Initially, the competence of the cloning procedure is very low, as showed by the detail that majority embryos subsequent from the cloning procedure do not live. To recover cloning competence, additional developments are essential in various intricate steps of nuclear transfer, such as the enucleation procedure for oocytes, the definite transfer of a nucleus to this enucleated oocyte, and the stimulation procedure that initiates the cloned oocytes to commence dividing. Furthermore, cell cycle synchronization between donor cells and recipient oocytes must be achieved.

Reprogramming and Generation of Induced Pluripotent State (iPS) cells

Surrounded by exciting reports in the past few years, regarding the positive revolution of adult somatic cells into pluripotent stem cells concluded genetic "reprogramming" have been available. Reprogramming is a method that encompasses de-differentiation of adult somatic cells (such as fibroblasts) to yield patient-specific pluripotent stem cells. This procedure is especially inspiring since it permits pluripotent stem cells to be gained deprived of the utilization of embryos. Also, cells created by reprogramming are genetically matching to the somatic cells utilzed (and consequently to the patient who donated these cells) and should not be rejected.

24 genes were studied that were supposed to be significant for embryonic stem cells and recognized 4 key genes that, when presented into the correspondent fibroblasts through retroviral vectors, caused in drug-resistant cells. These were Oct3/4, Sox2, c-Myc, and Klf4. The subsequent iPS cells controlled the illustrious development features of self-renewing embryonic stem cells, articulated genes precise for embryonic stem cells, and produced embryonic bodies *in vitro* and teratomas *in vivo*.

Once iPS cells were inserted into mouse blastocysts, they donated to a diversity of cell kinds. On the other hand, while iPS cells designated in this method were pluripotent, they were not indistinguishable to embryonic stem cells. Contrasting embryonic stem cells, chimeras made from iPS cells did not effect in full-term pregnancies. Gene appearance shapes of the iPS cells presented that they influenced a separate gene appearance signature that was dissimilar from that of embryonic stem cells. Furthermore, the epigenetic state of the iPS cells was anywhere between that originate in somatic cells and that originate in embryonic stem cells, signifying that the reprogramming was unfinished.

These consequences were better expressively sated in July 2007. Fibroblasts were infected with retroviral vectors and designated for the initiation of endogenous Oct4 or Nanog genes. Consequences from this investigation presented that DNA methylation, gene appearance outlines, and the chromatin state of the reprogrammed cells were comparable to those of embryonic stem cells. Teratomas encouraged by these cells incorporated differentiated cell kinds demonstrating all three embryonic germ layers. Most prominently, the reprogrammed cells from this trial could form viable chimeras and give to the germline-like embryonic stem cells, signifying that these iPS cells were totally reprogrammed. It was perceived that the number of reprogrammed colonies enlarged when drug choice was introduced later (day 20 rather than day 3 post-transduction). This proposes that reprogramming is a slow and steady development and may clarify that preceding efforts caused in partial reprogramming.

Reprogramming of Human Cells

It has lately been discovered that reprogramming of human cells is expected. Human iPS cells were formed that equivalent to hES cells in terms of morphology, multiplying, gene expression, surface markers, and teratoma expansion. It was made known that retroviral transduction of the stem cell indicators OCT4, SOX2, NANOG, and LIN28 could generate pluripotent stem cells. Though, in both research, the human iPS cells were alike but not indistinguishable to hES cells.

While reprogramming is an reassuring phenomenon, the restricted knowledge of the mechanism fundamental it presently restricts the clinical applicability of the method, but the future possible of reprogramming is fairly hopeful.

Amniotic Fluid and Placental Stem Cells (AFPSC)

An alternative basis of stem cells is the amniotic fluid and placenta. Amniotic fluid and the placenta are identified to comprise numerous partially differentiated cell types resulting from the emerging fetus. Stem cell populations were secluded from these sources, called amniotic fluid and placental stem cells (AFPSC) that shows embryonic and adult stem cell markers. The undifferentiated stem cells increase widely short of feeders and double every 36 hours. Contrasting hES cells, the AFPSC do not form tumors *in vivo*. Lines upheld for over 250 population doublings reserved long telomeres and a standard karyotype.

AFS cells are generally multipotent. Clonal human lines confirmed by retroviral marking can be prompted to differentiate into cell types demonstrating each embryonic germ layer, containing cells of adipogenic, osteogenic, myogenic, endothelial, neuronal, and hepatic lineages. In this respect, they meet a normally known condition for pluripotent stem cells, deprived of implying that they can produce every adult tissue. Samples of differentiated cells resulting from AFS cells and showing dedicated functions comprise neuronal lineage cells secreting the neurotransmitter L-glutamate or expressing G-protein-gated inwardly rectifying potassium (GIRK) channels, hepatic lineage cells creating urea, and osteogenic lineage cells making tissue engineered bone. The cells could be found either from amniocentesis or chorionic villous selection in the emerging fetus, or from the placenta at the time of birth.

The cells could be well-preserved for self-use and utilized without rejection, or they could be banked. A bank of 100,000 specimens could possibly provide 99% of the US population with a perfect genetic match for transplantation. Such a bank may be simpler to create than with other cell sources, since there are around 4.5 million births per year in the USA.

Adult Stem Cells

Adult stem cells, particularly hematopoietic stem cells, are the greatest implicit cell type in stem cell biology. The occurrence of stem cells in the adult was initially distinguished though exploring the mechanisms by which the bone marrow could rejuvenate after exposure to radiation. Nevertheless, adult stem cell investigation rests an area of concentrated study, since their possible for treatment may be appropriate to a countless of deteriorating disorders, Figure 6.12.

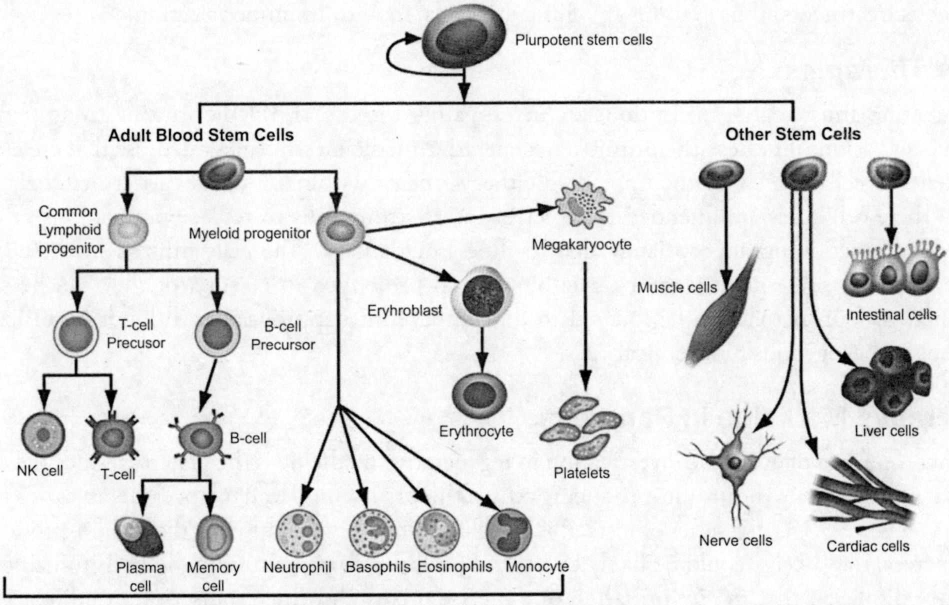

Figure 6.12: Adult Stem Cells

Courtesy: Maharaj Institute of Immune Regenerative Medicine

Throughout the past decade, adult stem cell populations have been established in numerous adult tissues other than the bone marrow and the gastrointestinal tract, counting the brain, skin, and muscle. Numerous other kinds of adult stem cells have been recognized in organs all over the body and are believed to assist as the main repair entities for their corresponding organs. The detection of such tissue-specific progenitors has unlocked up novel opportunities for investigation.

A distinguished allowance to the tissue-specificity of adult stem cells is the mesenchymal stem cell (MSC), also recognized as the multi-potent adult progenitor cell. This cell type is resulting from bone marrow stroma. Such cells can differentiate *in vitro* into many tissue kinds and can also differentiate developmentally if inserted into a blastocyst. Multi-potent adult progenitor cells can grow into a diversity of tissues containing neuronal, adipose, muscle, liver, lungs, spleen, and gut tissue, but particularly not bone marrow or gonads.

Moreover, stem cells resulting from adipose tissue may also be an autologous and self-rejuvenating cell basis. Adipose-derived stem cells (ADSCs) have been revealed to differentiate into a diversity of cell phenotypes, and meanwhile they are easily attained, they display excessive potential for future kinds of reconstructive surgery founded on tissue engineering and there have been numerous medical trials utilizing these cells.

Investigation into more differentiated kinds of adult stem cells has, though, advanced gradually, mostly because scientists have had excessive struggle in upholding adult non-mesenchymal stem cells in culture. More or less cells, such as those of the liver, pancreas, and nerve, have very low proliferative ability *in vitro*, and the functionality of some cell kinds is condensed after the cells are refined. Separation of cells has also been difficult, since stem cells are existing in exceptionally small numbers in adult tissue. Though the clinical usefulness of adult stem cells is presently restricted, great possible occurs for upcoming utilization of such cells in tissue-specific regenerative treatments. The benefit of adult stem cells is that they can be utilized in autologous treatments, thus evading any difficulties related with immune rejection.

Cellular Therapies

Several scientific and research institutions are investigating the use of 3D Bioprinting using stem cells as Bioink to treat various illnesses. The primitive regenerative medicine strategies are those that are established on the events of cells, which can be implanted either alone or within a kind of carrier material, such as a hydrogel. These cell cures are intended to inject or insert strong cells to replace populations of cells that are no longer functioning appropriately due to illness or damage. The cells utilized in these cures can be autologous cells subsequent from a tissue biopsy and prolonged in culture, or they can be stem cells from numerous sources that can be focused to differentiate into appropriate cell kinds by utilizing both endogenous and exogenous biochemical cues.

Regenerative Medicine in Pancreas

For instance, one area of extensive investigation in regenerative medicine is the pancreas, since the capability to substitute or regenerate the insulin-producing cells of this organ could lead to novel therapies or a treatment for diabetes, Figure 6.13. In a series of tests, it is established that regeneration of the insulin-producing cells of the pancreas, the β-cells, could be likely by utilizing cellular reprogramming methods utilizing a mouse model, they displayed that *in vivo* initiation of a precise mixture of three transcription influences (Ngn3, Pdx1, and Mafa) by use of adenoviral vectors led to the reprogramming of adult differentiated pancreatic exocrine cells into cells that carefully bear a resemblance to β-cells.

These cells were alike to inherent β-cells in size, shape, and ultrastructure, and they conveyed genes that are specific to β-cells as well. Remarkably, these cells secreted insulin and stated vascular endothelial development influence (VEGF), which permitted them to pattern the local vasculature in a way like to inherent β-cells. In fact, these reprogrammed cells were capable to partly ameliorate hyperglycemia in diabetic mice, signifying that reprogramming methods for curing illness may one day develop a reality.

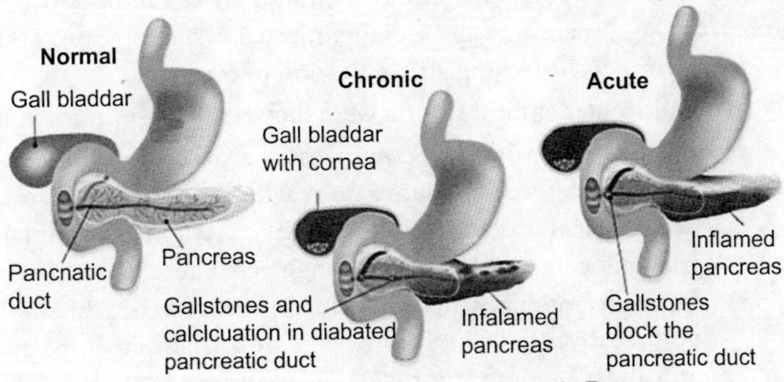

Figure 6.13: Normal, Acute, and Chronic Pancreas
Courtesy: Libra, Inc.

Degenerative Muscle Diseases

Degenerative muscle illnesses such as Duchenne's muscular dystrophy have overwhelming influences on value of life. To date, these genetic ailments have no appropriation cure. Early eagerness for gene treatment involvements has been mitigated by subjects of vector toxicity and insufficient gene transmission to mark muscle cells *in vivo*. Though, normal mechanisms of muscle reparation have recommended that cell-based treatment could yield gain of normal homing mechanisms to guide cells to the correct site. Several research and scientific institutions are investigating the use of 3D Bioprinting using stem cells as Bioink to treat degenerative muscle illnesses.

Trials utilizing the mdx mouse model, in which the dystrophin gene is mutated, designate that injection of usual muscle precursors and dermal fibroblasts into skeletal muscle may cause augmented appearance of dystrophin and enhanced functional results. Though, this cure choice necessitates additional investigations before it can be extensively used in the clinic. While numerous of these cell treatments are still in the trial phase, some are being rendered to the clinic and clinical trials are being executed.

Urine Flows Backwards

Vesicoureteral reflux (VUR; a disorder in which urine flows backwards from the bladder into the ureter and kidney) and stress urinary incontinence are two urologic circumstances that can effect from dysfunction of a specific sphincter muscle. Once severe, these circumstances are restored surgically. Though, cell-based treatments for both VUR and incontinence would be an significant substitute to surgical reparation of these circumstances. Preferably, such a treatment would be simply directed by injection and well endured by the patient. The injectable treatment must be non-antigenic, non-migratory, volume stable, and harmless for human use, and furthermore, it must be capable to transmit cells and help as a matrix *in vivo*.

Toward this purpose, long-term investigations were showed to regulate the properties of injectable chondrocytes for the behavior of VUR *in vivo*. Chondrocytes were selected since the use of autologous cartilage for the cure of VUR in humans would content all of the necessities for an ultimate injectable cell-based treatment. Chondrocytes resulting from an ear biopsy can be willingly full-grown and long-drawn-out in culture. Neo-cartilage development can be attained *in vitro* and *in vivo* by utilizing chondrocytes cultured on synthetic biodegradable polymers. In the VUR trials, chondrocytes were suspended in an alginate matrix and injected at the vesicoureteral sphincter. Early enough, regular cartilage substituted the alginate as the alginate gradually degraded. This system was then modified for the treatment of VUR in a porcine model.

These investigations demonstrates that chondrocytes can be effortlessly collected and joint with alginate *in vitro*, that the suspension can be easily injected cystoscopically, and that the elastic cartilage tissue shaped can remedy the VUR lacking any indication of obstacle.

Two multicenter clinical studies were showed by use of this engineered chondrocyte technology.

- First, patients with VUR were treated at 10 centers throughout the United States. The patients had a comparable achievement rate as with other injectable materials in terms of treatment. Cartilage development was not eminent in patients with treatment failure. Patients who were treated possibly had a biocompatible region of engineered autologous tissue present.
- Secondly, patients with urinary incontinence were treated endoscopically with injected chondrocytes at three diverse medical centers. Phase 1 trials presented an approximate achievement rate of 80% at 3 and 12 months postoperatively.

3D Bioprinting Organs

Tissue engineering plans are frequently mentioned to as "growing organs in the laboratory." In these plans, differentiated cells or stem cells are sown onto a biomaterial scaffold and this build is permitted to be established *in vitro* in a bioreactor for a short time prior embedding *in vivo*. These builds are intended to substitute a fabricating organ in its wholeness. In current years, it has been revealed that hollow organs, such as the urinary bladder, urethra, and blood vessels, can be effectively engineered in the laboratory, and these achievements are designated below, Figure 6.14.

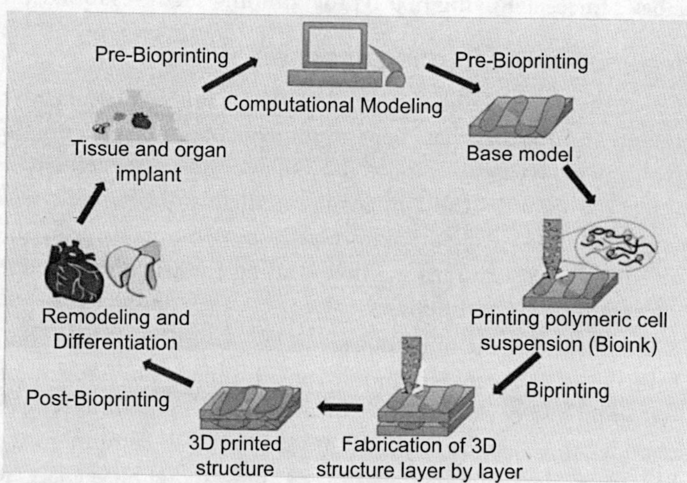

Figure 6.14: Schematic Representation of 3D Organ Bioprinting
Courtesy: Semantic Scholar

The urethra can be mended by utilizing tissue-engineered implants in numerous methods. It has been presented that numerous biomaterials deprived of cells, such as PGA and acellular collagen-based matrices from small intestine and bladder, can be utilized experimentally (in animal patterns) for the recreation of urethral tissue. Acellular collagen matrices resulting from bladder submucosa have been utilized experimentally and clinically.

In animal investigations, sections of the urethra were resected and substituted with acellular matrix implants in an "onlay" fashion. Histological checkup exhibited comprehensive epithelialization and advanced vessel and muscle permeation, and the animals were capable to invalid through the neo-urethras. These consequences were established in a scientific investigation of patients with hypospadias and urethral stricture illness. Decellularized cadaveric bladder submucosa was utilized as an only matrix for urethral restoration in patients with stricture illness and hypospadias.

3D Printing Tissue Engineered Bladder

Lately, tissue-engineered urethras that had been made from patients' distinct cells were utilized for tubularized urethral reconstruction. In this initial analysis, five youngsters who had urethral failings were chosen to receive treatments. A tissue biopsy was removed and cultured from each individual patient, and the muscle and epithelial cells subsequent from the biopsy specimen were stretched and planted onto tubularized polyglycolic acid: poly(lactide-co-glycolide acid) scaffolds to make neo-urethras for inserting. The patients then went through urethral reconstruction with the refabricated urethras.

Subsequent surgery, these patients were followed for up to 6 years. All of the patients recognized an upsurge in urinary movement rate, and sequential radiographic and endoscopic examination displayed that they well-maintained extensive urethral qualities all over the continuation period and did not advance attacks. Furthermore, urethral biopsies were attained in these patients, and these shown that by 3 months post-surgery, the Bio-fabricated implantations had advanced to an regular performance tissue architecture involving of a urothelial layer enclosed by a well-built layer.

Moreover, analogous methods have been utilized to make tissue-engineered bladder builds. Urothelial and muscle cells can be extended *in vitro*, embedded onto polymer scaffolds, and permitted to confer and form areas of cells. These values were utilized in some investigations in which tissue-engineered bladders were utilized to reestablish subtotal cystectomies in beagle dogs. The first medical practice in which engineered bladder tissue for cystoplasty rebuilding in patients was accomplished in early1999.

Also, a minor trial search of seven patients was documented, utilizing a collagen scaffold sowed with cells either with or without omentum sheet of fatty tissue, or a joined PGA-collagen scaffold sowed with cells and omentam sheet. The patients rebuilt with the bio-engineered bladder tissue fashioned with the PGA-collagen cell-sowed scaffolds exhibited augmented acquiescence, reduced end-filling burdens, augmented volumes, and lengthier dry times. While the knowledge is hopeful in terms of performance that engineered tissues can be inserted safely, it is just a jump start in terms of achieving the aim of bio-engineering completely well-designed bladders. Additional trials and scientific effort is being showed.

3D Bioprinting Bioink -Xenogenic Materials Replacing Blood Vessels

Xenogenic or artificial materials have been utilized as substitute to blood vessels for intricate cardiovascular cuts. Nonetheless, these materials characteristically absence of growth possibility, and may put the receiver in jeopardy for disorders such as stenosis - a constriction or narrowing of a duct, thromboembolization, or septicity. Tissue-engineered vascular grafts have been built by utilizing autologous cells and ecological scaffolds and have been utilized in dog and lamb models. The significant benefit of utilizing these autografts is that they degenerate *in vivo* and therefore make the novel tissue to be absent of the poor effect of foreign material. Conversion of these methods from the laboratory to the medical setting has commenced, with autologous vascular cells garnered, extended, and planted onto an ecological scaffold. The autologous build was utilized to substitute a stenosis pulmonary artery that had been formerly mended. Seven months after embedding, no indication of graft obstruction or aneurysmal deviations was recorded in the receiver. Moreover, additional group generated tissue-engineered blood vessels by utilizing the cell-sheet multilayer technique and then utilized these builds to ssuccessfully generate vascular contact points for hemodialysis in 10 patients. Lately, the same group finished a larger investigation on these engineered vessels for hemodialysis, which showed that the 1-month and 6-month patency of the grafts was 78% and 60%, respectively, which meets the permitted criteria for a high-risk patient group.

Congenital Tracheal Pathology

Few clinical treatment choices are presently obtainable for patients who feel pain from severe congenital tracheal pathology, such as stenosis - a constriction or narrowing of a duct, atresia - the often hereditary absence of a usual body opening, and agenesis - the incomplete development or total absence of a body par, due to the restricted obtainability of autologous transplantable tissue in the neonatal period. Tissue engineering in the fetal period may be a feasible substitute for the surgical cure of these prenatally identified congenital irregularities, since cells could be harvested and fully-fledged into transplantable tissue in analogous with the rest of maturation. Chondrocytes from both elastic and hyaline cartilage samples have been reaped from fetal lambs, extended *in vitro*, and then dynamically sown onto ecological scaffolds. The builds were then imbedded as substitute to tracheal tissue in fetal lambs. The resultant tissue-engineered cartilage was distinguished to undergo engraftment and epithelialization, though upholding its physical support and patency, Figure 6.15.

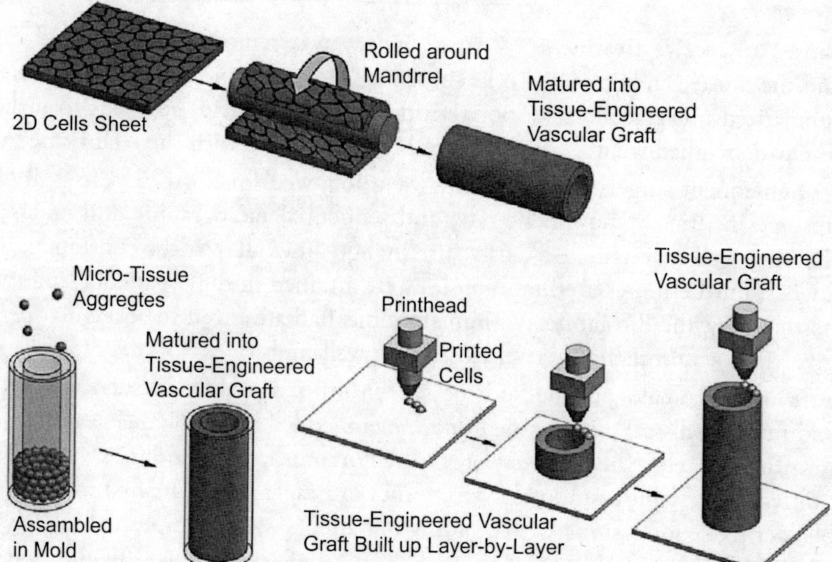

Figure 6.15: Tissue Engineered Vascular Graft Cell and Supporting Material are Deposited in a Layer-by-Layer Manner, Building up a 3D Construct

Courtesy: Semantic Scholar

Lately, this technology progressed into a human patient with end-stage airway illness. The cellular material and MHC antigens was detached from a human donor trachea and, utilizing a particular bioreactor to seed acellular matrix with chondrocytes and epithelial cells resulting from the patient to obtain the implant. This build was then utilized to substitute the patient's left main bronchus. There were no perioperative problems, and the left lung ventilated usually as soon as the implant was located. At 3 months after surgery, the patient's lung role was in the normal range for her age and sex, and she was capable to function generally. While lengthier continuation and greater investigation populations are desirable, this statement specifies that tissue engineering may be a new choice for patients with airway illness.

Nevertheless, while there has been stirring development with tissue engineering methods for hollow organs, the growth of approaches to produce superior, solid organs with more intricate histological building has been much more problematic. A number of issues must be recognized before completely functional, engineered organs such as liver and kidney can be arranged in the research laboratory. Initially, these organs comprise exceptionally intricate internal builds made up of many cell kinds organized in actual exact conducts, and simple cell-seeding methods may not be adequate for rebuilding these builds. Furthermore, the large bulk of these organs commands that the distribution of oxygen and nutrients to each part of the organ will be a test, except a technique for engineering a functional vascular network within the organ can be established.

Nevertheless, notwithstanding the trials, there have been some inspiring outcomes from numerous investigations. For example, the kidney encompasses numerous cell kinds and a compound functional structure that condenses it one of the most tough to rebuild, yet generating an elementary form of this organ that seemed to have at least the filtration characters of the inherent kidney was effective.

3D Bioprinting Tissue and Therapeutic Cloning

The fundamentals of both Bioprinting tissues, and therapeutic cloning were used in an attempt to generate genetically symmetrical renal tissue in a large animal model, the cow. Bovine skin fibroblasts from adult Holstein oxen were found by ear indentation and sole donor cells were secluded and microinjected into the perivitelline space of donor enucleated oocytes (nuclear transfer). The subsequent blastocysts were embedded into progestin-synchronized receivers to permit for additional *in vivo* development. After 12 weeks, cloned renal cells were reaped and extended *in vitro*. Next, the cloned renal cells were sowed on scaffolds comprising of three collagen-coated cylindrical silastic catheters.

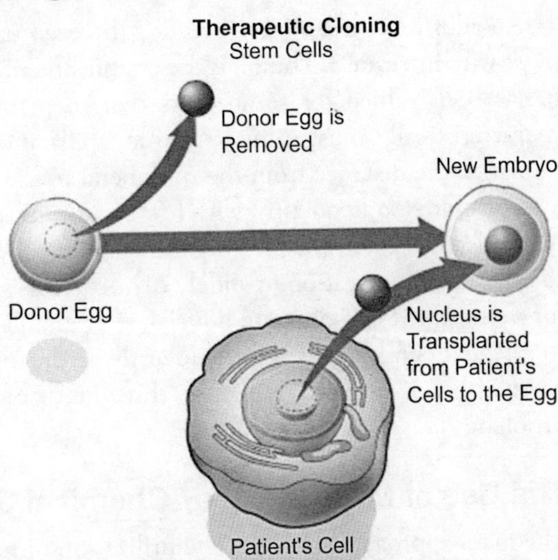

Figrure 6.16: Theraputic Cloning – Stem Cells

The ends of the three membranes of each scaffold were joined to catheters that terminated into a collecting reservoir. This generated a renal neo-organ with a mechanism for gathering the excreted urinary fluid. These scaffolds with the collecting devices were transferred intravenously into the same ox from which the inherent material initiated and then recovered 12 weeks after imbedding.

At this time, a yellow urine-like fluid was witnessed accumulating within the reservoir of the device. Chemical examination of this fluid, comprising urea nitrogen and creatinine levels, electrolyte levels, specific gravity, and glucose concentration, discovered that the inserted renal cells influenced filtration, reabsorption, and secretory abilities. Histological inspection of the recovered transplants exposed wide vascularization and self-organization of the cells into glomeruli and tubule-like builds. A clear endurance amid the glomeruli, the tubules, and the silastic catheter was distinguished that permitted the opening of urine into the collecting reservoir.

These investigations established that cells resulting from nuclear transmission can be effectively harvested, extended in culture, and relocated *in vivo* with the use of ecological scaffolds on which the sole suspended cells can shape into tissue builds that are genetically alike to those of the host. These investigations were the first shown the use of therapeutic cloning for renewal of tissues *in vivo*. Nonetheless, the dimensions of this device was small, and the test will be to generate a bigger device with functioning vasculature and innervations, so that it can substitute all of the countless metabolic functions of the kidney.

Regenerative Medicine – Transplanting to Clinic

The understandings of urethral, bladder, blood vessel, and tracheal substitution utilizing Bioprinted tissue deliver reassurance for future energies to engineer other organs in the research laboratory. These capabilities also, performers further understanding to unsolved complications. For instance, innervation of tissues and organs is significant for attaining complete functionality. In the canine engineered bladder trials, the scrutiny of positive S-100 staining was reliable in growing of neural builds into the neo-bladders, and bladder function was reestablished shortly after imbedding. Innervation of tissue-engineered builds has been perceived in other systems such as the small intestine. Not only is successful construction with the nervous system significant for the functionality of neo-organs, but also indication proposes that it can improve tissue renewal. The controlled release of neurotrophic factors is one possible method to encourage peripheral nerve renewal and synapse development with engineered tissue. Direct electrical stimulus has proven useful in muscle renewal and may have wider applicability.

An additional essential matter for the eventual success of laboratory-cultivated organs, predominantly those with intricate 3D build, is the establishment of satisfactory oxygen and the creation of new vasculature. It has been valued for some years that in metabolically lively tissues, the distance over which oxygen characteristically must diffuse from a capillary bed to reach a cell is about 0.1 mm, but that in clinical implants, the distance from the peripheral to the midpoint of the implant is possible to surpass that by at least 50-fold. Consequently, with few exclusions (e.g., cartilage), oxygen is rate-limiting for the capability of imbedded cells, and thus for organ engineering. Neovascularization, a complex morphogenetic procedure that permits the creation of widely divided vessels, even in an adult, must happen quickly and competently for a embedded neo-organ to flourish after establishment. Furthermore, special procedures may be essential to safeguard endurance of imbedded tissue during the original period after embedding, until a functional vascular bed is in place. Presently, three kinds of plans have been developed to resolve the oxygen source problem.

The Use of Mechanical or Chemical Sources of Oxygen

The main approach comprises the utilization of mechanical or chemical sources of oxygen that can sustain the build prior and immediately after implanting, until the neovascularization technique is made and can convey the neo-organ with adequate blood flow. An intra-tissue perfusion system take advantage of a collection of micro-needles to bring oxygen and nutrients and remove waste augments the capability and functionality of thick (1 mm) slices of liver tissue *in vitro* and might enable *in vivo* surrounding.

Also, the utilization of oxygen-carrying particles such as perfluorocarbons could excite the role of cells in culture and of encapsulated cells and organ builds inserted into animals. Research laboratory lately showed that a PLGA film incorporating an oxygen-generating system could avert the necrosis - the death of cells - of ischemic tissue over some days *in vivo*.

Generation of Competent Vascular Network

"prevascularization" plans intention to create neo-organs engineered with a preexistent channel structure to enable the creation of a capable vascular system. To complete this, endothelial extraction cells can be pre-seeded into the channels or may be employed *in vivo* by utlizing biochemical signals that are implanted in or freed by the scaffold. Though, there is still the query of how to generate channels in a way that will be understood as a natural vascular system by the body. One explanation would be to engage decellularized tissue as the scaffold. A new investigation established that perfusion of a complete heart with detergents produce an acellular build in which the inherent vascular channels continue to be whole. Perfusion-decellularization technology was autonomously developed, utilizing liver tissue and have established that the vascular tree of the whole organ scaffold rests patent and can be re-colonized with great numbers of endothelial cells.

Manufacture Scaffolds Using Laser Guided Wire Technology

Several technologies can be utilized to Bio-fabricate scaffolds with preshaped channels, possibly with cells incorporation, envisioned to excite neo-vascularization. For instance, laser guided "writing" (LGW) was used to form endothelial cells and excite their buildup into tubular vessels.

3D Ink-Jet-Based Bioprinting of Cells and Biomaterials

Ink-jet-based bioprinting of cells and biomaterials by thermal ink jet technology can deliver extraordinary guide of the fine build of engineered tissues, counting the creation of complicated vessel networks, Figure 6.17. Layer-by-layer ink jet printing is utilized to yield 3D builds comprising endothelial cells and presented that these grow functional microvascularization when embedded *in vivo*, as evaluated by magnetic resonance imaging MRI.

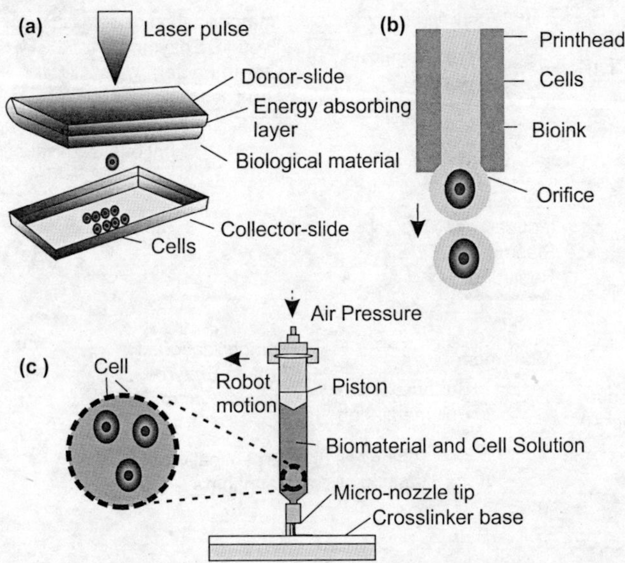

Figure 6.17: The Three Main Groups of 3D Bioprinting Technologies for Tissue and Organ Engineering: (a) Laser-based Bioprinting; (b) Inkjet-based Bioprinting; and (c) Extrusion-based Bioprinting

Courtesy: MGPI

Electrospinning of Living Cells with Biomaterials

Electrospinning of living cells with biomaterials provides comparable possibility to manufacture organ builds with pre-patterned vessels. Mathematical modeling of scaffolds considered to encompass a pre-existing arteriovenous loop displays by what means the delivery of an oxygen source within the scaffold can dynamically sustain additional neo-vascularization and tissue growth.

It is well recognized that development influences such as VEGF and FGF can encourage vascularization in engineered tissues. Current determinations have drawn-out this method by integrating further pro-angiogenic molecules into scaffolds, such as organ-specific ECM from liver to support sinusoidal endothelial cells. Artificial biomaterials well-thought-out to deliver signals frequently presented by the ECM will complement, and may finally replace, the utilization of the innate molecules.

Many of the medical examinations connecting bioengineered products have been put on hold since of the costs linked with the comprehensive technology. With a bioengineered product, costs are characteristically high since of the biological surroundings of the treatments involved, and as with any treatment, the cost that the therapeutic health care system can permit for an exact technology is restricted. Subsequently, the costs of Bioprinted products have to be dropped beforehand they can have an impact clinically. This is currently being known for many tissue-Bioprinted technologies. As the technologies develop over time, and the volume of the implementation is well-thought-out, costs will obviously reduce.

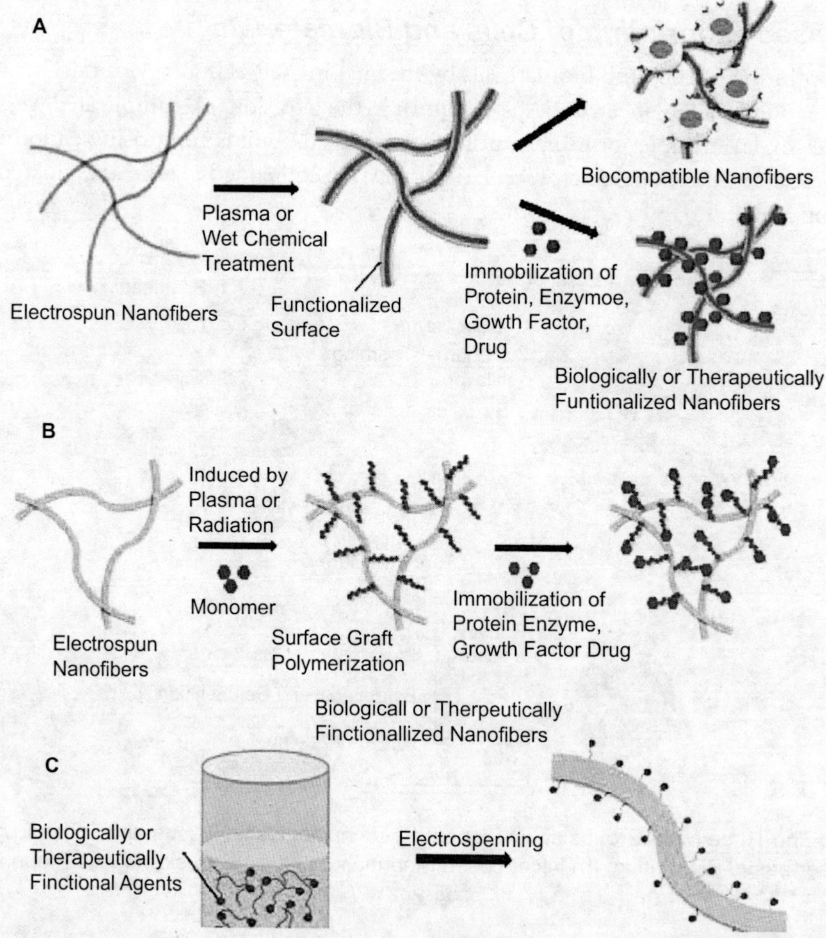

Figure 6.18: Electrospun Nanofibers - A: Plasma activation or wet chemical treatments - B: Plasma or radiation surface graft polymerization - C: Co-electrospinning

Courtesy: Elsevier

The Essence for Successful Application Bioprinting Tissue Technologies

Regenerative clinical development is presently happening experimentally for practically every kind of tissue and organ within the human body. For example, regenerative medicine incorporates the arenas of tissue engineering; cell biology, nuclear transfer, and materials science, scientists who have understood the ideal techniques of cell harvest, culture, expansion, transplantation, and polymer design are indispensable for the constructive use of these technologies to prolong human life. Numerous tissues are at dissimilar phases of growth, with some already being utilized clinically, a few in preclinical trials, and some in the finding phase. Existing development endorses that Bio-engineered tissues may have an extended clinical applicability in the future and may describe a achievable healing choice for those who would profit from the life-prolonging benefits of tissue replacement or renewal.

Bioprinted Cells Accelerate Effective Healing of Large Skin Wounds

Stem cells gained from amniotic liquid illustrate high proliferative capability in culture and multilineage differentiation possibility. Since of the absence of substantial immunogenicity, and the capability of the amniotic fluid-derived stem (AFS) cells to modulate the inflammatory reaction, had been examined if they could supplement wound curing in a mouse model of skin regeneration.

Bioprinting technology is utilized to heal full-thickness skin injuries in mice. AFS cells and bone marrow-resulted mesenchymal stem cells (MSCs) were re-suspended in fibrin-collagen gel and "3D printed" over the wound site. At days 0, 7, and 14, AFS cell- and MSC-driven wound closure and re-epithelialization was knowingly superior than closure and re-epithelialization in wounds preserved by fibrin-collagen gel only.

Histological investigation presented enlarged micro-vessel concentration and capillary diameters in the AFS cell-treated wounds associated with the MSC-treated wounds, while the skin preserved only with gel presented the lowermost amount of micro-vessels. Though, following of fluorescently categorized AFS cells and MSCs exposed that the cells continued briefly and did not perpetually assimilate in the tissue. These interpretations recommend that the augmented wound closure rates and angiogenesis may be due to transfer of secreted trophic factors, rather than direct cell-cell communications. Therefore, proteomic investigation was achieved, which presented that AFS cells secreted a number of development factors at concentrations greater than those of MSCs. In parallel, it is revealed that AFS cell-conditioned media prompted endothelial cell migration *in vitro*. Taken together, the consequences show that Bioprinting AFS cells could be an actual cure for large-scale wounds and burns.

3D Bioprinting May Heal 11 Million Injuries per Annum

Widespread burns and full-thickness skin injuries can be overwhelming to patients, even when treated. There are a projected 500,000 burns treated in the U.S. each year. The total mortality rate for burn injury was 5.9% between 2000 and 2018, and remedial costs for burn cures approach $3 billion per year. Universally, this statistic increases to 12 million wounds per year. Furthermore to burns, full-thickness long-lasting wounds establish a great patient base, and notwithstanding technological development of cures, healing rates continue below 50% successful. These non-healing long-lasting wounds are projected to affect 8 million people per year in the U.S., with yearly costs approaching $29 billion. Patients who suffer from either of these kinds of wounds value from speedy treatments that ends in complete closure and shield of the lesions. Especially, burn patients who have late cures often subject to widespread scarring that can effect in undesirable long-term physical consequence.

Allografts as Option to Heal Skin Wounds

Current developments have been made in treatments for less healing; though, the gold-standard still employed in the clinic is an autologous split-thickness skin graft. This includes take away a piece of skin from a secondary surgical operating site for the patient, elongating the skin, and reapplying the graft on the wound or burn. While this cure produces a rational medical consequence, if the wound is widespread, then the number and size of donor sites are restricted. Allografts are an added choice, but they suffer from the need of immunosuppressive drugs to avert immune rejection of the graft. These restrictions have consequently led to the growth of non-cellular dermal alternatives, which are most frequently consisted up of a polymeric scaffold. Samples contain Integra, and while such materials effect in better wound curative over untreated controls, they are costly to yield and consequence in fairly poor cosmetic outcomes.

Cellularized Graft-Like as Option to Heal Skin Wounds

Current developments in tissue engineering have produced to more intricate natural skin counterparts that may harvest more proper wound treatment selections for patients. Instances comprise cellularized graft-like products, such as Derma-graft, Apligraf, and TransCyte.

These products are usually poised of a polymer scaffold patch that is sowed with human fibroblasts and cultured *in vitro* previous to application. These grafts are also expensive to produce and, as allografts, can suffer from the same immunological disadvantages discussed previously.

Cell Spurting and Bioprinting to Heal Skin Wounds

Cell spurting and Bioprinting technique have recently been progressed for wound cure. In these approaches, cells are combined in a hydrogel carrier and located over the wound via a sputtering nozzle or print-head.

3D Bioprinting as Regenerative Medicine

Bioprinting has appeared as a flexible implement with possible in a diversity of tissue engineering and regenerative medicine applications. Bioprinting can be defined as robotic deposition that has the possible to construct organs or tissues. Over-all, bioprinting emoloys a computer controlled printing device, to precisely deposit cells, and biomaterials into exact 3D configurations in order to generate anatomically precise builds. These devices have the capability to 3D print cells - in the shape of cell combinations, cells encapsulated in hydrogels, or cell-seeded micro-carriers. The polymers that deliver structure or space-holding abilities are helping as the supporting "Bio-paper."

The cell basis employed in cellular treatments for wound medicinal treatment is a significant concern that has consequences in the cost, speed, and result of the cures. Human keratinocytes are possibly the ideal cell kind to utilize. Though, autologous and allogeneic keratinocytes undergo the same disadvantages as their autologous and allogeneic skin graft corresponding item, that is, secondary surgical sites and possible for rejection, respectively. This raises the issue; can cells that are helpful to wound remedial be utilized but may be immunoprivileged albeit they initiate from an allogeneic basis?

Mesenchymal Stem Cells MSC Treatment of Acute and Chronic Wounds

Mesenchymal stem cells (MSCs) have revealed healing possible for restoration and renewal of tissues impaired by injury or disease. Especially, MSC cure of severe and long-lasting wounds effects in quicker wound closure, augmented epithelialization, creation of granulation tissue, and angiogenesis. MSCs have lately been revealed to be also active for cultivating *in vivo* skin development, a method frequently employed to generate possibility for surgical grafts that generates tissue injury and demands tissue renewal.

Bioprinting Amniotic Fluid-derived Stem (AFS)

Amniotic fluid-derived stem (AFS) cells are a useful cell basis for applications in regenerative medicine because of their high propagation capability, multipotency, immunomodulatory movement, and absence of substantial immunogenicity. Contrasting embryonic stem cells, AFS cells do not form teratomas when injected into immune-deficient mice. Additionally, AFS cells continue steady and illustrate no signs of alteration in culture. The segregation of AFS cells is an easier procedure than that for separation of MSCs, and large numbers of AFS cells can be isolated and extended from as little as 2 ml of amniotic fluid. These cells propagate quickly with replication times of 30–36 hours, and do not need supportive feeder layers.

The immunomodulatory and high propagation characteristics of AFS cells propose that they can be employed as an "off-the-shelf" cell treatment product for wound curative. A murine large skin wound model and Bioprinting technology are utilized to bring AFS cells in a fibrin-collagen hydrogel to the wound site. AFS cells are expected to enable wound curative and could thus be additional well-thought-out for clinical use of extreme skin injuries.

Materials and Methods - Hydrogel Preparation

To create results that when printed shaped healthy gels, the constituents of fibrin were joined with collagen. Fibrinogen was liquefied in phosphate-buffered saline (PBS) to make a 50 mg/ml solution. Thrombin (Sigma-Aldrich) was dissolved in PBS to make a 20 IU/ml solution. Type I rat tail collagen was diluted in ice-cold PBS to give a 2.2 mg/ml solution, after which the pH was adjusted to 7.0 using 1 N NaOH. All solutions were then sterile filtered with a 0.4-μm syringe filter. To form gels, the fibrinogen and collagen solutions were mixed together in a 1:1 ratio by volume. That mixed solution was then in turn mixed with the thrombin solution, after which the gel started to cross-link within 25 seconds.

Cell Culture and Cell Labeling

Multipotent stem cells were secluded from human amniotic fluid. To attain a moderately homogenous subpopulation, the cells were long-drawn-out in culture from a single clone. AFS cells can multiply fast in culture deprived of feeder cells for many ways, while upholding chromosomal steadiness.

For picturing and following of the printed cells, AFS cells and MSCs were inherently branded with green fluorescent protein (GFP), utilizing a lentivirus vector transporting the GFP gene (Lenti-GFP), with a typical technique. Labeling competence (70%–80%) was confirmed by fluorescence-initiated cell categorization. The cells were then extended as required to spread enough numbers for bioprinting.

Bioprinting

In planning for printing, AFS "Amniotic Fluid Stem" cells and MSCs "Mesenchymal stem cells" were distinctly suspended in the fibrinogen/collagen solution at a density of 16.6 million cells per milliliter and located on ice. Although subjected to anesthesia, a sole full-thickness skin wound (2.0 × 2.0 cm) was surgically shaped with clippers on the mid-dorsal region of nu/nu mice.

A Bioprinting device was utilized for cell deposition using fibrin/collagen gels. The device contains of a way of moving in a three-axis movement ability in which is contained the core print head. The print head is invented of a set of pressure-driven nozzles over which the hydrogel solutions, with or without cells, are printed, and an extra set of high-pressure nozzles over which secondary solutions, such as cross-linking solutions, can be printed. The printable hydrogel solutions are contained in changeable cartridges in line with the back-pressure and the print nozzles. The printing procedure is then controlled exploiting software also advanced in-house.

For cure, the Bioprinter was utilized to deposit two layers of a fibrin-collagen gel by depositing a layer of thrombin, a layer of fibrinogen/collagen, a second layer of thrombin, a second layer of fibrinogen/collagen, and a final layer of thrombin. Equivalent volumes of the two solutions were used (0.3 ml), ensuing in 5 × 106 AFS cells or MSCs deposited per wound. Furthermore, gel-only cures, comprising no cells, were managed. Triple antibiotic was applied over each wound after gelation, tracked by a Tegaderm bandage.

Lastly, a made-to-order dressing was stitched in place in order avoid the Tegaderm from being detached. Wound dimensions was known by taking photographs directly after surgical operation and again at 7 and 14 days (n = 5). Animals were euthanized at 7 and 14 days, and the recreated skin was collected for histological examination. Moreover, a number of mice were euthanized at 1 and 4 days for cell follow up goals and to assess the constancy and delivery of bioprinted cells by confocal microscopy.

Wound Closure

Wound closure, tightening, and re-epithelialization ratios were considered utilizing snapshots and histological investigation of the injuries, acquired at the time of surgical operation, day 7, and day 14. The initial wound area was defined as A, the clearly re-epithelialized skin area was distinct as B, and the residual unclosed wound was well-defined as C. Ratio of wound closure was defined as C/A × 100%, percentage of shrinkage of the wound was defined as (A – B)/A × 100%, and percentage of re-epithelialization was defined as (B – C)/B × 100%.

Collected skin tissues were first rolled along a syringe needle past to being stable overnight in 4% paraformaldehyde. Samples were then cleaned in PBS three times for 30 minutes per wash, after which the samples were transported to 30% sucrose for an overnight maturation at 4°C. The rolled along tissues were then divided in half and flash frozen in Tissue-Tek OCT Composite lumps in liquid nitrogen. A cryotome was utilized to create 6-μm sections made up of the whole cross-sections of the rejuvenating wounds. These slides were kept at –20°C until histological events were achieved.

Units were discolored with hematoxylin and eosin for histology, and slides were pictured under light microscopy. Consideration was paid to the existence of blood vessels in the rejuvenated tissue and re-epithelialization of keratinocytes through the surface of the wound area.

Immunohistochemistry (IHC)

Immuno-histo-chemical (IHC) staining with "smooth muscle actin" (SMA) was utilized to envisage smooth muscle cells (SMCs) in established blood vessels in the rejuvenating skin, or with "von Willebrand factor" (vWF) to envisage endothelial cells in afresh shaped capillaries and lining the established blood vessels. For IHC, all incubations were approved out at room temperature unless otherwise stated. Slides were warmed at 60°C for 1 hour to upsurge attachment to the slides. Antigen recovery was accomplished on all slides and attained with incubation in proteinase K for 5 minutes.

Fluorescence Imaging of Cells

To examine if the deposited cells remain in the rejuvenating skin long-term after the bioprinting, GFP-transfected AFS cells and MSCs cells were utilized. Animals were euthanized on days 1, 4, 7, and 14 after cell bioprinting, and skin samples were harvested and equipped for histology. Samples were then cleaned three times in PBST, counterstained with DAPI, and washed three times before loading with Prolong Gold Antifade reagent. Sections were pictured utilizing a fluorescence microscope, and illustrative pictures were documented.

Bioprinting of Cells onto Skin Wounds

The bioprinting process, portrayed in Figure 6.19A, steadily produced a fibrin/collagen gel that delivered 100% coverage over the wound area and shaped a fitted seal with the skin at the boundaries of the excisional wound. To assess the dissemination of cells, brought to the wound by the bioprinting process, confocal microscopy of gel/tissue at day 1 was tilized to envisage the fluorescently considered AFS cells. Figure 6.19B illustrations an even delivery of cells in a illustrative picture spanning an approximately 900 × 900 μm2 area of the harvested tissue. Figure 6.19C illustrates a cross-sectional view (z axis) of the same sample, in which the vertical delivery of cells is constant across the sample.

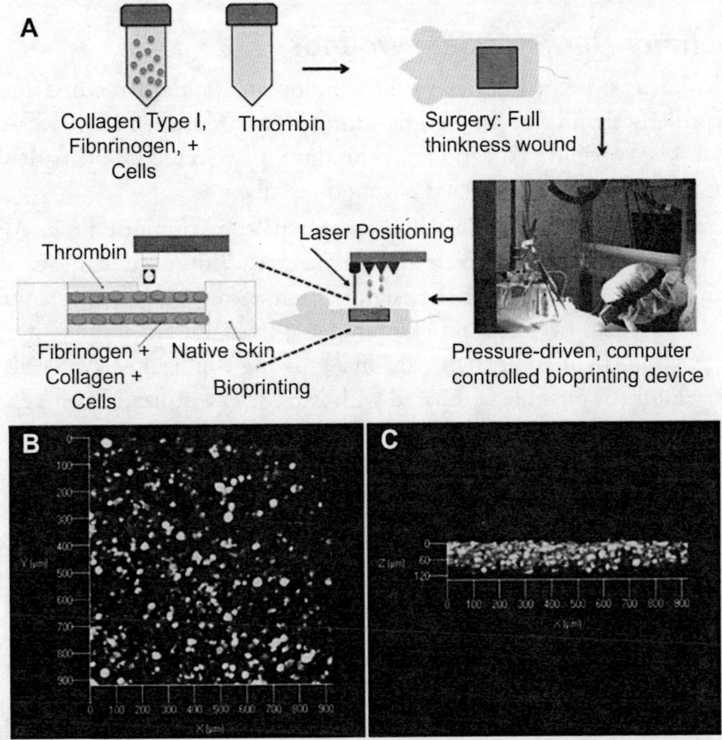

Figure 6.19: Stem Cells Remedy Skin Wound - A: Amniotic Fluid-derived Stem (AFS) Cells Generating Full-Thickness Skin Wound. B: Wounds Deposited Gels with Green Fluorescent Protein-tagged AFS Cells – C: Side View

Courtesy: US National Library of Medicine – National Institute of Health

Potential of Bioprinted AFS Cell Therapy

Extensive skin injuries, counting severe burns, can be overwhelming to patients, even when treated, and have a massive monetary weight on the medical institution. These patients necessitate a cure that outcomes in comprehensive closure and shield the wound. Although numerous developments have been made in remedies for such injuries, the autologous split-thickness skin graft is still the finest obtainable remedy.

The 3D Bioprinted AFS cell treatment method may have the possibility to deal with the scientific and medical necessity for more active remedy of burns and skin injuries. The Bioprinted banked cells with immunomodulatory Characteristics executing a rapid deposition procedure to sustain fast and comprehensive wound coverage is effective *in situ*. The AFS cells deposited in a collagen/fibrin gel augmented closure of full-thickness injuries in mice sooner than gel-only controls and were as active as MSC remedies. Furthermore, the AFS remedy convinced augmented neovascularization and blood vessel growth over the sequence of the exploration.

Founded on the examination of cell-situation media and an *in vitro* endothelial cell relocation assay, it is likely that the reaction of the AFS cells is contributed to secretion of a quantity of trophic factors significant for wound remedial and angiogenesis.

Optimal Treatment for Burns or Skin Wounds

An optimum cure for burns or skin injuries would employ an standard method for rapid application. Rapid reaction is particularly significant for victims with widespread burns. Moreover, the cure should not generate a secondary surgical operating location, it should not be rejected by the patient's immune system, and it desires to shape a healthy covering over the wound rapidly.

Multipotent and stretchable cells were initially segregated from amniotic fluid. AFS cells direct both embryonic stem cell and adult stem cell markers and can be extend for more than 250 channels. These cells can be encouraged to differentiate into cells that signify each germ layer, such as adipogenic, osteogenic, myogenic, endothelial, neuronal, hepatic, and chondrogenic lineages. Since their sites along the evolving timeline—they are "fledgling" than adult stem cells, in a growing condition—AFS cells may have enlarged differentiation and development possible compared with MSCs. The utilization of AFS cells is investigated in for numbers explanations:

- AFS cells are extremely stretchy, permitting actual cell banking, additional backup the basis for utilizing them as an standard treatment;
- AFS cells presented a degree of immunomodulatory movement and consequently might not be rejected in the wound atmosphere; and
- AFS cells, like other MSCs, were revealed to secrete development factors and cytokines that have an anti-inflammatory reaction and that support in wound curative and tissue rejuvenation.

AFS cells were similarly operative as bone marrow-derived MSCs at cumulative wound cover and re-epithelialization rates in full-thickness injuries. Though, AFS cells encourage robust angiogenesis associated with MSC-treated groups and gel-only groups. Not only were micro-vessel density and average vessel diameter greater, but also smooth muscle actin stains confirmed that these blood vessels were established and steady.

In gel-only and MSC groups, the existence of more extravasated red blood cells proposed fewer mature vessel creation than in the AFS group. Immunofluorescence of GFP-tagging AFS cells and MSCs exposed that both kinds of printed cells reduced in quantity and were typically gone by day 7. There are numerous possible fates, as well as apoptosis, clearance by macrophages, and migration to other regions of the animal.

Furthermore to their transiency, we did not perceive migration of the conveyed cells into the original tissue or localize with new vasculature. These consequences propose that secretion of trophic issues was perhaps accountable for the wound closure rates and encouraged neovascularization, as contrasting to direct cell incorporation into the new tissue.

AFS cells have been exposed to be proangiogenic in other investigations. Cytokines from AFS-acclimatized media encouraged endogenous restoration in ischemic full-thickness skin flaps in a rat model and encouraged neovascularization in a hind limb ischemic mouse model. Furthermore, a efficient *in vitro* assay established that AFS cells could effectively persuade migration of endothelial cells through Transwell membranes. However cell treatment delivers a dynamic cure with continued development factor discharged over time, acclimatized medium suggests only a one-time management of development factors that would probable consequence in a single eruption reaction of bioactivity that would weaken after a short time. Thus, AFS cell treatment can possibly effect in a better wound curative treatment than the application of concerted untrained medium.

MSCs are being discovered with augmented frequency as possible wound curative treatments for medical utilization application. A great bulk of work has established that topical transfer of these cells normally recovers wound curative rates and seems to entertain through communications with vascular cells and immunomodulation. For instance, use of autologous bone marrow-resulting cells upsurges neovascularization and fit extracellular matrix creation though lessening granulation tissue creation in a rabbit wound model. Effectiveness of MSC cures can be additionally augmented by selecting a suitable transfer vehicle. In one such instance, comparable to the investigation offered herewith, MSCs were transported to wounds in a collagen-based gel. The gel atmosphere augmented MSC secretion of trophic influences and appearance of record issues related with stem cells, though hastening the wound curative procedure.

The migration and incorporation of the GFP-tagged AFS cells and MSCs into the rejuvenated skin were not detected. Though the consequences of utilizing cells were considerably more effective than cell-free cures, it might still be helpful to surge cell migration to deliver physical support for the rejuvenating skin tissue.

3D Printing Body Parts Will Revolutionize Medicine

Bioprinting a human liver is comparable to building honeycombs. The masterworks of biomedical engineering are closely indistinguishable to tissue models from actual human livers, and they are built from real human cells, Figure 6.20. But as an alternative of developing them, scientists print them, just as they would a manuscript, or, more precisely, just as they would print a scale model.

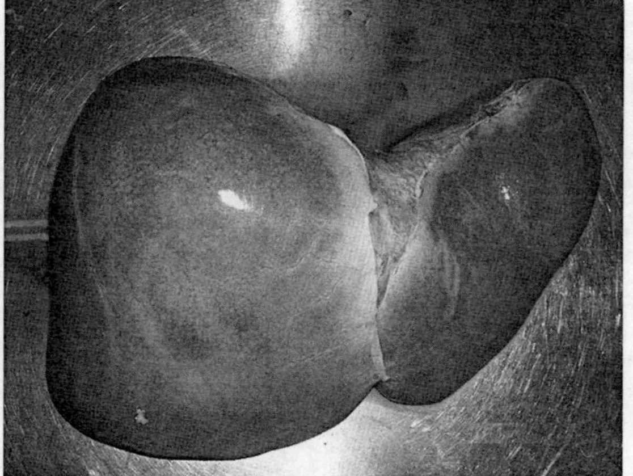

Figure 6.20: The World's First 3D Human Liver- 3D Bioprinted

Courtesy: public domain – Medical Daily LLC.

In past two decades, 3D printing has developed from a position manufacturing process to a $7 billion industry, accountable for the manufacture of all types of foods: toys, wristwatches, airplane parts, food. Nowadays scientists are occupied with applying comparable 3D Bioprinting technology to the field of medicine, speeding up an equally dramatic transformation. But it is much diverse, and much simpler, to print with plastic, metal, or chocolate than to print with a living cells.

It has been a fast moving mission; however, scientists and medical doctors are progressing well. In research laboratories around the world, bioengineers have commenced to print model body organs: heart valves, ears, artificial bone, joints, menisci, vascular tubes, and skin grafts.

From 2008 to 2019, the amount of scientific papers referencing bioprinting multiplied. Venture capital in the field has sky rocketed as well. Since 2007, the National Heart, Lung, and Blood Institute of the National Institutes of Health have awarded considerable endowments to bioprinting developments.

Three influences are leading the movement:

- more refined printers,
- developments in regenerative medicine, and
- advanced CAD software.

To pattern the liver tissue, merely the user may click "run program" with a mouse. Honeycomb-formed liver tissue is a lengthy way from an entirely operative organ, but it is a tangible phase in that path. Procurement to an entire organ employing a plug-and-play manufacturing approach may occur in our generation.

The very first bioprinters were not unduly costly. They look like low-cost desktop printers. The Bioprinters are in fact, have achievement similar to traditional printing role. In 2000, bioengineer and scientists had previously adapted inkjet printers to print fragments of DNA, in order to investigate gene sequencing. If an inkjet could print genes, the same hardware could print other biomaterials. The miniscule nano/microscale human cells are 10 micrometers, which is also the measurement of standard ink droplets, Figure 6.21.

Figure 21: 3D Organ Bioprinter
Courtesy: Fabion Biopriner

When a tinny dark silicon sheet fastened onto blank paper, and fed into the printer. As characters are typed the paper reeled out with printed characters. Likewise proteins can be applied in a similar way.

In 2000, Hewlett-Packard DeskJet 550C was reconstructed to print with E. coli bacteria, then to bigger mammalian cells. Then, the revolution of bioprinting commenced, that enabled other engineers to apply 3D printers to diverse medical encounters. They printed bone grafts from ceramic, dental crowns from porcelain, hearing aids from acrylic, and prosthetic limbs from polymer. They could print in 3D rather than just 2D.

When other bioprinting innovators adapted their printers, they deactivated the paper-feed mechanisms in their inkjets and supplemented them with an elevator-like base controlled by stepper motors; the base motion could translate up or down along the z-axis. Then, research laboratory could print one layer of cells, lower the base, and print another layer. Bioengineers went from sketching life on a flat canvas to constructing living figures.

Portable Printer for Regenerative Medicine

Numerous scientists for Regenerative Medicine have developed a transportable printer to graft skin directly onto burn sufferers. The capabilities to print cells in 3D have unlocked novel uses. Due to injuries variances from patient to patient; the depth is dissimilar; and they are very uneven. Therefore, charting the area, can govern how many cell layers are required for the subdermal tissue, as well as the epithelial area. The printer can supply cells more precisely and exactly.

Researchers could also print with numerous kinds of "Bioink." The spatial control over the depositing of cells has never been promising to this degree as scientists and engineers unlocked multiple dimensions of Bioprinting potentials to print a meniscus, the C-shaped slice of cartilage that cushions the knee and other joints. They used CT scans to generate a CAD file of a sheep's meniscus and extracted cells from the sheep to print a matching one. Nevertheless, the actual menisci are essentially hammered every day, all the time and they form up and develop stiff. So the beating that is in their atmosphere is truly a part of their development. Cells have precise supplies, depending on the tissue they are intended to become. In the case of a meniscus, it might mean emerging a bioreactor that can approximate pounding or use heat, light, or auditory pulses to stress the tissue into development.

For some tissues, even the simple ones, researchers don not know precisely what it earns to make the tissue perform similar a real tissue!

Most organs are highly refined structures with abundant cell kinds and intricate vasculatures initially fashioned to do very precise jobs. The liver alone achieves more than 500 functions. Like machines, bodies fatigue out and fail over time, occasionally or suddenly. Even when transplants are possible, donated organs cannot stay in step with demand. So as mechanical engineers started to shape early 3D Bioprinters, tissue engineers tried mounting replacement organs in research laboratory around the world. They started by pipetting cells into petri dishes by hand. Then, scientists started to plant those cells onto synthetic scaffolds. Made from biodegradable polymers or collagen, the scaffolds deliver a brief matrix for cells to cling to until they are strong enough to stand alone.

Scientists and doctors soon embraced 3D Bioprinters to create scaffolds more exactly. However, by hand placing the cells onto them continued a time-consuming and laborious procedure. Printed bladders can be made with just two cell types; kidneys, for example, contain of 30 types of cells. Engineering more intricate tissues will make it unaffordable to manually place diverse cell kinds into diverse sites that can duplicate the inherent tissue builds.

Printing with Stem Cells, Muscle Cells, and Vascular Cells

Researchers and doctors built special design bioprinters that are faster than modified inkjets and can print with numerous cell kinds—counting stem cells, muscle cells, and vascular cells. They also designed one printer to generate both the synthetic scaffold and tissue combined. Presently, they are utilizing it to yield complex ears, noses, and bones.

Scaffolds deliver tissues with mechanical steadiness, and they can be utilized to convey genes and development factors to emerging cells. But, as in the case of polymers, they can present foreign materials into the body and become the source of inflammation. Cell kinds also react in a different way to certain scaffold materials, and so the more intricate the organ, the more complex the necessary framework—and the more hard to forecast how the cells will migrate around it.

Scientists, in USA investigated morphogenesis, the procedure that controls cells shaping organs during embryonic growth. By positioning cellular combinations—miniscule spheres encompassing thousands of cells—into a circle. The operation was observed the spheres fuse and form new structures. The aggregates achieve this by working simultaneously, together. A molecule on one cell causes a receptor protein on the cell membrane to alter the shape, tugging on the cytoskeleton of a second cell. A flow of communication follows, ultimately reaching the nucleus and triggering an alteration in gene expression.

A team of scientists experimented with bioprinters instead of laying down aggregates by hand. The technology transformed the research. Using a bioprinter, it was easy to prove that aggregates containing diverse cell types also fuse, without any human interference or environmental signals.

Scientists and engineers ensured that tissue engineers cannot locate cells anywhere they would be in a finished organ. They must arrange cells founded on where they need to be to start forming an organ, as in an embryo.

Cells Instinctively Programmed Know Precisely its Functions

The cells inherently know exactly what to do as they have been performing their functions for thousands of years since the dawn of creation. *It was stated, "In the Beginning God Created --" Genesis 1:1.*

A tissue is an orderly, and very well systematized; according to very strict inherit rules, in cellular sets. A several micrometers of aggregate are already a piece of tissue. Those pieces bind together and exchange information.

It is already likely to construct tissue by piling heaps of cells along the z-axis. This method was practiced to cardiac cells; when fused, cells beat in unison, just like a heart. Biologically, though, there is quiet a main hindrance: It needs to flourish. An organ necessitates networks of blood vessels to dispense nutrients and oxygen, Figure 6.22. Deprived of this core function, cells will weaken and die.

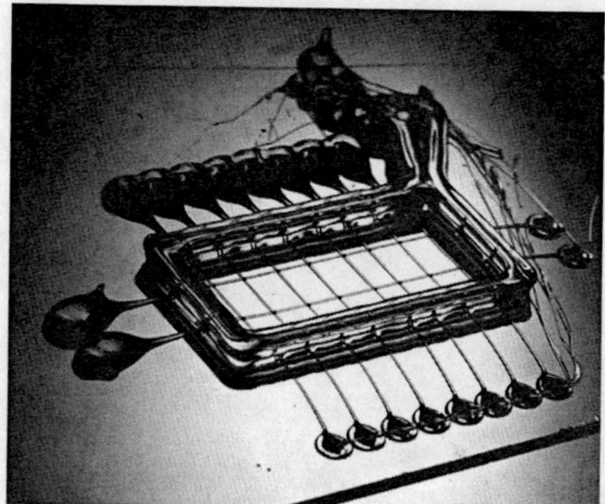

Figure 6.22: Printing Blood Vessels; *Courtesy: University of Pennsylvania*

Researchers and scientists have made vigorous vasculature by printing filler, such as hydrogel, among tubes of tissue cells. The filler can later be removed, leaving empty channels for blood cells. Therefore, a bioprinter was developed, which utilizes multiple arms moving in tandem, to deposit a vascular network and cellular aggregates at the same time.

The main obstacle is generating very small capillaries—the hair like blood vessels connecting larger vessels to cells. Once researchers can scale up the size and intricacy of the vascular system, progressing from biological parts and pieces to whole printed organs will become only a matter of time.

Bioprinters lacks biologically refined and reliable software. With an lifeless object like an automotive part, a 3D scanner can generate a CAD file in minutes and upload the design to a 3-D printer. Thus far, unfortunately, there is no medical equivalent. An MRI does not tell you where the cells are. Also, there are no devices to move cells around. It is beyond most computer software can undertake. Software model of a liver is more complex than a model for a jet plane.

Instead of printing a test tube out of polymer to perform chemistry in, test tube ought to be printed out of tissues. Nevertheless, Autodesk has teamed with "Organovo" to develop CAD programs that could be applied to bioprinting. As a first step, Autodesk plans to generate a contemporary cloud-based CAD shell to help modernize the design process. Ultimately, its objective is to assimilate the math that describes self-assembly and other cellular procedures into bioprinting software.

Every year, the pharmaceutical industry devotes more than $50 billion on R&D. According to the Food and Drug Administration, liver toxicity is the most common reason for a drug to be dragged out from clinical trials—as well as from the marketplace after it's been approved. There's still no reliable way to assess how a drug will affect the human liver before it is consumed—not even animal trials. The most important biological product will be 3D printing liver tissue for drug testing. There are some important class differences between animals like rats and humans.

Many researchers and scientists have attempted to resolve this problem by breeding mice with livers made up mostly of human cells. The investigation exhibited the mice projected how well a drug for treating hepatitis C would be metabolized by humans. Also, scientists have built minute liver models by means of micro-patterning, the same soft lithography method utilized to place copper wires onto computer chips. The problem is that micro-patterned structures are typically only a couple of cell layers thick, which limits the complexity researchers are facing.

Real Time Response Medical and Pharmaceutical Test

"Organovo," USA is selling its liver assay—a petri-dish-like well plate encompassing liver cells organized in a 3-D structure 200 to 500 microns thick, 2-5 times as thick as a human hair. The possible market is vast. It is well recognized that every drug consumed orally must pass through a liver, whether a painkiller, an anti-inflammatory, or a new cancer medicine. Instead of printing a test tube out of plastic to do chemistry in, it can be printed in test tube out of tissue, and achieve the chemistry in the tissue and look at the reaction in real time.

If Bioprinted assays deliver pharmaceutical investigators with healthier, faster data, the entire drug-discovery procedure will hasten. Furthermore, they could diminish the need for widespread animal testing. Bioprinted pancreatic tissue for therapy would be made up of only the endocrine cells proficient of producing insulin. If implanted in people, such tissue could control blood sugar and treat type 1 diabetes.

Bioprinters could also demonstrate priceless for medical schools. Students now train on cadavers. The processes of cutting out cancer would be priceless. Rather than printing healthy tissue, bioprinters could construct organs with tumors or other defects so that surgeons could rehearse before entering an operating room.

Whole, transplantable organs that function well will be the decisive challenge, but also, in the long run, change lives most profoundly. In the U.S., more than 578,000 people are currently on the national donor waiting list, which grows by 300 every month. It's not just an issue of supply versus demand. The odds of finding a suitable match are low. Bioprinting organs with cells full-grown from a patient's own body could ultimately aid doctors produce perfect matches at will.

Perhaps, scientists, and bioprinters could even enable bionic organs—body parts that do not just restore, but extends human capability. To that end, researchers at Princeton University have been experimenting with integrating electronics into bioprinting. They generated a matrix of hydrogel and bovine cells in the shape of an ear, incorporating silver nanoparticles to form a coiled antenna. The system could pick up radio frequencies beyond the range of normal human hearing. In a similar manner, bioengineers might one day incorporate sensors into other tissues—for example, creating a bionic meniscus that can monitor strain.

Bioprinters are already demonstrating scientists' remarkable mastery of biology and engineering.

REFERENCES

1. Cherry DK, Hing E, Woodwell DA, et al. National Ambulatory Medical Care Survey: 2006 summary. Natl Health Stat Report. 2008:1–39.
2. Pitts SR, Niska RW, Xu J, et al. National Hospital Ambulatory Medical Care Survey: 2006 emergency department summary. Natl Health Stat Report. 2008:1–38.
3. Miller SF, Bessey P, Lentz CW, et al. National burn repository 2007 report: A synopsis of the 2007 call for data. J Burn Care Res. 2008;29:862–870. discussion 871.
4. Peck MD. Epidemiology of burns throughout the world. Part I: Distribution and risk factors. Burns. 2011;37:1087–1100.
5. Kurd SK, Hoffstad OJ, Bilker WB, et al. Evaluation of the use of prognostic information for the care of individuals with venous leg ulcers or diabetic neuropathic foot ulcers. Wound Repair Regen. 2009;17:318–325.
6. Sen CK, Gordillo GM, Roy S, et al. Human skin wounds: A major and snowballing threat to public health and the economy. Wound Repair Regen. 2009;17:763–771.
7. Lesher AP, Curry RH, Evans J, et al. Effectiveness of Biobrane for treatment of partial-thickness burns in children. J Pediatr Surg. 2011;46:1759–1763.
8. Mirabella T, Cilli M, Carlone S, et al. Amniotic liquid derived stem cells as reservoir of secreted angiogenic factors capable of stimulating neo-arteriogenesis in an ischemic model. Biomaterials. 2011;32:3689–3699.
9. Brower J, Blumberg S, Carroll E, et al. Mesenchymal stem cell therapy and delivery systems in nonhealing wounds. Adv Skin Wound Care. 2011;24:524–532. quiz 533–524.

10. Sorrell JM, Caplan AI. Topical delivery of mesenchymal stem cells and their function in wounds. Stem Cell Res Ther. 2010;1:30.
11. Akela A, Nandi SK, Banerjee D, et al. Evaluation of autologous bone marrow in wound healing in animal model: A possible application of autologous stem cells. Int Wound J. 2012;9:505–516.
12. Rustad KC, Wong VW, Sorkin M, et al. Enhancement of mesenchymal stem cell angiogenic capacity and stemness by a biomimetic hydrogel scaffold. Biomaterials. 2012;33:80–90.
13. Marga F, Neagu A, Kosztin I, et al. Developmental biology and tissue engineering. Birth Defects Res C Embryo Today. 2007;81:320–328.
14. Norotte C, Marga FS, Niklason LE, et al. Scaffold-free vascular tissue engineering using bioprinting. Biomaterials. 2009;30:5910–5917.
15. Skardal A, Zhang J, McCoard L, et al. Dynamically crosslinked gold nanoparticle-hyaluronan hydrogels. Adv Mater. 2010;22:4736–4740.
16. Skardal A, Zhang J, McCoard L, et al. Photocrosslinkable hyaluronan-gelatin hydrogels for two-step bioprinting. Tissue Eng Part A. 2010;16:2675–2685.
17. Rahmanian-Schwarz A, Beiderwieden A, Willkomm LM, et al. A clinical evaluation of Biobrane® and Suprathel® in acute burns and reconstructive surgery. Burns. 2011;37:1343–1348.
18. Visconti RP, Kasyanov V, Gentile C, et al. Towards organ printing: Engineering an intra-organ branched vascular tree. Expert Opin Biol Ther. 2010;10:409–420.
19. Fedorovich NE, Alblas J, de Wijn JR, et al. Hydrogels as extracellular matrices for skeletal tissue engineering: State-of-the-art and novel application in organ printing. Tissue Eng. 2007;13:1905–1925.
20. Mironov V, Boland T, Trusk T, et al. Organ printing: Computer-aided jet-based 3D tissue engineering. Trends Biotechnol. 2003;21:157–161.
21. Maxson S, Lopez EA, Yoo D, et al. Concise review: Role of mesenchymal stem cells in wound repair. Stem Cells Translational Medicine. 2012;1:142–149.
22. Wang X, Li C, Zheng Y, et al. Bone marrow mesenchymal stem cells increase skin regeneration efficiency in skin and soft tissue expansion. Expert Opin Biol Ther. 2012;12:1129–1139.
23. De Coppi P, Bartsch G, Jr., Siddiqui MM, et al. Isolation of amniotic stem cell lines with potential for therapy. Nat Biotechnol. 2007;25:100–106.
24. Moorefield EC, McKee EE, Solchaga L, et al. Cloned, CD117 selected human amniotic fluid stem cells are capable of modulating the immune response. PLoS One. 2011;6:e26535.
25. Yoo JJ, Atala A, Binder KW, et al. Delivery System. U.S. patent application 20110172611. 2011 Jul 14; Kolambkar YM, Peister A, Soker S, et al. Chondrogenic differentiation of amniotic fluid-derived stem cells. J Mol Histol. 2007;38:405–413.
26. Delo DM, De Coppi P, Bartsch G, Jr., et al. Amniotic fluid and placental stem cells. Methods Enzymol. 2006;419:426–438.
27. Valli A, Rosner M, Fuchs C, et al. Embryoid body formation of human amniotic fluid stem cells depends on mTOR. Oncogene. 2010;29:966–977. 20. Yoon BS, Moon JH, Jun EK, et al. Secretory profiles and wound healing effects of human amniotic fluid-derived mesenchymal stem cells. Stem Cells Dev. 2010;19:887–902. Atala A, Lanza RP. Preface. In: Atala A, Lanza RPMethods of tissue engineering. San Diego: Academic Press; 2001.
28. Thomson JA, Itskovitz-Eldor J, Shapiro SS, Waknitz MA, Swiergiel JJ, Marshall VS, et al. Embryonic stem cell lines derived from human blastocysts. Science 1998;282:1145–1147.

29. Hasetine W. A brave new medicine. A conversation with William Haseltine. Interview by Joe Flower. Health Forum J 1999;42:28–30.
30. Dahms SE, Piechota HJ, Dahiya R, Lue TF, Tanagho EA. Composition and biomechanical properties of the bladder acellular matrix graft: comparative analysis in rat, pig and human. Br J Urol 1998;82:411–419.
31. Yoo JJ, Meng J, Oberpenning F, Atala A. Bladder augmentation using allogenic bladder submucosa seeded with cells. Urology 1998;51:221–225.
32. Piechota HJ, Dahms SE, Nunes LS, Dahiya R, Lue TF, Tanagho EA. *In vitro* functional properties of the rat bladder regenerated by the bladder acellular matrix graft. J Urol 1998;159:1717–1724.
33. Chen F, Yoo JJ, Atala A. Acellular collagen matrix as a possible "off the shelf" biomaterial for urethral repair. Urology 1999;54:407–410.
34. Kim BS, Mooney DJ. Development of biocompatible synthetic extracellular matrices for tissue engineering. Trends Biotechnol 1998;16:224–230.
35. Bergsma JE, Rozema FR, Bos RR, Boering G, de Bruijn WC, Pennings AJ. *In vivo* degradation and biocompatibility study of *in vitro* pre-degraded as-polymerized polyactide particles. Biomaterials 1995;16:267–274.
36. Li ST. Biologic biomaterials: tissue-derived biomateri-als (collagen). In: Brozino JDThe Biomedical Engineering Handbook. Boca Raton, FL: CRS Press; 1995. pp. 627-647.
37. Silver FH, Pins G. Cell growth on collagen: a review of tissue engineering using scaffolds containing extracellular matrix. J Long Term Eff Med Implants 1992;2:67–80.
38. Sams AE, Nixon AJ. Chondrocyte-laden collagen scaffolds for resurfacing extensive articular cartilage defects. Osteoarthritis Cartilage 1995;3:47–59.
39. Smidsrød O, Skjåk-Braek G. Alginate as immobilization matrix for cells. Trends Biotechnol 1990;8:71–78.
40. Lim F, Sun AM. Microencapsulated islets as bioartificial endocrine pancreas. Science 1980;210:908–910.
41. Gilding DK. Biodegradable polymers. In: Williams DFBiocompatibility of Clinical Implant Materials. Boca Raton, FL: CRC Press; 1981. pp. 209-232.
42. Mikos AG, Lyman MD, Freed LE, Langer R. Wetting of poly (L-lactic acid) and poly (DL-lactic-co-glycolic acid) foams for tissue culture. Biomaterials 1994;15:55–58.
43. Choi JS, Lee SJ, Christ GJ, Atala A, Yoo JJ. The influence of electrospun aligned poly (epsilon-caprolactone)/collagen nanofiber meshes on the formation of self-aligned skeletal muscle myotubes. Biomaterials 2008;29:2899–2906.
44. Lee SJ, Liu J, Oh SH, Soker S, Atala A, Yoo JJ. Development of a composite vascular scaffolding system that withstands physiological vascular conditions. Biomaterials 2008;29:2891–2898.
45. Pariente JL, Kim BS, Atala A. *In vitro* biocompatibility assessment of naturally derived and synthetic biomaterials using normal human urothelial cells. J Biomed Mater Res 2001;55:33–39.
46. Pariente JL, Kim BS, Atala A. *In vitro* biocompatibility evaluation of naturally derived and synthetic biomaterials using normal human bladder smooth muscle cells. J Urol 2002;167:1867–1871.
47. Amiel GE, Atala A. Current and future modalities for functional renal replacement. Urol Clin North Am 1999;26:235–246.

48. Amiel GE, Komura M, Shapira O, Yoo JJ, Yazdani S, Berry J, et al. Engineering of blood vessels from acellular collagen matrices coated with human endothelial cells. Tissue Eng 2006;12:2355–2365.
49. Yoo JJ, Park HJ, Lee I, Atala A. Autologous engineered cartilage rods for penile reconstruction. J Urol 1999;162:1119–1121.
50. Atala A. Autologous cell transplantation for urologic reconstruction. J Urol 1998;159:2–3.
51. Atala A. Bladder regeneration by tissue engineering. BJU Int 2001;88:765–770.
52. Atala A. Creation of bladder tissue *in vitro* and *in vivo*. A system for organ replacement. Adv Exp Med Biol 1999;462:31–42.
53. Cilento BG, Freeman MR, Schneck FX, Retik AB, Atala A. Phenotypic and cytogenetic characterization of human bladder urothelia expanded *in vitro*. J Urol 1994;152:665–670.
54. Oberpenning F, Meng J, Yoo JJ, Atala A. De novo reconstitution of a functional mammalian urinary bladder by tissue engineering. Nat Biotechnol 1999;17:149–155.
55. Scriven SD, Booth C, Thomas DF, Trejdosiewicz LK, Southgate J. Reconstitution of human urothelium from monolayer cultures. J Urol 1997;158:1147–1152.
56. Liebert M, Hubbel A, Chung M, Wedemeyer G, Lomax MI, Hegeman A, et al. Expression of mal is associated with urothelial differentiation *in vitro*: identification by differential display reverse-transcriptase polymerase chain reaction. Differentiation 1997;61:177–185.
57. Liebert M, Wedemeyer G, Abruzzo LV, Kunkel SL, Hammerberg C, Cooper KD, et al. Stimulated urothelial cells produce cytokines and express an activated cell surface antigenic phenotype. Semin Urol 1991;9:124–130.
58. Puthenveettil JA, Burger MS, Reznikoff CA. Replicative senescence in human uroepithelial cells. Adv Exp Med Biol 1999;462:83–91.
59. Freeman MR, Yoo JJ, Raab G, Soker S, Adam RM, Schneck FX, et al. Heparin-binding EGF-like growth factor is an autocrine growth factor for human urothelial cells and is synthesized by epithelial and smooth muscle cells in the human bladder. J Clin Invest 1997;99:1028–1036.
60. Nguyen HT, Park JM, Peters CA, Adam RM, Orsola A, Atala A, et al. Cell-specific activation of the HB-EGF and ErbB1 genes by stretch in primary human bladder cells. *In Vitro* Cell Dev Biol Anim 1999;35:371–375.
61. Harriss DR. Smooth muscle cell culture: a new approach to the study of human detrusor physiology and pathophysiology. Br J Urol 1995;75 Suppl 1:18–26.
62. Brivanlou AH, Gage FH, Jaenisch R, Jessell T, Melton D, Rossant J. Stem cells. Setting standards for human embryonic stem cells. Science 2003;300:913–916.
63. Richards M, Fong CY, Chan WK, Wong PC, Bongso A. Human feeders support prolonged undifferentiated growth of human inner cell masses and embryonic stem cells. Nat Biotechnol 2002;20:933–936.
64. Amit M, Shariki C, Margulets V, Itskovitz-Eldor J. Feeder layer- and serum-free culture of human embryonic stem cells. Biol Reprod 2004;70:837–845.
65. Reubinoff BE, Pera MF, Fong CY, Trounson A, Bongso A. Embryonic stem cell lines from human blastocysts: somatic differentiation *in vitro*. Nat Biotechnol 2000;18:399–404.
66. Reubinoff BE, Itsykson P, Turetsky T, Pera MF, Reinhartz E, Itzik A, et al. Neural progenitors from human embryonic stem cells. Nat Biotechnol 2001;19:1134–1140.

67. Schuldiner M, Eiges R, Eden A, Yanuka O, Itskovitz-Eldor J, Goldstein RS, et al. Induced neuronal differentiation of human embryonic stem cells. Brain Res 2001;913:201–205.
68. Schuldiner M, Yanuka O, Itskovitz-Eldor J, Melton DA, Benvenisty N. Effects of eight growth factors on the differentiation of cells derived from human embryonic stem cells. Proc Natl Acad Sci U S A 2000;97:11307–11312.
69. Zhang SC, Wernig M, Duncan ID, Brüstle O, Thomson JA. *In vitro* differentiation of transplantable neural precursors from human embryonic stem cells. Nat Biotechnol 2001;19:1129–1133.
70. Kaufman DS, Hanson ET, Lewis RL, Auerbach R, Thomson JA. Hematopoietic colony-forming cells derived from human embryonic stem cells. Proc Natl Acad Sci U S A 2001;98:10716–10721.
71. Kehat I, Kenyagin-Karsenti D, Snir M, Segev H, Amit M, Gepstein A, et al. Human embryonic stem cells can differentiate into myocytes with structural and functional properties of cardiomyocytes. J Clin Invest 2001;108:407–414.
72. Levenberg S, Golub JS, Amit M, Itskovitz-Eldor J, Langer R. Endothelial cells derived from human embryonic stem cells. Proc Natl Acad Sci U S A 2002;99:4391–4396.
73. Assady S, Maor G, Amit M, Itskovitz-Eldor J, Skorecki KL, Tzukerman M. Insulin production by human embryonic stem cells. Diabetes 2001;50:1691–1697.
74. Itskovitz-Eldor J, Schuldiner M, Karsenti D, Eden A, Yanuka O, Amit M, et al. Differentiation of human embryonic stem cells into embryoid bodies compromising the three embryonic germ layers. Mol Med 2000;6:88–95.
75. Gurdon JB, Elsdale TR, Fischberg M. Sexually mature individuals of Xenopus laevis from the transplantation of single somatic nuclei. Nature 1958;182:64–65.
76. Campbell KH, McWhir J, Ritchie WA, Wilmut I. Sheep cloned by nuclear transfer from a cultured cell line. Nature 1996;380:64–66.
77. Soloman, Sabrie "Sensors Handbook, 2nd Edition, McGraw-Hill Publishing Co., NY – ISBN 978-0-07-160570-0
78. Soloman, Sabrie "Sensors and Control Systems in Manufacturing 2nd Edition," McGraw-Hill Publishing Co., NY – ISBN 0-07-059626-3
79. Soloman, Sabrie "Affordable Automation," McGraw-Hill Publishing Co., NY – ISBN 0-07-059633-6
80. Soloman, Sabrie "Sensors Handbook, 1st Edition," McGraw-Hill Publishing Co., NY – ISBN 0-07-059630-1
81. Soloman, Sabrie "Sensors and Control Systems in Manufacturing 1st Edition," McGraw-Hill Publishing Co., NY – ISBN 978-07-10572-4

7 Rapid Prototyping – 3D Bioprinting Orthopedics

Rapid Prototyping - 3D Bioprinting Orthopedics

The field of tissue engineering research is the answer to the hopes of many in US and around the world. The bridging gap of hope between the scarce organ donor supplies, and the overwhelming abundant needs of organ transplantations is finally shortened if not totally answered through the advancements of tissue engineering. Nonetheless, constructing 3D vascularized organs continues to be the foremost technological obstacle to be diminished. Organ Bioprinting, which is characterized as biofabrication through digital control of additive manufacturing of 3-D cellular tissue constructs. Organ Bioprinting is benefiting of rapid prototyping technology to Bioprint cells, and bio-materials, Figure 7.1.

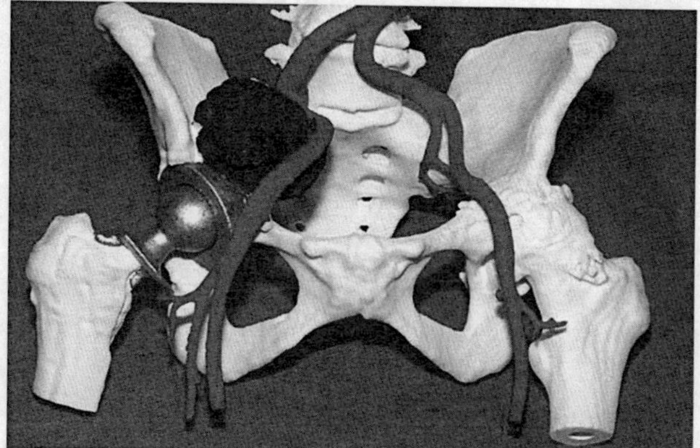

Figure 7.1: Personalized Orthopedics –3D Printing for Pre-operative Planning, Bespoke Implants - Precise Placement
Courtesy: Barker Brooks Communications Ltd., England

Constructing orthopedic devices rapid prototyping technologies has been until lately a highly rejuvenating operation that encompasses numerous users, comprising doctors, design engineers and rapid prototyping professionals. Current traditional systems constructing orthopedic parts through rapid prototyping do not trail the fundamentals of simultaneous engineering and design for manufacture. This leads to extensive communication between parties and delays in product realization time.

The outline for a unified expert system will allow a doctor to generate quickly and easily fully operational prosthetics and orthopedic implants. Necessary elements of the model acquisition procedure should contain volumetric segmentation of objects from a CT or MRI dataset and NURBS surface fitting to the boundary points. (**NURBS**) stands for "Non-uniform rational basis spline," which is a mathematical model commonly used in computer graphics for generating and representing curves and surfaces.

Finite element analysis and surface model alteration modules are also required, but should be delivered in an instinctive fashion for doctors who are not knowledgeable in computer aided design CAD. Preprocessing for rapid prototype building should be automatic, and should comprise optimal orientation, support structure generation and build simulation modules. The model should be forwarded to the rapid prototyping machine in a pre-sliced format for speed and accuracy.

Until lately, science had not advanced far enough to replicate the intricacies of the human skeleton. Rapid developments in the medical machining industry have led to an augmented number of available products for orthopedic utilizations. "*Protomatic Medical*" is at the forefront of these new developments, creating impressive medical prototypes that have been utilized by orthopedic doctors and surgeons all over the country. *Protomatic Medical* company, as well as many corporations continue to grow by developing more types of components required for artificial joint replacements, also known as orthopedic implants.

Thanks to the capabilities for rapid medical prototype testing of custom CNC orthopedic implants, *Protomatic Medical* has been capable to continually enhance the product quality while also refining the quality of human lives.

The widespread testing capabilities complement the medical industry innovations by helping speed time to clinical trials and commercial launch. Over the years, it has been established as a quality provider of orthopedic devices, forming long-lasting partnerships with customers who need high-quality precision medical products.

Operating a rapid schedule may take orthopedic implants from the design concept to medical device prototyping and on to production in record time. Utilizing the CAD/CAM systems, will yield high-quality accuracy of medical products quickly and efficiently.

3D Printing Rapidly Expanding in Medical and Dental Device Manufacturing

3D printing occupies a small but rapidly escalating role in medical and dental device manufacturing. Rendering to the 2013 Wohler's report, the 3D Printing industry had accumulative revenue of $2.204 billion in 2012. Medical and dental implementations were accountable for roughly a sixth of it at $361 million. Overall 3D Printing Industry is predictable to be $8.4 billion in 2020.

Implementation in medical devices includes rapid prototyping, Volatile Organic Compound (VOC), printing medical device components and surgical & dental guide mechanisms. Hospitals are starting to utilize it for producing 3D models of surgical areas to assist planning of complex maneuvers like skull base tumor removal. Other frontier implementations in the medical field which are still in research stage include Bioprinting the scaffold for intricate organs, organ systems, blood vessels & heart. It is being experienced to print skin for burn victims, print cartilage and bone for orthopedic implants.

Vital Controversial Issues

Like any new technology, 3D printing also faces vital controversial issues such as moral, religious, regulatory, and legal issues particularly bearing in mind the sensitive nature of the healthcare industry. Its cost, precision and effectiveness of large scale manufacturing also has a long way to go before it matches traditional high volume manufacturing. Therefore, while 3D printing in healthcare is interesting, it is likely to be somewhat controversial being confined to low volume lesser regulated areas in the near future. Nonetheless it could evolve into a intricate, interdependent eco-system of biotechnology firms, device OEM's, officials, insurers, patients, care givers, government and others.

3D printing is a procedure for making a physical object from a three-dimensional digital model. Additive manufacturing, usually utilized synonymously with 3D printing is a manufacturing performance where solid objects are manufactured by adding one or more specific material in successive layers. Although the term additive manufacturing is utilized interchangeably with 3D printing, it is not the only process involved. Even though, 3D printing is predominantly additive, subtractive manufacturing procedure like cutting, drilling, polishing, chiseling can also be utilized in 3D printing operations.

3D printing is done by a 3D printer utilizing a computer designed digital model. In additive manufacturing, the material/s is conveyed by one or more nozzles on a moving platform. This result in patterns laid one above the other, causing 3D objects. The digital model provides the inputs to the printer regarding the materials utilized, mobility of the platform and the quantity dispensed.

The materials utilized may be plastics like ABS or acrylic, metals including stainless steel, sterling silver, glass, ceramics, resins, sandstone or rubber. Most low cost 3D printers are filament printers where the extruder is heated up to melt plastics. The extruder then squirts out a thin string of molten plastic onto the moving platform's surface. More expensive 3D printers utilize a resin, and the most-expensive models utilize a powder that is melted or liquefied to bind to the layer below it.

Windows 8.1 Offers Support to 3D Printing

3D printing has emerged from a niche process or a hobby to a level where it has created interest among mainstream manufacturers and OEMs. Use cases are now sprouting in many industries including aerospace, defense, medical, automotive, architecture, consumer products among others. The growing importance of 3D printing processes can also be gauged by the fact that the latest version of operating systems like Windows 8.1 offers support to 3D printing. 3D Builder is a free 3D printing app which is bundled with Windows 8.1.

3D printing also plays a small but rapidly expanding role in medical and dental manufacturing. It is now seeing its first commercial use in areas like orthopedics, dermatology, disposables, surgical and in the manufacturing process of medical and dental devices. Additive manufacturing market landscape and applications especially with respect to the healthcare domain provides solutions that can be of interest to service organizations with the best practices applicable.

$8.4 Billion in 2020

According to the 2013 Wohler's report, the 3D Printing industry had collective revenue of $2.204 billion in 2012. Medical and dental implementations were accountable for about a sixth of it at $361 million. Generally 3D Printing Industry is projected to be $8.4 billion in 2020.

Applications in Organ Growth

The substantial progress made in the field of regenerative medicine over the past 30 years in its capability to Bio-fabricate living tissue substitutes. Traditional methodologies constructed on scaffolding and nano/micro/engineering are narrowed in their capability of yielding tissue constructs with accurate biomimetic characteristics. 3D bioprinting technology, bridges the deviation between synthetically engineered tissue builds and inherent tissues. Accordingly, 3D bioprinting offers extraordinary adaptability to supply cells and biomaterials with accurate control over their arrangements, spatial deliveries, and architectural precision, consequently accomplishing thorough and individualized review of the adequate form, structure, and configurations of target tissues and organs. Below it is briefly described current advancements of 3D bioprinting technology and related Bioinks proper for the Bioprinting procedure. The emphasis is on the implementations of tissue engineering technology in biofabrication of biomimetic builds of a number of typical tissues and organs, as well as blood vessel, heart, liver, and cartilage, summarized as follows:

- Assist planning of complicated surgical maneuvers like skull base tumor removal
- Precise arrangement of human embryonic stem cells – 3D printers can be used to squirt nano liters of stem cell culture which contains as little as 4 or 5 cells in precise arrangement over a scaffold to make complex organs and organ systems
- Organs and Organ Systems – There is potential for 3D printing of organs like liver and kidney. *Organovo* now expects to unveil a printed human liver of 500 micron thickness
- Printing blood vessels & heart tissue – The same process can be used to make blood vessels to be used for grafting purposes

- Printing skin – Printing skin involves designing a scaffold utilizing collagen like material and applying pluripotent stem cells over the scaffold

Application in Orthopedics

- Printing cartilage & bone – To print this kind of bone substitute, the 3D bio-printer generates a scaffold in the bone form and coats it with adult human stem cells, which can grow into many different tissue kinds. The printer's "ink" comprises of a polymer named polylactic acid and a gel-like material called alginate. This distributes the hard, mechanical strength of bone, along with a cushioning substantial for the cells.

Applications in Medical Devices

- Rapid prototyping – Implementations of 3D printing methods in the expansion of prototypes shrinkages the time to market and advances the quality of the final product. Utilizing 3D printing also progresses the synergy between numerous teams like industrial design team, hardware team, and software team thereby cultivating the probability of discovering the preeminent fitting.
- (VOC) Voice of Customer – 3D printing permits businesses to obtain valued user input at an early phase in design
- fabricating medical devices and mechanisms
- fabricating surgical guides – These are the "compasses" that surgeons utilizes to recognize skin surface markings over which incisions can be made to reach inner organs.

Rapid Prototyping

HCL provides rapid prototyping and low cost growth explanations for a large group of manufacturing industries including medical OEMs. The competences include formation of design data, modeling, and manufacturing utilizing CAD-CAM. It also includes software for CNC which are linked to 3D printers, VOC facilities linked to rapid prototyping.

These services make certain that the time to market is decreased and the product is closer to what the customer desires. Rapid prototypes aid the manufacturing design team and the engineering teams agree on an optimized design much more quickly. It also overcomes the supply chain management difficulty of traditional manufacturing for prototypes.

Rapid prototyping utilizing 3D printing also enables faster launch times for the product by enabling concurrent engineering as the whole product is not needed for a large portion of test cases.

Design Templates for Various Devices

Separately from the above mentioned services, HCL keenly pursues the advances of frameworks and solutions to deliver design pattern solutions for many common devices together with disposables and surgical instruments. These patterns, the generation of designs of these instruments can be much quicker than starting from scratch.

1. Advancements of software to classify bony defects by a 3D scan whose dimensions can be utilized to design a substitution
2. Software for 3-D–printed surgical guides which surgeons position over the surface of the body to lead their drills and knives as they bore and cut into flesh and bone.

By integrating inventive solutions around 3D printing, HCL brings flexibility in the novel product development workflow.

Limitations and Common Issues with 3D Printing

It would have been clear by now that 3D printing has some obvious benefits. Though, sensitivity to a number of considerations is needed when new boundaries are broken utilizing science and engineering. Following are the common issues that arise out of the emergence of 3D printing.

Cost and Time

3D printing cannot contend with mainstream traditional manufacturing technologies with respect to cost and time for manufacturing. It may take anywhere between a few minutes to several hours to manufacture a product utilizing 3D printing contingent upon the material utilized, size and intricacy. It is also possible that 3D printing companies might lock in consumers through a proprietary supply chain channel which might upsurge the cost of ownership.

Limitations of Material and Size

Today 3D printing is frequently done with only limited materials like plastics, resins and a few metals. Utilization of other materials though technically likely can be very costly and/or time consuming. In healthcare domain, care should be taken while generating implantable usage of bio compatible materials which would evade allergic responses while generating tissues or other implantable is narrowed right now to labs and proof of concepts. Further, it is particularly difficult and expensive to construct product which has more than one material in it. The consiguration of the printer also restricts the size of the product and it may be very difficult to yield parts of large industrial machines.

Competing Technologies

Competing technologies are now being presented which challenge 3D printing in at least several areas. One of the maor threats is "Laser Origami" which is claimed to be significantly faster and precise compared to 3D printing. It is assumed that laser utilizes a mixture of cuts and heating in order to separate and bend the plastic sheet into what the user wants to form.

Ethical and Intellectual Proprietary (IP) Considerations

The competence of artificially Biofabricating human organs and other body parts utilizing 3D bioprinting will elevate ethical, moral and religious queries about the part of man in operating the nature's own assortment process.

Most people do not yet comprehend that research is being directed at present, which involves not only printing non-human cells, but also integrating human cells with non-human cells.

The rapid advancements of 3D printing will also generate main contests in relation to intellectual property (IP) theft. It is documented that 3D printing is losing at least $100 billion per year in IP globally'

Insurance Considerations

Since 3D printing is upsetting industry, it is not simple to comprehend the future consequences beyond educated guesses. Since insurance industry characteristically requires historical data to analyze claims, they have a somber contest with respect to calculating the claims that may the result of this industry group.

The possible claims can be a product liability containing accidental injury/damage due to the 3D printed product or legal liability due to IP matters.

With a supply chain which may not include any physical products as designs and fabrication operations can be sent digitally, traceability becomes a difficulty too. This demands expert protection shelter as insurance.

There could also be an employer's responsibility angle, as the raw materials utilized are from time to time powder based, so there is potential for breathing issues for employees working the machines to long-tail employers' obligation.

Regulatory Considerations

The FDA presently deals with 3D-printed mechanisms in a similar manner it deals with traditionally made medical mechanisms. In order to to obtain FDA approval of aa mechanism, its inventors must either demonstrate the mechanism is corresponding to one already promoted in the market for identical utilization, or the mechanism must experience the procedure of reaching pre-market endorsement

Since 3D-printed products are manufactured utilizing a dissimilar fabrication technique than traditional medical mechanisms utilization, they could necessitate extra or diverse methods of testing. There are 2 FDA laboratories are observing the manners 3D printing may affect the way medical mechanisms are fabricated.

The FDA's operational performance and device utilization laboratory use computer-simulation technique to show how changes to a medical product's design could influence its safety and performance in numerous patient populaces.

The FDA's Research laboratory for Solid Mechanics emphasizes on how diverse printing techniques influence the strength and resilience of the materials utilized to make the mechnisms. The laboratory's discoveries will assist scientists to advance standards and set factors for scale, materials, and other acute features that underwrite to product safety and novelty.

3D Printing Technology - Disrupting Traditional Manufacturing Practices

3D printing technology is beginning to disrupt traditional manufacturing practices. Although it still cannot beat the cost and scale advantage of traditional manufacturing which is supported by considerable supply chains and just-in-time practices, it has started to bring fruit in certain niche areas particularly small scale manufacturing and prototyping, Figure 7.2.

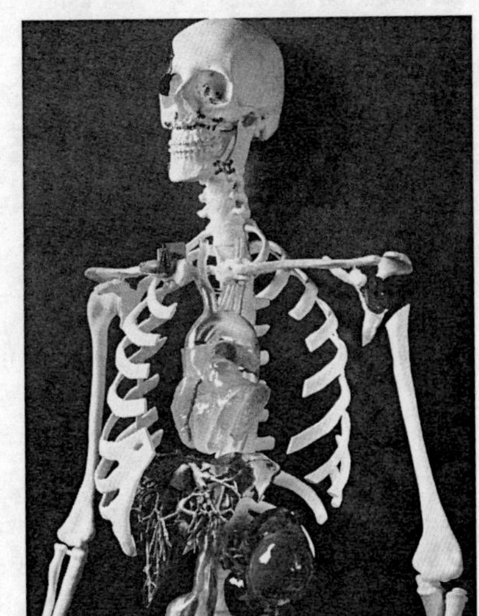

Figure 7.2: 3D Printing Assisting Orthopedic Surgery
Courtesy: Materialise

Not surprisingly, it has found application in healthcare too. There are now excitements about 3D Bioprinting of organs, organ systems, blood vessels etc. While these applications are at the cutting edge and have not yet reached the stage of commercialization, 3D printing is increasingly being used commercially in new product development of medical and dental devices. This is possible because of its ease of use and flexibility in rapid prototyping if there is an in-house manufacturing facility, which can produce prototypes inexpensively. It also promotes collaboration between the industrial design and engineering teams which usually work in tandem during the initial phases of new product development.

Therefore, companies which provide designing solutions as a service to OEMs can leverage the advantages provided by 3D manufacturing in its new product development and sustenance workflow. These companies can give specific services to medical device OEMs by providing design templates of various commonly used medical devices like disposables and surgical instruments. The service firms can also actively pursue development of software services for 3D OEM's so that they can support medical functions like identifying bony defects by a 3D scan whose dimensions can be fed into the computer to design a replacement.

Like any other new disruptive technology, 3D printing applications in healthcare also brings with it a unique set of challenges and considerations. These include challenges on ethical, moral and religious grounds and considerations with respect to insurance, intellectual property and regulatory issues. Good faith and thought therefore needs to be applied when such technology is applied to a field as sensitive as healthcare. While the intentions are noble, proper frameworks and guidelines should be evolved which makes sure that the considerations are addressed while not stifling innovation.

While 3D printing in healthcare is interesting, it is likely to be somewhat controversial being confined to low volume lesser regulated areas like prototyping, training, research in the near future. However it could evolve into a complex, interdependent eco-system of biotechnology firms, device OEM's, regulators, insurers, patients, care givers, government and others.

Orthopedic Prototypes -Biocompatible Implants

The development of a new orthopedic device from drawing board to manufacturing typically takes months. Involving high-precision design work, and subject to FDA approval, the process leaves little room for error. Bioprinting - Rapid prototypes (RP) help speed the refining of designs and verification of manufacturability, but high costs have usually precluded companies from building orthopedic prototypes in-house. Fortunately, recent RP technologies have eliminated this issue.

Corporation used RP sparingly, where time and budget permitted. In most cases, prosthetics development companies worked off CAD drawings to refine designs and then developed metal samples downstream, Figure 7.3, and 7.4. Often, this was sufficient, but often design flaws did not emerge until late in the development process, when design changes were expensive and time-consuming to make amend.

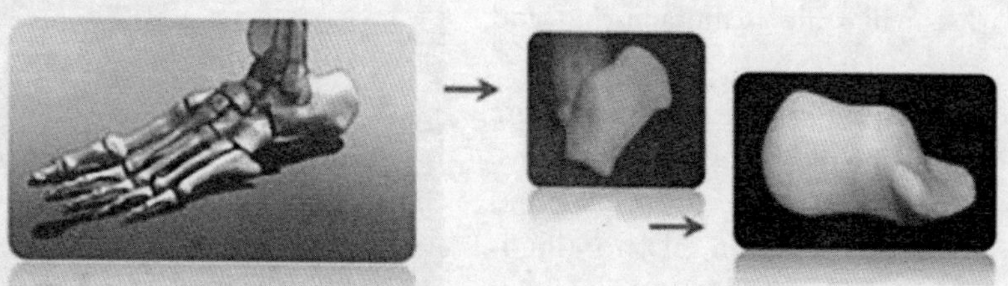

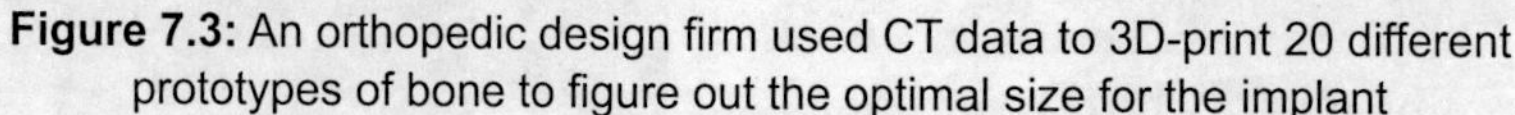

Figure 7.3: An orthopedic design firm used CT data to 3D-print 20 different prototypes of bone to figure out the optimal size for the implant

Figure 7.4: Optimal Size for Implant

Droplet-Based Deposition – Bioprinting Ink-Jet Technology

The advent of droplet-based deposition, derived from ink-jet technology, has allowed the development of compact, cost-effective 3D printers such as the Alaris30 from Objet, Billerica, MA, (objet.com) that fits in an office environment. The machine uses the company's photopolymer materials to build prototypes of orthopedic devices from CAD models, which in turn, can be generated from CT or MRI data. This allows fixtures to be customized for each patient. Since the fixtures fit well, they reduce patient discomfort, encourage healing, and last longer than traditionally developed implantable.

The 3D printer's jetting head slides back and forth along the X-axis, depositing a single, ultra-thin layer of photopolymer resin onto the build tray. Immediately after building each layer, bulbs alongside the jetting bridge emit UV light, curing and hardening each layer. This step eliminates the additional post curing required by other RP technologies. The internal jetting tray moves down slightly and the jet heads continue building the model, layer-by-layer, until it is complete, Figure 7.5.

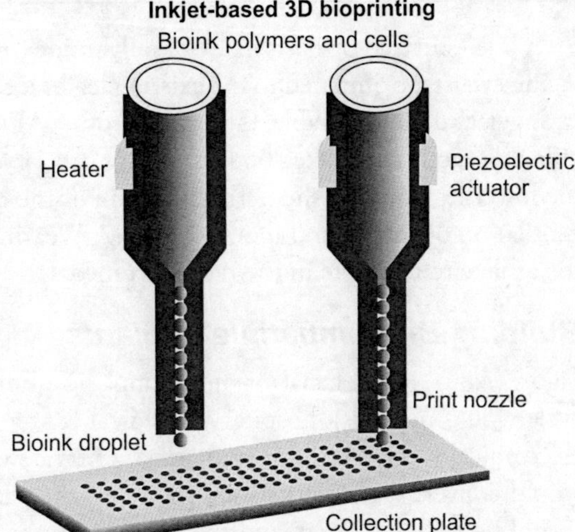

Figure 7.5: Schematic Diagram of Inkjet-based Bioprinting

Courtesy: Mohsen Soori

Bioprinter may print high-resolution prototypes in a few hours. Bioprinters yield prototypes for diverse project. This practice has increased the quality and manufacturability of the designs. For instance, prototypes can highlight possible areas of difficulty such as undercuts that are too small. This lets the business easily adjust the design before it reaches the patient for review or re-machining.

Additionally, sometimes patients modify design after viewing prototypes. Bioprinters provide patients a solid product to see and touch, rather than just CAD renderings or drawings.

Rapid prototyping in-house has been particularly helpful for determining the functional requirements for the small parts commonly utilized in orthopedics. In one instance, a patient requested with a new spinal device. Within a few days the CAD drawings were created and then printed the first 3D prototype. A few changes to the design necessitated the Bioprinting of a reviewed prototype one day later. With prototype in hand, the patient successfully communicated the design to investors and founded the start-up within very short period of time. Having a physical, working prototype was a compelling factor in obtaining the necessary capital investment, Figure 7.6.

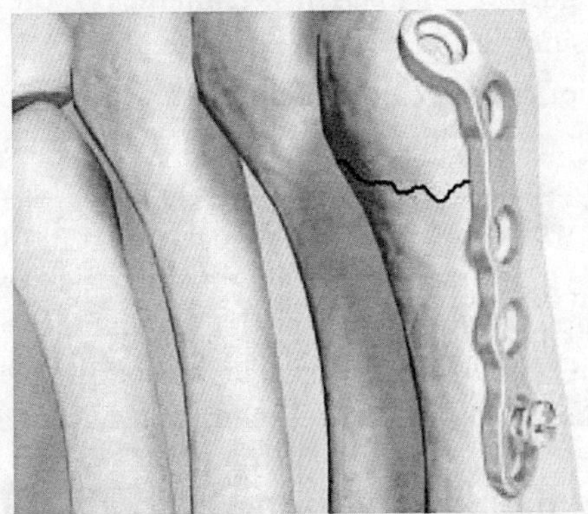

Figure 7.6: Orthopedic Plate Design

Courtesy: Materialise

In another instance, the orthopedic corporation partnered with a patient on the design of a plate for fractures. Utilizing a 3D CAD model generated from CT and MRI data, the designers printed 10 bones with diverse surface geometries and five prototype plates to study which plate would be most versatile and economical on diverse kinds of surfaces. This would have been too cost-prohibitive if a 3D printer was not utilized.

Functional Medical Devices

Another rapid technology called laser sintering allows the manufacturing of functional orthopedic devices and supplies. Manufacturer uses a high-end laser to fuse small particles of powdered materials. These include medical-grade stainless steel, titanium, cobalt chrome, and a variety of plastics based on polyamide, which is biocompatible and resistant to most chemicals. Additionally, a special PEEK material, considered one of the highest performing thermoplastic polymers, can now be processed by laser sintering.

A laser-sintered orthesis made of polyamide for disabled professional racing cyclists. Compared to the existing leg brace, the orthesis is lighter, stronger, and more durable. Generated from CAD data, the device is also tailored exactly to athlete's body, and thus can serve both as a support for the lower leg and as a shoe. The structure of the orthesis also optimizes ventilation of the leg and foot, Figure 7.7. Wearing the device may help the athlete to compete in the race for a medal.

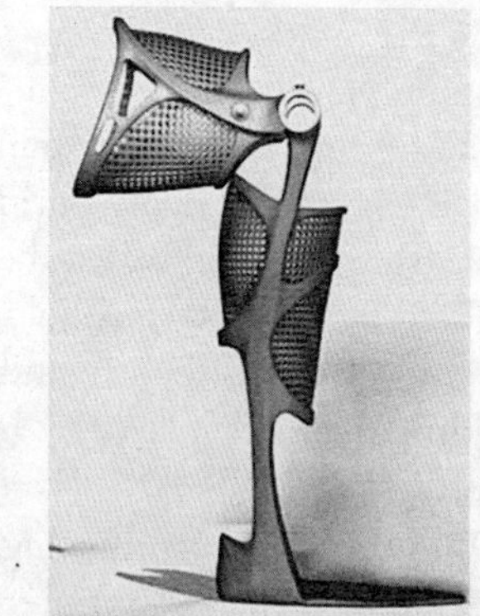

Figure 7.7: 3D Printed Leg Brace
Courtesy: Barceworks, Canada

Building Biocompatible Implants

Figure 7.8, - For a facial or an orthopedic implant, the 3D printing cycle would include a hospital or medical center creating the scan and electronically transferring the data to a service center. Here, software would convert scans into a solid model of the part to be printed. Parts would be printed, cured, and then sintered to provide strength. Finally, parts would be shipped to the hospital for implantation.

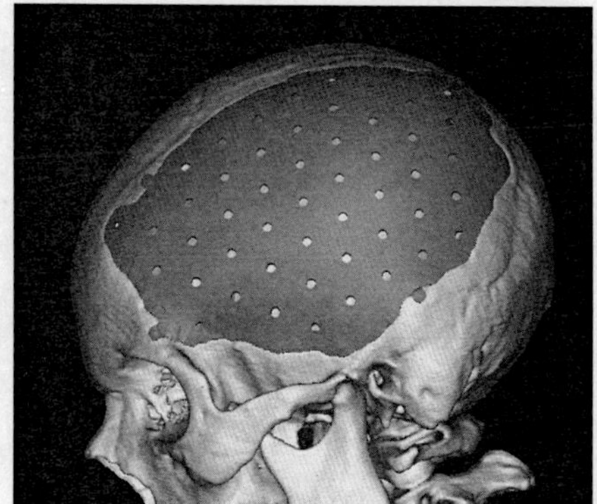

Figure 7.8: FDA Clearance - 3D-Printed Titanium Cranial/Craniofacial Implants
Courtesy: BioArchitects

An orthopedic company is experimenting with the 3D printing of pyrophoric metal powders to build orthopedic implants as well as engineered lattice structures. The process fuses titanium and magnesium materials, which in powder form are highly flammable. This technique is more cost-effective than laser sintering or electron beam melting.

Biomaterials can have either a biocompatible or bioresorbable quality depending on the function of the implant in the body. Biocompatible implants use titanium, to which the human body will not react. The material is used for functional implants that typically survive the life of the patient without degrading. Examples include hip implants and biomedical screws. Titanium is the primary metal used in biocompatible implant creation because it is corrosion-resistance and can interface with tissue.

Replacement of Pathologically Altered Tissues by Transplanting Living Cells

In contrast, bioresorbable implants utilize materials which degrade in the body over time. The body ultimately substitutes the material with living tissue. Present tissue-engineering approaches are motivated by the rebuilding or replacement of pathologically altered tissues by transplanting living cells in mixture with supportive scaffolds fabricating of many biomaterials. This opens a broad range of implementations for bioresorbable materials. Facial scaffolds to restore the jaw or skull are a characteristic instance. Magnesium is the main metal utilized in these implementations as it is effortlessly corroded by the body, letting the implant to be substituted by natural tissue.

3-D Printing

Additive manufacturing, usually called 3-D Bioprinting, can be utilized to fabricating substances vary from surgical guide mechanisms and medical device prototypes to batteries.

Many of the most inspirational implementations of the technology, nevertheless, can be found in new utilizations of the technology to deal with previously unfulfilled clinical requirements.

Printing Tissue

The fabrication procedure involves generating an object or objects via a manufacturing system that adds consecutive layers of materials, the "Bioink" of 3-D Bioprinting, versus the more traditional methods of utilizing molding and machining system. 3-D Bioprinting has illustrated potential in the field of growing tissue and organs, with Bioink comprised of living cells.

Prosthetic Limbs

Among the medical implementations of 3-D bioprinting that has seen the most progress is its utilization to product custom prosthetic limbs. One of biggest difficulty with traditional prosthetics is that they repeatedly do not perfectly fit patients' limbs. The capability of 3-D printing to make objects of virtually any form, utilizing specifications from a computer program, solves this issue.

One of the most touching instances of how 3-D printing has moved the standard of care can be found in the case of a young child who had been diagnosed with a disorder identified as arthrogryposis, which prohibited her from lifting her arms. At birth, she was only capable of moving her thumb and had difficulty learning how to walk.

To aid the little girl, scientists utilized 3-D bioprinting to custom-tailor exoskeleton. 3D Bioprinting system utilizing ABS polymer Bioprinted the exoskeleton enabled her to acquire control over her arms. The researchers were capable to yield a series of exoskeletons to fit the girl as she grew older. At the age of four, she referred to the exoskeleton as her "magic arms."

Figure 7.9: 3D Exoskeletons – 3D Bioprinted
Courtesy: Homayoon Kazerooni's Engineering Lab - Elena Zhukova

Custom Exoskeletons

3D Bioprinting is also well well-matched to making tailored made prosthetic hands. In recent times, a 12-year-old child tailored with a bright new prosthetic hand competent of gripping and holding objects. The cost of the 3D prosthetic was the fraction of a cost of customary prosthetics.

Personalized Airway Splint

3D Bioprinting has also demonstrated lifesaving in the instance of a baby who had trouble breathing. As a six-week old baby, would turn blue after stopping breathing. Doctors gave him a small chance of surviving.

At the University of Michigan scientists, doctors and researchers were gifted to save his life by developing a custom airway splint to substitute his crushed windpipe. A CT scan enabled the team to develop the device, Figure 7.10.

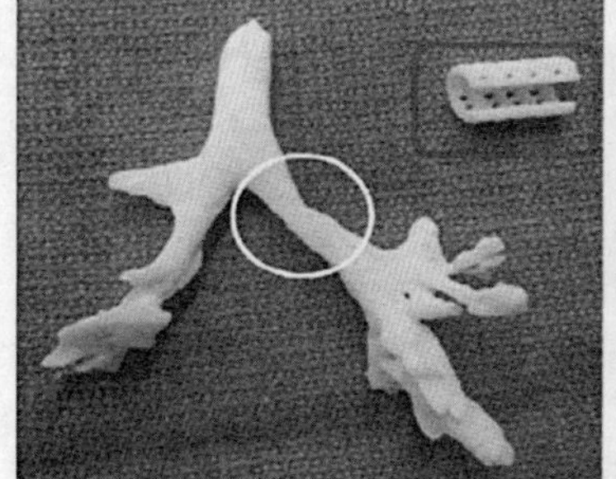

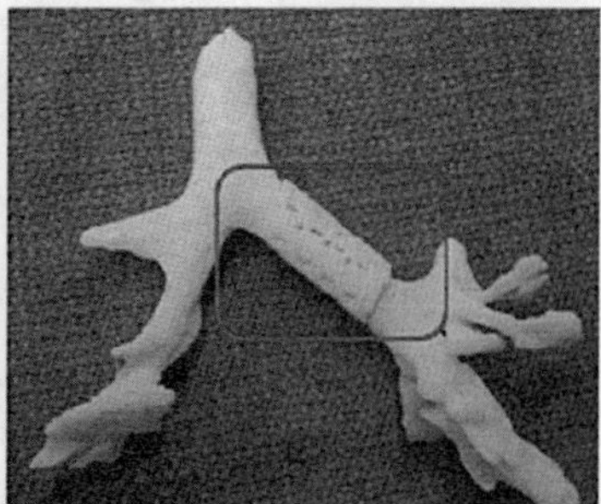

Figure 7.10: 3D-Airway Printed Splint
Courtesy: University of Michigan

Bone Scaffolds

It even has prospective to yield bone-like implants for reconstructive surgery that then perform as scaffolds for real bone cells, which multiply and take over the implant as real bone as the implant dissolves away. The designer has even developed a product known as Osteofab that was utilized earlier in the year to replace 75% of U.S. patient's skull. The product has received 510(k) clearance from the FDA.

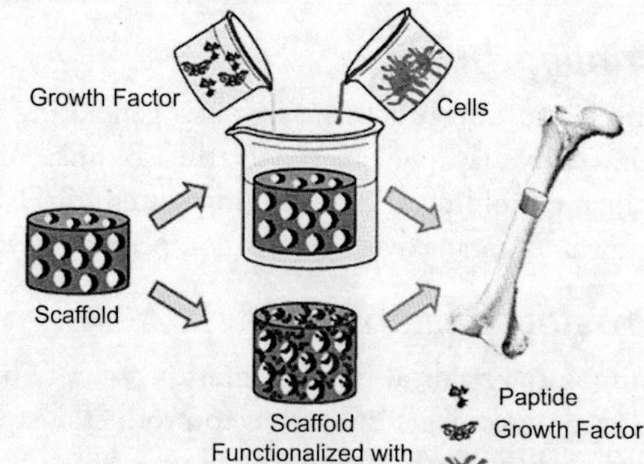

Figure 7.11: Figure 2. Schematic Representation of 3D Bioprinting Bone Tissue Engineering
Courtesy: Vallet-Regi

Cardiac Models

3D Bioprinting is also aiding designers better envision the parts of the body they are constructing devices for.

It can be utilized to yield extremely accurate models for cardiovascular implemetaations, as in the instance of the "Heart Print" service from "Materialise." Traditionally, engineers have utilized blown glass for rigid parts or silicone for compliant models, which offer limited accuracy. The "Heart Print" service takes advantage of CT or MRI data to generate realistic models, Figure 7.12. "It allows acquiring medical image data and select diverse regions of the body, whether be soft tissue or bone, and generate precise 3D illustrations from it."

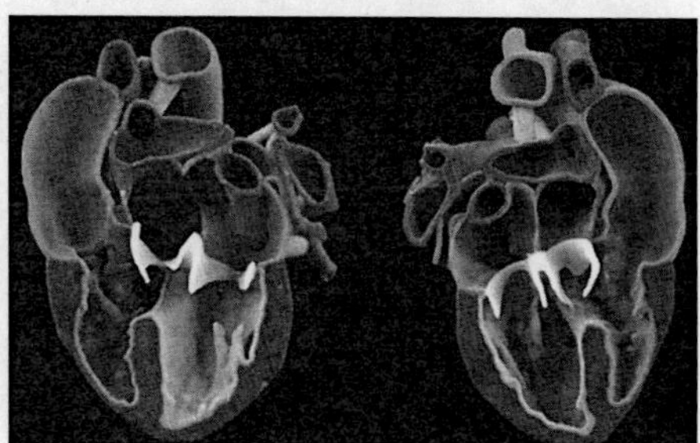

Figure 7.12: 3D Printed Heart Model with Congenial Abnormalities
Courtesy: 3D Systems

3D Printing Creates Unique Exoskeleton for Broken Bones

The upcoming innovation of health technology might not be developed by a technologist, scientist or roboticist, but by an artist, designer or dreamer. The notion of assimilating 3D scans of the broken limb with a 3D model to create an exoskeleton that guards the broken bone is an exclusive approach. The prototype is called Cortex.

For a broken hand, Cortex brings localized support for a fracture or broken bone, fundamentally shifting the pressure of the cast, in this instance the Cortex exoskeleton, to where the bone is truly broken versus immersing the entire limb in a heavy plaster or fiber glass cast, Figure 7.13.

Also, Cortex is a good for the environment —that it is recyclable which will help reduce the influence on medical excess which has been on the rise for the past 15 years.

Unlike other fiber glass and plaster casts, Cortex depends on 3D scanning and 3D printing to make the model for the cast.

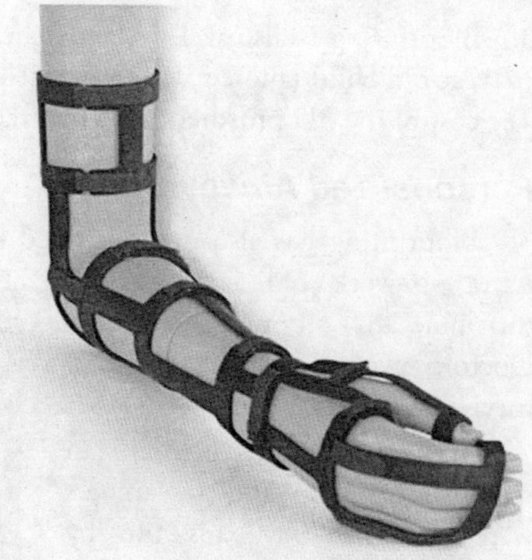

Figure 7.13: 3D Printed Cast Heals Broken Bones
Courtesy: Innovation Factory

Figure 7.14, Arthrogryposis, also recognized as arthrogryposis multiplex congenita, leads to "frozen" joints all over the body. When the fetus is not able to move in the womb, either due to unusual pressure on the uterus or some neurological disorder in the emerging child, muscles and joints are often immobile after birth.

Affecting about 1 in 3000 infants, the disorder is often identified before birth, and it destroys about 50% of infants born with it in the first year of life. Children who survive the first year of life with the disorder have to deal with exceptionally weak muscles and stiff joints, and frequently have to endure multiple surgeries to have any locomotion at all.

As an example, a new born baby survived infancy; by the age of three the child is basically immobile. The child could not move. Fortunately, a developer of a new technology: known as "WREX," the "Wilmington Robotic Exoskeleton."

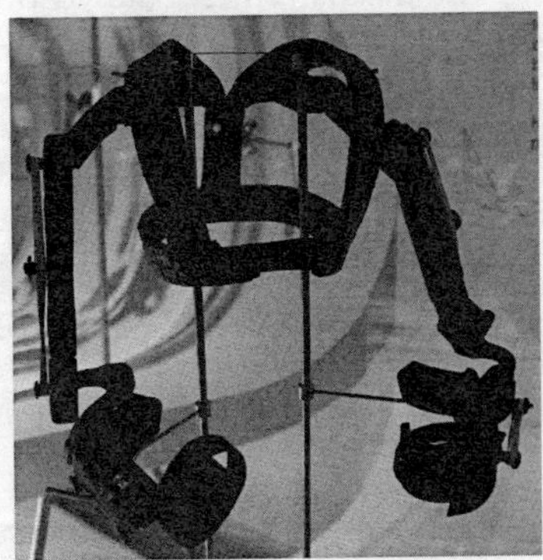

Figure 7.14: 3D Printed "WREX" Exoskeleton
Courtesy: Creative Commons

Suffer Muscle Failure

Many children agonize from diseases that rob them of muscle strength. Over and above arthrogryposis, Duchesne dystrophy and muscular dystrophy rob the children's ability to walk, play with toys, feed, and dress, and take themselves to the bathroom. Muscular illnesses can make it intolerable for a child to conduct simple normal movements. But the "Wilmington Robotic Exoskeleton" adds power to muscles that was not conceivable earlier.

The robotic exoskeleton is worn over the arms to give the child additional power. It is more of a piece of clothing than a burden of an extra attachment, Figure 7.4. The exoskeleton is made of plastic printed overnight with a D Bioprinter. Elastic suspenders grip the mechnism together and lift the arms and hands against the force of gravity.

As the WREX is made of polymer, it is lightweight and inexpensive. It can be printed out replacing the device as the child grows. The "Wilmington Robotic Exoskeleton" can strengthen the powers of muscles as it can give children the capability to move their arms in manners their own bodies merely cannot achieve.

3D Printing Robotic Exoskeleton Devices

Children are not the only people who can profit from robotic exoskeleton mechanisms. Another incidents as an example describing two adults who had arthrogryposis and one who had a spinal degenerative illness have been tailored with the mechanism. Even in maturity, it came to be possible for them to feed themselves, utilize a television remote, and work a keyboard for the first time in their lives. An unforeseen helpful side effect of the utilization of the mechaanism was liberation from the pain and rigidity of constricted muscles, and a greatly reduced incidence of falls.

WREX is a wearable mechanism. It is similar to a piece of clothing than a machine. The "exoskeleton," does not completely cover the arm. It is worn over a shirt and contains a few moveable plastic sleeves, spreading around the arm, powered by rubber bands.

Although the WREX could be mechanized, it is typically powered by the small residual forces of the wearer's arms, magnified by the elastic material dispersed in just the right places on the exoskeleton. It decreases the amount of power required in the shoulders, making movements of the arms slower, but more focused. The hands move without tremble and it may become feasible to pick up small objects.

The mechanism is tailored on a brace for users who can walk. It is tailored on a wheelchair for users who cannot walk. In children, utilization of the mechanism greatly upsurges participation in school—the child is capable to raise his or her hand. WREX can greatly increase reaching distance. Users who were not capable to pick objects off shelves and place objects on shelves sometimes gain these capabilities with the utilization of the mechanism.

The mechanism is useful for people who are paralyzed on just one side of the body by stroke. It can be utilized as a training mechanism that aids them relearn up, down, right, and left movments that their brains can no longer recollect. These learnt movements ultimately become second nature with practice on the mechanism.

The training mechanism is designed for at-home utilization, to strengthen the muscle movements learned with the help of a physical therapist. It may upsurge range of motion for the arms, but this range of motion is not limitless. Typically the mechanism can take the arms and hands into one of four positions.

Some children do not want to appear "handicapped" and reject wearing the mechanism in certain social situations. They have to utilize their own muscles, but occasionally they are more at ease with tough muscle movements than with the utilization of the mechanism when they are around their associates.

3D Printing Solutions for Digital Dentistry

The acceptance of digital technology is taking place fast in the dental industry and we see 3D Bioprinting as an essential role of this workflow tendency. The benefits of a process that incorporates oral scanning with digital design and digital Bioprinting over classical approaches are clear.

Production times are significantly condensed and bottlenecks are a thing of the past as many, unique parts are built in single Bioprint runs. By merging oral scanning, CAD/CAM design and 3D Bioprinting, dental labs can precisely and speedily yield crowns, bridges, stone models, and a variety of orthodontic appliances for dental offices, Figure 7.15. With a 3D printer executing the hard work, dental laboratories can remove the bottleneck of manual modeling and allow the business to expand and grow.

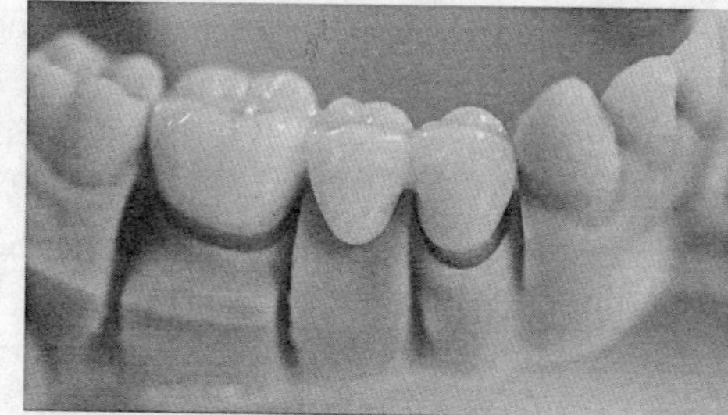

Figure 7.15: SLA 3D Printing Manufacture Temporary Teeth Restorations
Courtesy: University of California, Los Angeles (UCLA)

Digital Impressions: Virtually Perfect

Digital imprints signify cutting-edge technology that permits dentists to generate a simulated, computer-generated model of the hard and soft tissues in the mouth lasers and other optical scanning tools. The digital technique represents clear and highly precise dentations data in few minutes, without the necessity for classical dentation materials that some patients find difficult and disordered. Numerous patients find digital dentations a relaxed and more contented technique because classical dentation materials are evaded. The dentation data then is conveyed to a computer and utilized to generate rebuilding, frequently deprived of the essential for stone models.

Types of Digital Dentation Technology

There are two kinds of digital dentation technique presently accessible for dentists to utilize. One kind acquire the images as digital images, gives dentists and dental laboratories with a sequence of photographs; the other kind acquires photographs as digital video.

The photographs can be acquired lasers or digital scanning. Laser scanning utilizes focused light that is harmless and highly precise. It acquires all of the particulars and configurations of the teeth and gums while removing the patient's necessity to grip disdained, distasteful material in the patient's mouth. Digital optical scanners also are harmless and highly precise, but need teeth to be powder-coated with a unique spray before scanning to guarantee all parts of the dentation are registered correctly.

Benefits of Digital Dentations

Digital optical dentations considerably upsurge competence, productivity and precision, and make it possible for dentists to e-mail the virtual dentation to the laboratory, rather than send a classical dentation or stone model via regular mail. Also, digital dentations can be utilized to expedite same day dentistry rebuilding, thereby speeding up patient usage and reducing the necessity for many office visits.

Other benefits of digital denotations comprise:

- Enhanced image/dentation quality for better-fitting restorations
- Fewer chair time
- No necessity for distasteful dentation materials that made some patients to gag
- More comfortable, less anxious experience for patients and the dental team
- Reduced likelihood of dentation-taking errors and removal of material erroneousness for fewer restoration errors
- Patients incline to appreciate the new technique and state-of-the-art dental care, so they become more involved in, and better informed about, the treatment procedure because they can see their dentations on-screen chairside.
- The scan of the teeth being restored, as well as the opposing teeth and bite, can be accomplished in just three to five minutes.
- The digital dentation can be kept electronically indefinitely, which saves space, contributes to efficient recordkeeping, and supports a paper-free environment.
- Environmental dentistry and eco-friendly aspects include removing the necessity for disposable plastic trays and dentation materials, which otherwise would be polluting landfill space; digital data is removed with the "delete" button.

Digital Impressions versus Traditional Impressions

Traditional impression-taking potentially involves multiple materials and occasionally more steps. Because this is a highly delicate and skilled process, it is easier to introduce error throughout any of the numerous steps involved, either from the human element, or material defects such as voids, air bubbles, or improper setting or distortions.

Digital imprisoning devices eliminate much of the labor and guesswork associated with traditional impressions, and greatly reduce or eliminate errors. Simultaneously they increase time savings, both during the impression appointment and delivery of the restorations.

Digital impression devices, in most cases, eliminate the need for a return visit to the office considering the restoration can be made in the dentist's office rather than be shipped to a laboratory. If they are not made within the dental office, in most instances the restoration can be fabricated faster due to the elimination of the working time required with traditional materials, such as stone, and shipping times.

Digital impressions prevent the need for impression materials to be placed in the mouth for upwards of five minutes during conventional techniques. This allows patients who are fearful of gagging, or claustrophobic, to be more comfortable throughout the procedure. This advanced technology has allowed many patients to get their much needed dental work completed, which they may have avoided to prevent the use of conventional impressions.

3D Printing Digital Dentistry

With a physical impression, dentists rely on visual (usually with loupes) evaluations to determine if the impressions are ready to ship to the laboratory. A traditional impression is captured in the negative, making it more difficult to identify mistakes. If mistakes are identified, the dentist will need to take another impression, which means patients must undergo the procedure again; resulting in greater inconvenience and a longer appointment, as well as time lost, and added cost and material use for the dentist.

On the other hand, a digital scan enables dentists to see the "positive" image and magnify and evaluate it carefully. Errors can be corrected immediately before submitting the impression to the laboratory. With digital imprisoning, when the final impression is scanned and the bite registration obtained, a virtually articulated model of the preparation (one showing how the upper and lower teeth come together) is visible on the monitor.

The system also will indicate in color if any areas of inadequate tooth reduction are present. This allows dentists to modify the amount of tooth reduction immediately to ensure that proposed restorations will fit comfortably and function properly.

If an area is missed in the initial digital scan, some digital impression systems (such as the Lava C.O.S.) allow the operator to "patch" the scan without retaking the entire impression.

With digital impression technology, additional scans can be layered onto the original virtual model for enhanced visual representation. A repeat scan to capture a revised or voided image improves the previous virtual model, without introducing new errors. Only the previously missing data is added to the model. The more scans that are added, the more accurate the virtual model; this is exactly the opposite of a traditional physical impression.

Digital Dentation Procedure

With a digital dentation system, the dentist or dental assistant acquires an image of a tooth/teeth preparation. With some digital dentation systems, once the domain to be treated is anesthetized and removing the saliva and blood, the teeth are lightly dusted with specially formulated titanium dioxide powder in order to scan both configurations and the bite dents. Other systems (such as Cadent iTero and the dentation technology that go together with the E4D Dentist System for in-office CAD/CAM restorations) permit dentists to generate a powder-free, 3D image of the patient's teeth.

The digital dentation characteristically is acquired an intraoral wand that is place into the patient's mouth and moved over the surface area of the tooth or teeth. Most digital dentation systems utilize a chairside monitor to exhibit the dentation image as it is acquired, Figure 7.16.

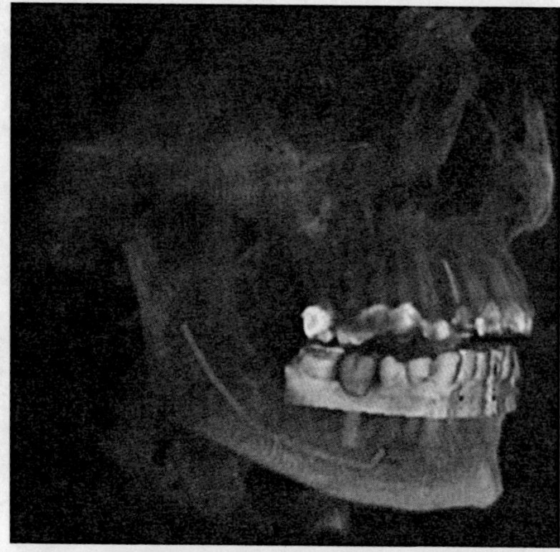

Figure 7.16: CAD/CAM 3D Printed Implant – Guided Placement with GALILEOS CAD/CAM Software

Courtesy: Compendium

Furthermore, most digital dentation systems also depend on point-and-click acquire, which necessitates the photographs to be patched together to generate the last digital ddentation. Though, the "Lava C.O.S" system is the only one that utilizes 3D real-time video capture to show live images on a touch-screen monitor. This live video capture generates the final digital dentations without the necessity to piece together data.

It takes almost 90 seconds to capture a digital dentation of prepared teeth. A dentation of the teeth in the opposite arch takes almost 45 seconds.

As the dentist evaluates the electronic real-time photograph, the dentist can enlarge and manipulate it for enhanced detail to guarantee that any possible faults are identified and corrected on-screen before forwarding the digital dentation electronically to the dental laboratory or in-office dental CAD/CAM system. The delivery workflow will vary contingent upon the digital dentation system the dental practice works with. Current digital dentation systems accessible for the dental office are digital dentation only or digital chairside CAD/CAM systems.

Digital impression systems — such as Cadent iTero, 3M ESPE Lava Chairside Oral Scanner (C.O.S.) and Sirona CEREC Connect — digitally record the tooth preparations and adjacent teeth in the arch or quadrant, as well as the opposing teeth and bite registration. The data is transmitted electronically to the dental laboratory for use in restoration fabrication.

Digital chairside systems — such as D4D Technologies E4D Dentist System and Cirona CEREC Acquisition Center (AC) — also record digital impressions, but the recorded data is used within the dental office to design, mill and deliver the final restoration in one appointment for same day dentistry.

The iTero and Lava Cos are the only digital system that can be used for all types of dental restorations, including bridges, crowns, inlays and on lays, and veneers. The other digital impression systems are designed for use with select types of ceramic and composite systems.

Training for Digital Impressions

Dental doctors and subordinates may take digital dentation. Some preparation in the utilization of the instruments is essential. Constructors of the various digital dentation systems characteristically provide in-office training of two to three days for installation, incorporation and certification, as well as supplementary offsite instruction at designated training localities.

Digital Impression Costs

After the initial investment - typically around $21,000 to $24,000 for equipment and training – digital scanning devices offer dental practices with numerous cost savings. They decrease the need for classical dentation materials, which cost approximately $30 per dentation. Meanwhile files are sent electronically, there are no mailing costs; the higher accuracy and ease of digital scanning mean there are fewer or no required adjustments or remakes, representing added savings in time, efficiency, and material and labor costs.

In instances where the restoration is not considered cosmetic in nature, the cost of digital imprisoning may be covered under some dental insurance plans or out-of-pocket deduction programs.

For a growing number of dental lab owners, digital dentistry is already here. With 3D printing as part of their business strategy, dental labs can speed part production while improving quality and precision.

Combining Oral Scanning, CAD/CAM and 3D Printing

By merging oral scanning, CAD/CAM design and 3D Bioprinting, dental labs can precisely and quickly yield crowns, bridges, stone models and a variety of orthodontic applications. This is contributed to 3D bioprinting enabling the competences of CAD design to the production stage, Figure 7.14.

With 3D Bioprinter accomplishing most of the work, dental labs remove the bottleneck of manual modeling and allow the business advance.

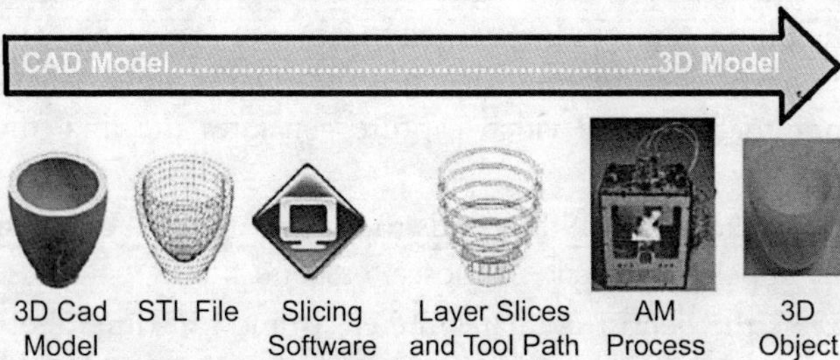

Figure 7.17: CAD/CAM Design to 3D Bioprint
Courtesy: Saudigazette

3D Printing – Significant Impact on the Medical Field

Lately 3D Bioprinting technology has stirred up the conventional trend, but there are countless number of people who are still not aware of this notable technology. Clearly, 3D Bioprinting is being carefully examined by scientists, researchers, designers, futurists, and hobbyists. No doubt, it will revolutionize our lives; 3D Bioprinting is already reshaping them. In the long run, 3D printing may have the most influence in the medical field, where fusion of living cells instead of plastic material in a 3D printer has led to bioprinting, Figure 7.18.

3D bioprinting has already transformed all the branches of the medicine. Moreover, it touches upon a contentious topic of artificial organs, and cloning.

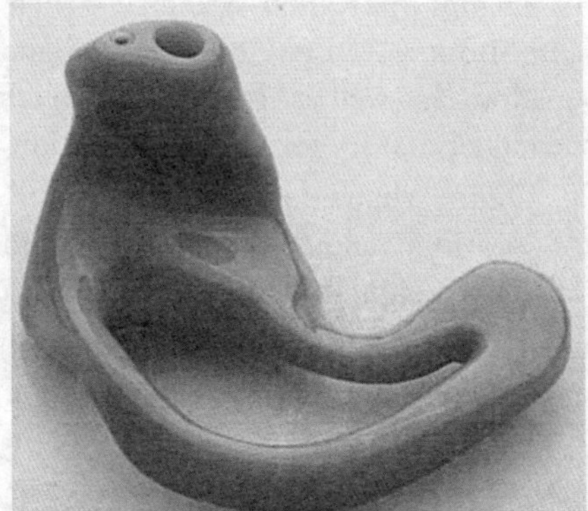

Figure 7.18: 3D Printed Hearing Aid

3D Printing Advancing Hearing Aids

Countless number of people do not comprehend that they have by now become an intwined part of 3D printing revolution by merely wearing hearing aids.

Interestingly enough, 98% of hearing aids, about 17 million users are 3D printed today. Manufacturers started to implement hearing aid 3D printing technology in 1998. 3D printing hearing aids is a significant improvement to patients and manufacturing alike. The process has been efficiently condensed to 3 steps stated as follows:

- scanning,
- modeling, and
- printing

A single 3D printer of a certain type of additive manufacturing is capable to produce 30 hearing aids in a single 90 minutes.

The hearing aids worldwide are efficiently fabricated 3D printers. The foremost motive for this makeover is owing to the fact that 3D printers have rehabilitated a manual, labor-intensive industry into a simple one.

The former procedure to construct hearing aids entailed of many vital steps extended from creating cast molds to changing them into ear dentations to pruning the finishing shell. Manufacturers of hearing aid engaged artificers and set up simple manual -shops to perform the operation, which spent slightly more than a week.

The hearing aids shaped with this operation were typically similar. Their similarity was echoed in adjustment and fitting issues for the finished product. Custom tailor adjustment and fittings are no longer an issue in the realm of 3D printing of hearing aids. The operation of carving and molding the finished product can competently be finalized within a day. 3D printing presented more science and technology into what was mostly a cumbersome art. Hearing aids operation is shrunk into a mere 3 steps, which are scanning, modeling, and printing. The the audiologist, in the new digitized process, scans the ear a 3D scanner to generate an ear dentations laser technique. The digital scan, generates approximately 100,000 to 150,000 pixel points of reference digital cameras. The digital scan is then sent to the individual to generate the necessary templates and geometric shapes to the specific ear dentations.

During the scanning step, the technician will perform several verification tests and alternatives tests of several geometric patterns for customizing the hearing aid for a specific group of customers. Consequently, shells are 3D printed resin and are adjusted with the vital acoustic openings and electronics.

The manufacturers of the 3D printers are rapidly fabricating the shells, once the sells are modeled. For instance, "EnvisionTEC's" printers can competently print 65 hearing aid shells or 47 hearing aid molds, all within 60 to 90 minutes at the most. The 3D printing speed aids manufacturers scale and adjust the demand to match and balance the supply. Furthermore, the digital file assists modelers adjust and reiterate adjustments to reutilize ear dentations to correct for flaws. 3D printers empower rapid prototyping and manufacturing to advance intricate an complex products. The amalgamation of specialized 3D printers with scanners at audiologists allowed simplification of the manufacture of hearing aids at any part of the world. This evolution permitted caring for countless patients worldwide.

Also, more than 90% of hearing aids are fabricated by "Stereolithography Printers," or SLAs, which is a type of 3D printers. 2% of hearing aids are not manufactured with 3D printers due to medical complication.

Cost Considerations of Hearing Aids

Characteristically, the utilization and introduction of 3D printing technology in a labor-intensive industry stimulates efficiency and decreases costs. This is particularly significant in an industry where product costs have up surged by 8% annually on an average within the past 20 years. The $2 billion hearing aid industry in 2014 is predicted to grow $4 billion between 2017 and 2019.

Though 3D printing machines decrease labor charges and enhance efficiency, they upsurge capital investment costs for hearing aid manufacturers. The utilization of 3D printers in the hearing aid business has motivated to bring down manufacturing costs for "In-The-Ear" or "ITE" hearing aids. The computerized automated process in manufacturing hearing aids has saved time and effort for manufacturers. The cost hearing aid 3D printers are significantly less than most industrial grade 3D printers.

While additive manufacturing technology has streamlined production and presented steadiness into the manufacturing procedure, 3D printing operations have also increased capital asset costs.

Depending on the product build size and intricacy, the price of a hearing aid 3D printer can vary from anywhere between $20,000 and $150,000. The outline of 3D printers in the production operation of the hearing aids has the possibility to drive down prices.

Digital Dentistry - 3D Printing

3D printing's influence to the dental industry has been game-changing. Scientists, and dental researchers personnel who work in various dental laboratories, conveyed that before the 3D printing has become the public view, dentists have been it for 10 years, to fabricate dentures that may not be fabricated in any other method.

3D printing aids to advance quality and speeds up the denture manufacture. Technology allows the patient to receive transparent 3D printed teeth for day-to-day use. The dentist 3D printer is now printing out a new dental implant, as well as dental crowns, bridges, stone models and a diversity of orthodontic implants, Figure 7.19.

Figure 7.19: 3D Printing - Dental Industry
Courtesy: 3der

Onsite 3D Printing

The world of dentistry and tooth restoration dates back a few thousand years to the Indus Valley Civilization where indication of tooth drilling has been discovered in a Neolithic necropolis. In 1840, the first dental institute opened in the US, and over the last 160 years, preventative dental care has progressed considerably.

Today current dentistry is on the threshold of a revolution with assistance from 3D printing and correlated technologies. A patient may step into a dentist's office, requiring a new crown. A machine may be chosen that captures a 3D digital scan of the teeth and loads the scan in the computer. Computer codes and commands are then sent to a 3D printer, which is placed at the dental office to print the patient's crown onsite, relatively in short time. The dentist may complete the task without patient leaving the office.

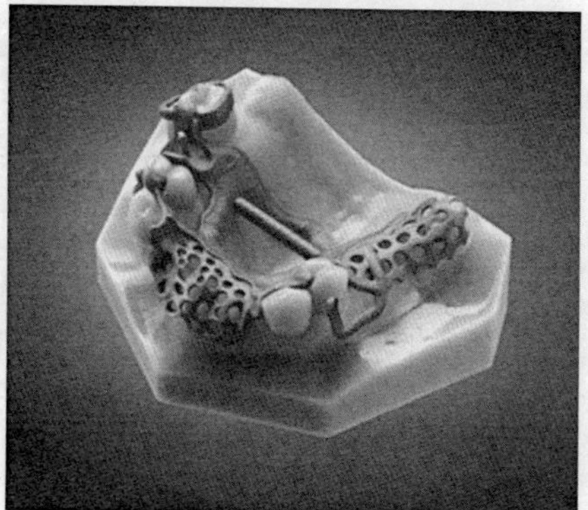

Figure 7.20: 3D Printing - Just-in-Time Dentures
Courtesy: NeoMetrix Technologies

This real-world practice is an excellent example of how 3D printing may be the catalyst for main changes in the medical field, leading to impressive progress and a $5.75 billion 3D printing industry by 2020.

For a mounting number of dental laboratory proprietors, digital dentistry is currently at hand. With 3D printing as a fundamental segment of their business strategy, dental and orthodontic laboratories can accelerate production, Figure 7.20.

3D printing translates the efficiencies of digital design to the manufacturing stage. By merging oral scanning, CAD/CAM design and 3D printing, dental laboratories can precisely and rapidly yield crowns, bridges, stone models and a range of orthodontic implants. Dental laboratories may eliminate the bottleneck of manual modeling with a 3D printer and enables the business grow.

3D Printing Body Parts and Bone

Earlier this year, an American patient received a radical surgery in which 75% of his skull was replaced with a 3D Bioprinted implant. This material was not only biocompatible but also a bone-like, Figure 7.21. It is intended to perform between 300 and 500 implants per month to patients in the U.S.

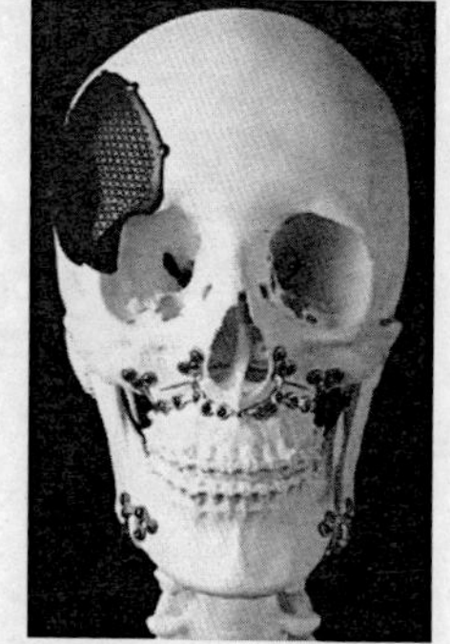

Figure 7.21: 3D Printed Patient-Specific Implants Replacing Damaged Bones – Using CT Scan - Virtual 3D Model of Patient's Face
Courtesy: Materialise

3D Printed Jaw Implant

In an evolutionary first in the medical arena, scientists have produced a method for 3D printing to manufacture an operational lower jaw implant that saved their patient from a massive contamination. The advent of printed implants can be correlated to an evolution in medical innovation.

The patient was an 83-year-old female who was anguished from a major infection in her mandible. Classical cures, such as taking away the lower jaw, would cause in greatly declined quality of life. Happily, a medical group of doctors combined with various biomedical engineers to develop an innovative treatment 3D printing, Figure 7.22.

Digital and computational technology generated a veritable revolution in the medical orthopedic field. Training and effective utilization of such technology is desirable. Doctors and engineers together work together to advance and design computer and printing procedures to innovate the technology. The 3D Bioprinted mandible was fabricated a titanium powder in only a few hours. Within 1 day after surgery, the patient had normal operative speech, swallowing and movement.

3D Printing and Growing a Bionic Ear

The scientists and researchers developed a bionic, artificial ear. With the help of 3D printing, the team created a skeletal structure which is seeded with cartilage cells, and 10 weeks later, a fully ear is fashioned.

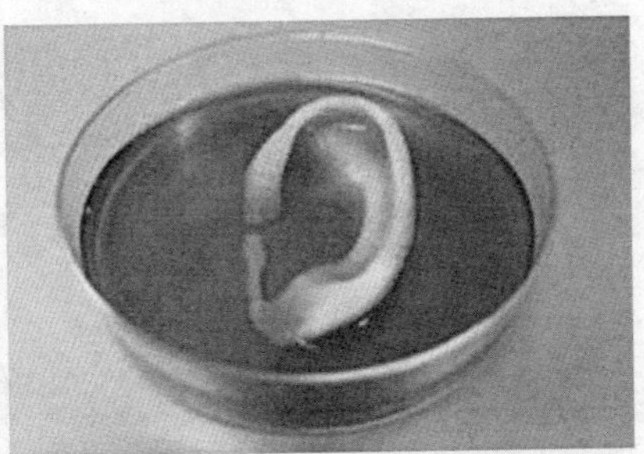

Figure 7.22: 3D Printing and Growing an Ear

Courtesy: Wake Forest Institute, USA

3D printing organs and stem cells is currently being researched. 3D printing an ear material to synthesize human cartilage is also being developed. Researchers have been working to develop a living Bioink that can be utilized to 3D print the cartilage for a human ear. .To make the ear, a patient's twin daughter's ears were scanned. Researchers utilized a 3D printer to consstruct a plastic mold based on the scan. Those printers, similar to a home inkjet, lately have also been adapted to experiment with making chocolate and even kidneys.

Researchers and scientists are then injected a liquid of collagen, living cartilage cells, and culture medium. An ear was formed in a several minutes as the mold was injected.

To examine whether their ear-mold would develop living, useful ear cartilage, the researchers inserted samples under the skin on the back of laboratory rats. In three months, cartilage cells took over the collagen, making for a solid-yet-flexible chunk of cartilage that retained its precise shape and size.

3D Printing Stem Cells Used in Artificial Organs

3D printing permitted scientists and researchers to harvest clusters of embryonic stem cells. The scientists utilized the method of valve-based printing in order to keep these cells in high level of viability, to stay precise to produce spheroids of identical size and to maintain their pluripotency that addresses to differentiation into any other cell type, Figure 7.23.

While artificial organs are almost at hand, yet the accomplishment thus far is extremely significant for drug testing purposes when artificial human tissue or even printing cells directly inside the body, Figure 7.24.

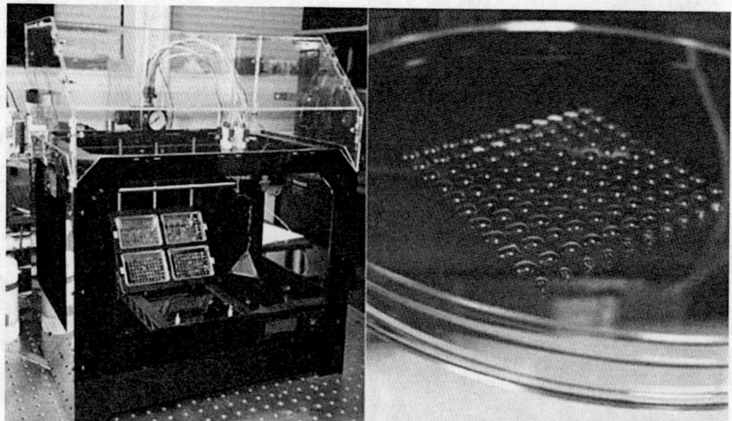

Figure 7.23: 3D Printing Stem Cells Used in Artificial Organs
Courtesy: Heriot-Watt University in Edinburgh

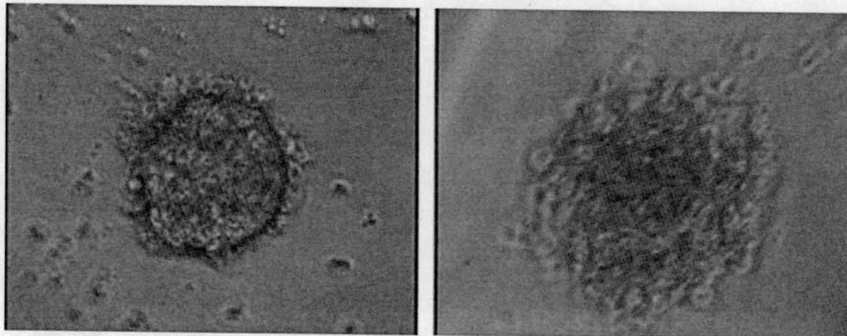

Figure 7.24: Aggregated Embryonic Stem Cells after 24 hours (Left) and after 48 Hours (Right)

Accordingly, the printed human embryonic stem cells anticipate very high cell viability. After printing the stem cells, it is discovered that the stem cells reserved their key biological occupation, which is their potency—their capability to be converted into any other kind of cells in the body. Also, the key constituent of the printer is essentially its motion control.

Stem cells aggregate are essentially tiny balls—they can be called tiny spheres, or tiny balls—made up of hundreds of stem cells. Stem cells incline to aggregate together. They would not endure living in single cells, especially human embryonic stem cells—they like to aggregate together and formulate these 3D objects, or micro-tissue, which is called "spheroid cell aggregates." They are very tiny, frequently smaller than 1 millimeter in diameter.

Essentially, we utilize the liquid substance to generate the gradient of sizes of cell aggregates. There are stem cell culture media. But one with human embryonic stem cells, the other without human embryonic stem cells. In the process, combine them and print them together. Then droplets can be generated with different cell numbers in each droplet. Therefore, in each droplet one can form cell aggregates with different sizes.

Invasive Procedure

An invasive procedure is one that infiltrates or disrupts the skin or enters a body cavity. Instances of invasive events comprise those that involve perforation, an incision, a catheterization, or other entry into the body. Surgical procedures are characteristic medical invasive procedure.

Non-invasive Procedure

A medical procedure is firmly distinct as non-invasive when no disruption in the skin is generated and there is no exchange with the mucosa, or skin break, or internal body cavity beyond a usual or artificial body orifice. For instance, deep palpation and percussion is non-invasive but a rectal examination is invasive. Similarly, examination of the ear-drum or inside the nose or a wound dressing change all fall outside the strict definition of non-invasive procedure. There are many non-invasive procedures, fluctuating from simple observation, to particular forms of surgery, such as radiosurgery.

For centuries, physicians have employed many simple non-invasive methods centered on physical limitations in order to evaluate body function in health and disease (physical examination and inspection), such as pulse-measuring, the auscultation of heart sounds and lung sounds (the stethoscope), temperature examination (thermometers), respiratory examination, peripheral vascular examination, oral examination, abdominal examination, external percussion and palpation, blood pressure measurement (the sphygmomanometer), change in body volumes (plethysmograph), audiometry, eye examination, and many others.

The detection of the first modern non-invasive techniques based on physical methods, electrocardiography and X-rays, dates back to the end of the 19th century. Subsequently, non-invasive techniques – which penetrate the body nonetheless, but by electromagnetic or molecule radiation rather than a blade – have incessantly widened the scope of medical technology. Non-invasive techniques frequently utilized for analysis and therapies.

3D Bioprinting of Tissues and Organs – None Invasive Procedure

Additive manufacturing, 3D Bioprinting, is driving major innovations in numerous areas, such as engineering, manufacturing, art, education and medicine. Current developments have empowered 3D printing of biocompatible materials, cells and supporting elements into intricate 3D efficient living tissues.

3D Bioprinting is being implemented to regenerative medicine to satisfy the necessity for appropriate tissues and organs for transplantation. Likened with non-biological printing, 3D Bioprinting includes additional intricacies, such as the selection of materials, cell kinds, development and differentiation factors, and practical encounters related to the sensitivities of living cells and the construction of tissues, Figure 7.25. Addressing these intricacies needs the combination of technologies from the fields of engineering, biomaterials science, cell biology, physics and medicine. 3D bioprinting has now been utilized for the generation and transplantation of several tissues, including multilayered skin, bone, vascular implants, tracheal splints, heart tissue and cartilaginous structures. Other implementations comprise developing high-throughput 3D bioprinted tissue models for research, drug discovery and toxicology.

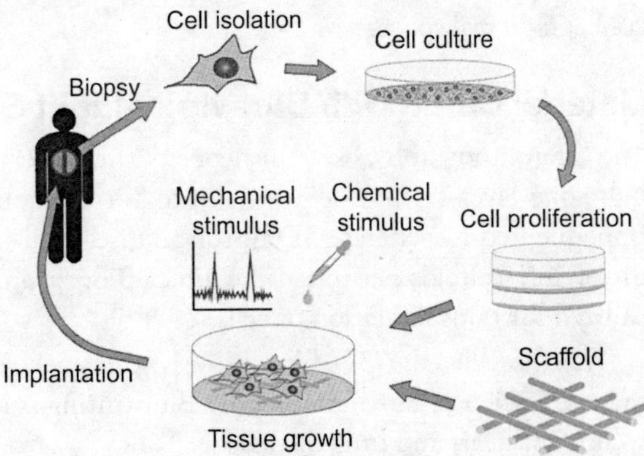

Figure 7.25: Typical Preparation Process for Bioprinting 3D Tissues

Courtesy: Springer Nature Switzerland AG

Damaged Tissue

Imaging of the impaired tissue and its surroundings can be utilized to monitor the design of bioprinted tissues. Biomimicry, tissue self-assembly and mini-tissue constructions are design approaches utilized individually and in combination. The choice of materials and cell source is essential and specific to the tissue form and function. Common materials comprise synthetic or natural polymers and decellularized ECM. Cell supplies may be allogeneic or autologous. These elements have to incorporate with bioprinting systems such as bio-inkjet, micro extrusion or laser-assisted printers. Some tissues may necessitate a period of maturation in a bioreactor before transplantation. Alternatively the 3D tissue may be utilized for *in vitro* practices.

Woodblock Printing

The invention of woodblock printing, and the successive progress of the industrial-scale printing press in the 15th century, facilitated rapid replica of text and images and the propagation of information. Printing had a radical effect on society, affecting education, politics, religion and language across the globe. Over the past few decades, printing technology has progressed from 2D printing to an additive procedure in which successive layers of material are distributed to form 3D outlines.

3D Printing Transforming Science and Education

The production of 3D constructions with intricate configurations by printing is being implemented both to permit rapid prototyping and fabrication in industry and to the making of customer tailored products in the home, such as bicycle parts, jewelry and electrical components. In addition to implementations in the industrial and consumer sectors, 3D printing is converting science and education. For instance, archeologists and anthropologists yield duplications of exceptional artifacts or fossils that can be held, shared and circulated. Just as "Watson and Crick" modeled the structure of DNA a ball-and-stick model, 3D Bioprinting is now being utilized to model intricate particles and protein exchanges, and to fashion customized laboratory tools. 3D Bioprinting allows students to design, visualize, hold and assess their concepts in real space.

Material Cured with Ultraviolet Light Sequentially printing Layers

The "sterolithography," system deposits thin layers of a substance that may be hardened with ultraviolet light one layer at time sequentially to form a solid 3D structure. This same printing process was later implemented to generate 3D dissolvable scaffold from biological materials. The advance development of dissolving scaffolds permitted the direct Bioprinting of biological substances into 3D scaffolds that van be utilized for transplantation of cells.

Additionally, the 3D bioprinting extensively employed tissue engineering serving cell biology and materials science. Furthermore, 3D Bioprinting is utilized to yield vital medical devices, such as stents and splints for heart and lung diseases.

3D Bioprinting, Biomimicry, Autonomous Self-assembly and Mini-tissue Building Blocks

In order to fabricate 3D structures 3D bioprinting technique, a layer-by-layer is precisely positioned of biological materials, bio-chemicals and living cells, with spatial regulator of the positioning of active element, is utilized. There are many approaches to 3D bioprinting, including biomimicry, autonomous self-assembly and mini-tissue building blocks. Researchers, and scientists are developing these approaches to fabricate 3D active living human constructs with biological and mechanical characteristics suitable for clinical renovation of tissue and organ occupation. One important challenge is to modify technologies specifically developed to print molten polymers and metals to the Bioprinting of sensitive, living biological substances. Though, the crucial challenge is to replicate the intricate micro-architecture of extracellular matrix (ECM) elements and many cell kinds in adequate resolution to recapitulate biological task.

The implementations of 3D bioprinting to tissue and organ engineering:

- Study the key strategies for Bioprinting tissue constructs.
- Define the diverse kinds of bioprinters and their influence on the Bioprinted tissue construct.
- Deliberate the steps of Bioprinting a tissue, the restrictions of current technologies and the challenges for forthcoming research

3D bioprinting is founded on three central approaches:

- biomimicry,
- autonomous self-assembly and
- mini-tissue building blocks.

Biomimicry

The cell-culture methods and nanotechnology application to 3D bioprinting technology often deals with the biofabrication of duplicate replicas of cellular and extracellular elements of a tissue or organ.

This can be accomplished by replicating precise cellular functional elements of tissues, for instance, mimicking the diverging patterns of the vascular tree or biofabrication physiologically precise biomaterial kinds and gradients. For this method to prosper, the duplication of biological tissues on the micro-scale is essential. Thus, an understanding of the microenvironment, as well as the specific preparation of functional and supporting cell kinds, gradients of soluble or insoluble factors, composition of the ECM as well as the nature of the biological forces in the microenvironment is desired. The progress of this knowledge-base will be significant to the success of this approach and can be drawn from basic research in fields of engineering, imaging, biomaterials, cell biology, biophysics and medicine.

Autonomous Self-Assembly

Another approach to replicating biological tissues is to utilize embryonic organ development as a guide. The early cellular elements of a developing tissue harvest their own ECM elements, appropriate cell signaling and autonomous organization and patterning to yield the desired biological micro-architecture and function.

A "scaffold-free" type of this approach utilizes self-assembling cellular spheroids that experience fusion and cellular organization to mimic developing tissues. Autonomous self-assembly depends on on the cell as the primary driver of histogenesis, guiding the composition, localization, functional and structural characteristics of the tissue. It needs an intimate knowledge of the developmental mechanisms of embryonic tissue genesis and organogenesis as well as the capability to influence the environment to drive embryonic mechanisms in bioprinted tissues.

Mini-Tissues

The concept of mini-tissues is relevant to both of the above strategies for 3D bioprinting. Organs and tissues comprise smaller, functional building blocks or mini-tissues. These can be defined as the smallest structural and functional component of a tissue, such as a kidney's fine tubule. Mini-tissues can be biofabricated and assembled into the larger construct by rational design, self-assembly or a blend of both.

There are two major strategies:

- first, self-assembling cell spheres (alike to mini-tissues) are assembled into a macro-tissue biologically inspired design and organization;
- second, precise, high-resolution duplicates of a tissue unit are designed and then permitted to self-assemble into a functional macro-tissue.

Instances of these approaches comprise the self-assembly of vascular building blocks to form diverged vascular networks and the utilization of 3D bioprinting to precisely replicate active living tissue units to generate "organs-on-a-chip," which are upheld and connected by a microfluidic network for utilization in the examination of drugs and vaccines or as in *in vitro* models of disease.

Mixtures of the above strategies are likely to be required to Bioprint a intricate 3D biological structure with many functional, structural and mechanical elements and characteristics. The main steps in the bioprinting procedure are imaging and design, selection of materials and cells, and bioprinting of the tissue build. The Bioprinted build is then transplanted, in some instances after a period of *in vitro* maturation, or is kept for *in vitro* analysis.

Noninvasive Medical Imaging Technology – Cell 3D Structure and Function

An indispensable necessity for replicating the intricate, heterogeneous architecture of active living tissues and organs is a complete understanding of the structure and organization of their elements. Medical imaging technology is a crucial tool utilized by tissue engineers to deliver data on 3D structure and function at the cellular, tissue, organ and organism levels. These technologies comprise most noninvasive imaging modalities, the most common being:

- Computed tomography (CT) and
- Magnetic resonance imaging (MRI).
- Computer-aided design and computer-aided manufacturing (CAD-CAM) tools an
- and Mathematical modeling

These are also utilized to assemble and digitize the intricate tomographic and architectural data for tissues.

CT imaging, utilized for both diagnostics and interventional events, is deependent on the variable absorption of X-rays by diverse tissues. The X-ray source rotates around the object, and as the X-ray beam penetrates the body, sensors measure the conveyed beam intensity and angle, and document the data as an accumulation of pixels that signify a small volume (voxel) of tissue. This imaging modality yields closely spaced axial slices of tissue architecture that, after surface rendering and stereolithographic editing, fully describe the volume of tissue.

A second approach, MRI, also can deliver high spatial determination in soft tissue, with the advantage of increased contrast resolution, which is useful for imaging soft tissues in close proximity to each other, without exposure to ionizing radiation. MRI utilizes nuclear magnetic resonance: a strong magnetic field affects a small fraction of nuclei in the tissue being imaged to align themselves with the magnetic field.

Alterations to energy states of nuclei yield radiofrequency signals, which can be measured with receiver coils. The contrast of biological builds can be greatly increased with the utilization of contrast agents such as barium or iodine for CT scans and iron oxide, gadolinium or metalloproteins for MRI scans. These agents attenuate X-rays or improve magnetic resonance signals that are normally utilized to highlight builds, such as blood vessels, which otherwise would be problematic to delineate from their surroundings.

Transformation of Analytical Anatomy into Synthetic Anatomy

When raw imaging data have been assimilated from these photographic modalities, the data must be managed tomographic reconstruction to yield 2D cross-sectional images. 3D anatomical representations can be fashioned for further analysis or adjustment. This procedure has been characterized as the conversion of "analytical anatomy" into "synthetic anatomy." One method to generate computer-based 3D models of organ or tissue architectures is to utilize CAD-CAM and mathematical modeling procedures. The 3D anatomical representation yields views of organ anatomy while holding the image-voxel information that can be utilized for volume rendering, volumetric representation and 3D image representation. Rebuilt images or models can be observed in many ways, comprising as outline stacks, as wire-frame models, shaded models or solid models with variable lighting, transparency and reflectivity.

Patients' Own Organ May Not be Desirable Due to Disease or Injury

If the goal is to yield a precise duplicate of the imaged organ or tissue, 2D cross-sections or 3D representation can be utilized directly for bioprinting implementations. Alternatively, a direct copy of a patients' own individual organ may not be necessary (due to disease or injury) or might not be economically possible for large-scale production. In these circumstances, computer-based models may completely or partly contribute to anatomical structural design, analysis and simulation.

Additionally, computer modeling can support in forecasting mechanical and biochemical characteristics of bio-fabricated tissue constructs. To date, CT and MRI data have been utilized most frequently in regenerative medicine to deliver precise measurements of tissue dimensions to aid the design of a bioprinted construct.

Numerically Controlled Bioprinting Systems

The completed tissue or organ model is interfaced with numerically controlled bioprinting systems for prototyping and bio-manufacturing. This is achieved by reversing the 2D to 3D reconstruction, such that the 3D-rendered model is divided into thin 2D horizontal slices (with custom mae size and orientation) that are imported into the bioprinter system. The anatomical and architectural data confined in the 2D horizontal slices delivers the bioprinting device with layer-by-layer deposition instructions. Distinctions in the accessible bioprinting technologies also affect tissue and organ design. Some bioprinting systems deposit a continuous bead of material to form a 3D construction. Other systems deposit multiple materials in short interrupted or defined spaces. Tissue design must take into account the capabilities and properties of the bioprinting systems.

Tissue bioprinting strategies

Components of Inkjet, Microextrusion and Laser-Assisted Bioprinters

The central technologies utilized for deposition and patterning of biological materials are inkjet, micro extrusion and laser-assisted printing, Figure 7.26. Diverse properties of these technologies, Table 7.1, should be measured in light of the most significant features in 3D bioprinting, which are area resolution, cell viability and the biological biomaterials utilized for Bioprinting, Figure 7.27.

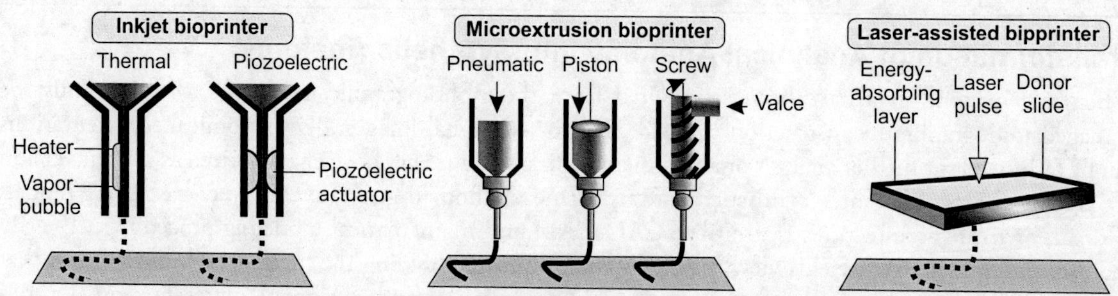

Figure 7.26: Left: Thermal Inkjet Bioprinters – Middle: Microextrusion Bioprinters – Right: Laser-assisted Bioprinter

Table 7.1: Comparison of bioprinter types

	Bioprinter type		
	Inkjet	**Micro-Extrusion**	**Laser Assisted**
Material viscosities	3.5–12 mPa/s	30 mPa/s to >6 × 107 mPa/s	1–300 mPa/s
Gelation methods	Chemical, photo-crosslinking	Chemical, photo-crosslinking, sheer thinning, temperature	Chemical, photo-crosslinking
Preparation time	Low	Low to medium	Medium to high
Print speed	Fast (1–10,000 droplets per second)	Slow (10–50 µm/s)	Medium-fast (200–1,600 mm/s)
Resolution or droplet size	<1 pl to >300 pl droplets, 50 µm wide	5 µm to millimeters wide	Microscale resolution
Cell viability	>85%	40–80%	>95%
Cell densities	Low, <106 cells/ml	High, cell spheroids	Medium, 108 cells/ml
Printer cost	Low	Medium	High

Inkjet Bioprinting

Bio-Inkjet Bioprinters, also recognized as drop-on-demand Bioprinters, are the most frequently utilized kind of Bioprinter for both non-biological and biological implementations. Certain amount of droplet volume deposited to predefined locations. The first Bio-inkjet printers utilized for bioprinting implementations were reengineered and adapted versions of commercially available 2D ink-based printers.

The ink filling the cartridge was replaced with a biological Bioink material, and the paper was substituted with an electromechanical micro-elevator stage to deliver and control the z axis. The Z axis is the dimension in addition to the X and Y axes. Now, inkjet-based bioprinters are custom-tailor machine to deal with printing biological materials at much greater resolution, precision and speed. Bio-Inkjet printers utilize thermal or acoustic waves to force ejecting drops of liquid onto a predetermined location, where a substrate is placed securely, which supports the final can build.

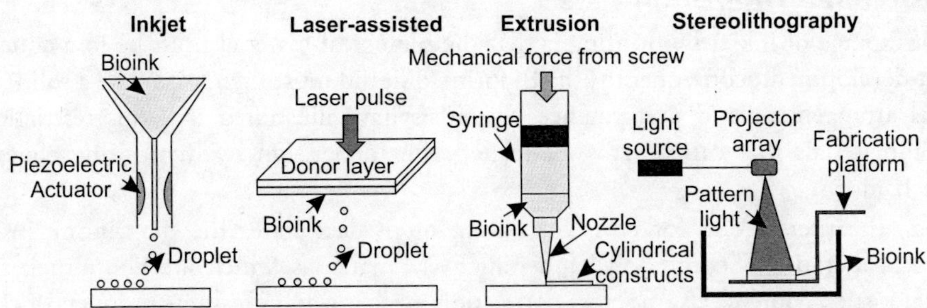

Figure 7.27: 3D Bioprinting Technologies for Regenerative Medicine

Courtesy: Hapres Co., Ltd.

Electrically heating the Bioprinter generates thermal inkjet at the head nozzle to create pulsating pressure to force droplets from the nozzle. The pulsed heating ranges from 200 °C to 300 °C, which does not have a much influence either on the consistency of the biological particles, such as DNA, or on the stability or post-Bioprinting function of living cells. The short duration of the pulsating heat, (~2 µs) results in an overall temperature rise of only 4-10°C in the printer head nozzles. The benefit utilization of thermal high speed Bio-inkjet printers is low cost and wide market availability. Nonetheless, the risk of exposing Biological cells and biomaterials is summarized as follows:

1. Thermal and acoustic wave stress,
2. Low droplet directionality,
3. Non-uniform droplet configurations,
4. Frequent blockage of the nozzle and undependable cell encapsulation

All pose substantial drawbacks for the utilization of these printers in 3D bioprinting.

Several inkjet Bioprinters encompass a piezoelectric crystal that constructs an acoustic wave inside the print head to disrupt the flow of liquid into droplets at regular short periods of time. Subjecting a voltage to a piezoelectric material induces a fast response in shape change, which consequently produces the pressure desirable to expel droplets from the nozzle. Other inkjet Bioprinters utilize an acoustic wave force correlated to an ultrasound domain to expel liquid droplets from an air-liquid boundary. Ultrasonic properties, such as pulse, time interval and repeatability, can be modified to govern the size of droplets and the rate of expulsion.

Acoustic Wave Inkjet Printers

The benefits of implementing acoustic wave technique on the inkjet Bioprinters comprise the ability to generate and govern identiThe benefits of implementing acoustic wave technique on the inkjet Bioprinters comprise the ability to generate and govern identical droplet size and expulsion directionality as well as to circumvent exposure of cells to heat wave and stress. Furthermore, the sheer stress forced on cells at the nozzle tip wall can be evaded by an open-pool nozzle-less expulsion system. This decreases the possible loss of cell capability and function, and circumvents the problem of nozzle blockage. Acoustic wave ejectors can be combined as multiple ejectors in an adjustable array format, enabling concurrent Bioprinting of multiple cell and Biomaterial kinds. Nonetheless, there linger some apprehensions concerning the 15–25 kHz frequencies utilized by piezoelectric inkjet bioprinters and their possibility to induce impairment of the cell membrane and destruction. Inkjet bioprinters also have limitations on Biomaterial viscosity. Ideally, viscosity is <10 centipoise. It is owing to the excessive force required to eject drops utilizing solutions at higher viscosities.

Drawback of Inkjet Bioprinting

One vital disadvantage of inkjet bioprinting is that the biological material must be in a aqueous form to permit droplet development; consequently, the Bioprinted liquid must then formulate a solid 3D structure with structural arrangement and performance. Scientists have illustrated that this restriction could be resolved by Biomaterials that can be cross-correlated after deposition by Bioprinting chemical, pH or ultraviolet mechanisms.

Nonetheless, the prerequisite for cross-correlating often decelerates the bioprinting procedure and includes chemical alteration of commonly happening ECM materials, which alters both their chemical and material characteristics. Moreover, some cross-correlation mechanisms necessitate products or circumstances that are toxic to cells, which consequence in reduced capability and performnce.

Another limitation come across by users of inkjet-based bioprinting technology is the trouble in accomplishing biologically applicable cell densities. Often, low cell concentrations, scarcer than 10 million cells/ml are utilized to ease droplet creation, evade nozzle blockage and decrease shear stress. Higher cell concentrations may also constrain some of the hydrogel cross-correlating mechanisms.

Inkjet-Based Bioprinters Advantages

Although these drawbacks, inkjet bioprinters also compromise many benefits, including low cost, high resolution, high speed and compatibility with many biological materials. Additional benefit of inkjet Bioprinting is the possibility to present concentration gradients of cells, biomaterials or growing factors through the 3D build by changing drop densities or sizes. According to the accessibility of standard 2D inkjet Bioprinters, scientists in numerous labs can readily access, adapt and test with 3D inkjet–based bioprinting technology. Commercially accessible inkjet bioprinters are also relatively low cost, owing to their simple components and readily accessible design and control software. The extensive implementation of this technology by numerous entities has accelerated advances in the capacity of inkjet bioprinters to precisely deposit with high resolution and precision controllable droplet sizes with uniform cellular densities.

Droplet configuration and deposition rate can be governed digitally, and can vary from <1 pl to >300 pl in volume with rates of 1–10,000 droplets per second. Patterns of individual drops, each containing one or two cells, in lines ~50 μm diameter, have been Bioprinted. Future advances will carry on to adapt this technology to handle and deposit other biologically relevant biomaterials, in a manner that both facilitates their Bioprinting and provides the essential biological, structural and functional components of the tissue. Additional complexities, such as the necessity for multiple cell kinds and materials, will also have to be addressed.

Prominent instances of the inkjet bioprinting approach include the regeneration of living skin and cartilage *in situ*. The high Bioprinting speed enables direct deposition of cells and materials directly into skin or cartilage lesions. These implementations accomplished rapid cross-correlating of the cell-containing material via either a biocompatible chemical reaction or a photo-initiator and cross-correlator through exposure of the Biomaterial to ultraviolet light. The inkjet approach facilitated the deposition of either primary cells or stem cell kinds with uniform density throughout the volume of the lesion, and upheld high cell capability and function after printing. These studies demonstrate the possibility of inkjet-based bioprinting to regenerate living functional builds.

Combination of Electro-Spinning and Inkjet Bioprinting

Layered cartilage builds have also been bio-manufactured *in vitro* an amalgamation of electro-spinning and Bio-inkjet bioprinting. The hybrid electro-spinning–inkjet bioprinting technology allowed the Biomanufactured of a layered build that supported cell iving function and upheld appropriate mechanical and structural characteristics. Inkjet bioprinters have also been utilized to bio-fabricate bone builds, matured *in vitro* before implantation into mice. These builds sustained to mature *in vivo* and shaped highly mineralized tissues with similar density as depressed bone tissue.

Micro-Extrusion Bioprinting

The most economical non-biological 3D printers utilize micro-Extrusion. Micro-Extrusion bioprinters typically contain of a heat-controlled material-handling and dispensing system and stage mechamism, with one or both capable of movement along the x, y and z axes, a fiber optic light source to illuminate the deposition area and/or for photo-initiator activation, a video camera for x-y-z direct and control, and a piezoelectric humidifier. A few systems utilize numerous print heads to ease the sequential dispensing of several materials without retooling. Nearly 30,000 3D printers are sold worldwide every year, and academic institutions are progressively purchasing and applying micro-extrusion technology in tissue and organ engineering research. Industrial Bioprinters are significantly more costly but have greater resolution, speed, spatial controllability and more flexibility in the Bio-material they can print.

Micro-Extrusion printers perform by robotically controlled extrusion of a material, which is deposited onto a substrate by a micro-extrusion head. Micro-extrusion yields steam of continuous beads of material instead of liquid droplets. Tiny beads of Biomaterial are deposited in two dimensions, as directed by the CAD-CAM routines, the stage mechanism or micro-extrusion head is moved upward along the z axis, and the deposited layer aids as a basis for the following layer.

Pneumatic / Mechanical - Piston / Screw Bioprinting Dispensing Systems

A countless materials are well-suited for micro-extrusion Bioprinters, comprising Biomaterials such as hydrogels, biocompatible copolymers and cell spheroids. The most known methods to extrude biological substances for 3D bioprinting implementations are pneumatic or mechanical displacement mechanism. Mechanical dispensing systems might deliver more direct control over the material flow as of the delay of the pressurized gas volume in pneumatic systems. Screw-based systems might offer more spatial control and are thought to be beneficial for the dispensing of hydrogels with higher viscosities, though pneumatic systems could also be suitable to dispense high-viscosity materials.

Pneumatically mechanized printers have the benefit of having simpler drive-mechanism elements, with the force limited only by the air-pressure capabilities of the system. Mechanically driven mechanisms have smaller and more intricate parts, which provide greater spatial control but frequently at limited maximum force capabilities.

Micro-Extrusion techniques have a very extensive variety of fluid characteristics that are well-matched with the procedure, with a comprehensive array of biocompatible materials. Materials with viscosities fluctuating from 30 mPa/s to >6 × 107 mPa/s have been illustrated to be well-matched with micro-extrusion bioprinters, with higher-viscosity materials often giving structural support for the Bioprinted build and lower-viscosity materials giving a appropriate atmosphere for upholding cell sustainability and living function.

For Micro-Extrusion bioprinting, researchers, and scientists frequently utilize Biomaterials that can be thermally cross-correlated and/or hold sheer-thinning characteristics. Several biocompatible Biomaterials can flow at room temperature, which lets their extrusion together with other biological elements, but cross correlate into a steady material at body temperature. Alternatively, materials that flow at physiologically appropriate temperatures (35–40 °C), but cross correlate at room temperature may also be beneficial for bioprinting implementations. Materials with shear-thinning characteristics are usually utilized for micro-extrusion implementations. This non-Newtonian material performance allows a reduction in viscosity reacting to upsurges in shear rate. The high shear rates that are current at the nozzle during biofabrication permit these materials to flow through the nozzle, and upon deposition, the shear rate reduces, producing a sharp upsurge in viscosity. The high resolution of micro-extrusion systems allows the bioprinter to precisely manufacture intricate builds designed utilized CAD software and enable the patterning of many cell kinds.

The key benefit of micro-extrusion bioprinting technique is the capacity to deposit very high cell densities. Accomplishing physiological cell densities in tissue-engineered organs is the main puroise for the bioprinting applications. Some researchers have utilized solutions comprised only of cells to generate 3D tissue builds with micro-extrusion Bioprinting.

Tissue Spheroids Replicating Mechanical Properties of Tissue ECM

Multi-cellular cell spheroids are dropped and permitted to self-assemble into the anticipated 3D build. Tissue spheroids are thought to hold material characteristics that can reproduce the mechanical and functional characteristics of the tissue ECM. Contingent on the viscoelastic characteristics of the building blocks, the apposed cell combinations fuse with each other, creating a unified macroscopic build. One benefit of the self-assembling spheroid strategy is possibly quicker tissue organization and the capability to direct the development of intricate builds. This methodology demonstrates potential in permitting the generation of an intra-organ branched vascular tree in 3D thick tissue or organ builds by patterning self-assembling vascular tissue spheroids, in 3D Bioprinted organs. The most known technology utilized for scaffold-less tissue spheroid bioprinting is mechanical micro-extrusion.

Cell capability after micro-extrusion bioprinting is lower than that with inkjet-based bioprinting; cell mortality rates are in the range of 40–86%, with the rate lessening with increasing extrusion force and increasing nozzle gauge. The lessened capability of cells deposited by micro-extrusion is likely to result from the shear stresses perpetrated on cells in viscous fluids. Dispensing pressure may have a more considerable effect on cell capability than the nozzle diameter. Although cell capability can be maintained low pressures and large nozzle sizes, the drawback may be a major loss of resolution and print speed. Maintaining high capability is essential for accomplishing tissue performance. Although many studies report maintenance of cell capability after printing, it is important for researchers to demonstrate that these cells not only endure, but also perform their essential performance in the tissue build.

It is a challenge for any user to increasing the print resolution and speed of micro-extrusion bioprinting technology. Non-biological micro-extrusion printers are able of 5 μm and 200 μm resolution at linear speeds of 10–50 μm/s. Utilization of better biocompatible materials, such as dynamically cross-correlated hydrogels that are mechanically resilient during printing and that develop secondary mechanical characteristics after printing might aid to hold cell capability and performance after printing. Single-phase, dual-phase and continuous-gradation scaffolds are also being constructed comparable principles. Moreover, developments in nozzle, syringe or motor-control systems might lessen print times as well as permit deposition of multiple diverse materials simultaneously.

Micro-Extrusion Bioprinters for Aortic Valves, Branched Vascular Trees, *In Vitro* Pharma-kinetic, and Tumor Models

Bioprinters of micro-extrusion have been utilized to manufacture many tissue kinds, containing aortic valves, branched vascular trees and *in vitro* pharma-kinetic as well as tumor models. Although the manufacturing time can be slow for greater-resolution intricate builds, have been manufactured that range from clinically pertinent tissue sizes down to micro-tissues in microfluidic chambers.

Laser-Assisted Bioprinting (LAB)

Laser-Induced Forward Transfer technology is the foundation for creating the "Laser-Assisted Bioprinting (LAB) applications. It is developed to successfully transmit biological material, such as peptides, DNA and cells. While it is not well known as inkjet or micro-extrusion Bioprinting, LAB is progressively utilized for tissue- and organ-engineering implementations. A characteristic LAB instrument contains of an equal interval pulsed laser beam, a beam focusing system, a "ribbon," which has a donor transport support typically made from silica that is protected with a laser-energy-absorbing layer (e.g., gold or titanium) and a layer of biological material (e.g., cells and/or hydrogel) arranged in a liquid solution, and an accepting substrate opposite the ribbon. LAB performance focused laser pulses on the absorbing layer of the facing ribbon to produce a high-pressure bubble that propels cell-holding substances to the accumulator substrate, Figure 7.28.

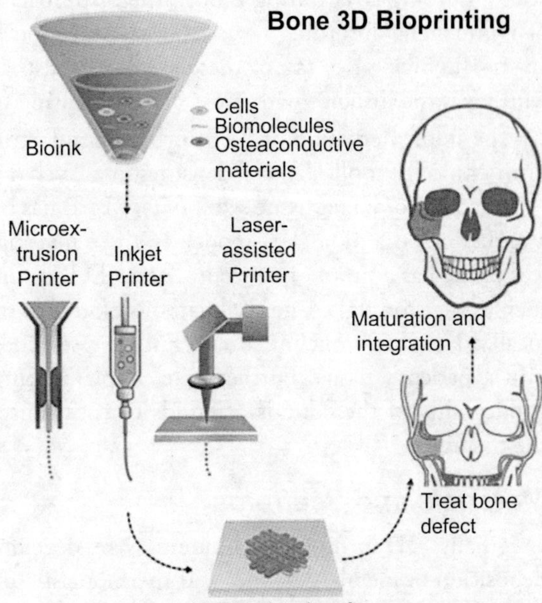

Figure 7.28: Laser Assisted Bioprinting - 3D Bioprinting Fabricating Bone Constructs, Utilizing Micro-extrusion, Inkjet, and Laser assisted Bioprinting – Bioinks: Hydrogels, Cells, Osteoconductive Materials, and Growth Factors
Courtesy: Alexandros G. Sfakianakis

The sensitivity of LAB is traditionally known as the resolution. It is influenced by many factors, stated ass follows:

a. The laser energy supplied per unit area,
b. the surface tension,
c. the wettability of the substrate,
d. the air gap between the ribbon and the substrate, and
e. the thickness and viscosity of the biological layer

As the LAB does not have a nozzle, the difficulty of clogging with cells or Biomaterials that adversely affecting other bioprinting technologies is avoided. LAB is well-matched with a variety of viscosities (1–300 mPa/s) and can Bioprint mammalian cells with negligible effect on cell capability and performance. LAB can drop spheres of cells at a density of up to 108 cells/ml with micro-scale resolution of a single cell per drop utilizing an equal interval laser pulse at a recurrence cyclic rate of 5 kHz, with speeds up to 1,600 mm/s.

Notwithstanding these benefits, the high resolution of LAB requires fast gelation kinetics to attain high configuration fidelity, which outcomes in a relatively low overall flow rate107. Planning of each individual ribbon, which is often required for each printed cell or hydrogel kind is time-consuming and may become burdensome if multiple cell kinds and/or materials have to be co-deposited. Since the nature of the ribbon cell coating, it can be problematic to precisely target and locate cells.

Several of these challenging tasks might be avoided by cell-recognition scanning technology to allow the laser beam to choose an individual cell per pulse. This so-called "aim-and-shoot" technique that could guarantee that each printed droplet sphere contains a predefined number of cells. Though, statistical cell Bioprinting can be accomplished utilizing a ribbon with very high cell concentrations, evading the necessity for such particular cell targeting.

Lastly, metallic remains are current in the final Bioprinted build owing to the vaporization of the metallic laser-absorbing layer during Bioprinting. Approaches to evade this contamination comprise the utilization of nonmetallic absorbing layers and adapting the Bioprinting operation to not necessitate an absorbable layer. The high price tag of these systems is also a concern for fundamental tissue-engineering research, while as is the instance with most 3D Bioprinting techniques, these costs are quickly dwindling.

The implementation of LAB to bio-manufacture a cellularized skin build established the potential to print clinically applicable cell densities in a layered tissue build, but it is uncertain whether this system can be scaled up for larger tissue sizes. *In vivo* LAB has been utilized to deposit spherical nano-hydroxyapatite in a mouse calvaria 3D defect model. In these investigations, a 3 mm diameter, 600 μm–deep calvarial holes was packed as a proof of concept. Laser 3D Bioprinting has been utilized to manufacture medical devices, such as a custom tailor, noncellular, bioresorbable tracheal splint that was grafted into a young patient with localized tracheobronchomalacia. Future researches might utilize materials that can directly incorporate into a patient's tissue. Furthermore, implementing the patients' own individual cells may facilitate the applicability of these kinds of builds to contribute to both the structural and functional components of the tissue.

Materials and Scaffolds

Originally, 3D printing techniques were developed for non-biological implementations, such as the deposition of metals, ceramics and thermoplastic polymers, and mostly involved the utilization of organic solvents, high temperatures or cross-correlating agents that are not well-matched with living cells and biological materials. Consequently, one of the main challenges task in the 3D bioprinting field has been to find Bio-materials that are not only well-matched with biological materials and the printing operation but can also deliver the desired mechanical and functional characteristics for tissue builds.

Materials presently utilized in the application of regenerative medicine for repair and regeneration are mainly founded on either instinctively resulting polymers, including alginate, gelatin, collagen, chitosan, fibrin and hyaluronic acid, frequently isolated from animal or human tissues) or artificial molecules (polyethylene glycol). The benefits of ordinary polymers for 3D bioprinting and other tissue-engineering implementation is their resemblance to human ECM, and their intrinsic bioactivity. The benefit of artificial polymers is that they can be custom-made with precise physical aspects to suit particular implementations.

Heavy burdens in the utilization of artificial polymers comprise poor biocompatibility, toxic degradation products and loss of mechanical characteristics during degradation. Even so, artificial hydrogels, which are both hydrophilic and absorbent, are suitable for 3D bioprinting regenerative-medicine implementations owing to the simplicity of governing their physical characteristics during synthesis.

As the diversity of biological materials for medical implementations upsurges, the list of required traits for printable materials has developed more particular and intricate.

Ideal material properties for bioprinting

The choice of appropriate Biomaterials for utilization in bioprinting and their presentation, in a particular implementation contingent upon several characteristics are listed below:

- **Printability:** Characteristics that enable handling and deposition by the bioprinter may include viscosity, gelation approaches and rheological aspects.

- **Biocompatibility:** Materials should not induce unwanted local or systemic reactions from the host and should contribute actively and controllably to the biological and operational elements of the construct.
- **Degradation kinetics and byproducts:** Degradation rates should be matched to the capability of the cells to yield their own ECM; degradation byproducts should be nontoxic; materials should demonstrate suitable swelling or contractile characteristics.
- **Structural and mechanical properties:** Materials should be selected founded on the essential of mechanical characteristics of the construct, extending from rigid thermoplastic polymer fibers for strength to soft hydrogels for cell compatibility.
- **Material biomimicry:** Engineering of anticipated structural, operational and dynamic material characteristics should be founded on knowledge of tissue-specific endogenous material compositions.

Materials must have appropriate cross-correlating mechanisms to enable bioprinter deposition, must be biocompatible for transplantation over the long-term, and must have appropriate swelling characteristics and short-term steadiness. Short-term steadiness is essential to uphold original mechanical characteristics, guaranteeing that tissue structures such as pores, channels and networks do not fail. As Bioprinted tissues grow *in vivo*, they should be amenable to remodeling, enabling the shaping of structures determined by cellular and physiological requirements. Most prominently, materials must support cellular attachment, proliferation and functionality.

Printability

A vital aspect of a convenient biomaterial is its accurate and precise ability to deposit viable functional cells. Several kinds of bioprinting technology as inkjet have restrictions on Biomaterial viscosity, whereas others, such as micro-extrusion, may need the Biomaterial to have particular cross-correlating mechanisms or shear-thinning characteristics. Operational factors, such as nozzle gauge, determine the shear stress to which cells are exposed, as well as the time essential for the Biomaterial to be deposited to formulate a 3D structure. For instance, inkjet Bioprinting necessitates biomaterials with a fast cross-correlating time to enable the layering of a intricate 3D structure. Micro-Extrusion, though, can implement highly viscous Biomaterials to uphold a 3D form after deposition, with final cross-correlating happening after manufacture.

The selection of Biomaterial may too be influenced by the capability of the Biomaterial to protect cell capability during the Bioprinting operation. Thermal inkjet Bioprinting and LAB both involve the localized heating of the biomaterial to deposit spherical drops of cells. Biomaterials with either low thermal conductivity or the capability to cushion the cells during transfer may upsurge cell capability and function after bioprinting. Although post-printing cell capability can range markedly centered on Bioprinter specifications, material characteristics, resolution and cell kinds, inkjet bioprinting studies typically quote cell capabilities in excess of 85%, micro-extrusion Bioprinting studies report viability ranges of 40–80% and LAB studies report viability in excess of 90%.

Biocompatibility

The Beginning of tissue engineering technology, has progressively led to the creation of biocompatibility implementations. Tissue engineering has led a type of Bio-revolution from necessitating an implanted material to coexist with the endogenous tissue without provoking any unwanted local or general effects in the host, to imbedded Biomaterials being predictable to inertly permit or vigorously yield wanted properties in the host. Biocompatibility in bioprinting comprises the anticipation of a living and manageable influence to the biological and efficient mechanisms of the build. This could comprise interface with endogenous tissues and/or the immune system, supporting suitable cellular activity and easing of molecular or mechanical sensing systems, all of which are vital for positive transplantation and purpose.

Degradation Kinetics and Byproducts

It a well-known beneficial fact that scaffolds material degrades giving ways to the embedded cells to secrete proteases, and consequently yield ECM proteins that outline the new tissue. The materials degradation dynamics necessitates understanding to govern the tissue regeneration effectively. As a result, the details of degradation should be considered tto learn the effect of degradation to the health of the new cells. There are some features of degradation that must be taken into considerations, stated as follows:

1. The capability to regulate degradation rates, in an ideal world balancing the rate of degradation with the capability of cells to substitute the materials with their own ECM proteins. This is represents a great challenge as materials with appropriate operational and mechanical properties for a providing given tissue may not balance the capability of the cellular elements to substitute the material upon degradation.
2. Degradation outcome substances are also vital as they frequently characterize the biocompatibility of any degradable material.
3. The degradation products must be nontoxic, naturally absorbed and quickly disposed from the body. Toxic products may contain small amount proteins particles and molecules, as well as nonphysiological pH, temperature or other parameters that mau be damaging to cell sustainability and function. For instance, large-molecular-weight polymers, which are originally inactive may be broken down into fewer than 5 monomer units namely oligomers, or monomers that can be detected by cells and cause inflammation and other harmful effects.
4. Swelling causing inflammation and shrinkage features of materials is particularly of fear in the manufacture of tissue-engineering products. Excessively swelling materials can possibly cause absorption of liquid from the nearby tissues, and shrinkage may cause the narrowing of pores or vessels that are vital for cell relocation and nutrient distribution.
5. Furthermore, it is significant to comprehend these reactions when implementing many materials with different swelling or shrinking behavior as this could possibly cause loss of layer coherence or distortion of the final build.

Structural and Mechanical Properties

Maintaining of a 3D structure requires essential adequate materials to resist the surrounding forces to cell structure. The appropriate materials act as an anchoring point for mechanical influence for continued function of the build. Materials must be prudently designated contingent upon the obligatory of mechanical characteristics of the build. Diverse structural necessities will require diverse tissue kind ranging from skin and liver to bone. One alternative to bypass this retraction is the utilization of sacrificial materials that can deliver the essential physical and mechanical characteristics over a assumed period of time. This sacrificial material either may be utilized at the time of Bioprinting to permit adequate cross-correlating to happen in the build or otherwise could be combined into the build, operating until the endogenously formed materials can adequately carry out this operation. With this method, maintenance must be considered to construct a material with precise structural and degradation characteristics while evading possible foreign body reactions or toxic degradation byproducts in the build.

Material Biomimicry

The significance of Biomimicry for biocompatibility has only lately been investigated. The capability to integrate biomimetic elements into a Bioprinted build may have an active result on the cell supplement, relocation, proliferation and operation of both internal and external cells. It is well recognized that materials have a substantial effect on cell attachment as well as cell configuration and form, and these values may be beneficial in governing the proliferation and differentiation of cells in a scaffold. The accumulation of surface ligands to a material has the possibility to upsurge cell attachment and proliferation on the material substrate. The existence of nanoscale characteristics such as ridges, steps and grooves also influences cell attachment, proliferation and cytoskeletal assembly. The 3D surroundings in a tissue-engineered build can affect cell form and affect the differentiation procedure. Nanoscale features of materials can influence cell adhesion, cell positioning, cell capability of moving by independent means, surface antigen that stimulates the production of an antibody, cytoskeletal condensation and modulation of intracellular indicating trails that control transcriptional activity and gene manifestation.

A biomimicry methodology to construct materials with specific physiological behavior necessitates an appreciation of the inherently taking place in tissue-specific arrangement and localization of ECM elements in the tissue of concern. New progress in tissue decellularization techniques could deliver intact ECM scaffolds for comprehensive analysis of ECM configurations, localization and biological occupations. This procedure contains the the destruction of cells by disruption of the bounding membrane, allowing the cell contents to escape, and elimination of the cellular elements of a tissue, typically via perfusion with deionized water or mild detergents, though leaving behind the tissue-specific ECM. The capability to replicate matching ECM scaffolds utilizing a bioprinting method would be beneficial in tissue engineering and regenerative medicine.

Intensive investigation in tissue decellularization emphasizes on striking a stability between total elimination of cellular elements and upkeep of the fine vascular and other tissue builds. Moreover, some toxicity has been perceived when cells are developed on decellularized tissue scaffolds, possibly due to the holding of the decellularization detergent within the ECM. In mammals, there are more than 300 ECM proteins as well as several ECM-modifying enzymes, ECM-binding development parameters and other ECM-associated proteins. The most ample and implicit proteins are the collagens, proteoglycans and glycoproteins. These proteins deliver strong suit and space-filling performance, bind growth factors, control protein developments, encourage cell adhesion, partake in cellular signaling and may have extra behavior. A "scaffold-free" method to bioprinting may be another stimulating take on the idea of material biomimicry. As cells yield and deposit the tissue ECM, bioprinted self-assembling cellular spheroids may yield an ECM surroundings best matched for their own performance. Engineering these dynamic ECM mechanisms into materials provides additional control over cell behavior. One difficulty is to advance approaches to integrate these materials into builds bioprinting technology, guaranteeing that the materials have appropriate degradation times and byproducts, and that these materials have well-comprehended and governable structural and operational biological properties in the build.

Cell Sources

The selection of cells for tissue or organ Bioprinting is vital for precise operating of the fabricated build. Tissues and organs comprise many cell kinds with particular and vital biological performance that must be reviewed in the transplanted tissue. Furthermore to the main functional cell kinds, most tissues also comprise cell kinds that deliver supportive, structural or barrier performance, are involved in vascularization or deliver a niche for stem cell upkeep and differentiation. Present options for Bioprinting cells involve either the deposition of many principal cell kinds into patterns that faithfully signify the inherent tissue or Bioprinting stem cells that can proliferate and differentiate into vital cell kinds. Cells selected for Bioprinting should closely mimic the physiological state of cells *in vivo* and are anticipated to uphold their *in vivo* performance under optimized circumstances.

In the least cell kind selected for Bioprinting should be capable to grow into adequate numbers for Bioprinting. Accurate control of cell proliferation *in vitro* and *in vivo* is significant for bioprinting. Too little proliferation may cause the loss of capability of the transplanted build, whereas too much proliferation may cause hyperplasia or a form of cell death necessary to make way for new cells namely apoptosis. Determinations to control cellular proliferation in the transplanted build are crucial to accomplish physiological ratios of living and supporting cells. In addition, the timing of cellular proliferation is significant. Originally, a high cellular proliferation rate may be necessary to populate the build, but over the long term, proliferation must be upheld at a rate appropriate to accomplish tissue homeostasis, though without hyperplasia.

Some methodologies to solve this difficult encompass viral transfection or utilization of small molecules to induce cell proliferation and avert aging. *In vivo*, endogenous stem cells behavior to substitute fatally differentiated cells following usual cell turnover or injury. For a bioprinted build to uphold function long-term, after transplantation, bioprinted tissues must be capable to uphold cellular balance, self-renew and react to tissue impairment or injury. Better understanding of the nature and configuration of endogenous stem cells and their niches will be beneficial in engineering tissues that can uphold their long-term performance after relocation.

Autologous Sources of Cells

As with any relocated tissues or organs, rejection of bioprinted builds by the host immune system is a likely difficulty that can be avoided by an autologous source of cells or by tolerance-induction approaches. Autologous sources of cells may be gained from biopsies, from the generation and differentiation of autologous stem cells or through reprogramming approaches. Though, if a patient is already ill or has either genetic or metabolic disorders, it may not be conceivable for the patient to undergo an invasive surgical procedure, and the isolated cell kinds may not yield the anticipated performance in the bioprinted build.

Embryonic Stem Cells and Induced Pluripotent Stem Cells

Many primary cell kinds are hard to isolate and culture and their finite lifespan is a limitation for the long-term performance of any bioprinted builds. Stem cells are a promising cell kind for tissue-engineering implementations owing to their ability to proliferate in an undifferentiated but multipotent state (self-renewal) and their capability to generate multiple functional tissue-specific cell phenotypes. Embryonic stem cells and induced pluripotent stem cells are capable of indefinite self-renewal and have established their permanence by upholding their undifferentiated state for over 80 passages. The capacity of pluripotent stem cells to generate large numbers of cells highlights the possible of these cells for bioprinting (and other) therapeutic implementations.

Other kinds of stem cells—such as adult stem cells from bone marrow and fat or perinatal stem cells from amniotic fluid or placenta—are believed to have a more limited multipotent differentiation potential but are measured safer for clinical transplantation and have the possibility for autologous implementations. With recognized protocols for the isolation, growth and differentiation, mesenchymal stromal cells (MSCs) may also be a hopeful cell source for bioprinted builds. Clinically pertinent numbers of MSCs have been successfully generated *in vitro* for clinical trials, and future advances in cell-culture performances are likely to make use of other stem cell populations for bioprinting clinical implementations an accurate option.

The cells utilized for bioprinting implementations must be vigorous enough to endure the bioprinting procedure and endure physiological pressures once transplanted, including physical forces such as shear stress and pressure as well as biological stressors including existence of toxins, enzymes and nonphysiological pH.

Several published bioprinting studies utilize cell lines that are recognized to be capable of substantial proliferation and are very robust, such as fibroblasts or transformed cell lines. Although these cells are appropriate for proof-of-concept investigations, it is significant to understand that bioprinting skill may have to be adapted to integrate cell kinds that are more sensitive to forces such as shear stress or culture surroundings as well as the time it takes to make the build. Advances in cell culture techniques as well as in reprogramming and directed differentiation approaches will be significant for yielding highly proliferative, functional, non-immunogenic and robust cell populations that are appropriate for bioprinting implementations.

Several of the difficulties in front of the 3D bioprinting field relate to precise technical, material and cellular properties of the bioprinting procedure, Table 2. While the field is at a crucial stage, it has already thrived in generating multiple tissues at human scale that are approaching the performance essential for transplantation, Figure 7.29. Technological difficultiess include the essential for improve, speed and compatibility with biologically pertinent materials. As scientists retreat from the alteration of preexisting technology and start to construct 3D bioprinters to handle particular biological elements, the variety of well-matched materials can be comprehensive, and methods to deposit materials and cells with cumulative accuracy and specificity can be established. The speed of manufacture must be augmented to fabricate builds of clinically relevant configurations. One way to attain this would be to generate miniature living tissue blocks that could be scaled to a clinically pertinent configuration by a macro-scaffold to join blocks. Commercialization may necessitate scalable automated robotic technologies that integrate each of the elements of the biofabrication production line. This may embrace not just the bioprinting device but also the fabrication of materials, cells and other supportive mechanisms.

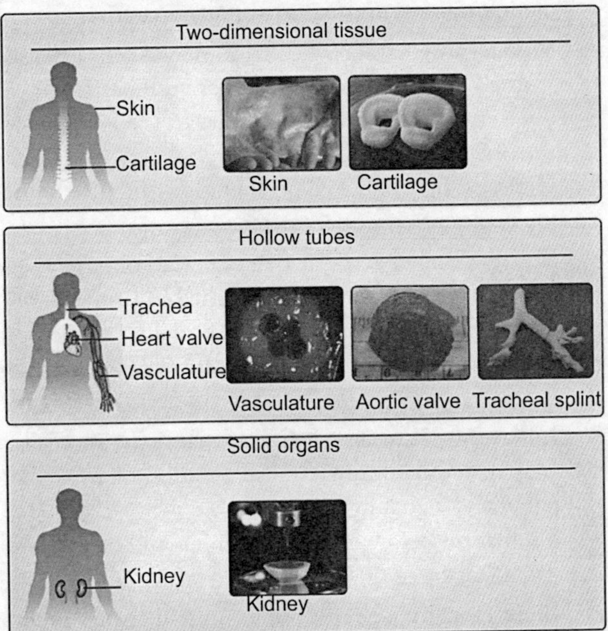

Figure 7.29: 3D Bioprinting Tissue Engineering - Regenerative Medicine - Cartilage Substitutes - Using Inkjet Bioprinting - either *in vitro* or *in situ*

Courtesy: Wake Forest Institute

Figure 7.29 - Skin and cartilage substitutes developed utilizing inkjet bioprinting systems, capable of fabricating tissues either *in vitro* or *in situ*. A vascular graft construct manufactured utilizing micro-extruded and fused cellular vascular rods8 and a micro-extrusion-bioprinted aortic valve fabricated with dual cell types, aortic root sinus smooth muscle cells and aortic valve leaflet interstitial cells. A laser bioprinted bioresorbable airway splint and an early stage kidney prototype, manufactured utilizing micro-extrusion bioprinting. All of these bioprinted tissues required integration of multiple components for the fabrication of functional, appropriately sized tissue constructs. Bioresorbable airway splint created with a three-dimensional printer.

Table 7.2: Issues of Concerns Regarding Scientific and Practical Applications

Area	Focus for future research
Bioprinter technology	• Well-matched with physiologically pertinent materials and cells Augmented resolution and speed • Scale up for commercial implementations Merging bioprinter technologies to overcome technical difficulties
Biomaterials	• Intricate arrangements or gradients to accomplish anticipated operational, mechanical and supportive characteristics • Adapted or constructed to enable bioprinter deposition, while also displaying anticipated post-printing characteristics • Utilize of decellularized tissue-specific ECM scaffolds to investigate ECM configurations, and/or as printable material
Cell sources	• Well-regarded and replicated source of cells essential Combinations of cell phenotypes with particular perfomance • Greater comprehension necessary of the heterogeneous cell kinds exist in the tissues • Direct control over cell proliferation and differentiation with small molecules or other parameters
Vascularization	• Well-developed vascular tree essential for large tissues • May have to be constructed in the bio-printed build • Capillaries and micro-vessels essential for tissue perfusion • Appropriate mechanical characteristics for physiological pressures and for surgical connection
Innervation	• Innervation is necessary for normal tissue survival • May be inducible after transplantation pharmacologic or development parameter signaling • Simulation before transplantation could be accomplished bioreactors
Maturation	• Time necessary for assembly and maturation • Bioreactors may be utilized to uphold tissues *in vitro* • Deliver maturation parameters as well as physiological stressors • Possible for pre-implantation testing of builds

Tables 7.2 represents suggested future research in the fields of Bioprinter Technology, Biomaterials, Cell Sources, Vascularization, Innervation and Maturation.

REFERENCE

1. Howard J. Brown. "Additive Versus Subtractive Manufacturing" accessed from http://www.dmass.net/2012/01/17/additive-versus-subtractive-manufacturing/ on 25th February 2014
2. What is 3D printing? Accessed from http://3dprinting.com/what-is-3d-printing/ on 25th February 2014
3. What are the materials used to print 3D objects? Accessed from http://www.3ders.org/3d-printing-basics.html#materials on 25th February 2014
4. Microsoft Windows 3D Builder. Accessed from http://apps.microsoft.com/windows/en-in/app/3d-builder/75f3f766-13b3-45e9-a62f-29590d5781f2 on 25th February 2014
5. Wohler's Report 2013. Accessed from http://wohlersassociates.com/2013report.htm on 25th February 2014
6. Edinburgh scientists use 3D printing to produce stem cells. Accessed from http://www.bbc.co.uk/news/uk-scotland-edinburgh-east-fife-21328109 on 25th February 2014
7. Lucas Mearian. "The first 3D printed organ -- a liver -- is expected in 2014". Accessed online from the website http://www.computerworld.com/s/article/9244884/The_first_3D_printed_organ_a_liver_is_expected_in_2014 on 25th February 2014 on 25th February 20149.
8. Kate Yandell. "Organs on Demand". Accessed at http://www.the-scientist.com/?articles.view/articleNo/37270/title/Organs-on-Demand/ on 25th February 2014
9. Kelsey Campbell-Dollaghan "How 3D Printers Are Cranking Out Eyes, Bones, and Blood Vessels". Accessed from http://gizmodo.com/how-doctors-are-printing-bones-eyes-noses-and-blood-1474983505 on 25th February 2014
10. Richard Darell. "Laser Origami: The Faster Alternative To 3D Printing" Accessed from http://www.bitrebels.com/technology/laser-origami-alternativ-3d-printing/ on 25th February 2014
11. Mary-Ann Russon. "3D Printers Could be Banned by 2016 for Bioprinting Human Organs" Accessed online from http://www.ibtimes.co.uk/3d-printers-could-be-banned-by-2016-bioprinting-human-organs-1434221 on 25th February 2014
12. Rhiannon Williams. "3D printing human tissue and organs to 'spark ethics debate'". Accessed online from http://www.telegraph.co.uk/technology/news/10604035/3D-printing-human-tissue-and-organs-to-spark-ethics-debate.html on 25th February 2014
13. Zurich Insider. "Insurance implications of 3D printing". Accessed from http://insider.zurich.co.uk/market-expertise/insurance-implications-of-3d-printing/ on 25th February 2014
14. Tanya Lewis. "3D-Printed Medical Devices Spark FDA Evaluation". Accessed From http://www.livescience.com/39339-how-fda-regulates-3d-printed-devices.html on 25th February 2014. Kruth, J.-P. Material incress manufacturing by rapid prototyping techniques.
15. CIRP Annals-Manufacturing Technology 40, 603–614 (1991).
16. Hull, C.W. et al. Method of and apparatus for forming a solid three-dimensional article from a liquid medium. WO 1991012120 A1 (Google Patents, 1991).
17. Malone, E. & Lipson, H. Fab@ Home: the personal desktop fabricator kit. Rapid Prototyping J. 13, 245–255 (2007). Article
18. Allard, T., Sitchon, M., Sawatzky, R. & Hoppa, R. Use of hand-held laser scanning and 3D printing for creation of a museum exhibit. in 6th International Symposium on Virtual Reality, Archaelogy and Cultural Heritage (2005).

19. Shimizu, T.S. et al. Molecular model of a lattice of signalling proteins involved in bacterial chemotaxis. Nat. Cell Biol. 2, 792–796 (2000).
20. Bailey, M.J., Schulten, K. & Johnson, J.E. The use of solid physical models for the study of macromolecular assembly. Curr. Opin. Struct. Biol. 8, 202–208 (1998).
21. Symes, M.D. et al. Integrated 3D-printed reactionware for chemical synthesis and analysis. Nat. Chem. 4, 349–354 (2012).
22. Gonzalez-Gomez, J., Valero-Gomez, A., Prieto-Moreno, A. & Abderrahim, M. A new open source 3D-printable mobile robotic platform for education. in Advances in Autonomous Mini Robots 49–62 (Springer, 2012).
23. Hull, C.W. Apparatus for production of three-dimensional objects by stereolithography. US 4575330 A (Google Patents, 1986).
24. Nakamura, M., Iwanaga, S., Henmi, C., Arai, K. & Nishiyama, Y. Biomatrices and biomaterials for future developments of bioprinting and biofabrication. Biofabrication 2, 014110 (2010).
25. 11.Zopf, D.A., Hollister, S.J., Nelson, M.E., Ohye, R.G. & Green, G.E. Bioresorbable airway splint created with a three-dimensional printer. N. Engl. J. Med. 368, 2043–2045 (2013).
26. 12.Michelson, R.C. Novel approaches to miniature flight platforms. Proc. Inst. Mech. Eng. Part G J. Aerosp. Eng. 218, 363–373 (2004).
27. 13.Reed, E.J., Klumb, L., Koobatian, M. & Viney, C. Biomimicry as a route to new materials: what kinds of lessons are useful? Philos Trans A Math Phys. Eng. Sci. 367, 1571–1585 (2009).
28. Huh, D., Torisawa, Y.S., Hamilton, G.A., Kim, H.J. & Ingber, D.E. Microengineered physiological biomimicry: organs-on-chips. Lab Chip 12, 2156–2164 (2012).
29. Ingber, D.E. et al. Tissue engineering and developmental biology: going biomimetic. Tissue Eng. 12, 3265–3283 (2006).
30. Marga, F., Neagu, A., Kosztin, I. & Forgacs, G. Developmental biology and tissue engineering. Birth Defects Res. C Embryo Today 81, 320–328 (2007).
31. Steer, D.L. & Nigam, S.K. Developmental approaches to kidney tissue engineering. Am. J. Physiol. Renal Physiol. 286, F1–F7 (2004).
32. Derby, B. Printing and prototyping of tissues and scaffolds. Science 338, 921–926 (2012).
33. Kasza, K.E. et al. The cell as a material. Curr. Opin. Cell Biol. 19, 101–107 (2007).
34. Mironov, V. et al. Organ printing: tissue spheroids as building blocks. Biomaterials 30, 2164–2174 (2009).
35. Kelm, J.M. et al. A novel concept for scaffold-free vessel tissue engineering: self-assembly of microtissue building blocks. J. Biotechnol. 148, 46–55 (2010).
36. Kamei, M. et al. Endothelial tubes assemble from intracellular vacuoles *in vivo*. Nature 442, 453–456 (2006).
37. Alajati, A. et al. Spheroid-based engineering of a human vasculature in mice. Nat. Methods 5, 439–445 (2008).
38. Huh, D. et al. Reconstituting organ-level lung functions on a chip. Science 328, 1662–1668 (2010).
39. Sonntag, F. et al. Design and prototyping of a chip-based multi-micro-organoid culture system for substance testing, predictive to human (substance) exposure. J. Biotechnol. 148, 70–75 (2010).
40. Gunther, A. et al. A microfluidic platform for probing small artery structure and function. Lab Chip 10, 2341–2349 (2010).

41. Mankovich, N.J., Samson, D., Pratt, W., Lew, D. & Beumer, J. III. Surgical planning using three-dimensional imaging and computer modeling. Otolaryngol. Clin. North Am. 27, 875–889 (1994).
42. Pykett, I.L. et al. Principles of nuclear magnetic resonance imaging. Radiology 143, 157–168 (1982).
43. Megibow, A.J. & Bosniak, M.A. Dilute barium as a contrast agent for abdominal CT. AJR Am. J. Roentgenol. 134, 1273–1274 (1980).
44. Zagoria, R.J. Iodinated contrast agents in neuroradiology. Neuroimaging Clin. N. Am. 4, 1–8 (1994).
45. Johnson, W.K., Stoupis, C., Torres, G.M., Rosenberg, E.B. & Ros, P.R. Superparamagnetic iron oxide (SPIO) as an oral contrast agent in gastrointestinal (GI) magnetic resonance imaging (MRI): comparison with state-of-the-art computed tomography (CT). Magn. Reson. Imaging 14, 43–49 (1996).
46. Wolf, G.L. Current status of MR imaging contrast agents: special report. Radiology 172, 709–710 (1989).
47. Matsumoto, Y. & Jasanoff, A. Metalloprotein-based MRI probes. FEBS Lett. 587, 1021–1029 (2013).
48. Mironov, V. et al. Biofabrication: a 21st century manufacturing paradigm. Biofabrication 1, 022001 (2009).
49. 35.Horn, T.J. & Harrysson, O.L. Overview of current additive manufacturing technologies and selected applications. Sci. Prog. 95, 255–282 (2012).
50. Sun, W. & Lal, P. Recent development on computer aided tissue engineering–a review. Comput. Methods Programs Biomed. 67, 85–103 (2002).
51. Hollister, S.J. Porous scaffold design for tissue engineering. Nat. Mater. 4, 518–524 (2005).
52. Peltola, S.M., Melchels, F.P., Grijpma, D.W. & Kellomaki, M. A review of rapid prototyping techniques for tissue engineering purposes. Ann. Med. 40, 268–280 (2008).
53. Hutmacher, D.W., Sittinger, M. & Risbud, M.V. Scaffold-based tissue engineering: rationale for computer-aided design and solid free-form fabrication systems. Trends Biotechnol. 22, 354–362 (2004).
54. Klebe, R.J. Cytoscribing: a method for micropositioning cells and the construction of two- and three-dimensional synthetic tissues. Exp. Cell Res. 179, 362–373 (1988).
55. 41.Xu, T. et al. Complex heterogeneous tissue constructs containing multiple cell types prepared by inkjet printing technology. Biomaterials 34, 130–139 (2013).
56. Xu, T., Jin, J., Gregory, C., Hickman, J.J. & Boland, T. Inkjet printing of viable mammalian cells. Biomaterials 26, 93–99 (2005).
57. Cui, X., Boland, T., D'Lima, D.D. & Lotz, M.K. Thermal inkjet printing in tissue engineering and regenerative medicine. Recent Pat. Drug Deliv. Formul. 6, 149–155 (2012).
58. Cohen, D.L., Malone, E., Lipson, H. & Bonassar, L.J. Direct freeform fabrication of seeded hydrogels in arbitrary geometries. Tissue Eng. 12, 1325–1335 (2006).
59. Iwami, K. et al. Bio rapid prototyping by extruding/aspirating/refilling thermoreversible hydrogel. Biofabrication 2, 014108 (2010).
60. Shor, L. et al. Precision extruding deposition (PED) fabrication of polycaprolactone (PCL) scaffolds for bone tissue engineering. Biofabrication 1, 015003 (2009).
61. Barron, J.A., Wu, P., Ladouceur, H.D. & Ringeisen, B.R. Biological laser printing: a novel technique for creating heterogeneous 3-dimensional cell patterns. Biomed. Microdevices 6, 139–147 (2004).
62. Guillemot, F. et al. High-throughput laser printing of cells and biomaterials for tissue engineering. Acta Biomater. 6, 2494–2500 (2010).

63. Guillotin, B. et al. Laser assisted bioprinting of engineered tissue with high cell density and microscale organization. Biomaterials 31, 7250–7256 (2010).
64. Xu, T., Kincaid, H., Atala, A. & Yoo, J.J. High-throughput production of single-cell microparticles using an inkjet printing technology. J. Manuf. Sci. Eng. 130, 021017–021017 (2008).
65. Soloman, Sabrie "Sensors Handbook, 2nd Edition, McGraw-Hill Publishing Co., NY – ISBN 978-0-07-160570-0
66. Soloman, Sabrie "Sensors and Control Systems in Manufacturing 2nd Edition," McGraw-Hill Publishing Co., NY – ISBN 0-07-059626-3
67. Soloman, Sabrie "Affordable Automation," McGraw-Hill Publishing Co., NY – ISBN 0-07-059633-6
68. Soloman, Sabrie "Sensors Handbook, 1st Edition," McGraw-Hill Publishing Co., NY – ISBN 0-07-059630-1
69. Soloman, Sabrie "Sensors and Control Systems in Manufacturing 1st Edition," McGraw-Hill Publishing Co., NY – ISBN 978-07-10572-4

The Digital Revolution – 3D Bioprinting Bio-manufacturing

Powering the 4th Industrial Revolution

3D printing is transmuting the $12 trillion engineering and Manufacturing industry and powering the "4th Industrial Revolution." The 3D printing technology is an advanced practical technology that is disturbing the $12 trillion manufacturing industry. "3D printing will thrust the creation of goods nearer to the consumer, where control is given the consumer. The technology is democratizing manufacturing on a comprehensive measure and sanctioning goods to be economically customized to consumers' prerequisites. Many countries around the world is toiling heard to catch up with U.S. as the world leader followed by Germany, South Korea and Italy.

The most passionate adopters of the technology are the aerospace and medical industries, but the list of businesses espousal the technology is mounting. They include Pfizer, Johnson & Johnson, Merck, Boeing and Airbus, Ford and other automakers, consumer electronics manufacturers like Foxconn and brewers like Heineken. The estimates the market for 3D printing/3D Bioprinting should grow at a brisk 13% rate from current $9.87 billion to $31 billion by 2029, Figure 8.1.

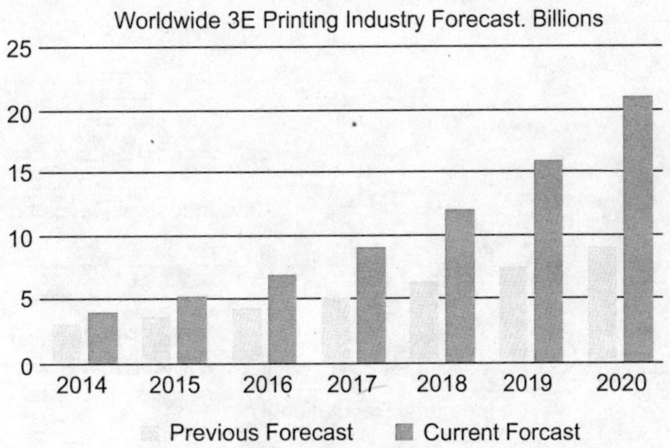

Figure 8.1: "Wohlers' Report" Issued in 2013 – Modified in 2019

The worldwide 3D printing industry has grown from $3.07 billion in revenue in 2013 to $12.8 billion by 2018, and expected to exceed $21 billion in worldwide revenue by 2020. The "Wohlers' Report 2013" had forecasted that the industry would grow to become a $10.8 billion industry by 2021. As far as 3D printing stocks are concerned, the industry's growth trajectory could offer a considerable runway for growth, assuming companies can grow alongside the industry and deliver on the potential for long-term fortunes.

3D bioprinting, as described also recognized as additive manufacturing or rapid prototyping, denotes to the bioprinting of physical 3D structures from computer-aided design (CAD) models in a layer-by-layer method. While the first successful illustration of additive manufacturing happened over 30 years ago, it is only in current years that 3D bioprinting in progress transforming the process of manufacturing mechanical parts. 3D bioprinting has now developed into a more than $16 billion industry and is forecast to top $21.8 billion by 2021.

3D Bioprinting – the 3rd Industrial Revolution

Advances in computer technology, decrease in hardware costs, expiration of important patents, open-sourcing and collective improvement of pertinent software tools and algorithms, and the occurrence of novel bioprinting materials considerably contributed to the great rate at which 3D bioprinting technology has advanced. 3D bioprinting is now being employed to manufacture fully useful organs suitable for end use and is no longer only set aside for prototyping resolves. 3D bioprinting technology has the possible to radically decrease product growth time and efficiently generate customized and geometrically difficult structures on-demand across a extensive diversity of diverse applications. This has fascinated the thoughts of engineers, scientists, and economists to the point where numerous experts have proclaimed 3D bioprinting technology as the "third industrial revolution," Figure 8.2.

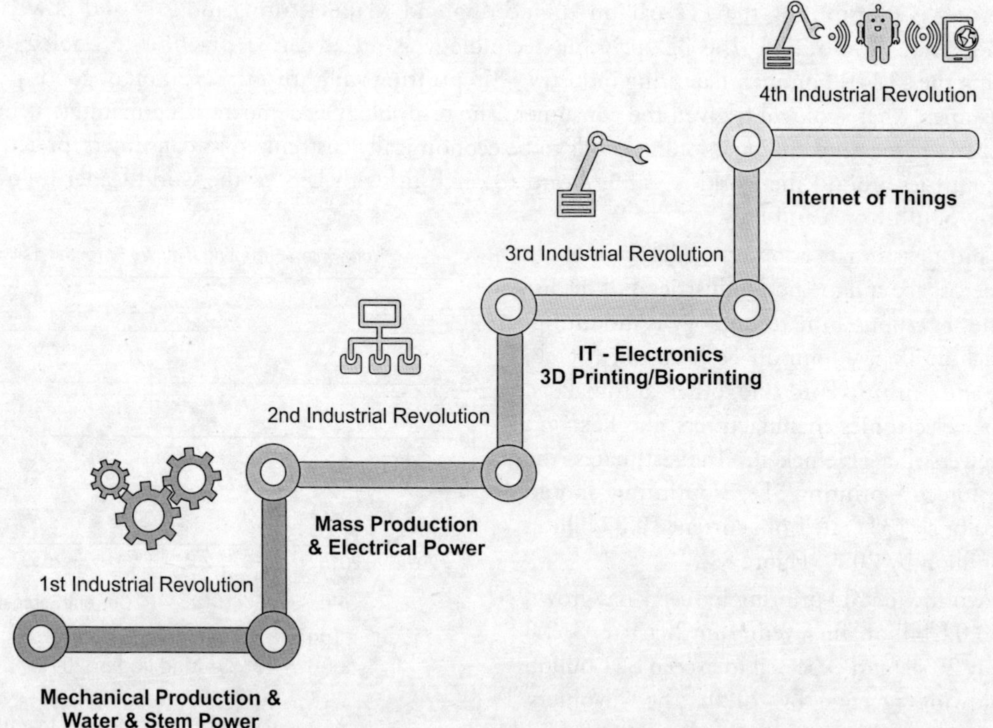

Figure 8.2: 3D Printing – The 3rd Industrial Revolution
Courtesy: Roboyo- DISQUS

The formation of biological structures from digital models is a rapidly developing application of 3D bioprinting that could possibly have transformative consequences on the field of tissue engineering. This specific application is now regularly being mentioned to as 3D bioprinting. Moreover the decisive purpose of printing organs for repair or replacement, the utilization of 3D bioprinting to generate practical tissue constructs is a shorter-term application that can deliver biologists with a new tool for investigative the pathology of particular diseases and enabling the detection of novel therapeutics.

Nowadays it takes an average of $1-4 billion and 12-14 years to advance a drug. Drugs can fail for a diversity of details, but in several cases it is owing to an improper effectiveness or toxicity valuation made during pre-clinical examination. Present pre-clinical analysis platforms can take a number of diverse methods; though, the most pervasive are 2D cell cultures and animal models, neither of which fully models the human physiology. 2D cultures can fluctuate considerably from the intricate 3D intercellular connections happening in organs. Animal models deliver a influential system level viewpoint but are frequently deprived prognosticators of the human drug reaction. Current investigation has established that 3D cell co-cultures better signify human *in vivo* environments compared to the normal 2D cultures presently being employed in the drug discovery procedure. Furthermore, 3D cell co-cultures have the likelihood to substitute animal models. Utilizing 3D co-cultures in the drug screening procedure could contribute to substantial developments in the projecting precision of the pre-clinical drug discovery process by gving the pharmaceutical industry with 3D tissue models that better mimic *in vivo* environments. These improvements could initiate a essential change in the pharmaceutical industry, permitting the improvement of completely novel therapeutics and permitting pharma to examine drugs they may have abandoned in the past due to a absence of suitable models and allowing pharma to examine drugs they may have abandoned in the past due to a lack of suitable models.

Bioprint 3D Heterogeneous Human Tissues on-Demand

The capability to quickly design and then print 3D heterogeneous human tissues on-demand could permit progressions in the fields of tissue engineering, drug discovery, and regenerative medicine. Biological tissues are characteristically difficult heterogeneous structures made up of diverse regions. Each region may be collected of several key components such as cells, extracellular matrices, and complex vascular networks that are organized in precise 3D configurations.

Figure 8.3: 3D Bioprinted Successfully Functional Blood Vessel
Courtesy: Regenerative Medical Research Center

The structure of an artery, Figure 1 contains a central lumen and a set of concentric walls with diverse constructions, cell arrangements, and morphologies. In order to engineer a physiologically relevant tissue that mimics native behavior, different regions of a structure may necessity to be printed utilizing diverse characteristics. Control over composition, structure, and parameters such as patterning and porosity is indispensable as these characteristics have been revealed to influence biological reaction, Figure 8.3.

Designing and then printing cell-laden build with difficult configuration, several materials, heterogeneous tool path planning and printing characteristics in an automated way causes a challenge in tissue engineering. Providing new design methods and tools that permit the formation of 3D tissue models as well as a platform for sharing the software-described tissue models is predictable to help biologists and tissue engineers in preceding the field of *in vitro* tissue model progress.

Achieving Tissue Model Objectives

The main aims of Bioprinting are to grow an automated procedure and user-friendly software tool series for designing and 3D printing heterogeneous tissue constructions as well as a platform for relating and distributing tissue models. Accomplishing these goals will first necessitate an in-depth understanding of typical 3D printing values and an investigation of the most collective file formats and open source software tools presently being utilized by the 3D printing community. Then a technique will be advanced for designing 3D biological buildings, taking into account the characteristic heterogeneity and difficulty of inherent tissues. Algorithms will be designed and carry out to code tissues in 3D with programmable geometrical and material characteristics. Then, the performance of the established algorithms will be assessed by examining implementation times.

Software will then be established to ease control and organize of print-head purpose and stage motion. A language for relating and sharing tissue models must be advanced. The result of design selections and printing characteristic on the general printing procedure can be examined. To confirm the functionality and validate the abilities of the established software tool series, numerous test structures can be designed and then 3D printed.

The Physical Structure of 3D Bio-Printing

The elementary idea behind 3D printing is that a CAD object is transformed into a physical build by successively stacking a set of 2D layers until the whole build is recognized. Each deposited 2D layer characterizes a cross-sectional slice of the 3D build and has a related finite layer thickness. The 3D printing procedure thus produces builds that are approximations of CAD models where a layer is the elementary construction unit. Decreasing the layer thickness progresses the quality of the subsequent 3D build at the cost of increasing the time required to build the entire structure. Within each layer, various printing control parameters are deployed in order to diverge the steadiness, porosity, and composition of the printed builds to setup the particular application necessities.

Fused Filament Fabrication (FFF)

Of the numerous diverse 3D printing deposition approaches, presently one of the most collective procedures is an extrusion-based technology recognized as Fused Filament Fabrication (FFF). The method FFF works is that a solid thermoplastic filament is heated into a liquid polymer that is extruded over a nozzle and patterned rendering to a computer-controlled motion arrangement.

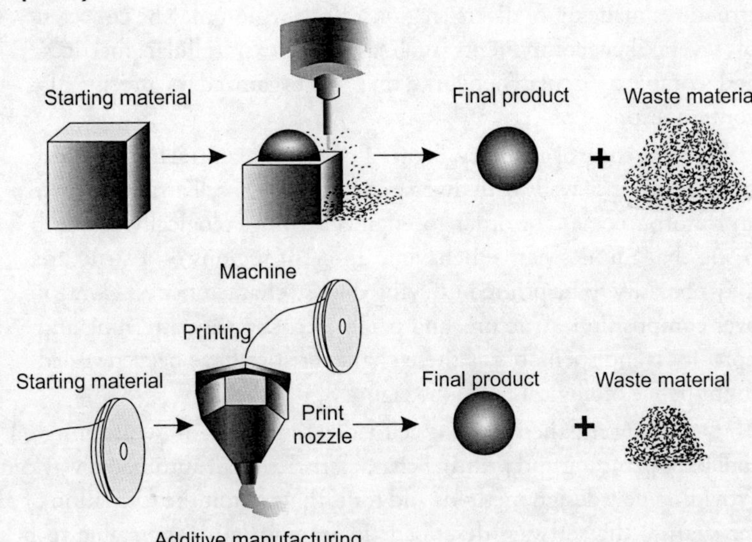

Figure 8.4: Additive Manufacturing versus Subtractive Manufacturing
Courtesy: US Government Accountability Office

In difference to 3D printers, which apply an additive manufacturing procedure, subtractive manufacturing machines such as Computer Numerical Controlled (CNC) machines begin with a solid chunk of material and utilize cutting tools to carve the preferred object by eliminating the additional. Because of this essential modification, 3D printing postures some advantages over subtractive manufacturing procedures. Some instances of these advantages contain a decrease in material depletion; improved geometrical difficulty doesn't inevitably obscure the manufacture procedure, and greater easiness in programming and software planning of build instructions. A diagram portraying the change in procedure between additive manufacturing and subtractive manufacturing is displayed in Figure 8.4.

With the fast reduction in the cost of 3D printers in current years, a flourishing online community of 3D printing supporters has arisen. Through this open community, an overall 3D printing tool chain has been embraced and users vigorously partake in the cooperative growth of different computer-aided design (CAD) and computer-aided manufacturing (CAM) software tools, such as OpenSCAD, ReplicatorG, Slic3r, and Printrun. Creativities such as the RepRap and Fab@Home projects give whole hardware provisions for users to build fully operational individual 3D printers. Websites such as Thingiverse serve as platforms for hosting and sharing 3D printing design files.

3D Modeling, Slicing, Control, and Print

The commonly acknowledged 3D printing software workflow for converting a 3D model in the virtual world to its real world comparable entails of three steps, Figure 8.5:

- modeling,
- slicing, and then
- printer control to print the actual structure.

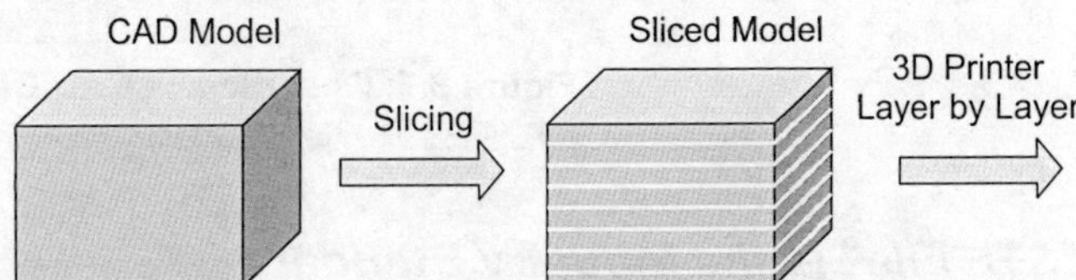

Figure 8.5: CAD Model Slicing and Printing
Courtesy: SPIE

While the precise employment may diverge slightly and extra elements such as initialization processes and post-processing functions may be merged for diverse printing uses, the software workflow and the related file formats illustrated in Figure 3 are collective across almost all 3D printers. Modeling includes fully stipulating the surface configuration of the anticipated 3D build utilizing CAD software such as SolidWorks. The surface configuration is most frequently defined and exported as a StereoLithography (STL) file. Slicing includes the utilization of computer-aided manufacturing (CAM) tools to transform a 3D CAD model into machine information implicit by the particular 3D printer applied in the printing procedure. The created slicing data is frequently defined utilizing G-Code. Printer control is the procedure whereby machine-specific instructions are transferred to the 3D printer in order to generate the entire build, Figure 8.6.

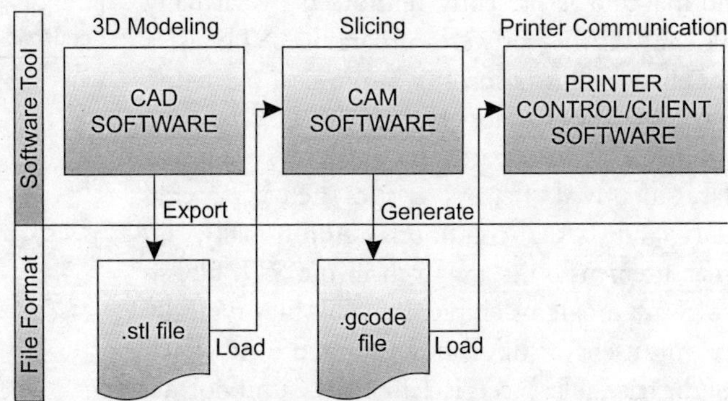

Figure 8.6: Software tools and file formats used in the general 3D printing software tool chain

Defined below, the particulars of the diverse software basics and file formats utilized in the overall 3D printing procedure. The first section clarifies the procedure of designing a computer model of a wanted physical build utilizing CAD software and delivers a background on the industry-standard tessellation file format. The second section defines the procedure of creating toolpaths essential to understand a wanted physical build utilizing CAM software and delivers an impression of one the most frequently utilized numerical control languages.

3D Model Design Tessellated Representation of Physical Structure "STL"

Product design is the first step in any systematic fabricating procedure and includes abstracting an idea and converting it into an entirely well-defined visualization. This also incorporates to the 3D printing procedure, which typically begins with designing a 3D computer model of a physical build. Models can be designed by a user through the application of 3D CAD modeling software, gained or adapted from online sharing sources, or reverse-engineered utilizing application precise technologies such as 3D scanning or medical imaging. The throughput of this step is a 3D model that is most usually defined and signified utilizing the STL file format, Figure 8.7.

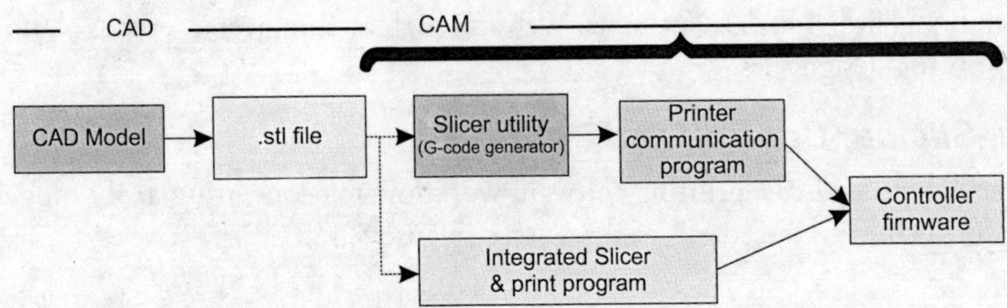

Figure 8.7: Open-Source CAD/CAM Software Stack

Courtesy: PicoPrint

"STL File" is the Industry Standard

Generated by "3D Systems Inc.," in 1987, the STL file format has since become the industry normal and is presently reinforced by virtually all 3D CAD software programs. STL is a tessellated illustration of a physical structure whereby the surface configuration of a 3D model is approximated by a list of triangular facets deprived of characteristics such as material data and CAM conditions. Additionally, no measurement units are given in the STL file, so software programs employing STL files typically permit users to stipulate the wanted units. Each of the triangular facets is defined by a unit normal and three vertices as illustrated in Figure 8.8. The unit normal and each of the three vertices are defined by X, Y, and Z coordinates.

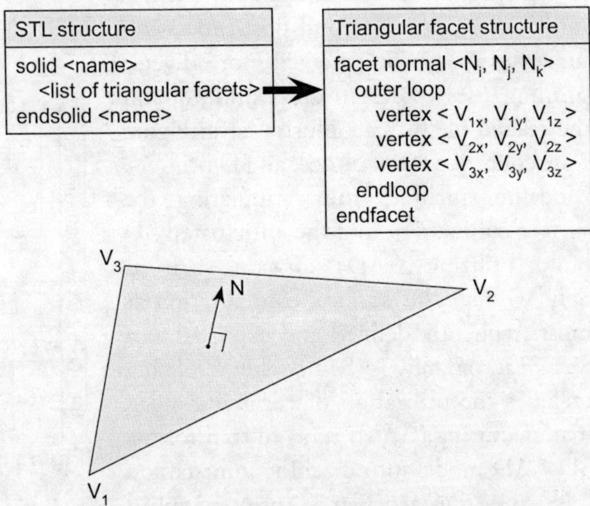

Figure 8.8: Transferring STL file Structure to Triangular Facet Structure

STL files originate in two different storage formats: ASCII and binary. Both formats utilize a ".STL" extension. The ASCII format is human-readable and is frequently kept for debugging and instructional purposes though the binary format is more frequently utilized owing to its compact file size. An instance of an ASCII STL file exported utilizing SolidWorks that defines a tetrahedron is illustrated in Figure 8.9. While both the ASCII and binary forms approximate the tetrahedron utilizing 4 triangular facets, the file sizes are 722 bytes and 284 bytes, correspondingly.

Generating Tool Path - 3D Model Slicing

Subsequent the conclusion of the essential modeling and tessellating tasks employing 3D CAD software and exporting the conforming STL file, the next step in the 3D printing workflow includes CAM procedure planning. This contains creating the essential printer commands and sending these commands to the 3D printer.

Meanwhile 3D printers build up physical substances over the consecutive deposition of material in a layer-wise method, tessellated objects defined in STL files must first be sectioned into a set of geometrically and dimensionally precise 2D horizontal layers (slices) with related user-defined layer thicknesses. This procedure is regularly denoted as slicing within the 3D printing community.

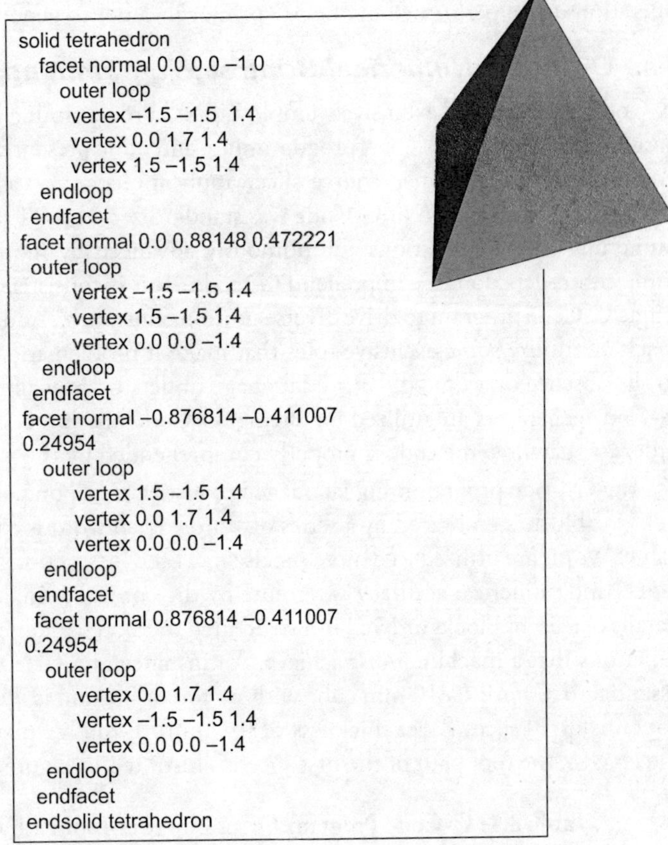

```
solid tetrahedron
  facet normal 0.0 0.0 –1.0
    outer loop
      vertex –1.5 –1.5 1.4
      vertex 0.0 1.7 1.4
      vertex 1.5 –1.5 1.4
    endloop
  endfacet
facet normal 0.0 0.88148 0.472221
  outer loop
      vertex –1.5 –1.5 1.4
      vertex 1.5 –1.5 1.4
      vertex 0.0 0.0 –1.4
    endloop
  endfacet
facet normal –0.876814 –0.411007
0.24954
  outer loop
      vertex 1.5 –1.5 1.4
      vertex 0.0 1.7 1.4
      vertex 0.0 0.0 –1.4
    endloop
  endfacet
  facet normal 0.876814 –0.411007
0.24954
  outer loop
      vertex 0.0 1.7 1.4
      vertex –1.5 –1.5 1.4
      vertex 0.0 0.0 –1.4
    endloop
  endfacet
endsolid tetrahedron
```

Figure 8.9: ASCII STL File Describing a Tetrahedron

The Slicing Process

The slicing procedure is accomplished by first intersecting the 3D model with a set of imaginary slicing planes whose common normal is in the vertical build direction, Figure 8.10. From the intersection points that are shaped between a specified imaginary slicing plane and the 3D surface depiction, a contour (boundary) is drawn out for each slice. After the borders are well-defined for each slice, tool paths that the print head desires to follow in order to fill the space surrounded by each slice's border with raw material allowing a precise pattern are created. The throughput of the slicing procedure is printing commands.

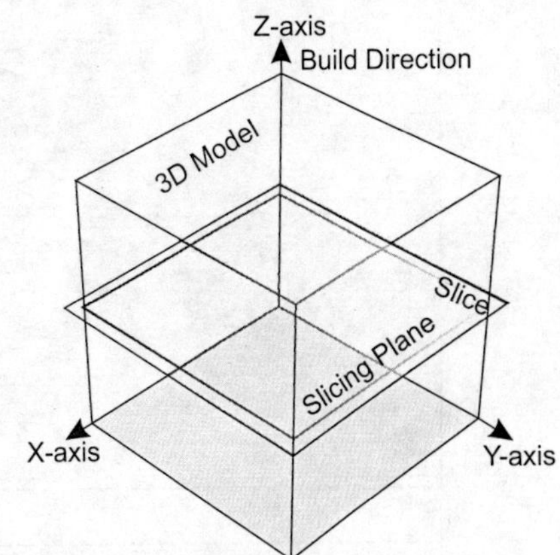

Figure 8.10: Intersection of slicing plane and 3D model

The whole slicing procedure, also occasionally mentioned as tool path creatoin, can be achieved over the utilization of slicer software. With the upsurge of 3D printing, some diverse slicer uses have been collaboratively established, optimized, and are presently obtainable online as open-source applications. Some instances of the more collective slicer programs comprise Slic3r, Skeinforge, Repsnapper, and Cura. Slicer applications characteristically accept an STL file as an input and permit the user to describe numerous diverse CAM provisions such as print head speed, layer thickness, and fill density and patterns in order to create the specified set of printing commands. Some slicer applications employ added algorithms to add extra fill, borders, and layers when required in order to guarantee structural steadiness and develop the quality of the printed object. Assumed a precise STL input file, the output of the slicer software is a file comprising a set of printing commands defined utilizing numerical control (NC) code. An NC program is fundamentally a structured arrangement of code blocks stipulating the precise motion profiles and complementary operations to be performed by the 3D printer in order to generate a physical structure.

The G-Code – Numerical Control Programming Language – Code Generator

G-Code is the greatest extensively employed NC Programming language, utilized comprehensively over the past few decades for CNC part programming, and more presently for desktop 3D printing tool path creation. For this cause, many open source slicer applications are occasionally also denoted as *G-Code generators.* Though a precise version of G-Code was standardized in the 1980s with ISO 6983, there presently happen numerous diverse executions autonomously advanced by motion controller manufactures. Furthermore, some controllers do not comprehend G-Code. This inconsistency in motion controllers makes it hard for a single G-Code program to drive diverse motion controllers. Additionally, numerous additive manufacturing machines utilize some exclusive roles that make it problematic to standardize a precise NC language. Due to the absence of a united control language understood by all controllers, in various circumstances where G-Code generators are utilized for slicing, G-Code post-processing or interpretation is an essential step in order to guarantee the code is properly comprehended by the machine.

The G-Code programming language is constructed upon consecutive lines of code named blocks. Each G-Code block is embraced by a series of words. Each word is contained of a letter followed by a numerical value. Words are utilized to denote precise machine instructions or influences. The whole list of permissible letters and numerical accuracy is resolute by the controller manufacturer. A G-Code program is a file that entails of a list of blocks ordered in a structured method to define the meticulous motion and supplementary functions that a machine must achieve. An instance of a G-Code program relating the motion tool paths essential to 3D print a 10 mm cube with no top or bottom is illustrated in Table 8.1. The G-Code program contains of 50 slices (layer thickness of 0.2 mm) in the XY-plane and the related new slice movements in the Z-axis. The tool path of the first slice is illustrated in Figure 8.11.

Table 8.1: G-Code Program Describing the Tool Path of a Cube with No Top or Bottom

3D Model Sliced		
G-Code Program		
	G-Code Block	Description
1	G28	Home taxes
2	G90	Use absoluts coordinates
3	G21	Set unit to millimeters

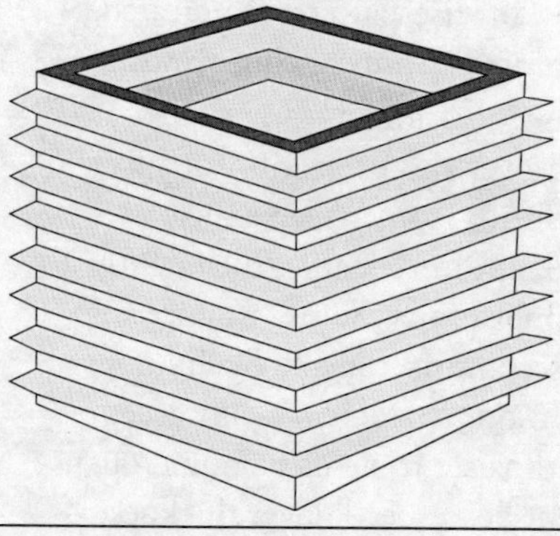

4	G1 20.200 F800.000	Move to next layer (0)
5	G1 X5.050 Y14.950	Move to most per mazer point
6	G1 X5.050 Y5.050	Perimeter point
7	G1 X14 950 Y5.050	Perimeter point
8	G1 X14.950 Y14.950	Perimeter point
9	G1 X5.050 Y14.950	Perimeter point
10	G1 70.400	Move to next layer (1)
11	G1G X5.050 Y5.050	Perimeter point
12	G1 14.950 Y5.050	Perimeter point
13	G1 X14 950 Y14.950	Perimeter point
14	G1 C5.050Y14.950	Perimeter point
.	.	.
.	.	.
.	.	.
250	G1 Z10.000	Move to next layer (49)
251	G1 X5.050 Y0.050	Perimeter point
252	G1 X14.950 Y5.050	Perimeter point
253	G1 X14.950 Y14.950	Perimeter point
254	G1 X5.050 Y14.950	Perimeter point

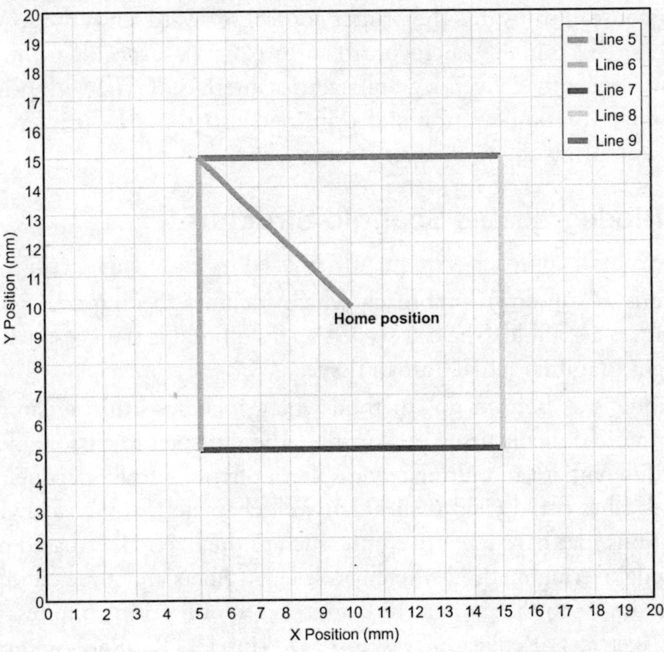

Figure 8.11: Tool Path of First Layer of Sliced Cube with No Top or Bottom

Preparatory G-Code Instructions

In the instance G-Code program illustrated in Table 8.1, the primary few commands (lines 1-3) are introductory G-Code instructions created for a specific 3D printer hardware outline. These introductory commands are encompassed to achieve functions such as preparing all axes and configuring the suitable units. The residual lines of code (lines 4-254) are all G1 commands. G1 commands are the most frequently employed commands created by open source slicer programs. The G1 command signifies a straight line movement from the present point to the quantified point at a specified speed. For instance, on line 4 in Table 8.1, the G1 command initiates the printer to move in a straight line from the present position to the definite Z coordinate (0.200 mm) at the programmed speed (1800 mm/minute) in order to begin printing a new layer of material. On line 5, the G1 command initiates the printer to move in a straight line from the present position to the X and Y coordinates (5.050 mm, 14.950 mm) which resembles the first boundary point.

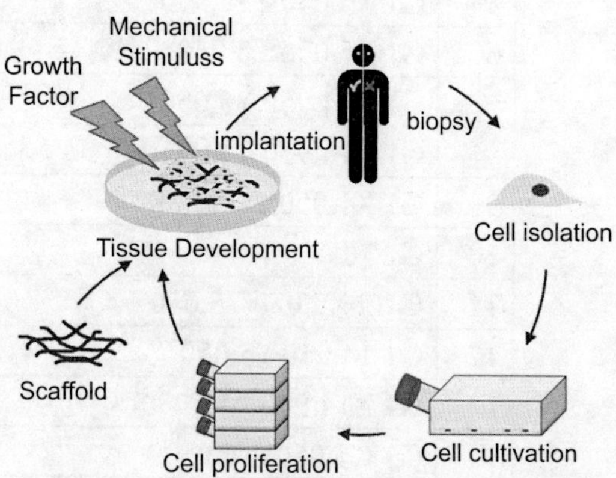

Figure 8.12: Biocompatible – Biodegradable Scaffold Structure in Tissue Engineering
Courtesy: Medgadget

Upon positive design the model of the anticipated physical structure utilizing 3D CAD software, exporting the resulting STL tessellation file, slicing the model and creating the corresponding G-Code toolpath file, the last step in the 3D printing software workflow includes sending the G-Code file to the 3D printer via some communication link such as USB or Ethernet. This communiqué task is held by the 3D printer's client (printer control) software. The printer control software sends the sequence of commands to the firmware fitted on the 3D printer's microcontroller to perform a specific printing procedure. The 3D printer control software frequently delivers a graphical user interface (GUI) with printing-related functions such as *initialization*, direct motion control, and printing start/stop. The printer firmware construes the established commands and prints the wanted structure.

Biocompatible - Biodegradable Scaffold Structure

There are presently few central approaches frequently utilized in tissue engineering, Figure 8.12.

The first technique is a "top-down" approach which includes seeding cells onto a biocompatible or biodegradable build named a scaffold. The cell-seeded scaffold build is then gestated and cultured until the cells propagate and begin to mimic actual human tissue.

The second technique is a "bottom-up" method which includes utilizing simple cell-laden structure blocks to construct up more difficult tissue pre-cursors. These tissue pre-cursors must then be cultured to permit for the cells to differentiate and ultimately display a physiologically-pertinent function. Bottom-up tissue engineering approaches could yield tissues that well characterize *in vivo* environments, particularly for biological constructions that have repetitive units such as the liver. Fiber-based printing is one instance of a bottom-up tissue engineering method whereby cell-laden fibers are modeled into builds. This method to tissue engineering is similar to the popular FFF approaches utilized in countless industrial 3D printing systems such as the "MakerBot" "Replicator Desktop 3D Printer", "RepRap," and others.

Inkjet – Extrusion Bioprinting

Majority bioprinting systems depend on two kinds of printing techniques:

- inkjet printing and
- extrusion printing.

Inkjet-based bioprinting systems permit for high-resolution and micro-scale printing. There are some difficulties connected to inkjet-based bioprinting contain high force of ejected cells, blockage at the orifice, restricted to printing substance with low viscosities, sedimentation of cells in inkjet cartridge, and long build times essential to print macro-scale structures. Although extrusion-based bioprinting systems allow the printing of big structures, they are naturally restricted to printing macro-scale structures and materials with adequately high viscosities. Additionally, most extrusion-based systems depend on an isolated orifice for each material, Figure 8.13.

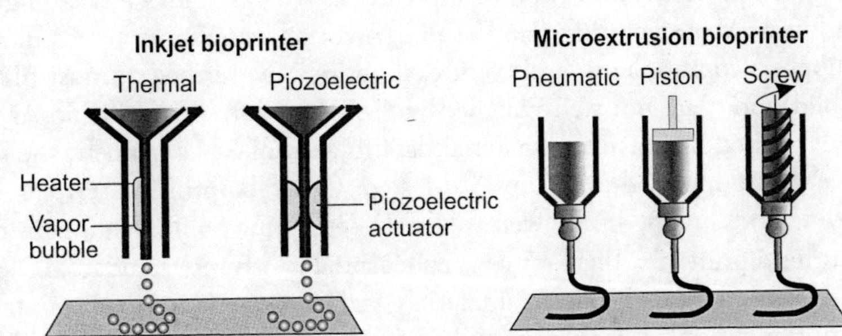

Figure 8.13: Inkjet – Extrusion (Robotic Dispensing) Bioprinting

The established bioprinting system utilizes a microfluidic print head to deposit a spherical hydrogel fiber that is chemically cross-linked on-chip via coaxial flow concentrating. utilizzing this method, material is placed into the printer as a liquid, but dispensed as a solid fiber, permitting for numerous on-chip programmatic microfluidic processes prior to fiber development. Creating cell-laden hydrogel fibers utilizing this method frequently mentioned as microfluidic spinning. Spreading this notion to 3D printing for the maanufacture of fiber-based biological build has also been verified before, but with very restricted automation and deprived of the capability to design difficult heterogeneous builds founded on user-defined 3D CAD models. The scientists depend on manually programming a set of motion commands, deprived of the utilization of user-defined 3D CAD models, to print simple models. Furthermore, though the scientists determines the utilization of numbers of substances, material switches during the printing procedure are held manually and are not automatically activated founded on pre-defined CAD provisions and material data.

Conventionally, 3D printing systems were utilized to generate objects that are standardized in terms of material arrangement and CAM provisions. This describes why STL – which is the most shared file format well-matched with nearly every 3D printing system – only defines the surface configuration of a build, and does not comprise any material data or CAM qualifications. STLs have been utilized in many investigation concentrated on printing homogenous biological build. Open source slicers such as RepRap and Slic3r have been employed to create the tool paths from an input STL file. In current years, 3D printers able of synthesizing multiple material objects have arisen due hard work from industry, academia, and a prosperous community of 3D printing supporters.

Heterogeneous Bioprinting Materials

Aimed at applications that necessitate the utilization of many materials such as heterogeneous bioprinting, others have advanced software that takes multiple STLs as inputs, then allocates material data to each STL, and then compiles the set of STLs into a multi-material model. The multi-material model well-defined in CAD is then served into CAM software in order to create the set of printing commands. Utilizing this method, the same set of CAM conditions is imposed upon the whole build. This critically restricts the control over print settings and general functionality of the object. Others have suggested modifying STL file formats that contain supplementary data besides just configuration. One instance of this is the Additive Manufacturing File Format (AMF). Furthermore to requiring an object's configurations, AMF supports conformation and color descriptions. Nevertheless, most 3D CAD modeling programs, such as SolidWorks, are not well-matched with AMF. Moreover, utilizing AMF postures the same restrictions as the method of conveying material data to a set of STLs; namely, the same characteristics are forced upon the whole model during the CAM step. Other bioprinting systems depend on manually programming commands line-by-line as well as heavily depending on human involvement to enable material changeovers during a print job. There remains an essential to advance standard 3D printing design methods that permit users to effortlessly allocate materials and characteristics to diverse sections of a 3D model, create the printing procedure strategies, and then print these heterogeneous build in a mechanized manner.

Multi-Material – Versatile Bioprinters

Nonetheless high-end commercial 3D printers, such as those advanced by Stratasys and 3D Systems, are beginning to sustain multi-material printing, these systems often depend on exclusive file formats that are custom-made to exact printer hardware configurations. Most of these printing systems use inkjet technology and in-house rasterization algorithms to concurrently deposit multi-material liquid photopolymers that are cured utilizing ultraviolet light. While these printers are adept of generating objects with high resolution, building big structures utilizing this method can be exceptionally time overshadowing.

Filament-founded numbers of material 3D printers, such as those established by "MakerBot" and the multi-nozzle bioprinter defined in application a devoted extruder for each material. This kind of system shape typically leads to disturbances, breaks, and time postponements during the printing procedure as a consequence of print head switches. Decreasing disturbances produced by material switches is still an ongoing dificulties.

Multi-Material Structures Using single Nozzle Print Head

Specially tailored software provides the printing of multi-material builds utilizing a single nozzle print head. Numerous algorithms were established to apply the programmatic switching of materials on-the-the-fly deprived of postponements related with print head switches, permitting for the uninterrupted and coordinated interlinking of programmable multiple material fibers both within and through layers.

"Click - N Print" and "Script - N Print" Modes

Commercial bioprinting software advanced by "Organovo" can be employed. The software is accomplished of manufacturing builds consisted of several hydrogels and cell kinds. The software functions in two modes:

- "Click 'N Print" and
- "Script N Print."

In the "Click N Print" mode, a blank cross-section of stacked cylinders is showed on a GUI and users can design easy builds by allocating a material kind to each of the cylinders. Founding the design procedure around cylindrical building blocks instead of user-stated 3D CAD models significantly restricts patterning competences and control over the build's geometry and arrangement. The "Script 'N Print" mode is utilized to design more intricate builds, but necessitates users to manually program activities of the positioning stage and several print heads. This procedure can be particularly time consuming, error-inclined, and necessitates wide information and skill about motion control programming.

Innovative 3D bioprinting software was advanced that is well-matched with industry typical file formats and CAD/CAM tools for the manufacture of 3D printed biological builds with heterogeneous characteristics founded on user well-defined 3D models. A design technique is established that includes assembling difficult builds from easier building blocks; each structure block could be programmed and executed utilizing diverse characteristics such as fiber diameter, deposition/infill patterning, porosity, layer thickness, and conformation. A set of STL building blocks are sliced individually, rendering to a exclusive set of CAM conditions, to create a set of G-Code files. Then each G-Code file is allocated a material sort, 3D coordinates, and other exclusive characteristics well-matched with multifunctional print head procedures. The set of G-Code files is then amalgamated into a single file in order to create the heterogeneous layer-by-layer procedure planning commands. In this way, the procedure of allocating material data, print settings, and other exclusive characteristics is achieved during the CAM step. This permits us to decouple our exclusive capability to design and print building blocks under dissimilar circumstances from typical methods utilized in 3D CAD modeling, permitting us to influence on industry standards file formats and construct on open-source tools. We also concentrate on creating full automation and coordination between multifunctional print head procedures and phase motion as well as advance control systems that are printer independent and could be modified to other hardware applications. A widespread description language was established to create a design method to allow community growth of printed tissues.

Heterogeneous Tissue Implementation

Innovative 3D bioprinting system was advanced to incorporate tissue development software and microfluidics to generate custom-made 3D fiber-based heterogeneous tissue builds. Subsequent the printing procedure, cell-laden builds manufactured utilizing the bioprinting system must experience a procedure of cell culturing so that the cells differentiate properly.

3D Bioprinter Hardware Mechanisms

The 3D bioprinting mechanisms established in this development is primarily consisted of more than a few hardware mechanisms: a micro-fluidic print head, a pressure monitoring system, air pressure solenoid switch bank, a vacuuming head mechanism, a computer-controlled Multi-axes high precision positioning stage, and a specially tailored printing base. The print mechanism and the pneumatic solenoid switch bank were established. A diagram of the established bioprinting mechanism is illustrated in Figure 8.14.

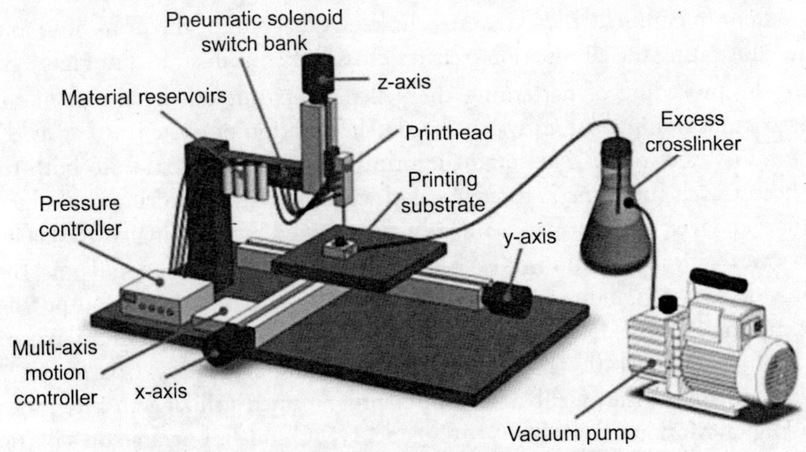

Figure 8.14: Diagram of the 3D Bioprinting Mechanisms

The print head is encompassed of a microfluidic chip proficient of giving out cell-laden hydrogel fibers. A flow of aqueous sodium alginate is coaxially concentrated by adjoining it with aqueous calcium chloride (cross-linker) inside the print head. In this way, sodium alginate is laden into the printer as a liquid but give out as a gelled fiber (calcium alginate). According to this, performing different on-chip microfluidic procedures such as mixing and sequencing preceding to gelling are likely. By altering the pressure ratio between the aqueous sodium alginate and the cross-linker, the fiber diameter could be diverse from 70-300μm. The on-chip flow rates are structured utilizing a pressure-driven flow controller (Fluigent MFCS-4C). To allow deposition of several materials, such as hydrogels laden with diverse kinds of cells, pneumatically controlled valves were integrated for each fluid channel in the print-head. A dedicated pneumatic solenoid switch was employed to activate the on-chip valve for each fluid channel. By changing the valves' shape, the kind of material being give out as well as the alginate gelation procedure could be controlled. The microfluidic print head is manufactured from PDMS (poly-methylsiloxane) using a soft lithography process. The molds used in the procedure are 3D printed utilizing a commercial 3D printer *"Objet 24"*.

Cross-linker Exits the Print Head Orifice

Since the cross-linker departures the print head orifice with the rapidly gelling alginate, it will pool and interfere with the already printed fibers. To solve this problem, the present example of the system must print materials on a porous substrate and vacuum must be useful to eliminate extra cross-linker during the printing procedure. Two diverse printing substrates with porous membranes were employed. The first kind of printing substrates utilized was typical well plate enclosures "Corning® Costar® Transwell® cell culture inserts, 8.0μm pore size." These are extensively utilized and well-matched with typical drug screening technology. The second kind of printing substrates utilized was specially designed and 3D printed. The special printing substrate comprises a porous surface on which fibers are patterned and loaded. A passage feed from the printing substrate is linked to a vacuuming system to eliminate the extra cross-linker. A diagram of the print head and the specialized printing substrate illustrating the elimination of extra cross-linker is illustrated in Figure 8.15.

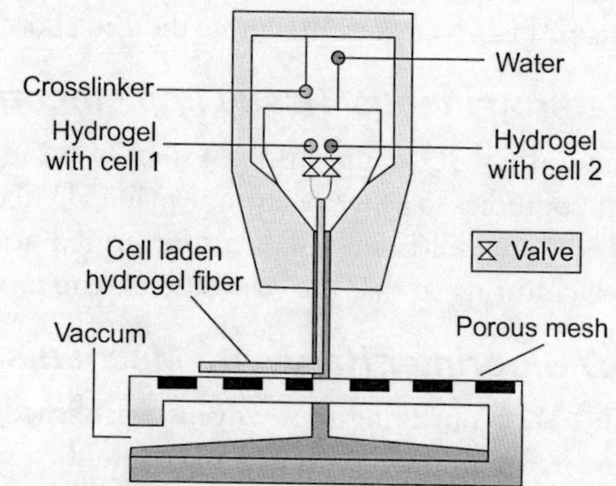

Figure 8.15: Schematic of print head and excess cross-linker removal mechanism

To mechanize the procedure of patterning the gelled fibers put out by the print head, a computer controlled multi-axis positioning system was utilized. The positioning system is consisted of X, Y, and Z axes. The print head is fixed to the Z axis and the printing substrate is held onto both the X and Y axes. Thus, the print head travels up in the Z-direction with each new layer allowing to the user-specified layer thickness. Within each layer, adjacent patterning is accomplished by coordinated travels in both the X and Y directions. The mechanical resolution of each of the three linear actuators is 1μm. Three autonomous motor drivers provide each of the three stepper motors with the necessary pulse output signals in order to deliver the necessary travel to the diverse lead screws. For coordinated travel, a multi-axis motion controller was employed "Galil" DMC-2140." The "Galil" travel controller utilizes a particular 32-bit Motorola 68331 series microcontroller. The "Galil" hardware is programmed utilizing an interpreted programming language called "Digital Motion Controller DMC" code. DMC code is operated on a particular "Real Time Operating System RTOS" that is uploaded on the "Galil" hardware.

Switching between Different Materials during Bioprinting

Over and above supporting embedded programming and operating programs entirely on the "Galil" motor controller, host-centric programming is correspondingly supported. An application programming interface "API" can be utilized to direct instructions from the host computer to the controller. For message between the host computer and "Galil's" motion controller, an Ethernet connection was utilized. In addition to synchronized-motion, this microcontroller assists as the 3D bioprinters' main processing component and is utilized for the control and synchronization of outside actions, such as swapping between diverse materials throughout a print job. Restrictions for each axes are detected using optical limit switches. A system level drawing is illustrated in Figure 8.16.

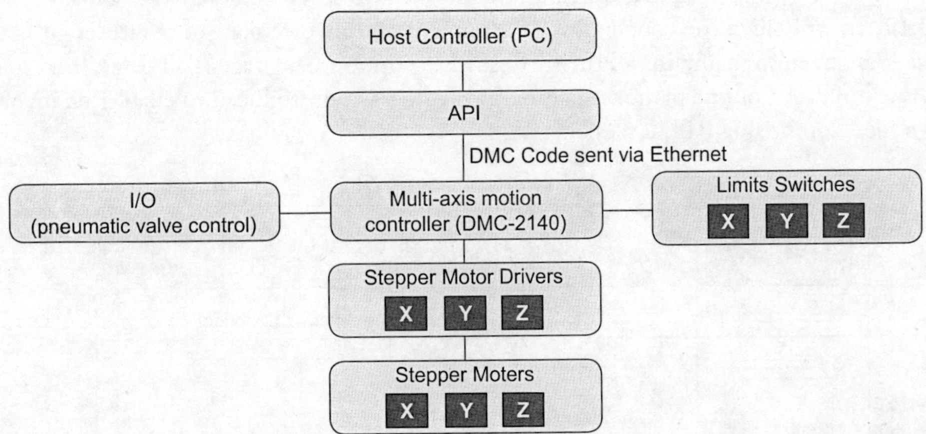

Figure 8.16: System Level Diagram Swapping between Diverse Materials throughout Print Job

3D Bioprinting Custom Software

The specially advanced special software can sustain the construction and 3D bioprinting of cell-laden hydrogel build with complicated designs, programmable configurations, and several printing characteristics such as fiber diameter, layer thickness, infill patterning, and porosity. The special software can be constructed around the impression of an element-based architectural design technique. This design technique is utilized in areas such as software engineering to progress large software-motivation systems as well as in aerospace engineering to construct aircrafts made of numerous diverse particular subsystems. Bottom-up tissue engineering approaches, such as the fiber-based method utilized by the 3D bioprinter, are founded around the same element-based design technique. In the setting of tissue printing, it is called this design method the "Tissue Building Block Design (TBBD)" technique. Utilizing the TBBD technique, intricate tissues could be accumulated from a set of easier building blocks as illustrated in Figure 8.17. These building blocks may be located in 3D either vertically or across. Then the organized building blocks are printed layer-by-layer to recognize the wanted 3D build.

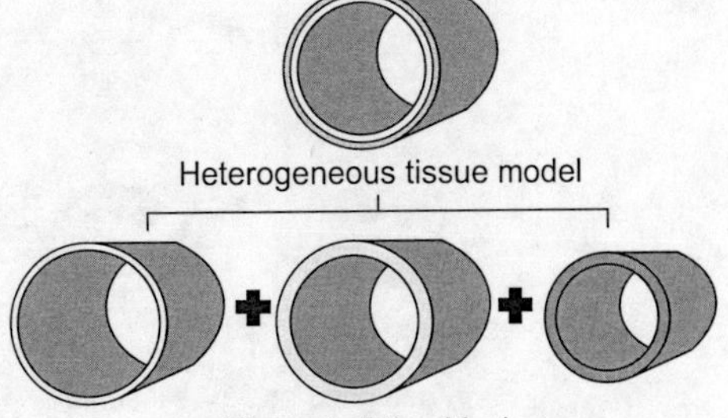

Figure 8.17: 3D Bioprinting - Assembling of Tissue Building Blocks

An illustration of specially established software tool chain with the diverse software tools, file setups, and online platform is illustrated in Figure 8.18. The software tools comprise CAD Software, CAM Software, Tissue Designer, and Printer Control Software. The created files contain STL files, G-Code files, tool path files, material files, and PTDL files. The workflow starts with designing diverse building blocks utilizing 3D CAD software. Each building block is signified by one STL file. Then each STL file is sliced individualistically employing CAM software permitting to particular CAM conditions and transferred as a G-Code file. After that, the set of G-Code files (set of sliced building blocks) are combined into a single tool path file and a single material file utilizing Tissue Designer. At last, the created tool path and material files are interpreted by the Printer Control Software and instructions are directed to the 3D bioprinter in order to perform the identified print job. An instance illustrating the diverse steps correlated in 3D printing a heterogeneous tissue build designed utilizing the bioprinting specially developed software is illustrated in Figure 8.19. To embolden the cooperative development and involvement of tissue design files between users, Tissue Designer also maintains a Printed Tissue Description Language (PTDL) utilized to define the designed tissue build. An online platform named OpenVitro was introduced to enable the involvement of tissue design files, comprising PTDL files.

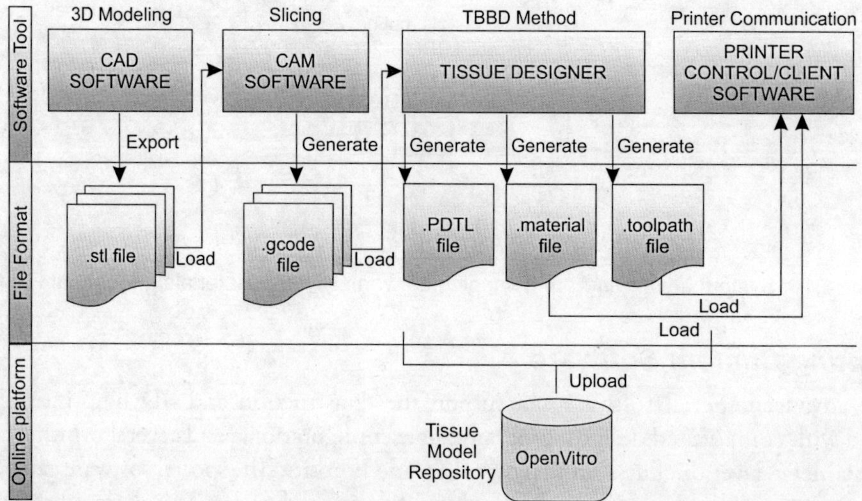

Figure 8.18: 3D Bioprinting Custom Software - Tools, File Formats, and Online Platforms used in our

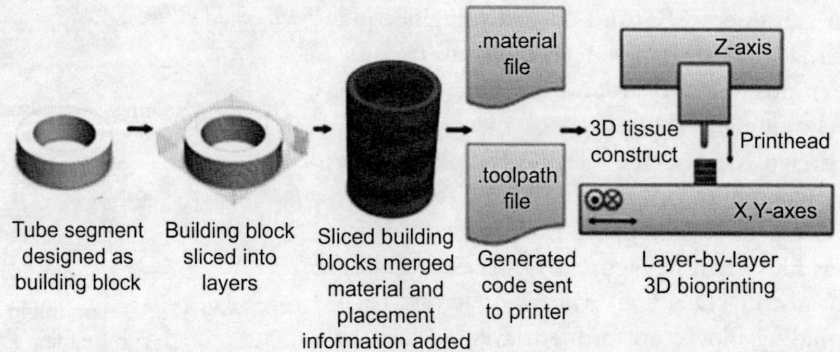

Figure 8.19: 3D Bioprinting Steps of Software Tool Chain – Tissue Building Block

The particulars of diverse elements of the 3D bioprinting software "tool-chain" are defined below. The procedure of designing builds utilizing the TBBD technique is stated. The employed open-source slicer and the pertinent CAM provisions is also stated below. Also, new algorithm utilized to unify separate tissue building blocks is also stated below. Moreover, the particulars motivating Tissue Designer, PTDL, and OpenVitro are described. The printer control, synchronization, and communication are stated below.

Tissue Building Block Design Technique

The TBBD technique purposes to permit for the design of a varied tissue model consisted of a set of tissue building blocks. Each building block could be a set of diverse material projects, CAM qualifications, and other constraints that share building blocks with one another, such as 3D positioning and print importance. Tissue building blocks can be designed utilizing any typical 3D CAD software that maintains the creation of STL files. SolidWorks, one of the most widespread 3D CAD programs, was employed for the design of most building blocks in this work. To generate a varied tissue structure, an individual building block must first be designed and exported as a STL file for each region that needs a diverse surface configuration, material type, or set of CAM specifications. Utilizing SolidWorks, an assembly file on behalf of the surface configuration of the whole tissue build may be designed and the discrete parts demonstrating the diverse building blocks may be exported into discrete STL files. As formerly stated, meanwhile STL files only hold surface configurations, CAM specifications and material allocations are allocated to each building block at following stages.

For instance, the varied structure in Figure 8.20 needs discrete STL files defining the surface configuration of each building block. Utilizing the TBBD technique, the varied structure is generated by amalgamating the two different building blocks. Each building block will be separately sliced permitting to a set of CAM specifications and then materials and other characteristics will be allocated in later stages of the 3D bioprinting specially developed software.

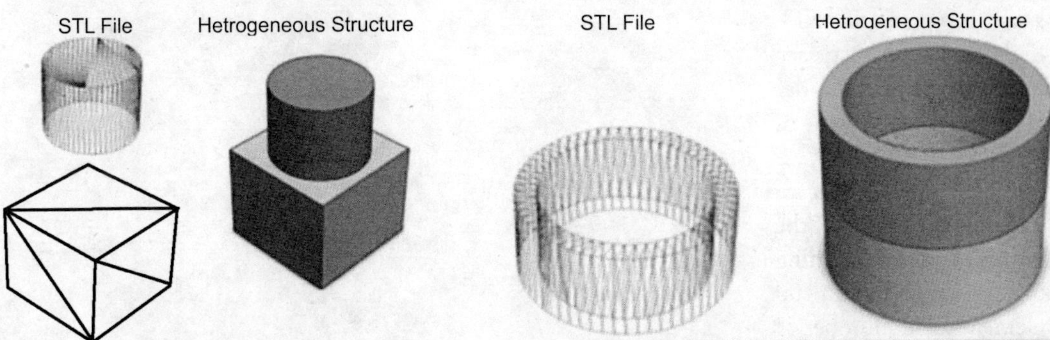

Figure 8.20: 3D Building Blocks with Different Surface Geometries

Figure 8.21: 3D another Building Block with Identical Surface Geometries

In the circumstance where two building blocks of a tissue build have the same surface configurations, then only one STL file is required. An instance of this situation is illustrated in Figure 8.21. The anticipated varied tissue build is consisted of two stacked tube segments, with the material kind being the only change between them.

Slicing Tissue Building Block

As formerly described, the 3D printing procedure is a layered fabricating method that depends on slicing a 3D CAD model into a set of stacked horizontal layers and the creation of the essential tool paths to fill them. In order to create a set of slices from a building block, an open-source slicing tool (Slic3r Version 0.9.8) was employed. The input of the slicer software is a building block STL file. Typically, slicer tools permit the operator to rotate the imported STL file to the wanted positioning. Furthermore, most open-source slicer programs are bundled with many purposes and settings precisely enhanced for the normally utilized FFF printing technique.

For the reason of the 3D bioprinting custom software, the open-source slicing tool is only utilized to create the slices and the related tool paths required to fill them. This data is defined utilizing G-Code. Thus, any other G-Code creator such as SkeinForge, Repsnapper, and Cura may be utilized in its place of Slic3r. In other words, the 3D bioprinting specially developed software is not dependent on any one specific slicer software. Furthermore, diverse slicer programs could be utilized to separately slice each building block providing the user additional design flexibility. The CAM specifications that are of most consideration during the slicing procedure and their correlated explanations are summarized in Table 8.2.

Table 8.2: CAM specifications for the slicing process

CAM Specification	Description
• Layer thickness (mm)	The vertical distance between each consecutive layer
• Number of perimeters	The number of outlines or contours per layer
• Fill density (%)	The percentage of the area within a perimeter that is filled with material
• Fill pattern	The specific geometry that is used to fill the area within a perimeter
• Extrusion width (mm)	The width of the deposited fiber within a layer

As an instance, an STL file relating the surface configuration of a cylinder with a diameter and height of 10 mm is sliced, Figure 8.22. The layer thickness is set to 100 μm, subsequent in the creation of a sliced building block that contains of 100 slice elements. The number of perimeters is set to 2. The fill density is set to 40%, which resembles to 60% porosity. The extrusion width is set to 100 μm. A drawing illustrating the 50th slice of the cylinder with diverse fill patterns is illustrated in Figures 8.22.

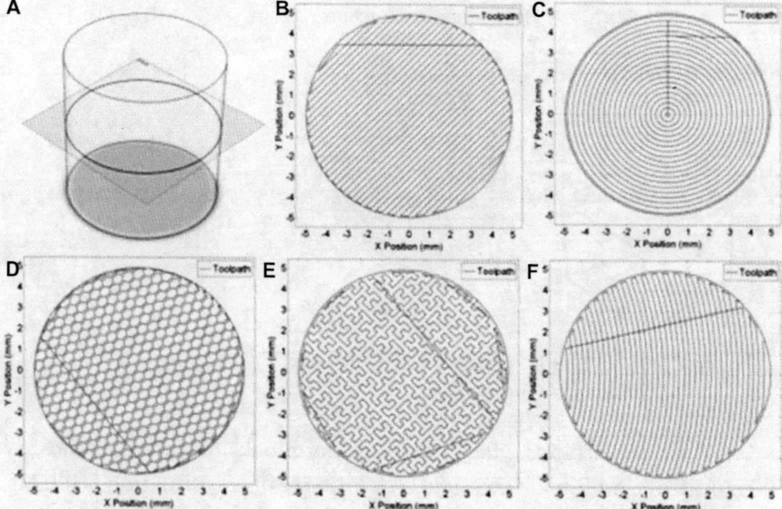

Figure 8.22: (A) Slice of a Cylinder (B) Rectilinear Fill Pattern (C) Concentric Fill Pattern (D) Honeycomb Fill Pattern (E) Hilbertcurve Fill Pattern (F) Archimedeanchords Fill Pattern

Diverse fill patterns could be utilized to accomplish complicated core tissue architectures that could influence biological reaction.

The G-Code file created from the slicing procedure contains of successions of G1 instructions. As formerly stated, a G1 instruction signifies linear interpolation between the present absolute location and the identified absolute location. Meanwhile, the created G-Code program defines a tool path that contains of a succession of straight line positions. A characteristic created G-Code file contains of a organized list of three diverse kinds of G-Code block patterns illustrated in Table 8.3. Type I G-Code block is well-defined as a block that designates a straight line travel within a layer while depositing substance. A Type II G-Code block designates a straight-line travel within a layer deprived of depositing substance. A Type III G-Code block designates a layer modification.

Table 8.3: Different Types of Defined G-Code Blocks

G-Code Block Type	Code	Description
Type I	G1 X_{Number} Y_{Number} FNumber E_{Number}	Move from the current XY position to the specified position (X_{Number}, Y_{Number}) at the specified feed rate (F_{Number}) while extruding the specified length of filament (E_{Number})
Type II	G1 X_{Number} Y_{Number} F_{Number}	Move from the current XY position to the specified position (X_{Number}, Y_{Number}) at the specified feed rate (F_{Number}) without extruding
Type III	G1 Z_{Number} F_{Number}	Move from the current Z position to the specified Z position (Z_{Number}) at the specified feed rate (F_{Number})

To track an incremental tissue expansion method, a sliced building block could be bioprinted. Then the development could be additionally adapted founded on observed structural and operational results. When the building block design is enhanced and the bioprinted build is displayed to exhibit the anticipated performance, it could be incorporated with formerly enhanced building blocks and bioprinted as one varied structure. For a more quick tissue improvement procedure or for experimentations with endpoints that rely wholly on the communication between varied elements, the sliced building blocks could be combined with other sliced building blocks and then the whole set could be bioprinted at the same time.

Design Algorithm of Tissue Building Block

The 3D bioprinting specially developed software: integrating a set of sliced building blocks utilizing the advanced TBBD Algorithm. If building blocks are stacked on top of each other, such as the configuration illustrated in Figure 8.21, a consecutive bioprinting method whereby an assumed building block is totally bioprinted before beginning to bioprint the succeeding building block may work. Nevertheless, this method will not work if, for instance, the building blocks were organized closely side-by-side. In this situation, a crash would happen amid the bioprint head and the first bioprinted building block when it begins bioprinting the second building block. With the intention of 3D bioprint heterogeneous structures consisted of several building blocks organized side-by-side, such as concentric tubes required to be capable to combine slices from diverse sliced building blocks as the structure is built up layer-by-layer.

The input of the TBBD algorithm is a set of sliced building blocks signified by a set of G-Code files as well as a material number, bioprinting precedence, and XYZ parameters for each building block. Utilizing specially developed software, a multi-material structure could supposedly be consisted of an unlimited number of material kinds. Nevertheless, in a everyday sense, the number of materials is restricted by the number of material channels reinforced by the single-nozzle microfluidic bioprint head used in our bioprinting system. The microfluidic bioprint head architecture only maintains a maximum of two materials.

To define the issue of amalgamating slices from diverse sliced building blocks at the same Z level, a priority-based preparation arrangement is utilized. A fixed, user-specified bioprinting importance is allocated to each sliced building block during the design phase. Bioprinting significances are utilized in preparation the order in which slices from diverse building blocks at the same Z level is bioprinted. Higher importance slices are bioprinted before lower importance slices. Algorithms could be established in the future to automatically assign bioprinting importance founded on precise design guidelines or application-specific necessities. With the purpose of position a set of building blocks in 3D space, each building block is provided an x-center position, y-center position, scaling factor, and z-offset. The x-center and y-center locations are employed to position the building block in the XY plane. For instance, if x-center and y-center positions are equal to zero, the building block will be centered at the source of the printing substrate. The scaling factor is employed to scale the building block in the XY plane. The z-offset constraint is utilized to show the z-position of the first slice of a building block. Therefre, the z-offset parameter is utilized to stack building blocks on top of each other.

- The TBBD algorithm is distributed into three main modules:
- G-Code Interpreter,
- Slice Scheduler, and
- Output Code Generator, Figure 8.23.

For prompt handling and effective memory usage, the algorithm can be written utilizing C++. The G-Code Interpreter analyses and parses through the input and stores it into diverse classes of dynamically generated objects. The Slice Scheduler is responsible for scheduling the order in which slices are bioprinted. The Output Code Creator accepts slice objects from the Slice Scheduler and yields the tool path and material code recognized by the 3D bioprinter. Each of these modules will now be clarified in additional clarification.

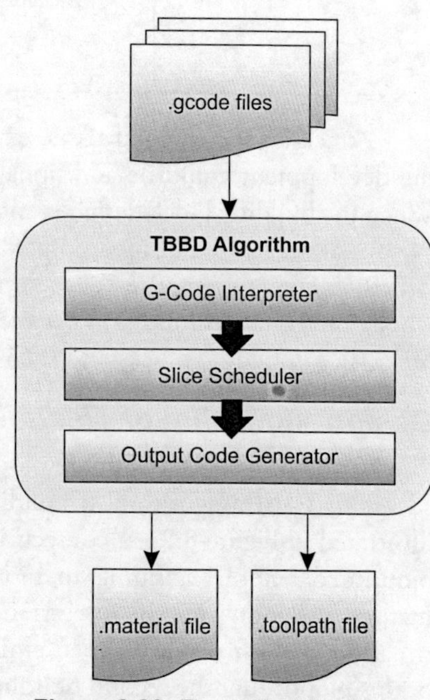

Figure 8.23: Tissue Building Block Design TBBD Algorithm

The G-Code Interpreter is established utilizing Flex and Bison. Flex and Bison are two powerful tools implemented to advance interpreters and compilers. Flex and Bison work together to discover patterns in organized input and yield a reaction founded on the known pattern. This procedure is divided up into two stages: lexical analysis, which is done utilizing Flex, and parsing, which is completed utilizing Bison. Lexical analysis contains describing the input into substantial units named tokens and unfolding each token utilizing a regular detail. The tokens are well-defined as the diverse G-Code words in a G-Code block. A grammar is detailed in Bison and describes the associations between a sequence of tokens and their meaning. The distinct grammar stipulates the instructions of how the distinct tokens make up the diverse G-Code blocks. In runtime, Bison parses the G-Code utilizing the well-defined grammar.

The G-Code Interpreter structures the input giving the class diagram illustrated in Figure 8.24. The G-Code Interpreter begins with generating a GCodeFile object for each input G-Code file "building block" and stores the related material number, bioprinting importance, x-center, y-center, scaling factor, and z-offset. The G-Code Interpreter then scans each G-Code file line-by-line and senses whenever a sequence of tokens matches the well-defined grammar. Each time a valid G-Code block is sensed, a function is called to store the numerical portions of the G-Code words and the kind of the G-Code block in a GCodeBlock object. The generated GCodeBlock object is an addition to a current Slice object or a new Slice object. A new Slice object is generated each time facing a Type III G-Code block, Table 3, which shows a layer modification. A data member called z_level in the Slice class is utilized to retain track of the absolute height of the slice. In other words, z_level is equivalent to the z-offset plus the numerical portion of the "Z" G-Code word of a Type III G-Code block; in this manner, building blocks that are sliced with diverse layer thicknesses can be combined. This procedure remains until the end of the effective G-Code file. Keeping the G-Code in a structured manner as opposite to just a sequence of instructions enables the software powerful, flexible, and scalable. A basic flowchart illustrating how one effective input G-Code file and the related constraints are processed and stored utilizing the G-Code Interpreter is illustrated in Figure 8.25.

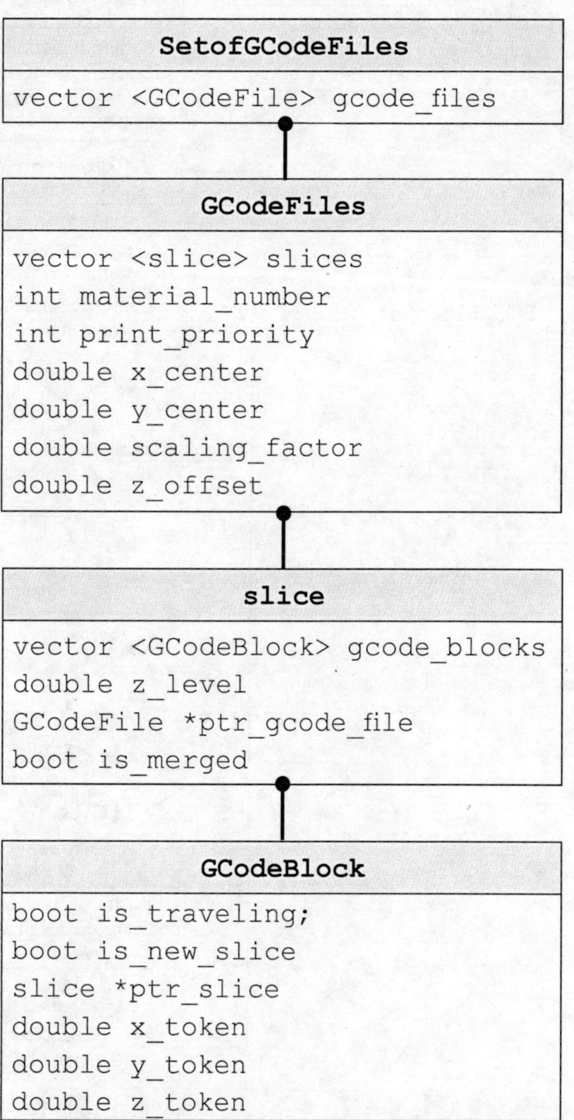

Figure 8.24: Class diagram showing composition relationship

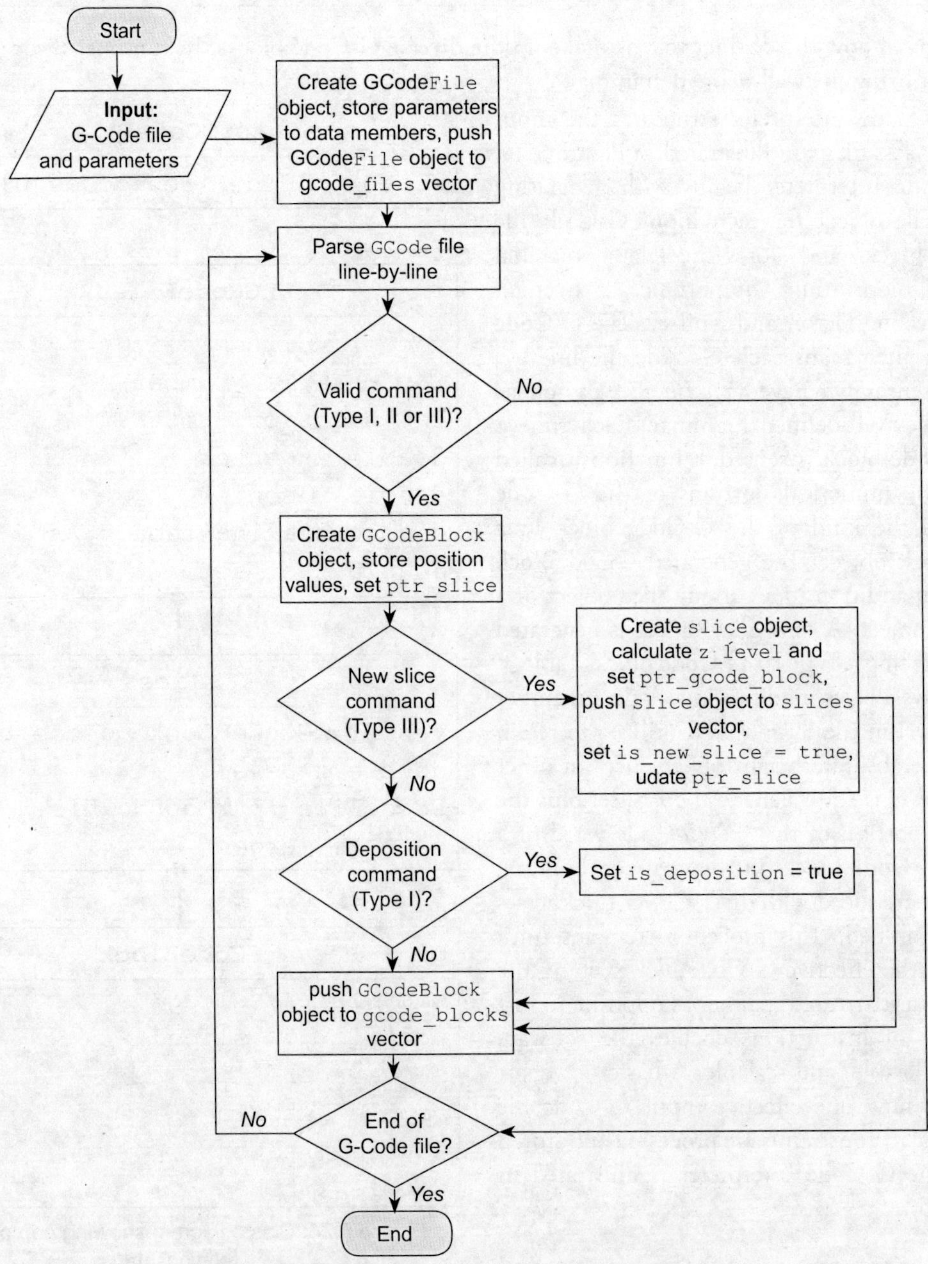

Figure 8.25: Flowchart of G-Code interpreter module

The Slice Scheduler is accountable for defining which "Slice object" is delivered to the Output Code Creator to be bioprinted next. Meanwhile the slices are properly organized within each building block, the z level and bioprint importance of each Slice object at the front of the vectors are compared. Amid these compared Slice objects, the one with the lowest zlevel and highest bioprint importance is scheduled to be bioprinted next. If the z level of the present slice object is equivalent to the z level of the preceding slice object, then the two slices are combined. In other words, the bioprint-head will not travel up in the Z direction and both toolpaths are joined. This is done by setting is merged flag value true. This flag notifies the Output Code Creator to disregard any Type III (layer change) G-Code blocks for merged slices. As an example, a graph illustrating the toolpath of two combined slices, each slice generated from isolated building blocks is illustrated in Figure 8.26. A flowchart of the Slice Scheduler, is illustrated Figure 8.27.

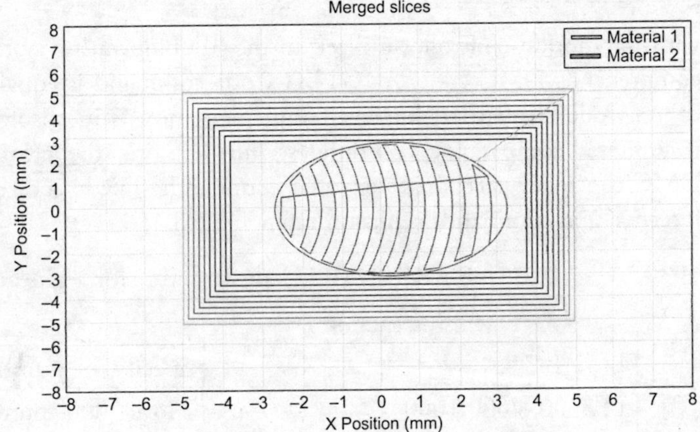

Figure 8.26: 3D Printing - Merged Slices

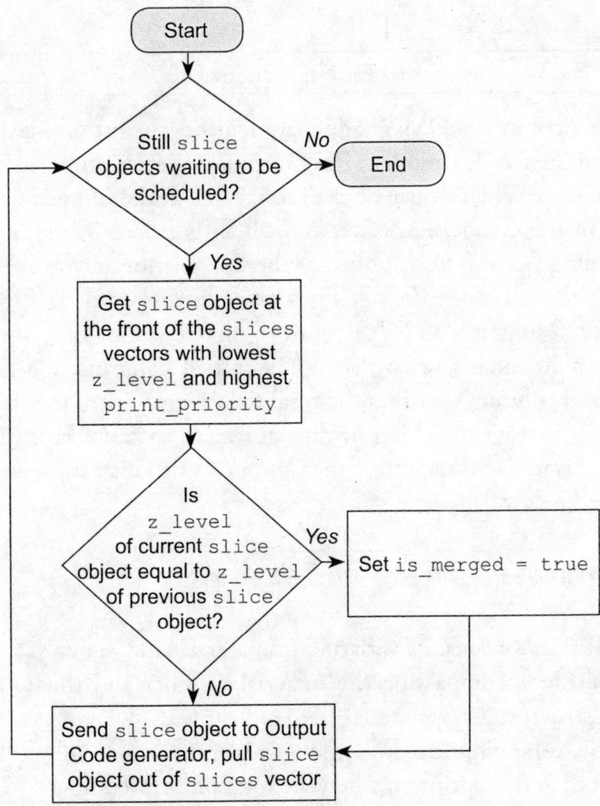

Figure 8.27: Flowchart of Slice Scheduler

The Output Code Creator is primarily accountable for generating two files utilized to drive the 3D bioprinter: a tool path file and a material file. As formerly stated, due to the absence of a universal G-Code accepted by motion control manufactures, G-Code interpreting or post-processing is an essential step. Additionally, some motion controllers do not comprehend any form of G-Code altogether. This is the case with the motion controller utilized in the 3D bioprinting system. The motion controller employed only comprehends DMC code. A "G1" G-Code command is equal to a "LI" DMC command. Additionally, the G-Code created by the slicer implementation is in absolute coordinates, but the "Galil" controller necessitates comparative coordinates. Furthermore, the created G-Code is in millimeters, while the DMC controller necessitates encoder counts. The DMC code structure for directing a sequence of linear interpolation commands is illustrated in Table 8.4.

Table 8.4: DMC Code Structure for a Linear Interpolation Sequence

DMC Instruction	Description
LM ABC	Specify the number of axes
LI 3000, 2000,1000	Move 3000 counts in X, move 2000 counts in Y, and 1000 counts in Z from the current XYZ position
. . .	. . .
LE	End of sequence

The created tool path file comprises DMC code that identifies the list of linear interpolated movements necessary to bioprint the designed 3D model. The Output Code Creator pauses until it obtains a Slice object from the Slice Scheduler. When a Slice object is established, the Output Code Creator loops over all the GCodeBlock objects in the Gcode blocks vector "which is stored in the Slice object" and translates G-Code to DMC. The scaling factor is also applied to the XY position travels at this stage. If the is merged flag value is true, the Z movement of the GCodeBlock object is disregarded. If the is traveling flag value is true, it is likely to question a function to halt depositing material, but bioprint-head presently does not sustain this. The material file contains of two spaces delimited columns: the first column identifies the material type and the second column identifies the total fiber length (μm) that has to be bioprinted "from the start of the bioprint job" before switching to the equivalent material kind. To calculate each of these values, the total fiber length is updated each time the Output Code Creator processes a GCodeBlock object utilizing the equation stated below.

$$L_{fiber} = \sum_{i-1}^{n} \sqrt{x\_displacement_i^2 + y\_displacement_i^2 + z\_displacement_i^2}. \quad (1)$$

Once the Output Code Creator senses a variation in material kind, a new row is put in the material file encompassing the material number and the present fiber length. The structure of a material file is illustrated in Figure 8.28. For instance, once the total bioprinted fiber length is equal to 438483 μm, material 1 will be placed and bioprinted; when the total bioprinted fiber length is equal to 589769 μm, material 2 will be placed and bioprinted; when the total bioprinted fiber length is equal to 666751 μm, material 3 will be placed and bioprinted; when the total bioprinted fiber length is equal to 779290 μm, material 4 will be placed and bioprinted.

The computed total fiber length is also utilized to approximate the build time for a particular bioprint job. The build time is roughly equal to the total fiber length divided by the bioprint speed.

material1	438483
material2	589769
material3	666751
material4	779290

Example material file

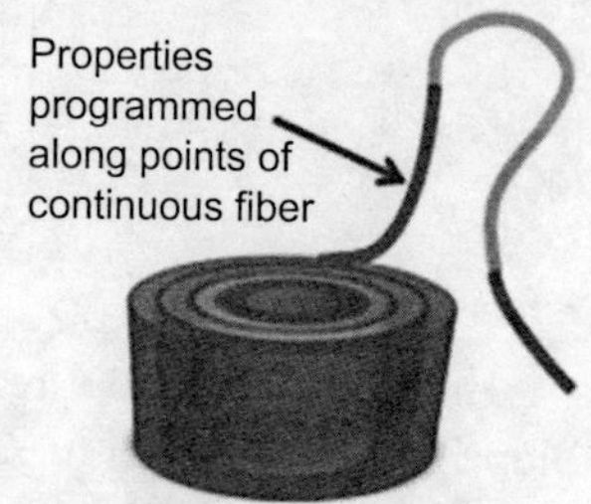

Figure 8.28: Structure of Material File

$$Build\ time \approx \frac{Total\ \ fiber\ length}{Speed}. \tag{2}$$

The projected build time permits the user to better assess costs prior they are sustained and better plan bioprinting experiments.

Detecting Problems before Starting the Bioprinting Process

Furthermore to creating the material and tool path files, a folder encompassing the point clouds of each layer is outputted. The point clouds are plotted and illustrated to examine the created tool paths, layer-by-layer, before bioprinting. This could be significant as users could sense glitches before beginning the bioprinting procedure, which could possibly save time and material costs. Illustrations of issues that may be perceived by visualizing the point clouds include overlying toolpaths as well as trailing fibers produced by Type II G-Code blocks. Tissue Designer, PTDL, and OpenVitro.

Bioprinter Control, Synchronization and Communication

Control over on-chip performance and harmonization between bioprint head task and stage motion is vital for bioprinter automation. To achieve this harmonization, the microcontroller is employed to create output signals controlling on-chip valves' structures founded on particular events. In this manner, the bioprint head could be programmatically arranged.

Instances of present bioprint head tasks embrace chip initialization and cleanup, the starting and halting of fiber creation, and material swapping. Future bioprint head roles could embrace material mixing and concentration gradient generators.

Initializing the Bioprint Head before Bioprinting

To initialize the bioprint head before bioprinting, a purpose was established that involves of a timed-sequence on-chip valve shapes. The primary task of the initialization process is to fill the diverse on-chip channels with the suitable materials and approve correct chip role prior to bioprinting. The pressure values for all the material inputs can be set to a maximum of 75 mbar. Table 5 delivers a thorough process. All the valves to permit materials to pass through, which are water, cross-linker, and both hydrogel pre-cursors, can be unlocked for five seconds. This is achieved to purge the bubbles by permitting liquid stream through all the channels. The water-valve is then locked, and the residual valves are kept unlocked for two seconds to ensure that a fiber is being created. The exiting channel is then flushed from any residue by closing all the valves excluding for water for fifteen seconds. Finally, all the valves are locked. After the conclusion of the initialization/cleanup process, the bioprint head should be prepared for bioprinting, Table 8.5.

Table 8.5: On-chip Valve Configurations for Initialization/Cleanup Procedure

#	Duration (Seconds)	Water	Cros-slinker	Alginate 1	Alginate 2
1	5	OPEN	OPEN	OPEN	OPEN
2	2	CLOSED	OPEN	OPEN	OPEN
3	15	OPEN	CLOSED	CLOSED	CLOSED
4	–	CLOSED	CLOSED	CLOSED	CLOSED

Accounting for the Deposited Fiber

Considering the quantity of fiber deposited during a travel sequence, the entire distance moved from the beginning of the tool path is uninterruptedly stored and updated in real-time on the microcontroller's on-board memory. By observing this value and observing the delays and implementation times related with on-chip processes, exclusive bioprint head operations could be implemented at controlled points along an uninterruptedly deposited fiber. Though our bioprint head only supports material swapping, as the bioprint head is further established and other characteristics such as concentration and temperature control, cell counting and mixing are incorporated into the hardware, uninterrupted fibers with more progressive and spatially-controlled characteristics could be bioprinted.

The bioprint head chores should be performed a priori so that properties could be used in a controlled means along diverse points of a uninterruptedly bioprinted fiber. When bioprinting multi-material builds, the bioprint head requirements to guarantee that the identified material is being deposited at the exact time. Because the procedure of swapping materials happens within the chip a space away from the tip of the nozzle, there is an inescapable delay related with material swapping that should be considered. Meanwhile, material swapping requirements to be activated on-chip proceeding to a material swap point within a tool path to permit for the preceding material within the channel to exit the nozzle and the new material to fill the channel.

With the understanding that the fiber departs the bioprint head nozzle at a rate equivalent to the stage speed, the distance "channel length" it takes within the bioprint head to alter materials should be constant. If the fiber exits the nozzle at a rate that is faster than the stage speed, the fiber will coil within the channel, producing the effective channel length to be better than the actual length.

In addition disturbing the channel length, coiling within the channel may also cause eventual chip blockage. If the fiber exits the nozzle at a rate that is slower than the stage speed, the fiber will be strained. This will also distress the channel length and may also lead to other difficulties such fiber damage. These three fiber exit velocity circumstances are summarized in Figure 29.

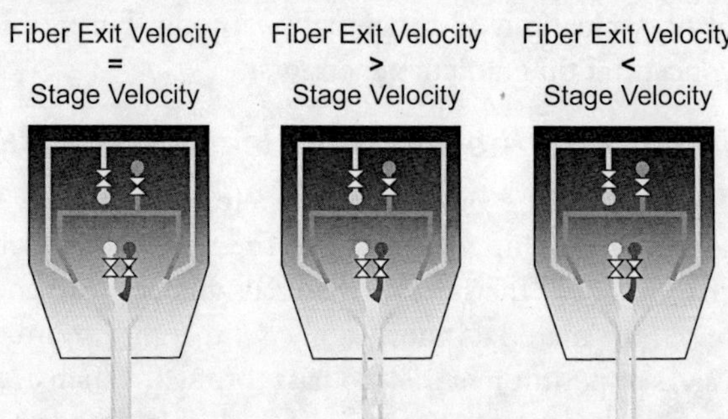

Figure 8.29: Fiber exit velocity versus stage velocity

The established algorithm creates a material file with some switch points. These switch points relate to the length of the fiber to be deposited and the material type for each building block. Though only two instructions are presently supported "materialA and materialB", additional commands could be defined to accommodate for tasks other than material switching. By equating the distance moved "which is continuously updated on the microcontroller's on-board memory" and equating it with the distinct switch points, exclusive chip situations could be invoked at the start of a building block's tool path. In this way, suitable characteristics could be used to each building block. The capacity to programmatically relate several properties to diverse regions of a 3D bioprinted tissue pre-cursor could help guarantee that the post-bioprinting cell culturing procedure allows cells to differentiate suitably and display a biological function.

To describe how material swapping is synchronized with the travel tool path utilizing the method stated above, an instance is offered in Figure 8.30. The 3D build illustrated in the example is a multi-material coaxial tube. A number of key points are considered along the tool path and the equivalent valve shapes are summarized. To permit fiber gelation during the whole tool path, the water valve stays locked whereas the cross-linker valve stays unlocked. At point 1, material 2 valve is unlocked whereas material 1 valve is locked, permitting for the deposition of purple fiber. Typically, an initialization structure is bioprinted some distance away to guarantee that the channel is primed with the suitable material and that a constant fiber is being deposited by the first point of the wanted build's tool path. By point 2, the initialization build should have bioprinted effectively and the tool path starts traveling in the direction of the first point of the build. The valve shapes stay unaffected from point 1 to point 5. At point 5, the residual arc length for the inner purple circle is equivalent to the channel length. Consequently, material 2 valve is unlocked and material 1 valve is locked. By point 6, all material 2 should have departed the channel and material 1 is deposited. These valve configurations remain unchanged from point 6 to finish. After the tool path is finalized, all valves are locked, Figure 8.30.

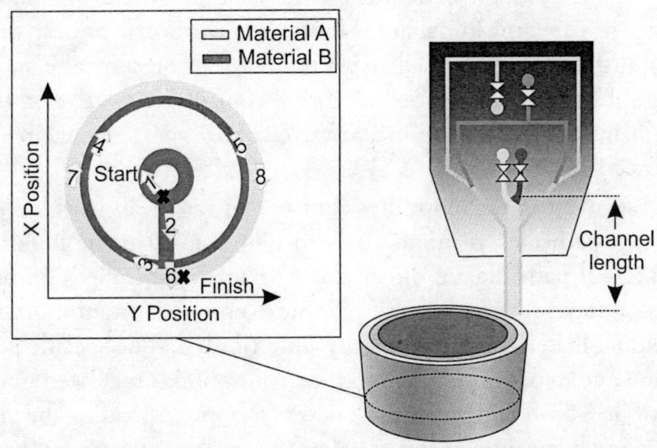

#	Water	Crosslinker	Material A	Material B
1	CLOSED	OPEN	CLOSED	OPEN
2	CLOSED	OPEN	CLOSED	OPEN
3	CLOSED	OPEN	CLOSED	OPEN
4	CLOSED	OPEN	CLOSED	OPEN
5	CLOSED	OPEN	OPEN	CLOSED
6	CLOSED	OPEN	OPEN	CLOSED
7	CLOSED	OPEN	OPEN	CLOSED
8	CLOSED	OPEN	OPEN	CLOSED

Figure 8.30: Example of synchronization between material switching and tool path

Bioprinter control software was established to facilitate control over bioprinter function. A GUI of the software can be integrated. The bioprinter control software properties can be divided into four classifications:

- operation controls,
- moment controls,
- valve controls, and
- bioprint controls.

Real-time control of the on-chip channel pressure values is controlled by the pressure controller software. Many buttons and inputs are applied to deliver users with valuable bioprint-related functions. A terminal console window may be utilized to exhibit output messages such as those involving to linking and delinking of the bioprinter, system and software faults, bioprint status, and on-chip valve configurations.

Setup and initialization begins by first creating a linking with the bioprinter. Then the positioning stage is reset by moving all three axes to the limits and then returning back to the home set position. The "Initialize Chip" button calls the initialization process outline in Table 8.5.

Motion control characteristics permit direct control of the 3D bioprinter. Each axis could be positioned individually by regulating the distance and the speed. Home and waste bin locations can be well-defined and saved. This permits users to set these two locations at the start of a bioprinting experiment and back to these special positions at any time. The home location is characteristically the starting location of the bioprint job and the waste bin location is frequently the location where the bioprint head yields waste material during initialization. To set the initial home location of the nozzle prior to bioprinting, the bioprint head was traveled down until contact was made with the bioprinting substrate and the nozzle was in the center of the bioprinting substrate. Furthermore to motion controls, each valve on the bioprint head could be switched on or off utilizing the GUI. Motion and valve controls are valuable in analyzing bioprint head functionality and for correcting purposes.

The bioprint controls segment of the bioprinter control software deals with bioprint jobs. The software considers material and tool path files as inputs so as to bioprint 3D tissue builds. The multi-axis move instructions stated in the tool path file are directed to the microcontroller's on-board buffer which can store up to 512 incremental travel sections. It is significant to ensure constant motion during a bioprint job with the purpose of avoiding material building up asymmetrically in one specific position of the build. To guarantee nonstop motion and to permit for the bioprinting for builds that need more than 512 movement segments, the microcontroller's on-board buffer must never be emptied pending the tool path's last movement is performed. Else, motion will be stopped at once after the implementation of the last incremental travel segment, even if the whole tool path has yet to be completely performed. Meanwhile, it is significant to not send instructions if the on-board buffer is full to evade buffer overflow. To resolve this producer-consumer communication difficulty, Figure 8.31, an assigned thread is provided the chore of observing the status of the on-board buffer. As travels are performed and space becomes accessible in the buffer, instructions are sent to the controller.

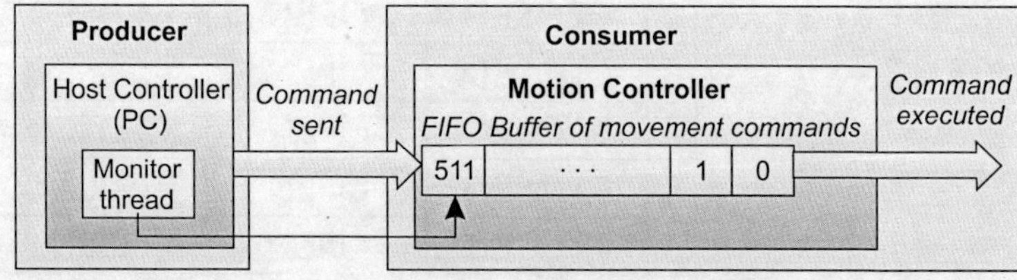

Figure 8.31: Sending and executing move commands using buffer and monitor thread

To manage material swapping, the bioprinter control software first reads the input material file and stores the material switch points as well as the equivalent material kind in a queue of material objects. An assigned thread is utilized to observe an on-board register that stores the total distance moved from the start of a bioprint job. If the material switch point of the material object stored at the front of the queue minus the channel length equivalents the distance traveled, then output signals are directed from the microcontroller to the pneumatic solenoid switch bank in order to switch to the identified material. As materials are switched, material objects are pulled from the front of the material queue.

To authenticate this process and permit the software to be well-matched with diverse bioprint head architectures, the channel length is one of the inputs of the bioprinter control software. This flowchart for this harmonization procedure is illustrated in Figure 8.32.

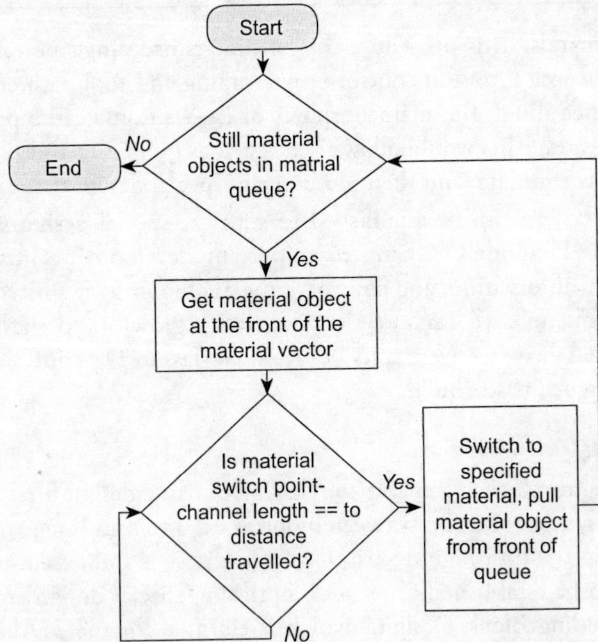

Figure 8.32: Flowchart for synchronization amid bioprint head and toolpath.

With the accumulation of new hardware modules, Figure 8.33, other constraints could be presented to consider the task-specific delays. For instance, several kinds of materials such as collagen may require being bioprinted under controlled-temperature situations to avert early gelation and bioprint head blockage. New control constraints could be presented consider the time involves to reach the temperature set point and this time could be synchronized to travel events in a comparable method to that utilized in dealing a priori material swapping. Furthermore, new tags that describe constraints such as temperature and concentration could be supplementary to consider for these added building block characteristics and bioprinting qualifications. In this manner, the bioprint head could be assumed as a programmable lab-on-a-bioprinter.

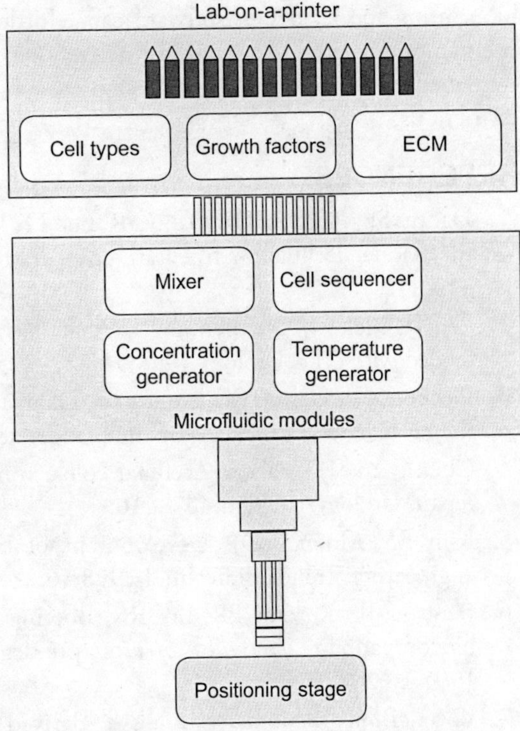

Figure 8.33: Lab-on-a-bioprinter schematic

3D Bioprint Customizable Tissue Structures

Proceeding to the growth of the 3D bioprinting special design software, simple bioprinting builds were narrow. To bioprint a wanted build, a user would have to manually program, line-by-line, and the essential bioprinting commands. This procedure can be time consuming, monotonous, and error-prone. Furthermore, gen about the exact motion control programming and tool path creation theory is essential to yield fine procedure scheduling. The main purposes of is to automate the process of bioprinting 3D heterogeneous tissue structures. This would allow users, such as biologists and tissue engineers, to rapidly design, visualize, graphically simulate, and then 3D bioprint special design tissue builds.

Novel 3D bioprinting system can be established to generate special design 3D assorted tissue builds founded on user distinct 3D models. Owing to important developments made in the functionality, effectiveness, incorporation, automation, and control of the 3D bioprinting procedure, tissue improvement could be significantly augmented. To reassure the cooperative growth and sharing of tissue design files between users, "Tissue Designer" software and "Bioprinted Tissue Description Language" also can be advanced to define the designed tissue build.

Sharing Tissue Design Files

An online platform can be introduced to enable the sharing of tissue design files. Particulars of the control over on-chip conduct and synchronization between bioprint head role and stage travel as well as bioprinter communication is clarified. The thorough descriptions of the design and execution of the 3D bioprinting special design software can be established. The 3D bioprinting special design software can be structured about a novel "Tissue Building Block Design," which is essential for the 3D bioprinting special design software. Each tissue building block could be individually modeled utilizing CAD software and then sliced allowing an exclusive set of CAM conditions and allocated exclusive characteristics. The established 3D bioprinting special design software can effortlessly incorporate CAD and CAM, although the manual programming method does not instinctively expedite this. Bioprinting difficult diverse builds are no longer bottlenecked by the capability to design these builds in software, but frequently by hardware and material restrictions.

REFERENCES

1. Dahms SE, Piechota HJ, Dahiya R, Lue TF, Tanagho EA. Composition and biomechanical properties of the bladder acellular matrix graft: comparative analysis in rat, pig and human. Br J Urol 1998;82:411–419.
2. Yoo JJ, Meng J, Oberpenning F, Atala A. Bladder augmentation using allogenic bladder submucosa seeded with cells. Urology 1998;51:221–225.
3. Piechota HJ, Dahms SE, Nunes LS, Dahiya R, Lue TF, Tanagho EA. *In vitro* functional properties of the rat bladder regenerated by the bladder acellular matrix graft. J Urol 1998;159:1717–1724.
4. Chen F, Yoo JJ, Atala A. Acellular collagen matrix as a possible "off the shelf" biomaterial for urethral repair. Urology 1999;54:407–410.
5. Kim BS, Mooney DJ. Development of biocompatible synthetic extracellular matrices for tissue engineering. Trends Biotechnol 1998;16:224–230.
6. Bergsma JE, Rozema FR, Bos RR, Boering G, de Bruijn WC, Pennings AJ. *In vivo* degradation and biocompatibility study of *in vitro* pre-degraded as-polymerized polyactide particles. Biomaterials 1995;16:267–274.
7. Li ST. Biologic biomaterials: tissue-derived biomateri-als (collagen). In: Brozino JDThe Biomedical Engineering Handbook. Boca Raton, FL: CRS Press; 1995. pp. 627-647.
8. Silver FH, Pins G. Cell growth on collagen: a review of tissue engineering using scaffolds containing extracellular matrix. J Long Term Eff Med Implants 1992;2:67–80.

9. Sams AE, Nixon AJ. Chondrocyte-laden collagen scaffolds for resurfacing extensive articular cartilage defects. Osteoarthritis Cartilage 1995;3:47–59.
10. Smidsrød O, Skjåk-Braek G. Alginate as immobilization matrix for cells. Trends Biotechnol 1990;8:71–78.
11. Lim F, Sun AM. Microencapsulated islets as bioartificial endocrine pancreas. Science 1980;210:908–910.
12. Gilding DK. Biodegradable polymers. In: Williams DFBiocompatibility of Clinical Implant Materials. Boca Raton, FL: CRC Press; 1981. pp. 209-232.
13. Mikos AG, Lyman MD, Freed LE, Langer R. Wetting of poly (L-lactic acid) and poly (DL-lactic-co-glycolic acid) foams for tissue culture. Biomaterials 1994;15:55–58.
14. Choi JS, Lee SJ, Christ GJ, Atala A, Yoo JJ. The influence of electrospun aligned poly (epsilon-caprolactone)/collagen nanofiber meshes on the formation of self-aligned skeletal muscle myotubes. Biomaterials 2008;29:2899–2906.
15. Lee SJ, Liu J, Oh SH, Soker S, Atala A, Yoo JJ. Development of a composite vascular scaffolding system that withstands physiological vascular conditions. Biomaterials 2008;29:2891–2898.
16. Pariente JL, Kim BS, Atala A. *In vitro* biocompatibility assessment of naturally derived and synthetic biomaterials using normal human urothelial cells. J Biomed Mater Res 2001;55:33–39.
17. Pariente JL, Kim BS, Atala A. *In vitro* biocompatibility evaluation of naturally derived and synthetic biomaterials using normal human bladder smooth muscle cells. J Urol 2002;167:1867–1871.
18. Amiel GE, Atala A. Current and future modalities for functional renal replacement. Urol Clin North Am 1999;26:235–246.
19. Amiel GE, Komura M, Shapira O, Yoo JJ, Yazdani S, Berry J, et al. Engineering of blood vessels from acellular collagen matrices coated with human endothelial cells. Tissue Eng 2006;12:2355–2365.
20. Yoo JJ, Park HJ, Lee I, Atala A. Autologous engineered cartilage rods for penile reconstruction. J Urol 1999;162:1119–1121.
21. Atala A. Autologous cell transplantation for urologic reconstruction. J Urol 1998;159:2–3.
22. Atala A. Bladder regeneration by tissue engineering. BJU Int 2001;88:765–770.
23. Atala A. Creation of bladder tissue *in vitro* and *in vivo*. A system for organ replacement. Adv Exp Med Biol 1999;462:31–42.
24. Cilento BG, Freeman MR, Schneck FX, Retik AB, Atala A. Phenotypic and cytogenetic characterization of human bladder urothelia expanded *in vitro*. J Urol 1994;152:665–670.
25. Oberpenning F, Meng J, Yoo JJ, Atala A. De novo reconstitution of a functional mammalian urinary bladder by tissue engineering. Nat Biotechnol 1999;17:149–155.
26. Scriven SD, Booth C, Thomas DF, Trejdosiewicz LK, Southgate J. Reconstitution of human urothelium from monolayer cultures. J Urol 1997;158:1147–1152.
27. Liebert M, Hubbel A, Chung M, Wedemeyer G, Lomax MI, Hegeman A, et al. Expression of mal is associated with urothelial differentiation *in vitro*: identification by differential display reverse-transcriptase polymerase chain reaction. Differentiation 1997;61:177–185.
28. Liebert M, Wedemeyer G, Abruzzo LV, Kunkel SL, Hammerberg C, Cooper KD, et al. Stimulated urothelial cells produce cytokines and express an activated cell surface antigenic phenotype. Semin Urol 1991;9:124–130.
29. Puthenveettil JA, Burger MS, Reznikoff CA. Replicative senescence in human uroepithelial cells. Adv Exp Med Biol 1999;462:83–91.

30. Freeman MR, Yoo JJ, Raab G, Soker S, Adam RM, Schneck FX, et al. Heparin-binding EGF-like growth factor is an autocrine growth factor for human urothelial cells and is synthesized by epithelial and smooth muscle cells in the human bladder. J Clin Invest 1997;99:1028–1036.
31. Nguyen HT, Park JM, Peters CA, Adam RM, Orsola A, Atala A, et al. Cell-specific activation of the HB-EGF and ErbB1 genes by stretch in primary human bladder cells. *In Vitro* Cell Dev Biol Anim 1999;35:371–375.
32. Harriss DR. Smooth muscle cell culture: a new approach to the study of human detrusor physiology and pathophysiology. Br J Urol 1995;75 Suppl 1:18–26.
33. Brivanlou AH, Gage FH, Jaenisch R, Jessell T, Melton D, Rossant J. Stem cells. Setting standards for human embryonic stem cells. Science 2003;300:913–916.
34. Richards M, Fong CY, Chan WK, Wong PC, Bongso A. Human feeders support prolonged undifferentiated growth of human inner cell masses and embryonic stem cells. Nat Biotechnol 2002;20:933–936.
35. Amit M, Shariki C, Margulets V, Itskovitz-Eldor J. Feeder layer- and serum-free culture of human embryonic stem cells. Biol Reprod 2004;70:837–845.
36. Reubinoff BE, Pera MF, Fong CY, Trounson A, Bongso A. Embryonic stem cell lines from human blastocysts: somatic differentiation *in vitro*. Nat Biotechnol 2000;18:399–404.
37. Reubinoff BE, Itsykson P, Turetsky T, Pera MF, Reinhartz E, Itzik A, et al. Neural progenitors from human embryonic stem cells. Nat Biotechnol 2001;19:1134–1140.
38. Schuldiner M, Eiges R, Eden A, Yanuka O, Itskovitz-Eldor J, Goldstein RS, et al. Induced neuronal differentiation of human embryonic stem cells. Brain Res 2001;913:201–205.
39. Schuldiner M, Yanuka O, Itskovitz-Eldor J, Melton DA, Benvenisty N. Effects of eight growth factors on the differentiation of cells derived from human embryonic stem cells. Proc Natl Acad Sci U S A 2000;97:11307–11312.
40. Zhang SC, Wernig M, Duncan ID, Brüstle O, Thomson JA. *In vitro* differentiation of transplantable neural precursors from human embryonic stem cells. Nat Biotechnol 2001;19:1129–1133.
41. Kaufman DS, Hanson ET, Lewis RL, Auerbach R, Thomson JA. Hematopoietic colony-forming cells derived from human embryonic stem cells. Proc Natl Acad Sci U S A 2001;98:10716–10721.
42. Kehat I, Kenyagin-Karsenti D, Snir M, Segev H, Amit M, Gepstein A, et al. Human embryonic stem cells can differentiate into myocytes with structural and functional properties of cardiomyocytes. J Clin Invest 2001;108:407–414.
43. Levenberg S, Golub JS, Amit M, Itskovitz-Eldor J, Langer R. Endothelial cells derived from human embryonic stem cells. Proc Natl Acad Sci U S A 2002;99:4391–4396.
44. Assady S, Maor G, Amit M, Itskovitz-Eldor J, Skorecki KL, Tzukerman M. Insulin production by human embryonic stem cells. Diabetes 2001;50:1691–1697.
45. Itskovitz-Eldor J, Schuldiner M, Karsenti D, Eden A, Yanuka O, Amit M, et al. Differentiation of human embryonic stem cells into embryoid bodies compromising the three embryonic germ layers. Mol Med 2000;6:88–95.
46. Gurdon JB, Elsdale TR, Fischberg M. Sexually mature individuals of Xenopus laevis from the transplantation of single somatic nuclei. Nature 1958;182:64–65.
47. Campbell KH, McWhir J, Ritchie WA, Wilmut I. Sheep cloned by nuclear transfer from a cultured cell line. Nature 1996;380:64–66.

48. Soloman, Sabrie "Sensors Handbook, 2nd Edition, McGraw-Hill Publishing Co., NY – ISBN 978-0-07-160570-0
49. Soloman, Sabrie "Sensors and Control Systems in Manufacturing 2nd Edition," McGraw-Hill Publishing Co., NY – ISBN 0-07-059626-3
50. Soloman, Sabrie "Affordable Automation," McGraw-Hill Publishing Co., NY – ISBN 0-07-059633-6
51. Soloman, Sabrie "Sensors Handbook, 1st Edition," McGraw-Hill Publishing Co., NY – ISBN 0-07-059630-1
52. Soloman, Sabrie "Sensors and Control Systems in Manufacturing 1st Edition," McGraw-Hill Publishing Co., NY – ISBN 978-07-10572-4

9 Organ Printing - Discovering Novel Treatments and Drugs

The Role of Mesenchymal Stem Cells - MSCs

Mesenchymal Stem Cells are defined as cells dedicated for the packing of fat. They are multipotent stromal cells that can differentiate into a diversity of cell kinds, comprising osteoblasts - bone cells, chondrocytes - cartilage cells, myocytes - muscle cells and adipocytes -fat cells, which give rise to marrow adipose tissue.

Mesenchymal stem cells (MSCs) is present in a diversity of tissues and organs of the human body from the time when their discovery in early 1970. The main function of the NSCs is to uphold and restore the corresponding tissue *in vivo*. MSCs can be effortlessly secluded from diverse tissues and can experience widespread self-proliferation prior to differentiation into numerous mesodermal cell kinds such as osteoblasts, chondrocytes, adipocytes, tenocytes, myocytes, and fibroblasts, Figure 9.1. Since this immense differentiation possibility, MSCs are a capable tool for regenerative medicine methods. They could represent a significant part in cellular treatment, tissue substitution and renewal. Mesenchymal stem cells can be relatively equated for their application and differentiation possibility to embryonic stem cells. The recognition of the induced pluripotent stem cells and the restrictions as well as the difficulties utilizing scaffolds for tissue restoration is imperative. Moreover, it is vital to ensure the isolation procedures for MSCs from the most collective sources specifically the bone marrow, adipose tissue and umbilical cord.

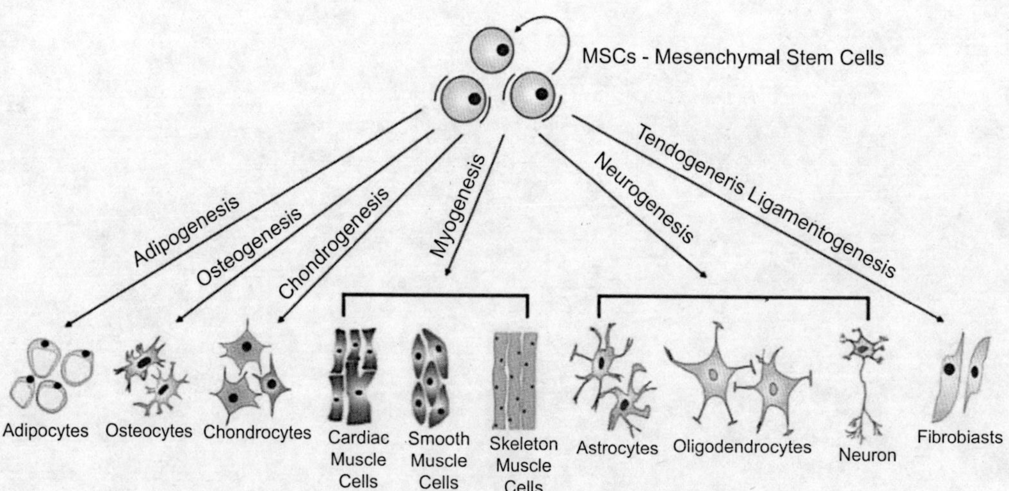

Figure 9.1: The Differentiation Possibility of Mesenchymal Stem Cells – MSCs

Courtesy: Springer eBook

Definitions - (*Google Scholar*)

Mesenchymal Stem Cells:	*They are multipotent stromal* ***cells*** that can differentiate into a *diversity of* ***cell*** *kinds, comprising osteoblasts (bone* ***cells****), chondrocytes (cartilage* ***cells****), myocytes (muscle* ***cells****) and adipocytes (fat* ***cells*** which give rise to marrow adipose tissue).
Adipocytes:	A cell specialized for the storage of fat, found in connective tissue.
Osteocytes:	*A bone cell, formed when an osteoblast becomes embedded in the matrix it has secreted.*
Chondrocytes:	*A cell which has secreted the matrix of cartilage and become embedded in it.*
Cardiac Muscle:	**Myocardium** - *the muscular tissue of the heart.*
Smooth Muscle:	*Muscle tissue in which the contractile fibrils are not highly ordered, occurring in the gut and other internal organs and not under voluntary control.*
Skeleton Muscle:	*An internal or external framework of bone, cartilage, or other rigid material supporting or containing the body of an animal or plant.*
Astrocytes:	*A star-shaped glial cell of the central nervous system.*
Oligodendrocytes:	*A glial cell similar to an astrocyte but with fewer protuberances, concerned with the production of myelin in the central nervous system.*
Neuron:	*A specialized cell transmitting nerve impulses; a nerve cell.*
Fibroblasts:	*A cell in connective tissue which produces collagen and other fibers.*

3D Bioprinting Considerations

Bioprinting advancements have materialized as a potent instrument for constructing tissue and organ builds in the arena of tissue engineering. This technology permits accurate positioning of cells, bio-substances and bio-particles in spatially distinct locations surrounded by confined three-3D builds, Figure 9.2. Numerous bioprinting techniques have been established and used for implementations in medicine, extending from researching cellular progress to building tissues and organs for applications, including blood vessels, myocardial tissue, heart valve, , trachea and. The implementations are based on intensive research focused on tissue regeneration utilizing bioprinting technology.

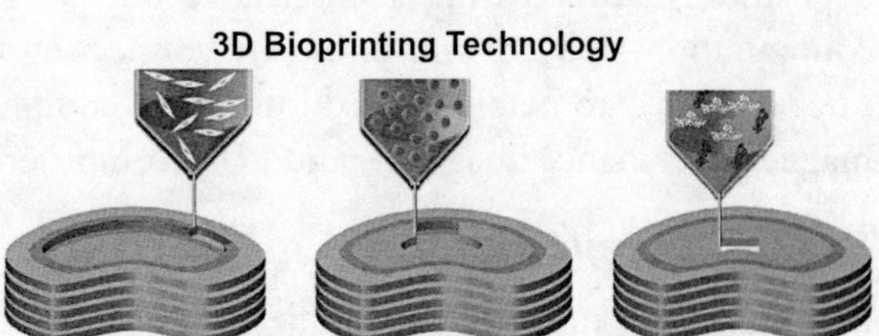

Figure 9.2: 3D Multi-Processes – Multi Function Bioprinting Techniques

Newly, bioprinting technology has substantially gained much consideration due to its capability to step above many engineering obstacles in the field of tissue engineering. The crucial mechanisms of tissue engineering comprise cells, scaffolds and biological particles, where cells are seeded onto dense biodegradable scaffolds, while tissue development is frequently prompted by the growth factors of many biomolecules. Many tissue engineering methodologies have been utilized to construct tissues and organs that may be implemented in patients, together with myocardial tissue, vessel, heart valve and trachea. Traditional tissue engineering methodologies has revealed its success in building a number of tissues clinically. However, the challenges are persisting to occur in constructing intricate and compound tissues and organs. These contain insufficient scaffold manufacturing techniques that could reproduce tissue micro-construction, restricted accessibility of biomaterials for tissue building and approaches to convey multiple cell kinds in their specific placements within the bio-scaffold, 9.3.

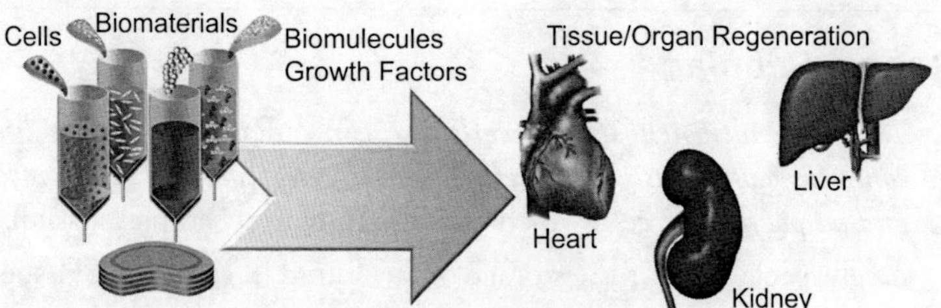

Figure 9.3: Biomaterials for 3D Tissue Bioprinting

Scaffolds for Tissue Redevelopment

In order to manufacture the perfect scaffolds for tissue redevelopment, numerous necessities must happen, as well as bio-compatibility, bio-degradability, porosity and physical sustenance. Many approaches have been used to construct scaffolds for tissue engineering implementations, in addition to solvent molding, particulate drainage, gas frothing, phase separation and freeze drying. While these approaches are capable to yield essential constructions, regulatory the size, micro-construction and inter- attachments of pores, which are required to carry oxygen and nutrients for cell existence, remains problematic. Furthermore, organic solvents employed to liquefy the bio-substances can continue to hold on the scaffolds, the rests of which can be lethal to cells. Thus, many scientists used additive manufacturing 3D Bioprinting techniques to manufacture scaffolds to step over the restrictions of traditional approaches.

Additive manufacturing is a fabrication procedure that yields intricate 3D constructions by carefully adding materials layer by layer through 3D Bioprinting technology such as stereolithography, fused deposition manufacturing and selective laser sintering. These technologies are able to govern the size, configuration, delivery and internally connecting the pores of the scaffold. Additionally, 3D Bioprinting can be utilized to manufacture biomimetic-configured 3D builds from medical imageries, such as MRI - magnetic resonance imaging – and CT - computerized tomography – utilizing CAD/CAM technologies.

Scaffold Limitations

While scaffolds manufactured utilizing 3D Bioprinting technique may provide a manageable construction, the effectiveness is frequently restricted according to problems in inserting multiple cell kinds, biomaterials and bioactive particles within the scaffold. 3D Bioprinting tools restricts these limitations by its capability to build 2D and 3D constructions with appropriate assignment of cells, biomaterials and biomolecules in distinct locations, Figure 9.4.

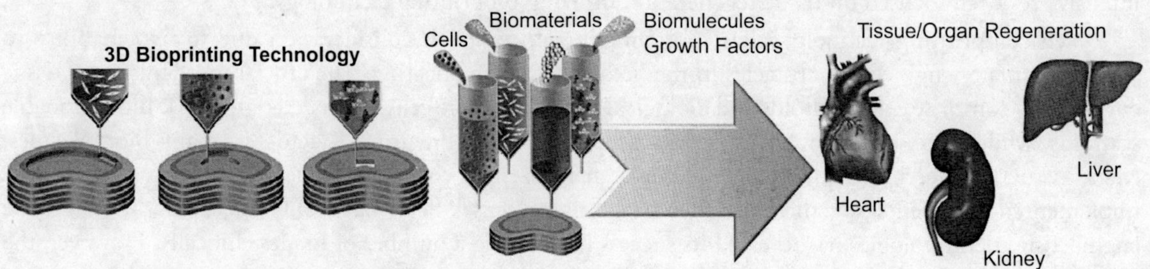

Figure 9.4: 3D Organ Bioprinting Procedure

3D Bioprinting is founded on additive manufacturing technology and it permits direct cell deposition in typical organic construction. Also, it can be integrated with CAD/CAM technology to manufacture a build that has an precise functional configuration. Consequently, 3D bioprinting technology is advancing the fabrication method for engineered tissue build,

The summary of the ideologies and obstacles of the most widely employed 3D bioprinting technologies, as well as jetting- and extrusion-based, and highlight representative study on tissue regeneration; comprising implementations in the arena of cardio-thoracic surgery are presented.

BIOPRINTING TECHNOLOGY

Inkjet Bioprinter

3D Inkjet bioprinters is also recognized as spherical drop-on-demand Bioprinters, which are the most normally utilized kind of bioprinter according to its inexpensive cost, good quality, relatively high speed, and consistent with numerous biological substances. 3D Inkjet Bioprinters utilize heat or acoustic waves to expel droplets of liquefied cells onto a specific structured scaffold, which can sustain the final configuration of the build. Thermal inkjet Bioprinters utilize short electric pulse to apply heat the print head source yielding a pulse of pressure to expels droplets from the dedicated orifice. The thermal temperature, which ranges from 200 C to 300 C, does not influence the permanence of the biomaterial or the cells capability. The benefits of heat inkjet Bioprinting are high speed, inexpensive cost, and widespread convenience. Nevertheless, there are some drawbacks to this methodology as well as inconsistent droplet directionality, unevenness of droplet size, danger of subjecting cells to physical stress, and recurrent blockage of the nozzle, Figure 9.5.

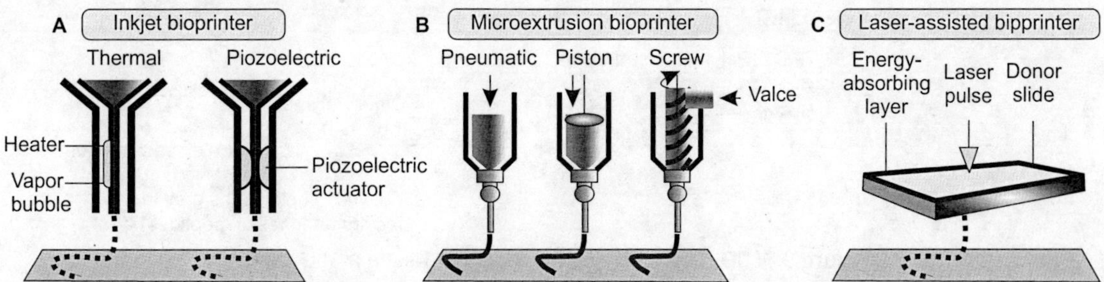

Figure 9.5: Most Used 3D Bioprinting Technologies

Acoustic wave Bioprinters utilizes piezoelectric quartzes to generate an acoustic wave inside the dedicated print head, segregating the liquefied bioink into spherical droplets at regular intervals. Acoustic wave inkjet Bioprinters have the capability to generate and govern an even droplet size and high droplet consistency by regulating the frequency and amplitude of the acoustic waves. This technique creates a sufficient amount of low energy that cells cannot be impaired and has been utilized with several kinds of cells. A common disadvantage of bio-inkjet bioprinting is the biomaterial should be liquefied to allow spherical droplet creation as well as to avert blockage of the nozzle. Inkjet bioprinters are restricted to low cell thicknesses to avert blockage and decrease shear stress. Instances of effective implementations of a bio-inkjet bioprinter contain the renewal of operational skin and cartilage *in situ*. Great capability and operation of cells after bioprinting, as well as affordable make inkjet bioprinters an irreplaceable tissue biofabrication engineering service, Figure 9.6.

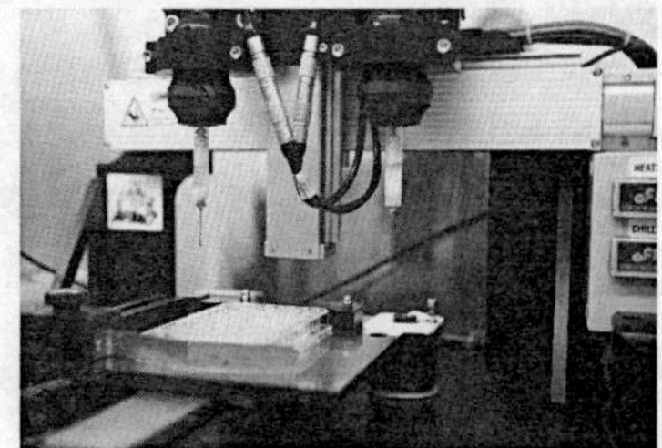

Figure 9.6: L'Oreal Teams with Organovo to 3D Print Skin Tissue
Courtesy: 3dprintingindustry

Micro-extrusion Bioprinter

Micro-extrusion bioprinters typically contain of a temperature-controlled biomaterial dispensing mechanism. It is a mechanism developed to translate the x, y, and z movements, a fiber-optic source of light subjected at the deposition domain for photo-initiator initiation, a video camera for x-y-z monitor and instruction, and a piezoelectric humidifier. Dissimilar from inkjet bioprinters, the microextrusion bioprinter creates a constant thread of bioink rather than numerous droplets of spherical bioink through implementing pressure, either air pressure or mechanical force order to expel bioink from a syringe. These threads are flown in 2D layers, as instructed by the CAD-CAM routines, and assisted as the foundation for the following layers while the stage is moving up in the Z-axis subsequent in the creation of a 3D build. Microextrusion bioprinters have a broader assortment of bioink as well as high viscosity Bio-substances such as hydrogels, well-matched copolymers, and cell spheroids. Thus far, cell endurance rate is lesser in comparison with inkjet bioprinters according to considerable stress. Microextrusion bioprinters have been utilized to manufacture an aortic valve, pronged vessels, and tumor representations, Figure 9.7.

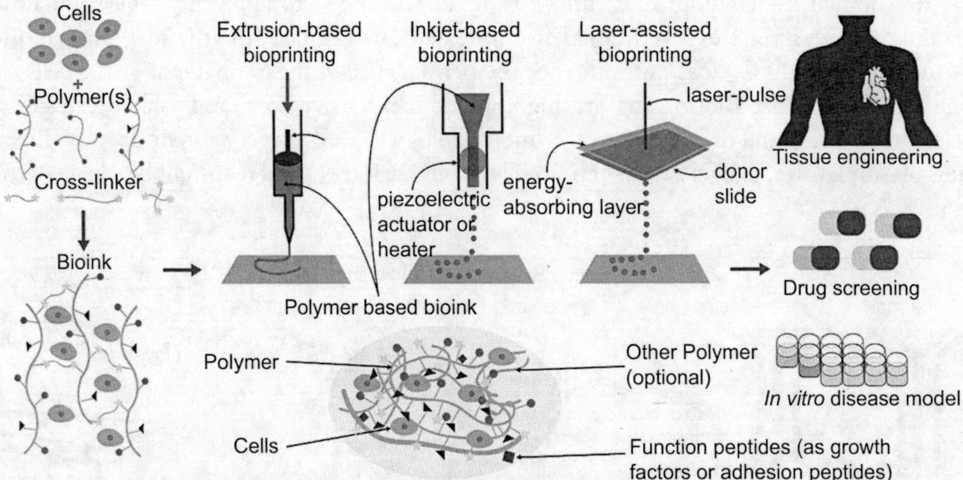

Figure 9.7: 3D Bioprinting Technologies of Tissue and Organs.

Laser-Assisted Bioprinter

LAB -The laser-assisted bioprinter- is not as common as inkjet or microextrusion bioprinters. However, its manufacturing implementations are growing. A LAB typically contains an excited laser pulses, a focusing mechanism, a ribbon provision usually made from glass protected with an energy absorbing layer – usually gold or titanium, and a layer of bio-substances comprising cells and hydrogel. The laser pulse focuses on the gold or titanium absorbing layer generating high pressure bubbles, impelling cells toward the substrate plate below. The laser pulsed energy is proportional to the droplet size. The LAB is a nozzle-free device accordingly; it causes the blockage of the nozzle to be evaded. LAB is well-matched with a widespread range of biomaterials with numerous viscosities. Furthermore, LAB can yield sensitive high quality 3D builds with high cell capability. Notwithstanding these benefits, the high expense of a LAB device thwarts its utilization for elementary studies.

3D Bioprinting Strategies

3D Bioprinting Approaches

3D bioprinting is founded upon three dominant methodologies:

1. biomimicry,
2. independent self-assembly, and
3. mini -tissue building-blocks.

Biomimicry contains the repetition of both intra-cellular and extra-cellular mechanisms of a tissue or organ. To effectively mimic an exact operational constituent of a tissue, an appreciation of micro-surroundings is necessary. This contains the particular preparation of operational and sustenance cell kinds, gradients of soluble and insoluble elements, and configuration of the ECM Biomimicry efforts to pattern a completely established tissue and its atmosphere.

Independent self-assembly utilizes the growth of embryonic organs as a director. During the initial phase of embryo development, cellular mechanisms yield their individual ECM mechanisms that provide the suitable signal for the development of operational biological micro-construction. Independent self-assembly trusts cells to direct the arrangement, localization, operational, and mechanical characteristics of the tissue. A scaffold-free tissue manufacture technique benefits of self-assembling cellular spheroids to mimic emerging tissue. This method would necessitate an in-depth comprehension of the atmosphere that initiates the growth of embryonic tissue beginning and organogenesis.

Mini-tissues are the operational building blocks of tissue and organs. Mini-tissues are manufactured independently and assembled into a bigger tissue through coherent design, self-assembly, or a mixture of both.

Imaging and Digital Design

To replicate intricate, heterogeneous operational tissues and organs originated in the human body, comprehending of configuration and arrangement of their mechanisms is a vital necessity. Medical computerized imaging elements are utilized to deliver data on the 3D construction and operation of these mechanisms at the cellular level. The greatest frequently employed approaches are CT -computed tomography- and MRI -magnetic resonance imaging-. CAD-CAM routines are also utilized to create intricate graphic designs for tissue. This is the initial phase in the path to generating a 3D tissue or organ pattern, Figure 9.8.

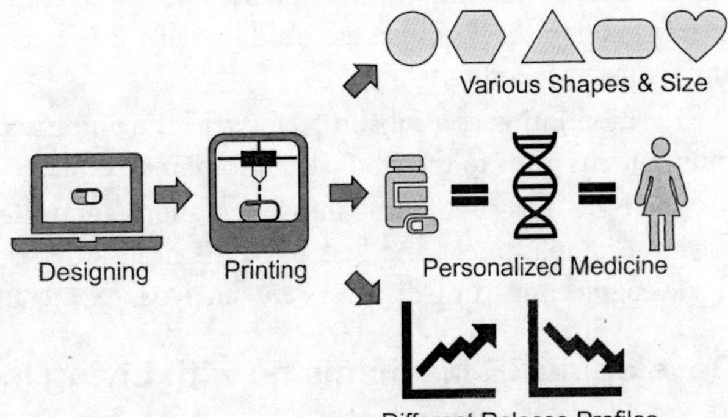

Figure 9.8: Precise 3D Bioprinting of Dosage Formation of Various Shapes

Courtesy: Asian Science and Pharmaceutical Sciences

Material Consideration

The collection of bioink substance is a significant concern as arranging to bioprint a tissue structure; substances must be appropriate for cross-correlation in order to enable bioprinter deposition and configuration preservation, as well as bio well-matched for long-standing relocation aims. There are innumerable conceivable bio substance appropriate for 3D bioprinting including: PEG -polyethylene glycol-, PLA -polylactic acid-, PGA -polyglycolic acid-, PLGA -polylactic-co-glycolic acid- and gellan gum. The election of substance is reliant upon anticipated cell and physical characteristics of the tissue being copied.

The newest developments in 3D bioprinting and living ink

Living devices out of 3D printer

The 3D Bioprinting refers to the process of making a physical biological object utilizing a 3D digital structure. Recently, it is well recognized that the 3D printing technique is utilizing lifeless substances such as resins, metals, and thermoplastics. Lately, within the past decade, the scientists have effectively studied the 3D Bioprinters for Bioprinting living mechanisms by inserting bacteria in a bio-compatible 3D Bioprinting ink.

Professional and well recognized panels have advanced an extraordinary 3D Bioprinting system by genetically adapting the cells for constructing a 'living tattoo.' A living tattoo is an effective consequence developed from a biological living ink, ink that is regenerative and maintainable to convert the carbon dioxide into the bioink by utilizing algae.

The 3D Bioprinting on active mechanisms is made productive by utilizing a technique in which the hydrogel inks are preset with the bacteria cells, acting as sensitive elements into the large-scale alive substances.

As a result, the 3D printing on active mechanisms has produced eminent success when combined with hydrogel, bacteria, algae nutrients and other microbes to withstand and uphold the cell practical application to serve a useful purpose.

The 3D printing living mechanisms, also known as 3D bioprinting systems have demonstrated as a godsend to the healthcare segment. The 3D Bioprinting technology is effectively utilized in the healthcare segments for both creation and implantation of numerous tissues as well as the skin, vascular implants, bones, heart tissue, tracheal splints, etc. Additionally, it is extensively utilized for the burn victims, pharmacology, and tissue engineering.

3D Bioprinting as a substitute answer for gaining access to organs that are bioficial, the 3D bioprinting multiplies its gains to the whole healthcare marketplace.

There are many additional implementations and utilization of the 3D bioprinting, which is accountable for its accomplishment and benefit to the healthcare segment. Other 3D bioprinting implementations are tissue engineering, drug discovery and analysis, regenerative medicine and research and development.

Developing 3D Mechanisms with Living Ink

The technology of 3D bioprinting has renovated the healthcare segment in countless ways. Getting driven by the benefits and advantages of the 3D printing and living ink, there are many emerging corporations taking the initiative of chart the path forward for the healthcare business.

The crucial constituents of the 3D bioprinting business were several national and International 3D bioprinting corporations taking intense strides for initiating the welfares of 3D printing and bioink into the healthcare manufacturing via partnerships, studies, contractual agreements, associations, growth, alliance and product advancements.

The Organovo Holdings Inc. has developed liver tissue for medicine trials, which has been accepted in healthcare sector. The company has launched kidney tissue for numbers of implementations. Many other products were advanced by many industries for healthcare such as 3D Bioprint thyroid gland, 3D Bioprinted surgical preparation frame guide for youngsters, Nano 3D Bioprinting breast cancer tissue, anthropoid nipple tissue, and Bioprint dura-mater **for brain surgery**.

There is a diverse utilization of the product as apiece of the utilization mainly for testing and investigation, diverse skin, brain and bone surgical operations, transfer of glands or organs, biofabrication of tissue, cancer exploration, pediatric care, and printed surgical frame guides.

The 3D Bioprinting of living mechanisms has caused a substantial influence on the healthcare segment. It is contributed to the effective investigation and progress of the substantial effort of many industries and academia to create particular tissue for the cure of lingering as well as individual illnesses, Figure 9.9.

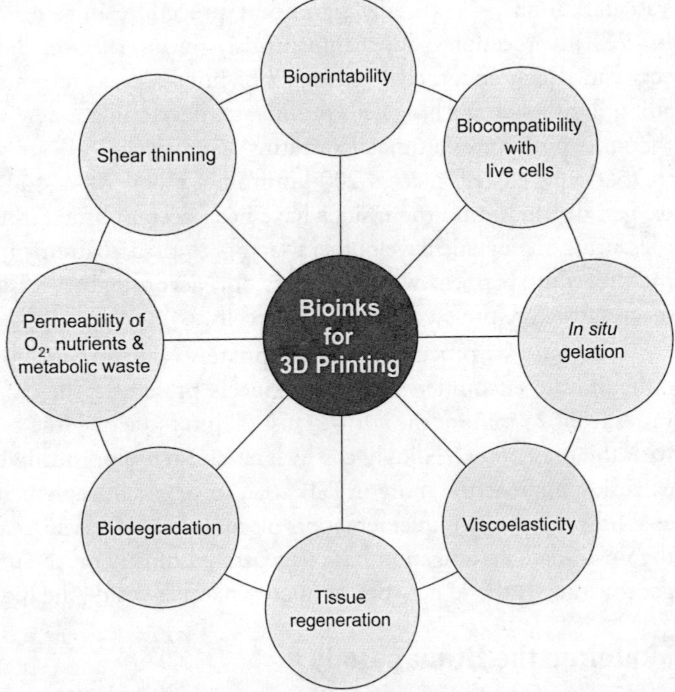

Figure 9.9: Selecting Bioink for 3D Bioprinting in Biomaterial Implementations

In addition to yielding bio-fabricated tissues, the 3D bioprinting practice has aided the world healthcare segment with tailor-made and individualized the medical products and well tested medicine. The 3D medical implementations are reforming healthcare with a novel wave of development in the cures and surgical procedure. 3D Bioprinting is the safest ways to bio-fabricate human tissue with minimum peril and expense.

Cell Types - Foundation of Tissue Engineering

The selection of cell kind is vital for a bio-fabricating living tissue or organ. Most tissue contains of number of cell kinds. Each cell kind has a diverse function that must be replicated in the transferred tissue. Most organs and tissues encompass cell kinds that are important for maintaining, physical, boundary tasks, vasculature interchange, and deliver cell proliferation as well as differentiation. Cells selected for bioprinting must mimic the biological state of innate cells to uphold their inherent purpose and endurance roles. As with many transfers, rejection of bioprinted portions by the host immune system is a likely difficulty. Consequently, utilizing similar foundations of cells may aid to avert immune rejection. Stem cells are a capable selection for bio-fabrication of tissue according to their capability to proliferate as well as self-differentiate. Additionally, they have the capability to create numbers of tissue-specific cell phenotypes. Embryonic stem cells and induced pluripotent stem cells are the furthermost capable cells for bioprinting implementations according to their immense possibility to differentiate to several diverse kinds of cells. As cells are constantly under physical pressure stress from the bioprinter, cells utilized in bioprinting must be vigorous enough to overwhelm that mechanical pressure and the non-living pH of the printing atmosphere.

Vascularization

Vascularization is a crucially significant procedure in safeguarding the capability and existence of any tissue. The procedure happens unsurprisingly *in vivo* over the procedure of the formation of new blood vessel in which endothelial cells react to inter-cellular responses by spreading toward the response. Cells adjoining these branching cells proliferate, developing a new vessel. More vessels divide off of this one in the same procedure, ultimately creating a massively intricate vessel linkage. It has been witnessed *in vivo* that virtually all cells place ≤ 200 μm from a blood vessel, so that it may uphold appropriate nutrient and oxygen distribution. Progressions have been accomplished in evolving microvascular networks *in vitro* by presenting angiogenic development factors to tissue patterns in the existence of endothelial cells. Though, the procedure happens within days, is not, accordingly a feasible choice for the existence of a bioprinted macro-structure pre-seeded with living cells.

Consequently, procurement appropriate vasculature in bioprinted tissue builds till one of the utmost difficulties to encounter in tissue engineers presently. The cross section diameter of human blood vessels varies from 25 mm for the aorta to just 20 μm for actual fine capillaries. The sensitive resolutions attainable with the present, exceedingly costly, laser-assisted bioprinting technique are 10's of μm. Inkjet and micro-extrusion approaches, more usually used in research implementations, print on the order of 100's of μm, enabling the direct implementation of complicated vascular networks presently unachievable. Therefore, the up-to-date investigation has concentrated mostly on the utilization of sacrificial gels and assemblies to generate an synthetic network of micro-channels inside the bioprinted pattern.

Modeling the Human Body

The Skull

Skull rebuilding is a collective process which can develop essential after extreme head shock or severe bacterial contamination. Other times cancers can assault the skull, or problems can ascend post cranial surgical operation. Conventional approaches include molding a polymer to the skull manually, usually polymethylmethacrylate (PMMA), which frequently marks an unacceptable, irregular form, or a poor fitting. Scientists developed a technique of generating synthetic skulls utilizing a 3D printer, where the drawing is reverse-engineered from patient CT-Computer Tomography images.

Methods

PMMA grafts, founded on computer designed patterns, were embedded into numerous female and male patients in 2018-2019. The patterns of the graft took place during the surgical operation. CT - Computer Tomography was used to scan of the patient's open skull. The scan was immediately utilized to design the essential configuration. The procedure was an imminent success.

The Aorta

Cardiovascular illness is a foremost reason of decease worldwide. Blood vessel transplants have been presented to be an effective technique of fighting this sickness; nonetheless, transplants are infrequent as a result of the absence of accessible donors in the patients' distressed region. Tissue engineering and bioprinting have given a prospective answer to this issue. Scientists pursued to bioprint a pattern of human abdominal aorta utilizing a micro-extrusion Bioprinter to deposit PEG hydrogel sowed with porcine aortic valve interstitial cells.

A 3D pattern of the human aorta was shaped utilizing CT medical image subdivision for engineering design of anatomy computer routines. It is ten and adapted to a stereolithography STL file due to the close compatibility with the 3D Bioprinting software. While the converted aorta mesh model formed in the STL file is not smooth, it however, signified by a series of triangular elements, Figure 9.10. To battle the absence of adequate bioprinting resolution from the model transformation to STL mesh format, mathematician established a precise model developed to make a greater printing resolution pattern of the human aorta established upon mathematical smoothing algorithms of the centerline curve within the aortic cylinder. Also, this mathematical model is planned to erect a structure that is self-supportive, enabling the build will uphold its form deprived of the employment of a scaffold.

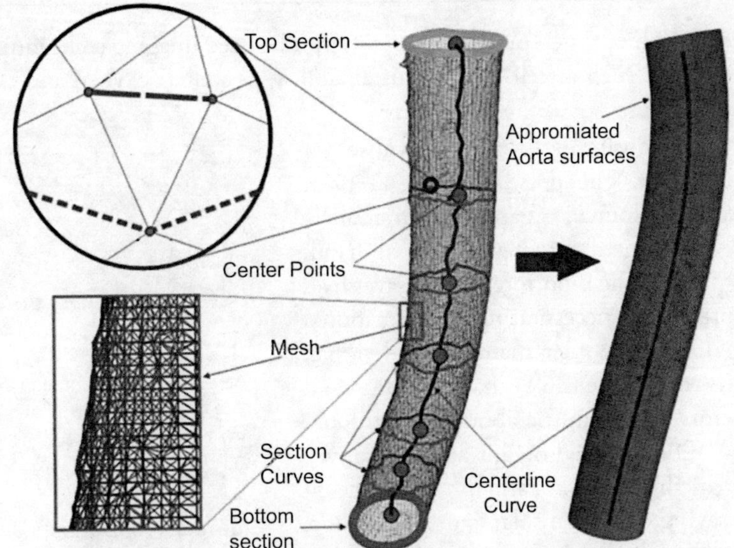

Figure 9.10: STL Files Illustrate the Mesh Configuration to Create Aorta Model

Self-supportive Aorta Structure

The creation of self-supportive aorta structure is a breakthrough in medical technology. The implementations of the mathematical algorithms have made the most of the contact area between cells on neighboring layers, easing the merging of cells. The accomplishment of this innovative employment of a mathematical algorithm intended to exploit printing effectiveness and cellular communication delivers encouragement for comparable approaches to be currently embraced by various healthcare providers.

There are some impediments which must be resolved to allow 3D bioprinting to extend its prospective achievement in tissue engineering. These include more better-quality vascularization approaches, refining the competences of CAD 3D modeling routines, up surging resolution, and rapid approaches of Bioprinting. Further development of CAD routines will enable for the augmented abilities for the custom tailor of 3D Bioprinted grafts or synthetic body parts.

Upsurge the printing quality through greater resolution of printed bioink layers will permit for better accuracy in forming bioprinted builds, particularly vascular networks which comprise micro vessel networks with cross sectional diameters as low as 20 μm. Also, rapid printing techniques will make it vital to uphold the sustainability of cell cultures during the printing procedure as enhanced techniques will ultimately guide the formation of superior and more intricate builds. Through present Bioprinting techniques, it is projected that the printing of an article the configuration of an adult human liver will consume more than 3 days. It is extremely significant that the cells be capable to endure this period. In order to deal with this subject, scientists applied innovative multi-nozzle head micro-extrusion Bioprinter. They were capable to competently reduce the duration of time it consumed to build the Aorta model. They have suggested that increasing the quantity of print heads is an actual realistic of upsurge print speed. However, increasing the size of the bioprinted models require more advanced imaging techniques. It is significant to monitor various parameters such as cell density, sustainability, as well as oxygen and nutrient distribution throughout the structure.

3D bioprinting is a good cause for enthusiasm. The possibility to 3D print a working human organ that can actually be transplanted into patients is at hand, Figure 9.9. The Bioprinted organ saves lives deprived of necessitating a contributed organ from another matching individual. It is comprehensible that patients and doctors are optimistic about that outlook. The 3D bioprinted organs actually convey remarkable gains. Patients may accept lifesaving organ transplants in a very short time, and deprived of excessive time to wait for a donor match, Figure 9.11. The 3D bioprint eliminates the years-long delay, as well as the guiltiness that originates from aiding from the demise of added individual. Moreover, the impression is that 3D printed organs are shaped from the patient's individual stem cells, removing the danger of rejection and the necessity for immunosuppressive drugs.

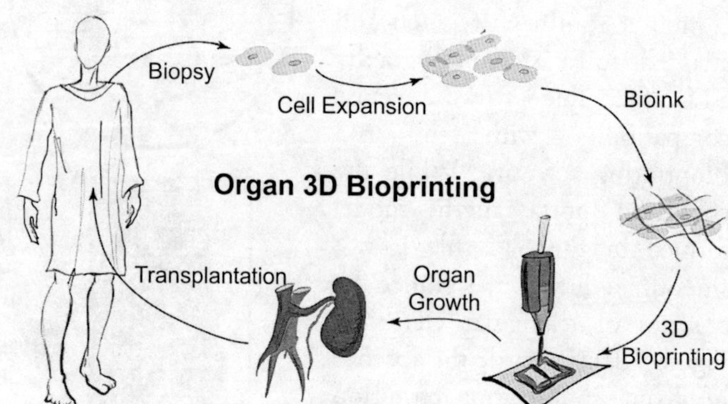

Figure 9.11: 3D Bioprinting Human Organ can be Transplanted into Patients

In fact, one perhaps would not grasp 3D Bioprinted, transplantable human organs for few more years. The 3D Bioprinting of an organ is beyond just 3D Bioprinting depositing layers of human cells into the configuration of a liver or a kidney; such organs should be capable to execute all of the diverse purposes of their normal corresponding organs. The Bioprinted organs must be competent of incorporating with the body's present functions, which includes the growth of nerves and blood vessels. Development is being accomplished in the expansion of 3D Bioprinted blood vessel complexes. However, within the past two years, the progression that experts and scientists have achieved towards 3D Bioprinted organs are unquestionably extraordinary, developing thyroid glands and ovaries being transplanted into lower animals, for instance.

Nonetheless, it is surprising to view comparable advancements and imagine we ought to transplant 3D Bioprinted organs into people by the following year!

Again, it's not as simple as one may imagine. The core fact is scientists, biologists, doctors, and researchers were able to 3D Bioprint living tissue at all is impressive, as the current at-hand technology is at present saving lives in the form of cardiac reinforcements and patches, for instance. Nonetheless, human lives are also being protected through a bioprinting application that receives less dramatic attention. 3D Bioprinted tissue is demonstrating to be actual resources of testing novel medications enabling drugs to be methodically evaluated and carried out to market more rapidly, all deprived of impairing animal trial subjects.

Medical Advancements

Medical science has developed insightful advancements in major innovations and continues to accelerate an array of cure possibilities for even the utmost deadly illnesses.

On account of medical doctor, scientists, biologists, and researchers who were occupied diligently to combat illnesses such as cancer, advance the patient's value of life, as well as branding our lives safer.

Medical Advancement in India - Gene Therapy

Healthcare, has substantially advanced from where it was 50 years ago. The past decade, in particularly, physicians, scientists, biologists, and researchers who have been working on personalized gene therapies for a rare disease, employing novel surgical techniques for full face transplants. Also, through 3D Bioprinting technology they have regrown organs and teeth.

Gene Therapy

Genomics, an innovation in healthcare, is making it conceivable to forecast certain conditions, identify, and cure illnesses more accurately. Over the past decade, scientist and researchers have achieved substantial improvements in genome engineering techniques, which allowed the editing of genome sequences much similar one edits software routines. At present scientists developed a powerful gene-editing tool that enabled them to easily modify DNA sequences and alter gene performance, Figure 9.12. Scientists claim that they could use the technique fight thousands of illnesses, extending from congenital blood disorders to several cancers.

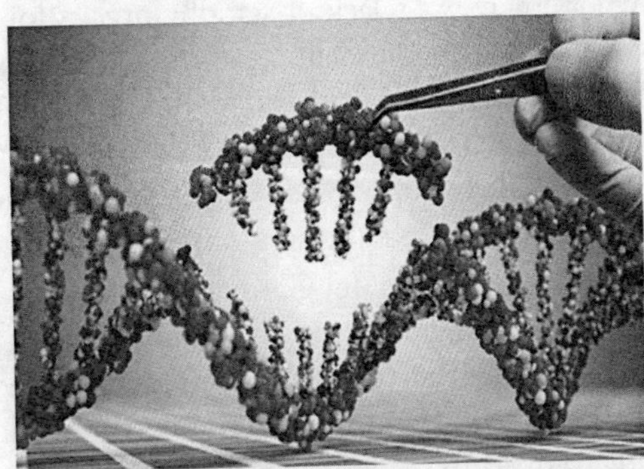

Figure 9.12: Genome Engineering Techniques Modifying DNA

Currently, scientist and many doctors worldwide are using the technology of gene altering therapy to utilize inside the human body for the first time in USA to treat an hereditary eye disorder that caused blindness.

Lately, US FDA has approved three gene treatments for severe and rare illnesses such as leukemia. It is one-time gene therapy for a genetic disorder. It had been established to return vision in patients living with an congenital retinal ailment that leads to total loss of sight.

Gene Therapy in Twin Babies

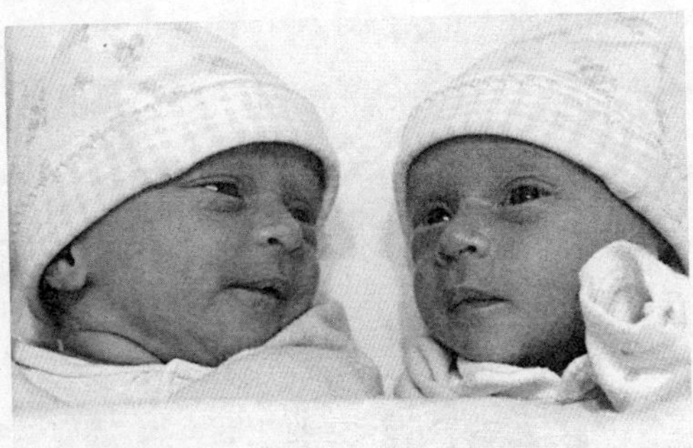

Figure 9.13: World's First Unethically Genetically-Edited Twin Babies by Chinese Scientists

HIV Gene Therapy

Sadly, as well as UNETHICALLY, **on November 25, 2018, Chinese scientists propagated to have generated the world's first genetically-edited twin babies.** Accordingly, severe global uproar sparked among scientists, clergymen, politicians, and the public prompting the World Health Organization to introduce a registry to trail explorations on human genome editing worldwide. The twin female infants had their DNA reformed to help avert them from contracting HIV. The utilized gene therapy tool was applied to the embryos that were generated to prevent HIV infection in the babies, Figure 9.13. The infant father was carrying of HIV, while the mother was not infected by the virus.

Researchers have been diligently occupied in creating a future where there is no lack of standby organs for transplantations. Scientists are confident that the heart, possibly, sometime soon the lung, might be made by 3D Bioprinters. In earlier 2019, Israeli scientists revealed the world's initial 3D-printed heart made by human tissue, generating human blood vessels, human ventricles and chambers, Figure 9.14.

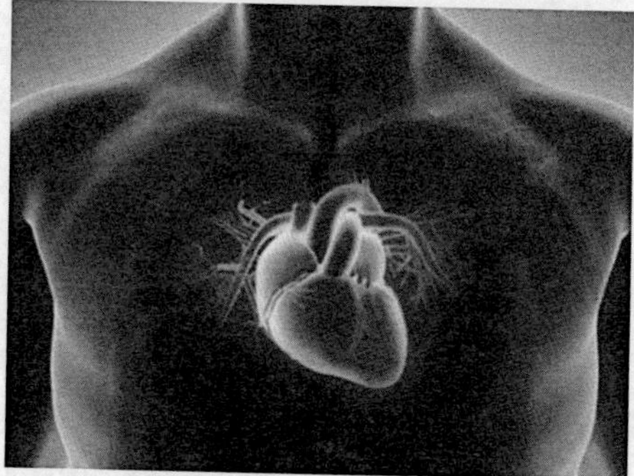

Figure 9.14: Implantable 3D Bioprinted Organs

Also, scientists in Germany generated transparent, see-through human organs utilizing novel 3D Bioprinting technology. The scientists asserted that the 3D Bioprinting development could assist 3D Bioprinting kidneys and liver for transplants. The drawback is the inability to 3D Bioprint intricate vascular networks as in airways and blood vessels of the lung or bile ducts in order to develop such organ replacements. Utilizing particularly-developed 3D Bioprinting technology, scientists generated a synthetic air pouch that anatomically and operationally mimics comparable constructions originated in the human lung, as well as the ability to throb or "inhale/exhale" deprived of overflowing.

Artificial Intelligence - Medical Diagnosis

Artificial Intelligence (AI) is changing medicine in many ways; possibly, it may even substitute human clinicians. Presently, healthcare employees are utilizing AI for genome therapy and research, CT, medical imaging, drug development, etc. Artificial intelligence based devices may analyze many circumstances, acquire and generate decisions much quicker than real medical workers. Artificial intelligence may identify ailments like doctors and foresee if a patient may pass away within months - such is the influence of AI.

Novel Methods to Treat Cancer

Scientists and researchers remain investigating greater means to inhibit, identify, cure and recover the value of life for cancer patients and survivors. Immunotherapy, a cancer cure that utilizes the body's immune system to battle the illness, has altered the method oncologists cure cancer. The biologic treatment has actually turned out to be a recognized pillar of cancer healing. Immune therapy has become important developments in the past few years, including the 2019's extraordinary treatment that can aid eliminate progressive breast cancer, Figure 9.15. The notable treatment constituted thrusting 90 billion cancer-killing immune cells into the body of a patient who had been wholly healed of progressive breast cancer. The treatment is the most highly personalized treatment conceivable.

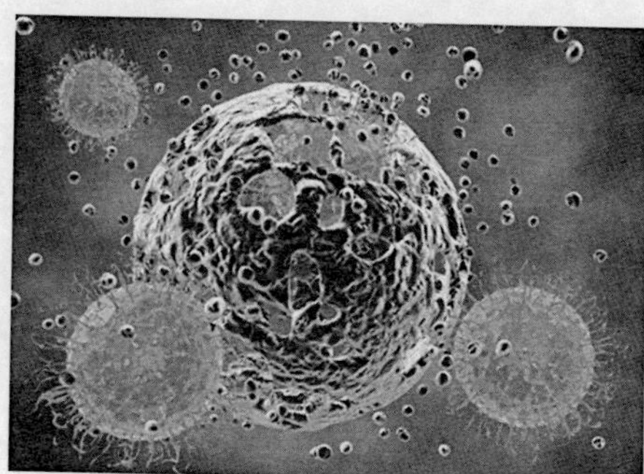

Figure 9.15: Body's Immune System Battling Cancer

Turning Cancer to Fat

2019, an amazing breakthrough, scientist and researchers at the University of Basel in Switzerland have devised a novel drug amalgamation, consisting of "diabetic drug," "Rosiglitazone" and "cancer drug Trametinib," that can halt metastasis in breast cancer by turning tumor cells into fat cells, Figure 9.16.

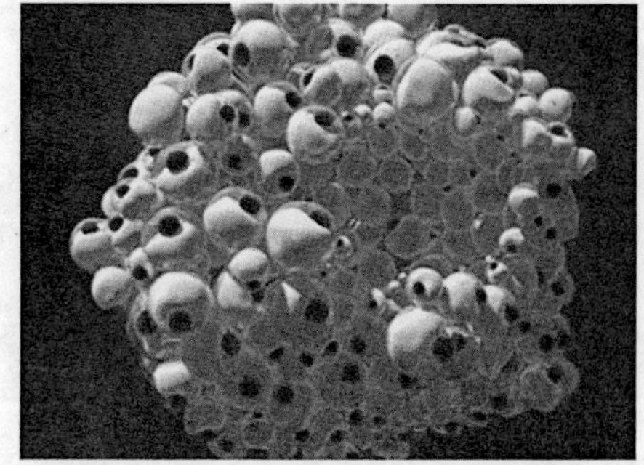

Figure 9.16: Metastasis in Breast Cancer Turned Tumor Cells into Fat Cells
Courtesy: Basil University - Switzerland

Life Saving Drugs

The world's initial drug strategy intended to halt cancer's deadly capability to evolve and progress resistance to treatment, novel treatments that affect no damage to healthy cells whereas aiming the tumor cells, and the novel drug - called "Ascimimib" - that can fight leukemia deprived of producing toxic side effects were several of the life-saving development in cancer cures.

The cancer cure technology established by a group of scientists and accepted by the FDA can identify the initial spread of cancer. The technology, which is appealed to be the quickest in the world, is predictable to revolutionize the initial analysis and managing of cancer patients in India. In 2018, John Hopkins Cancer Center developed a liquid biopsy called "CancerSEEK," which is a single blood test that can identify up to 8 kinds of cancers, Figure 9.17.

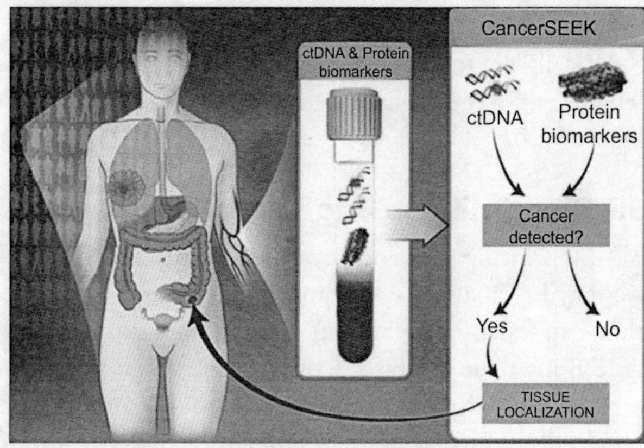

Figure 9.17: Single Test Reveals 8 Cancer Types Locations
Courtesy: Hopkin University USA

First Uterus Transplant Baby Born in India and Asia

A patient in India, after 17 months of receiving a transplanted womb from her mother delivered a baby girl through caesarean section, Figure 9.18. The baby, weighed at birt 1.4 kg at. It was first uterus transplant baby's birth in India and throughout whole Asia.

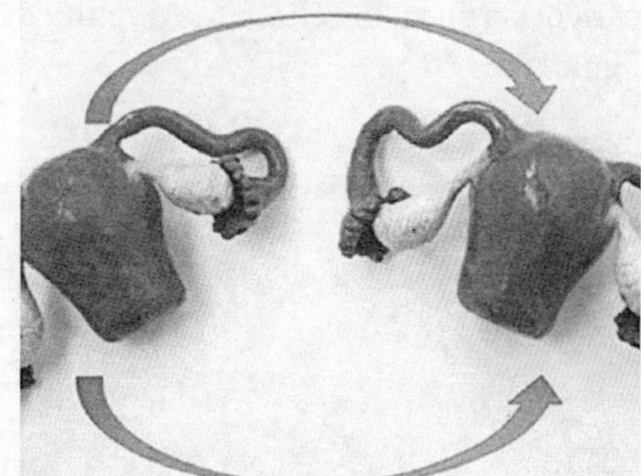

Figure 9.18: Transplanted womb Conceived a Baby

A Tablet Preventing HIV

Gilead Sciences' Truvada (TVD) has been established to cure and avert HIV. The blue, oval-shaped tablet, which is a fixed-dose mixture of "tenofovir, disoproxil and emtricitabine," which has led to an unparalleled decrease in new HIV instances universally when consumed daily. Truvada was permitted by the FDA to decrease the danger of sexually acquired HIV-1 in adults at high risk.

In January 2018, scientists, and researchers from the Massachusetts Institute of Technology- MIT, and Brigham and Women's Infirmary established a weekly capsule that can provide a week's worth of antiretroviral drugs in a solitary dose. This indicates individual patients may only require consuming a single tablet once a week in difference to HIV treatment that includes many medicines every day.

Tuberculosis Vaccine

The novel vaccine, which the scientists and researchers developed to combat against tuberculosis, is established to be partly averting an inactive infection from proceeding to active infection. The vaccine, M72/AS01E, has been established by pharmaceutical. The vaccine needs to be verified in greater trials before it can be licensed for use, Figure 8.19.

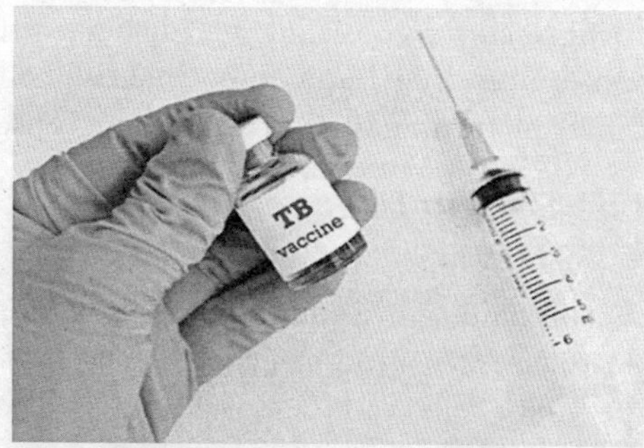

Figure 9.19: Injectable Tuberculosis Vaccine

Ebola Vaccine

Merck's Ervebo has developed the first vaccine permitted by the US FDA to aid stop the lethal Ebola virus sickness. The vaccine was established to be 100% effective in averting Ebola instances. Ervebo is a living, tempered vaccine that has been genetically engineered to comprise a protein from the Zaire "ebolavirus," and is managed as a single-dose inoculation, Figure 9.20.

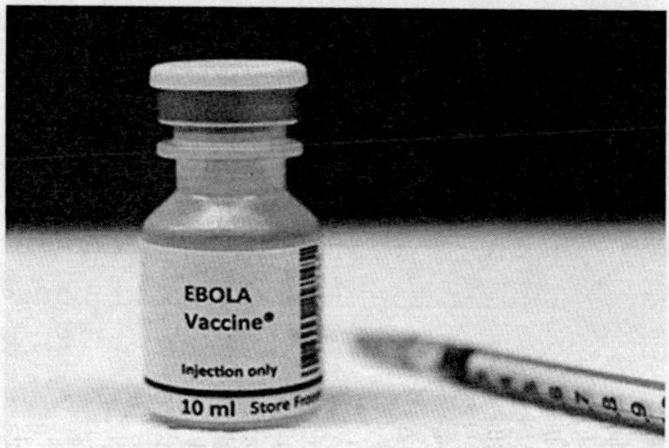

Figure 9.20: Injectable Ebola Vaccine

3D Bioprinting - Drug Testing

3D bioprinting is a good reason for enthusiasm. Typically, we may be capable to 3D Bioprint functional human organs that can essentially be transplanted into various patients, redeeming their survivals deprived of necessitating a donated organ from a functional organ. 3D bioprinted organs actually carry remarkable benefits. People could obtain lifesaving organ transplants immediately, without delay in years or even month or weeks for a donor match, eradicating the years-long wait lists in addition to the guilt that associates from profiting from the death of another person. In addition, the notion is that 3D Bioprinted organs are shaped from the patient's own stem cells, removing the danger of rejection and the necessity for immunosuppressive drugs, Figure 9.21.

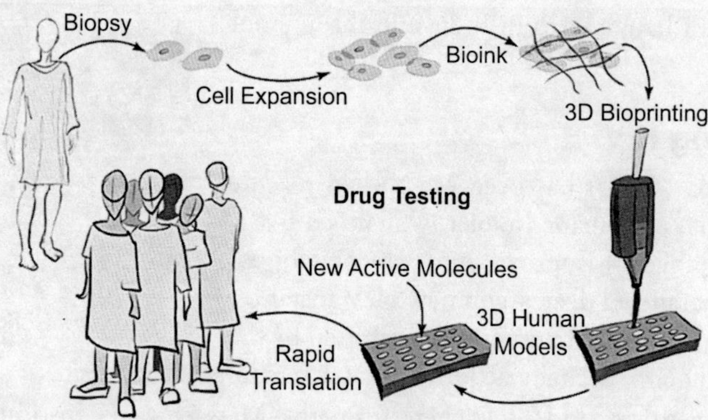

Figure 9.21: Drug Testing - of 3D Bioprinting Application

A group of researchers and scientists recently developed novel kind of Bioink that allows the 3D Bioprinting of cells and other biological constituents as part of a solitary production procedure. The Bioink is called "Mechanically Tunable Bioink" for 3D Bioprinting of human cells. These researchers and scientists believe that the potentials bioprinting provides the most significant utilization of the 3D Bioprinting technology, Figure 9.22.

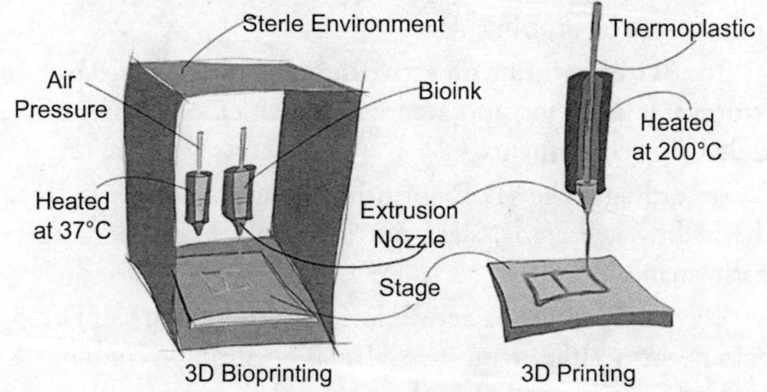

Figure 9.22: "Mechanically Tunable Bioink" for 3D Bioprinting of Human Cells

Utilizing present techniques, conveying a novel drug to marketplace has been projected to cost US$2.5 billion, at a span of more than ten years from start to finish. The likelihood of novel contestant drug to receive regulatory approval is low. In 2016, less than 10% were approved in USA.

The motivation behind the disappointment of numerous drugs is that --when a novel drug functions well on animals, its special effects may not necessarily interpret to humans. Human biology is very dissimilar from that of trial topics such as mice, so what mechanisms for a mouse may not at all times function for a human. With 3D bioprinted tissue, though, scientists and researchers can really generate the type of intricate human tissue originated in organs such as the heart, liver, kidneys, etc. and observe the consequences that a specific drug may have on those tissues privileged the human body.

Despite the fact the utilization of animals for investigation is not probable to be removed completely, the scientists and researchers endure 3D bioprinted skin tissue can remove the utilization of animals in the laboratory. The trials of cosmetics on animals have continuously been more contentious than trial for medical determinations. Presently, we have the capability to 3D print human skin. Accordingly, there is actually no necessity to try the chemical cosmetics elements on animals anymore. The European Union, in 2013, approved a law against cosmetics trials on animals. It is hoped that the 3D Bioprinted skin may be the motive for comparable laws to be approved on a worldwide level.

The more discrete kinds of tissue that scientists and researchers are capable to 3D Bioprint, the better the prospects become for more operative pharmaceutical testing – and possibly transplantable organs in the future. Bioprinting is not a panacea of all solution; dissimilar kinds of cells necessitate very diverse kinds of surroundings in order to operate correctly. Lately, scientists have offered new Bioink, which is taken out from marine algae that can be utilized to 3D Bioprint human stem cells in dissimilar, separate surroundings, without damage approaching to the cells. This work overlays the means toward the Bioprinting of intricate tissue-like constructions consisted of a variety of mechanically separate micro-domains that could possibly replicate normal mechanical features of operational tissues.

3D Bioprinting - Living Bioink

The 3D Bioprinting denotes to the procedure of creating a bodily entity utilizing a 3D digital model. Until today, it is well-recognized the 3D printing technology is employing lifeless materials such as resins, metals, and thermoplastics. However, recently, the scientists and researchers have positively investigated the 3D Bioprinters for printing live devices by embedding bacteria in a biocompatible 3D printing ink.

Expert professional research and scientific panel has established a unique 3D Bioprinting procedure by genetically adapting the cells for ingenuous a "live tattoo." In this event a living tattoo is an effective consequence from a living ink. The living ink is renewable and maintainable to convert the carbon dioxide into the ink by utilizing algae.

The 3D Bioprinting on active devices is prepared to be productive by utilizing a technique in which the hydrogel inks are incorporated with the microorganisms cells acting as penetrating elements into the large-scale living constituents.

Accordingly, the 3D Bioprinting on active devices is prepared to be made effective with the mixture of the hydrogel, bacteria, algae nutrients and other microorganisms to withstand as well as uphold the cell performance.

The 3D Bioprinting active devices, also known as 3D bioprinting devices have showed as a substantial gain to the healthcare business. The Bioprinting technology is effectively utilized in the healthcare business for both creation and transplantation of numerous tissues as well as the skin, heart tissue, bones, vascular grafts, tracheal splints, etc. The technology is also extensively utilized for the burn victims, pharmaceuticals, and tissue engineering.

3D Bioprinting is an alternative solution for accessing organs that are bioficial, the 3D bioprinting expands its advantage to the entire healthcare market.

The healthcare industry has profited with the development of 3D bioprinting in North America. Europe is the next leading region after North America.

There are additional applications and utilizations of the 3D bioprinting, which is accountable for its positive impact and advantage to the healthcare industry. As previously stated, there are many other 3D bioprinting implementations such as tissue engineering, drug discovery and testing, regenerative medicine and research and development.

The Organovo Holdings Inc., USA has industrialized "exVive3D - liver tissue" for drug testing, which has been approved in healthcare section by familiar reputable companies like L'Oréal and Merck. Organovo Corporation is currently preparing for the launch of "exVive3D - kidney tissue."

The other vital products established by this biomedical corporation are stated as follows:

- 3D print thyroid gland,
- a 3D-printed surgical planning guide for children,
- Nano3D printing breast cancer tissue,
- Human nipple tissue, and
- Bio-print dura mater for brain surgery.

There is a diverse utilization of the product according to their recognized usefulness but general the implementations of these products are mainly used for trials, testing and research, diverse skin, brain and bone surgical operations, transplant of glands or organs, manufacture of tissue, cancer exploration, pediatric care, and Bioprinted guides.

The 3D Bioprinting of active devices has accomplished a momentous impact on the healthcare industry. This is contributed to the effective research and development of the diligent universities and vital industries. The prospect to produce precise tissue for the treatment of chronic, as well as personalized diseases is finally at hand.

In addition to yielding the fabric tissues, the 3D bioprinting has also facilitated the world healthcare industry with custom-made medical products and pharmaceutical medicationss. The 3D bioprinting is transforming healthcare with a novel upsurge of progress in the cures and surgical operations. It is one of the harmless means to manufacture human tissue with substantial reduced jeopardy and cost. With a substantial growth in the universal market share, the 3D bioprinting will have an important influence on the healthcare business by contributing custom-made and personalization in the bio-medical implementations.

Medical devices will become affordable with enhanced productivity due to the economical ways to Bioprint many products. Also, the healthcare industry will benefit by the decentralization of design and biofabrication of numerous products with the use of 3D bioprinting. Additionally, depositing Bioink through the use of 3D bioprinting machines in the healthcare industry to fabricate tissues and organs will greatly enhance and mature successful implementations of organ manufacturing.

The scientists and researchers have presently revealed exceptional results by utilizing the 3D bioprinting for generating a heart valve, spinal disk, and artificial ear. All can be extended into much more growths within the coming months and years.

Medical Applications for 3D Printing

3D Bioprinting has been implemented in medical applications since the early 2000s, as the expertise was first utilized to manufacture dental implants and custom made prosthetics. Meanwhile, the medical implementations for 3D Bioprinting have grown significantly. Lately it is well recognized the utilization of 3D Bioprinting to yield the followings:

- bones, ears, exoskeletons, windpipes,
- a jaw bone, eyeglasses,
- cell cultures, stem cells,
- blood vessels, vascular networks, tissues, and
- organs,
- as well as novel dosage forms and drug delivery devices.

The present medical utilizations of 3D Bioprinting can be prearranged into numerous comprehensive categories as follows:

- Tissue and organ manufacture;
- Generating prosthetics, grafts,
- Functional models;
- Pharmacological research regarding drug detection, distribution, and amount forms.

Bioprinting Tissues and Organs

Tissue or organ demise due to old-age, illnesses, accidents, and birth deficiencies is a serious medical problem. Present cure for organ demise depends on typically on organ transplants from alive or departed donors. Nevertheless, there is a severe shortage of human organs obtainable for transplant. In 2009, 154,324 ailing individuals in the U.S. were waiting for an organ. Only 27,996 of them (18%) obtained an organ transplant, and 8,863 (25 per day) deceased during the wait. In 2014, roughly 120,000 persons in the U.S. were pending receive an organ transplant. The cost of organ transplant surgery as well as the follow-up procedures is also expensive. The cost varies from $300 to 350 billion in 2012. A further problem is that organ transplantation encompasses the often difficult mission of locating a donor who is a well matched tissue. This problematic issue could probably be eradicated by utilizing cells extracted from the organ transplant patient's individual body to construct a substitute organ. This would reduce the peril of tissue rejection, as well as the essential to take enduring immunosuppressant·

Treatments founded on tissue development and regenerative medicine is being followed as a possible answer for the organ donor scarcity. The customary tissue engineering plan is to separate stem cells from small tissue samples, combine them with growth factors, grow them in the laboratory, and seed the cells onto scaffolds that guide cell proliferation and differentiation into operating tissues. Though still in its beginning, 3D bioprinting provides extra significant gains beyond this customary regenerative technique, which fundamentally delivers scaffold provision alone, such as extremely accurate cell location and high digital control of speed, resolution, cell build, drop volume, and configuration of printed cells. Organ Bioprinting earns advantages of 3D Bioprinting technology to yield cells, biomaterials, and cell-laden biomaterials separately or in pair, layer by layer, directly generating 3D tissue-like builds. Numerous substances are obtainable to construct the scaffolds, based on the required strength, porosity, and kind of tissue, with hydrogels typically well-thought-out to be the best appropriate for yielding soft tissues.

Though 3D bioprinting systems may be laser printing, inkjet printing, or extrusion printing; however, inkjet-based bioprinting technology is most common. This technique deposits Bioink, droplets of living cells or biomaterials, onto a substrate base guided by digital routines to replicate human tissues or organs. Several print-heads may be utilized to deposit diverse cell kinds, such as organ-specific, blood vessel, or muscle cells, a necessary feature for manufacturing complete hetero-cellular tissues and organs. A procedure for bioprinting organs has occurred as follows:

1. Generate a blueprint of an organ with its vascular construction;
2. Generate a bioprinting procedure strategy;
3. Separate stem cells;
4. Differentiate the stem cells into organ-specific cells;
5. Formulate Bioink tanks with organ-specific cells, blood vessel cells, and support medium and place them into the Bioprinter;
6. Bioprint; and
7. Place the bioprinted organ in a bioreactor prior to transplantation.

Laser printers have also been utilized in the cell printing procedure, in which laser energy is utilized to stimulate the cells in a specific form, providing 3D control of the cellular atmosphere.

While tissue and organ bioprinting is still in its primary stages, numerous research have delivered evidence of validity. Scientists have utilized 3D bioprinters to generate a "knee meniscus," "heart valve," "spinal disk," other types of "cartilage and bone," and an "artificial ear." Biomedical scientists implemented inkjet 3D Bioprinting technique to renovate human joint cartilage. Also, scientists have utilized 3D bioprinting technology to deposit diverse cells within many biocompatible hydrogels to yield a synthetic liver. The utilization of a 3D Bioprinter and Computer Tomography imageries of a patient's airway allowed scientists to manufacture an accurately modeled, bioresorbable tracheal splint that was surgically implanted in an infant with tracheobronchomalacia.[7] The infant improved and while a full resorption of the splint has occurred.

Several of biotech corporations have concentrated on generating tissues and organs for medical investigation. It is likely to quickly study novel possible healing drugs on patient tissue, significantly reducing investigation costs and time. Therefore, researchers at Organovo are crating strips of Bioprinted liver tissue for this goal; shortly, researchers believe the material will be progressive enough to utilize in finding novel medicinal cures. Other scientists are working on methods to cultivate whole human organs that may be utilized for investigative goals during drug influence. An organ generated from a patient's own individual stem cells may also be utilized to study various cures to understand if a drug will be operative for that individual.

Vascularized Organs

Evidence of validity investigations with respect to bioprinting has been achieved effectively, but the organs that have been formed are minuscule and relatively modest. Also, those organs are frequently avascular, aneural, alymphatic, thin, or hollow, and are nurtured by the circulation from host vasculature. Though, as the thickness of the developed tissue surpasses 150–200 micro meters, it exceeds the restriction for oxygen diffusion between host and transplanted tissue. Accordingly, bioprinting intricate 3D organs will necessitate constructing accurate multicellular builds with vascular network incorporation, which has not yet been accomplished.

Most required organs for transplantation are dense and intricate, such as the kidney, liver, and heart. Cells in these large organ build may not uphold their metabolic operation without vascularization, which is typically delivered by blood vessels. Consequently, operational vasculature should be bioprinted into manufactured organs to stream the cells with oxygen/gas interchange, nutrients, growth factors, and waste-product elimination—all of which are required for development during perfusion. Though the predictable tissue development method is not presently able of generating intricate vascularized organs, bioprinting demonstrates potential in determining this serious restriction. The accurate allocation of multiple cell kinds is essential to manufacture dense and intricate organs, and for the concurrent build of the combined vascular or microvascular system that is vital for these organs to operate.

Thermal Inkjet Printers TIJ are believed to be the utmost capable for Bioprinting utilization. Though, numerous 3D Bioprinting methods and materials have been implemented effectively to generate vasculature as modest as a sole channel, as well as added intricate configurations, such as bifurcated or forked channels. Lately, Experts from a network of educational establishments, including the Harvard University, Massachusetts Institute of Technology, Stanford University, University of Sydney – all claimed that they bioprinted a operational and perfusable grid of vessels, an accomplishment that signifies an important step toward resolving this difficulty.

Prostheses and Implants

Grafts, Implants and prostheses can be fabricated in nearly any conceivable configurations assisted by the interpretation of the followings:

- x-ray, MRI, or
- CT scans into digital .stl
- 3D print files.

Consequently, 3D Bioprinting has been utilized productively in the health care segment to develop both standard and difficult custom made prosthetic limbs and medical implants, occasionally within hours. This methodology has been utilized to manufacture dental, backbone, and hip implants. Beforehand, prior implants could be utilized clinically; they had to be authenticated, which is extremely time-consuming.

The capability to rapidly yield personalized implants and prostheses resolves a strong and insistent difficulty in orthopedics, where typical implants are frequently not adequate for some patients, mainly in intricate instances.- Formerly, surgical doctors had to achieve bone graft surgeries or utilize blades and drills to adapt implants by flaking smithereens of metal and plastic to a anticipated form, dimensions, and fit. This is too true in neurosurgery; Skulls have uneven profiles, so it is tough to standardize a cranial implant. In sufferers of head injury, where bone is detached to allow the brain space to expand, the cranial plate that is later tailored must be flawless. While some plates are machine milled, more and more are produced utilizing 3D Bioprinters, which enables it much simpler to customize the fit and configurations.

There have been numerous other industrial and medical achievements concerning the 3D Bioprinting of artificial limbs and implants. A biomedical research group at the BIO-MED Research Institute in Belgium has effectively embedded the initial 3D Bioprinted titanium mandibular prosthesis. The implant was constructed through the use of a laser to consecutively liquefy thin layers of titanium fine particles. In 2013, "Oxford Performance Materials" obtained FDA endorsement for a 3D-printed polyetherketoneketone

(PEKK) skull implantation, which was the first positive implant occurred that year. Another company, "LayerWise," has developed 3D Bioprinted titanium orthopedic, maxillofacial, spinal, and dental implants. A structurally accurate 3D Bioprinted prosthetic ear capable of sensing electromagnetic frequencies has been bio-fabricated utilizing silicon, chondrocytes, and silver nanoparticles. There is an increasing and upward trend toward developing 3D Bioprinted implants out of a variety of metals and polymers. Recently, implants and grafts have been Bioprinted with live cells.

3D Bioprinting has caused a positive transitive consequence on hearing aid fabrication. Today, 97-99% of hearing aids that fitting into the ear are personalized utilizing 3D Bioprinting. Everyone's ear way is formed in a different way, and the utilization of 3D Bioprinting permits personalize hearing aids to be made competently and economically. The overview of personalized 3D Bioprinted hearing aids to the marketplace was enabled contributed to fewer controlling limitations. "Invisalign" supports are another successful profitable utilzation of 3D Bioprinting, with 50,000 Bioprinted every day. These transparent, detachable, 3D Bioprinted orthodontic supports are personalized and exclusive to each user. This product delivers a respectable instance of how 3D Bioprinting may be utilized competently and cost-effectively to enable sole, personalized, intricate items.

Surgical Preparation

The discrete alterations and intricacies of the human body brand the utilization of 3D Bioprinted models perfect for surgical groundwork, Figure 9.23. Requiring a perceptible model of a patient's structure obtainable for a surgeon to investigate or utilize to animate surgery is desirable to ddepend exclusively on MRI or CT scans, which may not as informative since they are observed in 2D on a flat screen. The utilization of 3D Bioprinted representations for surgical exercise is likewise desirable to exercise on cadavers, which cause difficulties with regard to obtainability and cost. Also, cadavers are often absent of the suitable pathology, so they deliver more of a teaching example in anatomy than a depiction of a operating patient.

Figure 9.23: CT Scan Creating Custom 3D Bioprinted Surgical Medical Model

Researchers at the "National Library of Medicine" generate digital files from clinical data, such as CT scans, that are used to make custom 3D-printed surgical and medical models.

3D Bioprinted neuro functional models can be predominantly supportive to neuro surgeons by giving a depiction of some of the most complex constructions in the human body, Figure 9.24. The complicated, sometimes masked relations between cranial nerves, vessels, cerebral constructions, and skull construction can be problematic to understand founded solely on radiographic 2D images. A minor mistake in circumnavigating this intricate structure can have possibly distressing consequences. An accurate 3D model replicating the relationship between a laceration

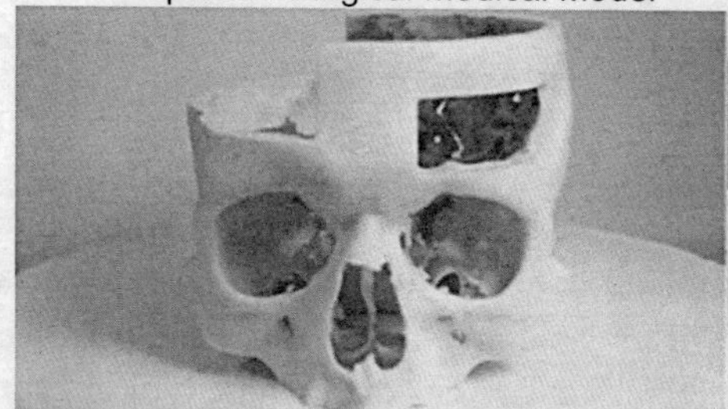

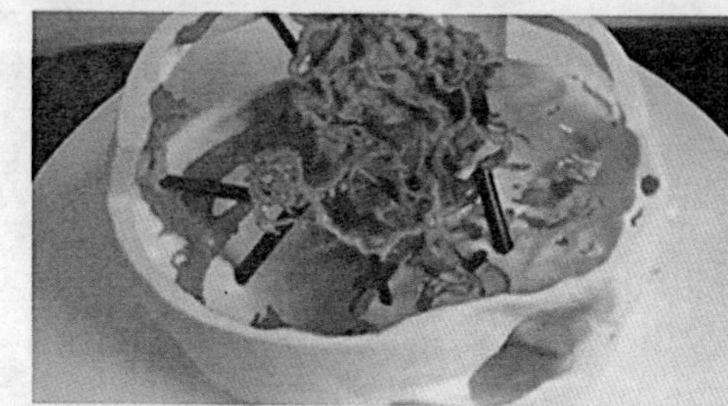

Figure 9.24: Depictions of Most Complex Constructions in Human Body Aiding Surgeons

and typical brain constructions can be supportive in defining the most harmless surgical strip and can also be valuable for the neurosurgeon to practice perplexing cases.- Intricate spinal malformations can also be investigated better over the utilizaation of a 3D model. High quality 3D functional models with the correct pathology for exercise doctors in executing colonoscopies are also vibrant, since colorectal cancer is the second important reason of cancer-related deceases in the U.S.

While still mainly investigative, 3D Bioprinted models have been utilized in many instances to increase understanding into a patient's precise structure prior to a medical operation. Groundbreaking surgical doctors at Japan's University Hospital have employed 3D Bioprinted representations to strategize liver transplantations. The surgeon utilize models of a patient's organs to decide how to best slice a donor liver with negligible tissue loss to place in the recipient's abdominal location. These 3D mockups are constructed of partly see-through, economical acrylic resin or polyvinyl alcohol—substances that have a water element and texture comparable to alive tissues, permitting an additional accurate infiltration by the surgical blades. Other surgeons have utilized a 3D Bioprinted model of a hardened aorta for medical surgery planning of plaque subtraction. An untimely blockage of infant's airway was also rebuilt in order to inveestigate aerosol drug distribution to the lungs. Also, orthopedic surgery trainee employed CT image scans and 3D modeling software routines to generate 3D Bioprint files on behalf of a patient's bones. The digital files were then directed to "Shapeways" to Bioprint custom made models utilized for strategizing surgery. The cost for 3D Bioprinting was a minuscule compared to the traditional cost to have custom models made, in addition to substantial saving in time to produce the object.

3D Bioprinted models can be valuable exceeding the surgical strategy.-Lately, a polypeptide chain pattern was 3D Bioprinted that it could be folded into secondary shape. Comparable replicas could be used to help the consideration of other kinds of biological or biochemical constructions, Figure 9.25. Pre- and post-understanding investigation outcomes have revealed that scholars are better capable to theorize molecular assemblies when such 3D replicas are employed.

Figure 9.25: 3D Bioprinted Representation of Influenza Hemagglutinin Trimer

3D-Printed Dosage Forms - Drug Delivery

3D Bioprinting technique are now being utilized in pharmaceutical investigations and manufacturing. They carry the promise to be transformative. Benefits of 3D Bioprinting embrace accurate regulation of droplet diameter and amount, high replication, and the capability to yield the dose forms with intricate drug-release properties.

Intricate drug fabrication procedures could also be made consistent through utilization of 3D Bioprinting to brand them simpler and more feasible. Also, 3D Bioprinting technology could be significant in the growth of modified medicine.

Personalized Drug Dosing

The motive behind drug improvement should be to upsurge effectiveness and reducing the danger of adversarial responses, an objective that can possibly be realized through the implementation of 3D Bioprinting to yield custom made medicines.

Consuming oral tablets are the most prevalent drug dosage method due to the ease of fabrication, pain evasion, precise medicating, and good patient acquiescence. Nevertheless, no feasible technique is obtainable that could regularly be employed to produce custom made compact dosage forms, such as tablets. Currently, oral tablets are arranged through well-recognized procedures such as combining, milling, and dry and wet granulation of powdered ingredients that are shaped into tablets through series of compression or molds. Each of these fabrication stages can present complications, such as drug dilapidation and form alteration, possibly leading to difficulties with construction or batch failures. Furthermore, these customary fabrication processes are inappropriate for generating custom made drugs and limit the capability to generate personalized dosage forms with extremely intricate configuratoins, new drug-release shapes, and continued stability.

Custom made 3D Bioprinted medications may mainly profit patients who are known to have a pharmacogenetic polymorphism or who utilize medicines with constricted therapeutic indices. Pharmacists could examine a patient's pharmacogenetic sketch, as well as other properties such as age, race, or gender, to regulate an ideal treatment dose. A pharmacist could then 3D Bioprint and dispenses the custom made treatment via an automated 3D Bioprinting system. If essential, the dose could be attuned further depending on medical reaction.

Also, 3D Bioprinting has the possibility to yield custom made drugs in completely novel preparations—such as tablets that contain several active ingredients, either as a lone mixture or as compound multilayer or multi-reservoir Bioprinted tablets. Patients who have several chronic ailments could have their medicines Bioprinted in one multi-dose method that is Bio-fabricated at the point of care. Offering patients with a precise, custom made dose of several medicines in a sole tablet could possibly advance patient acquiescence. Preferably, compounding pharmacies could distribute 3D Bioprinted drugs, since their clients are previously acquainted with buying custom made medicines.

Unique Dosage

The main 3D Bioprinting techniques utilized for medicinal fabrication is inkjet or inkjet powder-based 3D bioprinting. Either another substance or a fine particles is employed as the substrate is mainly what distinguishes 3D inkjet Bioprinting from powder-based 3D inkjet Bioprinting.

In inkjet-based drug bio-fabrication, inkjet Bioprinters are utilized to spray formulations of medicines and binders in small drops at accurate speeds, motions, and configurations onto a substrate. The most normally utilized substrates comprise diverse kinds of cellulose, coated or uncoated paper, microporous bio-ceramics, glass scaffolds, metal alloys, and potato starch films, among others. Researchers have further enhanced this technology by spraying unchanging "ink" drops onto a liquefied film that encapsulates it, creating microparticles and nanoparticles. Such matrices can be utilized to bring small hydrophobic molecules and growth factors. In powder-based 3D Bioprinting drug construction, the inkjet Bioprinter head sprays the "ink" onto the powder basis. When the ink interacts the powder, it solidifies and generates a hard dosage form, layer by layer. The ink can comprise active ingredients as well as binders and other inactive ingredients. After the 3D Bioprinted dosage form is dry, the hard substance is detached from the immediate loose powder substrate.

Complex Drug-Release Profiles

The formation of medicines with intricate drug-release outlines is one of the most investigated routines of 3D Bioprinting. Customary compacted forms are frequently prepared from a homogeneous combination of active and inactive ingredients, and are thus normally restricted to a simple drug-release property. Though, 3D Bioprinters can print binder onto a medium powder bed in layers characteristically 200 micro meters thick, generating an obstruction between the active ingredients to enable controlled drug release. 3D

Bioprinted dosage forms can also be Biofabricated in compound geometries that are porous and loaded with many drugs throughout, encircled by barrier layers that modulate release.

Implantable drug delivery strategies with novel drug-release properties can also be generated utilizing 3D Bioprinting. Dissimilar to customary general cures that can affect non-afflicted tissue; these strategies can be embedded to deliver direct cure to the area involved. Bone toxicities are one instance where direct cure with a drug implant is more needed than general treatment. Luckily, powder-based 3D Bioprinted bone scaffolding can be generated in high-resolution models with intricate configurations that mimic the natural bone extracellular matrix. The Bioprinting of treatments with custom made drug-release properties into such bone implant scaffolds has been investigated. One instance is the Bioprinting of a multilayered bone implant with a separate drug-release properties interchanging between rifampicin and isoniazid in a pulse release mechanism. 3D Bioprinting has also been utilized to print antibiotic micro-patterns on paper, which have been utilized as drug implants to eliminate *Staphylococcus epidermidis*.

In other investigation regarding drug-release properties, chlor-pheniramine maleate was 3D Bioprinted onto a cellulose powder substrate in quantities as minor as 10 to 12 moles to validate that even a minuscule quantity of drug could be released at a specified time. This investigation showed enhanced precision for the discharge of very small drug doses equated with predictably Biomanufactured medicines. Dexamethasone has been Bioprinted in a dosage form with two-stage release properties.

4D Bioprinting Revolutionize Drug Development

4D Bioprinting is a derivative of 3D bioprinting with an additional variable parameter. It embraces prominent potential in tissue regeneration and in bio-pharmaceutical usage of an extensive array of infirmities. As the arenas of biomedical engineering and bio-material-science have advanced, manufacturing techniques embraces including self-assembly, electrospinning nano-technology have progressed. Therefore, 3D and 4D bioprinting advancements are gradually maturing, giving novel treatment choices which could possibly come to be the upcoming human remedies. Moreover, the arrangement of technological advancement and the decrease in fabrication costs is powering the considerable utilization of bioprinting methods. As a resut, their implementation is greatly increasing, extending from printing elements, medical devices, and controlled biological constituents, vascularized tissues and operational human organs. Some significant considerations of therapeutic implementations are viewed to highlight the most relevant hurdles and challenges of 4D Bioprinting technology.

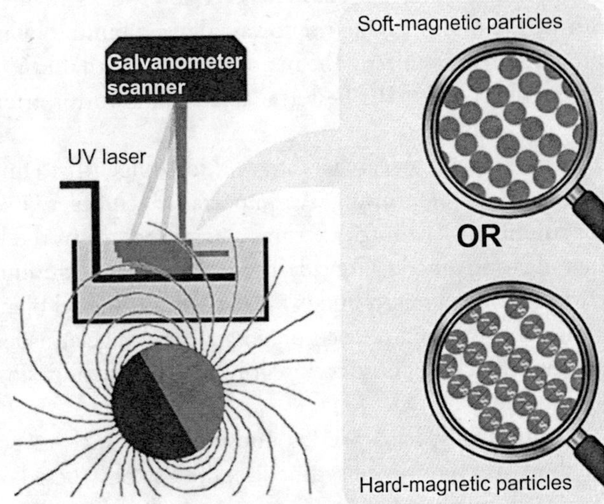

Figure 9.26: Schematic Illustration of 3D Bioprinting System Utilizing Gel with Soft and Hard Magnetic Powder

As discussed, 3D bioprinting is fundamentally established on the arrangement of additive manufacturing with alive organism and biomaterials to yield compound builds, while 4D bioprinting is described as an evolving method, where earlier build 3D formation is incorporated with "TIME" as the fourth variable parameter. Through the 3D constructions, Biologists, engineers, and scientists now capable to manufacture builds that have the capability to alter their shape in consequence to various inducements counting temperature, inflammation behavior, pH, concentration, humidity, electro-magnetic field, and light among others, Figure 9.26, producing a novel time in tissue engineering, as well as in other bio pharmaceutical arenas such as bio-actuation, bio-robotics and bio-sensing. Bio-manufacturing is explained as 'the automated rejuvenation of biologically living operational products with structural grouping from alive cells, bio-active particles, bio-materials, cell accumulations such as micro-tissues, hybrid cell-substance builds through bioprinting or bio-manufacture. Bio-Manufacture is established on 3 major manufacturing fundamentals embracing stated as follows:

- subtractive lithography,
- formative - drying in wells and
- additive manufacturing; being the latter the best way for biofabrication as it provides faster response, enables the tissue engineers to make complex structures, and leads to higher automation capability and reproduction.

As 4D bioprinting is a way of biofabrication, it also uses the 3 manufacturing principles above mentioned.

3D printing was in existence for 30 years, while the recent 4D biofabrication is an incorporation of a recent advancement stimulus to enhance the 3D bioprinting technique. The responsive bio-substance and cell grip forces to generate active builds. The absence of substance strength to sustain the force of fluids developed one of the major difficulties for 3D Bioprinting. In contrast, 4D methods are opening the door to precise and realistic bio-manufacture of structural tissues such as heart tissues, skeletal muscles, vessels, and glands.

Tissue Engineering - Drug Delivery

4D bioprinting grasps an abundant potential for implementations in tissue engineering and drug delivery. Concerning tissue engineering, 4D bioprinting is offering hopeful outcomes in the *in vitro* formation of biomimetic blood veins. The precise structural and biological properties of vasculature mandate significant obstacles in terms of vessels build and re-perfusion. Diverse cells containing endothelial cells, fibroblasts, and mesenchymal stem cells can be combined with hydrogels and then 4D Bioprinted into tube-shaped builds that mimic vasculature. After few weeks in media culture, cells journey, propagate and mature during the builds, giving rise to vascularized build in which endothelial-specific genes and traditional bond proteins are recognized. The major difficulty I still in the area of limited resolution to enable the production of vascular builds. The best available resolution for arteriole/venule builds and capillaries is 30 μm and 8 μm, respectively.

Scientists and researchers are working diligently to find new ways to develop and Bio-manufacture gland builds by 4D bioprinting. An important incomplete issue regarding oncological difficulty is the necessity for mimicking the intricate tumor surroundings in the laboratory in order to generate *in vitro* models for investigating cancers. Considering a self-folding technique method, researchers were capable to manufacture curled hydrogel micro-builds made of photo-model bilayers of poly- ethylene glycol- diacrylate (PEGDA), which shortly after, were co-polymerized with gelatin-derivatives. Actually, gelatin is recognized for its cell-stimulatory characteristics, integrin binding cell adhesion peptides and metalloproteinase-sensitive peptide arrangements.

Additionally, hard printed constructions, such as bone structure can be bioprinted. In a remarkable method, 3D Bioprinted synthetic scaffolds constructed of polycaprolactone, polylactic-co-glycolic acid and β-tricalcium phosphate were custom-made with a bone-mimicking cell-placed mineralized extracellular matrix to improve the biological properties of developed build. The bone implants were manufactured in a decanter bioreactor with human adenoidal inferior turbinate tissue-taken mesenchymal stromal cells to persuade a bone-like bony micro surrounding. The novel bone-like-vertebrate illustrated improved cell differentiation and calcium formation. Though, in spite of this progress, still much hard work are necessary to improve the mechanical and biomechanical characteristics of the 4D-printed implants.

A burst of scientific developments took place within the past few years experiencing the design and fabrication of other micro-structured tissues and organs. For instance, 4D bio-manufacture is being exploited in the bio-fabrication of cell-loaded hydrogels for bio-synthetic pancreas. In this field of synthetic pancreas, the cell-loaded builds must permit the cell capability of the surrounded cells while offering appropriate steadiness and absorbency characteristics. These limits can be enhanced by utilizing of 4D bio-manufacture technique. Additional methods embrace the manufacture of heart micro-tissues, form-memory tracheal-bronchial stents or cell-loaded hydrogels for 3D human cancer spheroids or builds. Remarkably, these minuscule tracheal-bronchial grafts could adapt their outline and shape contingent on a number of micro-environmental conditions, allowing a physiological version to the innate tissue. In actual fact, after inserting these implants in a baby with trachea-bronchomalacia, mechanical ventilation was evaded after 21 days of surgery.

A main emphasis of drug manufacturing industry has been the advance of medicines with the capability to discharge drugs in the most appropriate time and functional position in the body. The method to deliver spatial and temporal healing substances can be enhanced by utilizing 4D bioprinting technologies. For instance, 4D bioprinting allows the accurate control of the spatial distribution of diverse mechanisms that can self-fold or self-unfold to condense and release drugs or cells in a programmable way. For example, tiny oil drops capturing water-base droplets, called that can be bioprinted in water. The same methodology can be taken to grow other nano and micro-builds which react to other micro-environmental variations such as osmosis, temperature, humidity, light or electro-magnetic stimuli. Crosslinking the chemical polycaprolactone build with the capability of recovering its configuration was built for drug delivery goals. Once embedding in a rabbit lower jaw, which contains a single defected bone, the bone morphogenetic protein-2 loaded build recuperated its original shape and encouraged new bone regeneration when being stimulated by the inherit body temperature.

4D Bioprinting Taking a New Direction

The technologies of 3D and 4D bioprinting are founding new paths in drug advancements and other implementation in robotics and bio-sensing. The occasion to manufacture synthetic bio-builds with a form and function that bear a resemblance to the normal physiological tissues may transform the technique tissues will be redeveloped and medicines manufactured in the foreseeable future. The influence on drug fabrication, 3D bioprinting can expand in drug delivery methodology, accuracy on drug quantities and on-demand manufacturing of contractures. Accordingly, it is commonly understood that 4D bioprinting would advance those constructors.

Though, notwithstanding this development, the biomedical arena still confronts several challenges as well as the programmable scalability, cost, fabrication, and simplicity of technology implementations. Contrasting 3D bioprinting in which expenses are plummeting, the fabrication procedures for 4D are still difficult and costly. The utilization of sustainable cells, tissues, and intelligent materials is still high-priced and restricted. It is predicted a novel set of bio-inks for 4D bioprinting and novel intelligent-biomaterials that are receptive to several stimuli *in vivo* will be required to completely increase its healing prospects.

The Future of 4D Bioprinting

An all-embracing purpose of 4D Bioprinting is to expose new current therapeutic chances by allowing form-memory transporters that can be inoculated into the body and recover their original form in a post-injection situation to blend directly with the refined construction of innate tissues to enable optimal tissue regeneration. The architectural changes controlled by time can also produce flexible robots able of bringing their lading in a organized means. Nonetheless, this field requires more investigation with respect to materials utilization and the method to advance their mechanical characteristics to deal with the several necessities within the load-bearing and active atmosphere of the human body. Accordingly, the use of manufacturing procedures and data of numerous materials characteristics could help medical and biomedical scientists to deal with and resolve the present difficulties in the field. Lastly, it is predictable that in the near future high-resolution and novel stimulation mechanisms will be advanced which can react to intricate several-stimulus arrangements in harmony with the several combined biological processes *in vivo*. By means of the arenas of pharmaceutical expertise, biomedical research, bio-intelligent material science, and biology endure to prosper, scientists foresee that new and detailed stimulus systems and high-resolution several-stimuli mechanisms will be advanced, aiding the rise to a hopeful prosperity for 4D bioprinting.

REFERENCES

1. Schubert C, van Langeveld MC, Donoso LA. Innovations in 3D printing: a 3D overview from optics to organs. Br J Ophthalmol. 2019; 98(2):159–161.
2. Klein GT, Lu Y, Wang MY. 3D printing and neurosurgery—ready for prime time? World Neurosurg. 2018; 80(3–4):233–235.
3. Banks J. Adding value in additive manufacturing: Researchers in the United Kingdom and Europe look to 3D printing for customization. IEEE Pulse. 2018; 4(6):22–26.
4. Mertz L. Dream it, design it, print it in 3-D: What can 3-D printing do for you? IEEE Pulse. 2013; 4(6):15–21.
5. Ursan I, Chiu L, Pierce A. Three-dimensional drug printing: a structured review. J Am Pharm Assoc. 2013; 53(2):136–144.
6. Gross BC, Erkal JL, Lockwood SY, et al. Evaluation of 3D printing and its potential impact on biotechnology and the chemical sciences. Anal Chem. 2014; 86(7):3240–3253.
7. Bartlett S. Printing organs on demand. Lancet Respir Med. 2013; 1(9):684.
8. Science and society: Experts warn against bans on 3D printing. Science- 2018; 342(6157):439.
9. Lipson H. New world of 3-D printing offers "completely new ways of thinking:" Q & A with author, engineer, and 3-D printing expert Hod Lipson. IEEE Pulse. 2018; 4(6):12–14.
10. Cui X, Boland T, D'Lima DD, Lotz MK. Thermal inkjet printing in tissue engineering and regenerative medicine. Recent Pat Drug Deliv Formul. 2017; 6(2):149–155.
11. Hoy MB. 3D printing: making things at the library. Med Ref Serv Q. 2013; 32(1):94–99.
12. 3D Print Exchange. National Institutes of Health; Available at: http://3dprint.nih.gov. Accessed July 9, 2019.
13. Ozbolat IT, Yu Y. Bioprinting toward organ fabrication: challenges and future trends. IEEE Trans Biomed Eng. 2018; 60(3):691–699.
14. Bertassoni L, Cecconi M, Manoharan V, et al. Hydrogel bioprinted microchannel networks for vascularization of tissue engineering constructs. Lab on a Chip. 2019; 14(13):2202.
15. Centers for Disease Control and Prevention Colorectal cancer statistics. Sep 2, 2019. Available at: http://www.cdc.gov/cancer/colorectal/statistics. Accessed September 17, 2019.

16. Khaled SA, Burley JC, Alexander MR, Roberts CJ. Desktop 3D printing of controlled release pharmaceutical bilayer tablets. Int J Pharm. 2014; 461(1–2):105–111.
17. Plastics Today. FDA tackles opportunities, challenges, of 3D printed medical devices. Jun 2, 2014. Available at: http://www.plasticstoday.com/articles/FDA-tackles-opportunities-challenges-3D-printed-medical-devices-140602. Accessed July 9, 2019.
18. Food and Drug Administration Public workshop—additive manufacturing of medical devices: an interactive discussion on the technical considerations of 3D printing. Sep 3, 2019. Available: http://www.fda.gov/medicaldevices/newsevents/workshopsconferences/ucm397324.htm. Accessed September 17, 2019.
19. Soloman, Sabrie "Sensors Handbook, 2nd Edition, McGraw-Hill Publishing Co., NY – ISBN 978-0-07-160570-0
20. Soloman, Sabrie "Sensors and Control Systems in Manufacturing 2nd Edition," McGraw-Hill Publishing Co., NY – ISBN 0-07-059626-3
21. Soloman, Sabrie "Affordable Automation," McGraw-Hill Publishing Co., NY – ISBN 0-07-059633-6
22. Soloman, Sabrie "Sensors Handbook, 1st Edition," McGraw-Hill Publishing Co., NY – ISBN 0-07-059630-1
23. Soloman, Sabrie "Sensors and Control Systems in Manufacturing 1st Edition," McGraw-Hill Publishing Co., NY – ISBN 978-07-10572-4

10 3D Bioprinting Innovative Design

3D BIOPRINTING INNOVATIVE DESIGN

Healthcare, Education, Fashion, Food and Environmental Safety

Healthcare

The state of current 3D Bioprinting technology has set a prosperous and secure position in the history of Bio-manufacturing technology. This prominent position is highlighted by Today 7,000,000 working 3D printers acquiring US$3 billion in various industries, laboratories, academic and research institutions, medical and dental offices, as well as Universities and homes. Only two years earlier less than half of this quantity was in operation worldwide, Figure 10.1 illustrates the global 3D printing/Bioprinting revenues among large public companies.

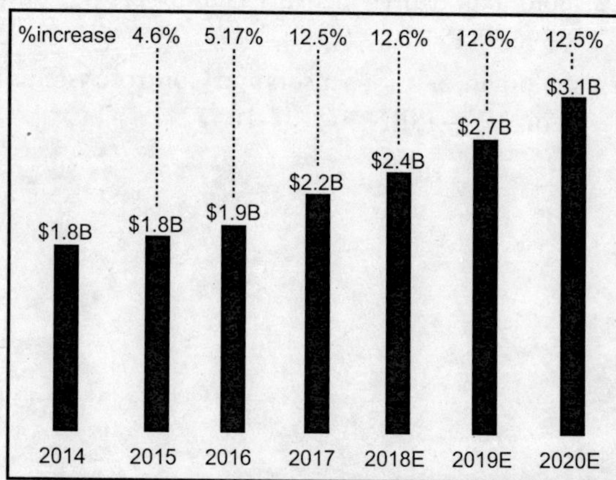

Figure 10.1: Growth in the 3D Printing Market

The progress of current 3D Bioprinters' capabilities is equally impressive when one assesses the expected breakthrough in the speed and the qualitative characteristics within the short and long range plans.

3D BIOPRINTING IN HEALTH CARE

Bone Reconstructive Surgery

In March 2014, surgeons used 3D Bioprinted parts to rebuild the face of a patient who had been seriously injured in an accident. Surgical uses of 3D Bioprinting therapies started in the mid-1990s with functional modelling for bone reconstructive surgery planning. Patient-corresponding implants were a natural addition, leading to truly personalized grafts that fit each exclusive individual, Figure 10.2. Virtual planning of surgery and guidance utilizing 3D Bioprinted, personalized instruments have been applied to many areas of surgery involving total joint replacement and "craniomaxillofacial reconstruction" with steady, Figure 10.3.

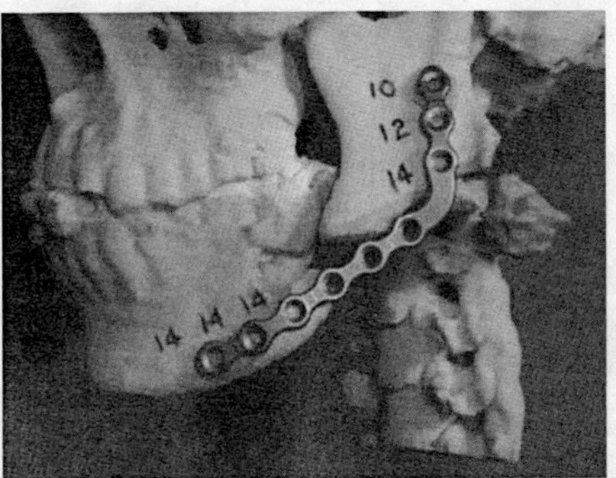

Figure 10.2: Stereolithographic Model of Craniofacial Region - Pre-bent Reconstruction.

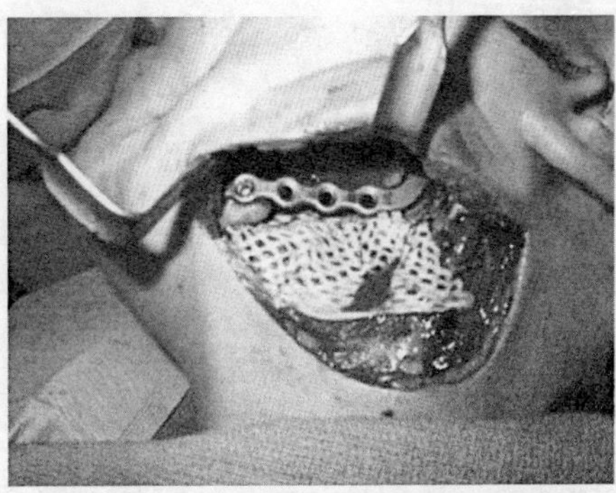

Figure 10.3: Intraoperative Placement of the Acellular Dermal Matrix Mesh after Fixation of the 3D Bioprinted Reconstruction Plate

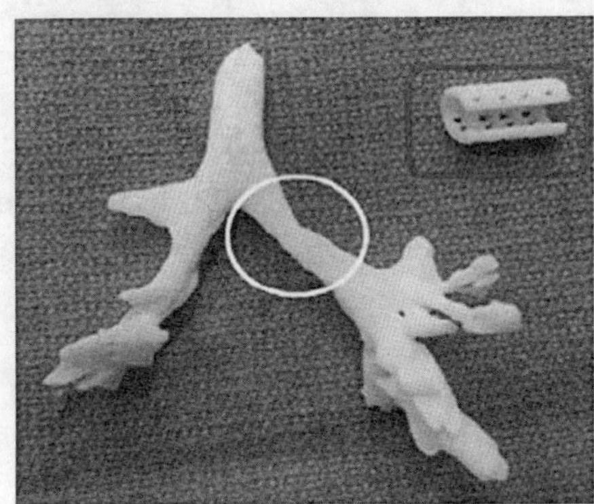

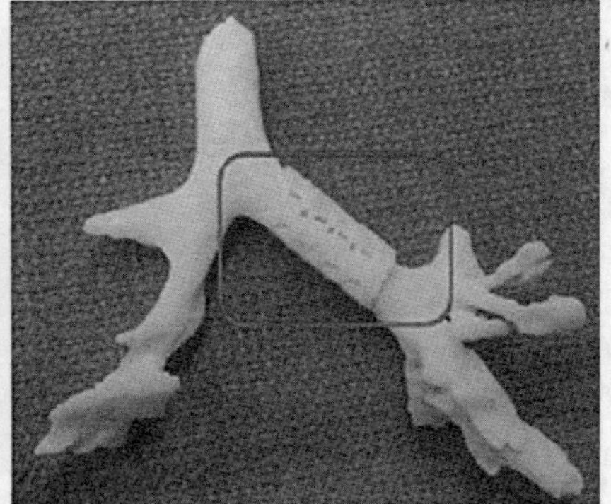

Figure 10.4: 3D Bioprinted Bioresorbable Trachial splint for a Newly Born Baby

One instance of this is the bioresorbable trachial splint to treat newborns with "tracheobronchomalacia." The use of additive manufacturing for serialized manufacture of orthopedic grafts, utilizing metals, is also increasing due to the capability to efficiently generate permeable surface structures that enable "osseointegration," Figure 10.4.

The hearing aid and dental industries are expected to be the largest area of future development utilizing the custom 3D Bioprinting technology, Figure 10.5. 99% of the hearing aids worldwide are manufactured through 3D Bioprinting.

Figure 10.5: Hearing Aids Used Stereolithography 3D Bioprinting

Living Tissues

As of 2012, 3D bio-printing technology has been implemented by many biotechnology organizations and academia for potential utilization in tissue engineering implementations in which organs and body parts are constructed utilizing inkjet methods. In this procedure, layers of living cells are deposited onto a gel environment or sugar matrix and slowly constructed up to form 3D structures containing vascular systems.

Lately, a heart-on-chip has been fashioned which matches characteristics of cells. A non-profit medical center is utilizing 3D Bioprinting to further its mission of generating the next innovation of medical technology to treat vascular illness, such as heart attack and stroke. They have established a way to 3D-Bioprint brain arteries to better comprehend brain aneurysms and stroke.

Medical Applications

Medical implementations for 3D Bioprinting are increasing rapidly and are predictable to transform health care. The genuine and potential medical applications can be organized into several broad categories, including:

- tissue and organ fabrication;
- creation of customized prosthetics, implants, anatomical models; and
- pharmaceutical research regarding drug dosage forms, delivery, and discovery.

Recently scientists define the utilization of 3D Bioprinting to yield bones reconstruction, ears, exoskeletons, windpipes, a jaw bone, eyeglasses, cell cultures, stem cells, blood vessels, vascular networks, tissues, and organs, as well as novel dosage forms and drug delivery devices.

The implementation of 3D Bioprinting in medicine can deliver many advantages, comprising:

- the customization and personalization of medical products, drugs, and equipment;
- cost-effectiveness; increased productivity; the democratization of design and manufacturing; and
- enhanced collaboration.

However, it should be warned that notwithstanding recent substantial and exhilarating medical improvements concerning 3D Bioprinting, distinguished scientific and regulatory challenges continue, and the most transformative implementations for this technology will require time to advance.

The categorization of healthcare-applications of 3D Bioprinting can be stated as follows:

1. Tissue and organ manufacture
2. Medical implants comprising tailored prosthetics and grafts

3D Bioprinting founded on imaging data from CT/MRI

3. Functional Models for Surgical Research and Planning
4. Drug quantity forms, delivery and discovery.
5. Dentistry

The categorization of healthcare application is detailed according to the current state of their descending level of importance worldwide. The list of importance may change in near future.

Bioprinting Tissues and Organs

Bioprinting - Heart on a Chip

In October 2016, Harvard scientists, and engineers introduced the world's first 3D-Bioprinted heart-on-a-chip with incorporated sensors. The heart-on-a-chip is manufactured completely utilizing multi-material 3D Bioprinting in a single automated process, assimilating six custom Bioprinting inks at micrometer resolution, Figure 10.6.

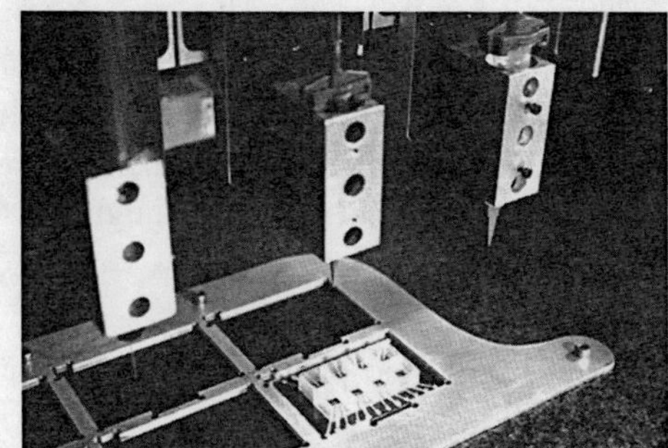

Figure 10.6: Heart-on-a-Chip, Made Entirely Utilizing Multi-material 3D Bioprinting in a Single Automated Procedure
Courtesy: Harvard University

Tissue or organ failure owing to old-age, heart ailments, accidents, and birth defects is a serious medical problem. Current cure for organ failure depend on typically on organ transplants from living or deceased donors. Though, there is a chronic shortage of human organs obtainable for transplant.

In 2009, 154,324 patients in the U.S. were waiting for an organ. Only 27,996 of them, 18% were donated organ transplant, and 8,863, 25 per day deceased while on the waiting list. As of early 2014, about 120,000 people in the U.S. were pending an organ transplant. Organ transplant surgery and continuation is also costly, costing more than $300 billion in 2012. An added problem is that organ transplantation embroils the often problematic charge of finding a donor who is a tissue match. This difficulty could possibly be abolished by utilizing cells cultured from the organ transplant patient's individual body to construct a substitute organ. This would lessen the jeopardy of tissue rejection, as well as the requirement to take lifelong immunosuppressant.

Solution - Organ Donor Shortage

Rehabilitations centered on tissue engineering and regenerative medicine are being followed as a possible solution for the organ donor unavailability. The traditional tissue engineering approach is to isolate stem cells from small tissue samples, blend them with growth factors, grow them in the laboratory, and seed the cells onto scaffolds that direct cell proliferation and differentiation into operative tissues.

While still in its infancy, 3D bioprinting bargains supplementary important benefits outside this traditional regenerative technique, which fundamentally delivers scaffold support only, such as, highly detailed cell assignment and extraordinary digital regulator of speed, resolution, cell concentration, drop volume, and diameter of printed cells. Organ 3D Bioprinting takes benefit of 3D printing technology to yield cells, biomaterials, and cell-laden biomaterials separately or in tandem, layer by layer, directly generating 3D tissue-like builds. Numerous materials are obtainable to construct the scaffolds, contingent on the anticipated strength, porosity, and kind of tissue, with hydrogels usually well thought-out to be most appropriate for generating soft tissues.

Droplets of Living Cells

Although 3D bioprinting systems can be laser-based, inkjet-based, or extrusion-based, inkjet-based bioprinting is most shared. This technique deposits "bioink," droplets of living cells or biomaterials, onto a substrate according to digital commands to duplicate human tissues or organs. Several Bioprint-heads can be utilized to deposit diverse cell kinds (organ-specific, blood vessel, and muscle cells), essential characteristics for Bio-manufacturing complete hetero-cellular tissues and organs.

A procedure for bioprinting organs has appeared as follows:

1. Generate a blueprint of an organ with its vascular construction;
2. Create a bioprinting procedure plan;
3. Segregate stem cells;
4. Differentiate the stem cells into organ-specific cells;
5. Formulate bioink reservoirs with organ-specific cells, blood vessel cells, and support medium and load them into the Bioprinter;
6. Bioprint; and
7. Place the bioprinted organ in a bioreactor prior to transplantation.

In bioprinting, there are three main types of printers that have been utilized.

These are stated as follows:

- Inkjet:
- Laser-assisted, and

Extrusion Bioprinters: Utilized in bioprinting for rapid and far-reaching products. One kind of inkjet printer, named drop-on-demand inkjet printer, prints materials in precise quantities, reducing cost and unused.

Laser printers have also been active in the cell Bioprinting procedure, in which laser drive is utilized to stimulate the cells in a specific array, if 3-D control of the cellular setting.

Bioprinting 2D and 3D Tissues:

Utilizing inkjet bioprinting schemes, scientists effectively printed skin and cartilage replacements, 2D tissues. In the meantime, other scientists achieved to utilize CT imageries of a patient's airway to laser bioprint a bioresorbable tracheal splint, 3D tissue. This splint was then positively embedded in a baby with tracheobronchomalacia, Figure 10.7.

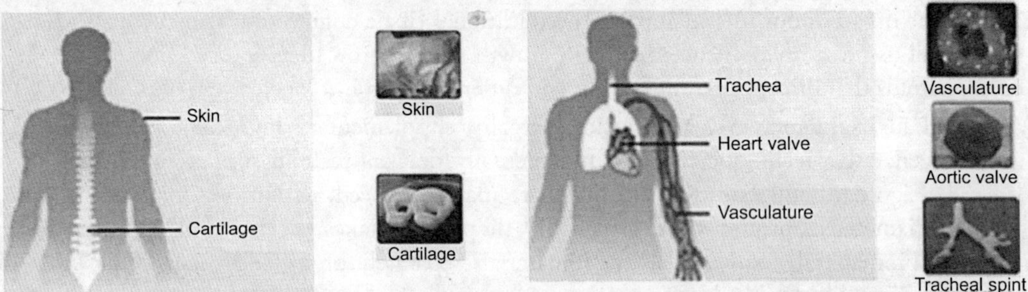

Figure 10.7: 2D Tissue **Figure 10.8:** Hollow Tube

While tissue and organ bioprinting is still in its early stages, many investigations have delivered evidence of perception.

Scientists have utilized 3D printers to generate a knee meniscus, heart valve, spinal disk, other kinds of cartilage and bone, and a synthetic ear. Researchers implemented inkjet 3D Bioprinting technology to restoration human articular cartilage. They have utilized 3D bioprinting technology to deposit diverse cells inside many biocompatible hydrogels to yield an simulated liver. Doctors issued a case study in the New England Journal of Medicine recording that utilization of 3D Bioprinter and CT imageries of a patient's airway permitted them to manufacture a exactly demonstrated, bioresorbable tracheal splint that was surgically embedded in a baby with tracheobronchomalacia, Figure 10.8. The baby recovered, and full resorption of the splint has to occur successfully.

Many biotech corporations have concentrated on generating tissues and organs for medical investigations. It may be likely to quickly monitor new possible healing drugs on patient tissue, significantly reducing investigation costs and time. Researchers at "Organovo" corporation are emerging bands of Bioprinted liver tissue for this strategy; they believe the material will be progressive enough to utilize in studying new drug behaviors. Other biologists are occupied with methods to produce whole human organs that can be utilized for understanding phases of drug discovery. An organ generated from a patient's individual stem cells could similarly be utilized to monitor drug behaviors to regulate if a drug will be useful for that individual.

Bioink Components

Bioprinting utilizes a computer-controlled 3D printing means to precisely deposit cells and biomaterials into the simulated organs. Bio-printing characteristically utilzes two inks. One is the biological material and the other is hydrogel that delivers the surroundings somewhere the tissue and cells cultivate.

Essentially, when a tissue design is designated, scientists produce "bio-ink" from the cells. Stem cells can familiarize effortlessly to tissues, so they are an striking choice for bioprinting diverse organs and bones. The cells are layered between water-based layers until the tissue is constructed. That hydrogel in between layers is occasionally utilized to seal spaces in the tissue or as supports to the 3D Bioprinted tissue. Collagen is additional material utilized to fuse the cells combined. This layer-by-layer method is comparable to the common 3D printing procedure, where products are constructed from the ground up.

Hydrogels for Bioprinting

Hydrogels have been distinct as two- or multi-constituent systems containing of a 3D set-up of polymer chains and water that plugs the space between macro-molecules. Contingent upon the characteristics of the polymer or polymers utilized, as well as on the nature and density of the network joints, such build in symmetry can comprise several volumes of water; characteristically in the swollen state, the mass fraction of water in a hydrogel is abundant than the mass fraction of polymer.

In preparation, to accomplish accurate swelling, it is normal to utilize artificial polymers that are water-soluble when in non-cross-linked form. Hydrogels come out as a type of biomaterial with good biocompatibility and now they are extensively utilized as the cell-laden materials for bioprinting. As the bio well-matched materials, hydrogels have been extensively utilized in 3D bioprinting/organ printing to load cell for tissue engineering. They have been extensively utilized as cell transporters and scaffolds in tissue engineering due to their physical resemblances to the natural extracellular matrix. Hydrogel alginates have developed as one of the most regularly employed materials in organ printing investigation, as they are extremely custom tailor, and can be fine-tuned to mimic certain motorized and biological characteristic of normal tissue. The capability of hydrogels to be custom-made to precise requirements permits them to be utilized as malleable scaffold materials that are matched for a diversity of tissue or organ assemblies and physiological circumstances. Amid these, intelligent hydrogels denote to a collection of hydrogels that is receptive to numerous peripheral incentives such as pH, temperature, light, electric, and magnetic field. Combining the possible of 3D Bioprinting and intelligent hydrogels is a stimulating new example in the manufacture of a efficient 3D tissue.

Generating materials that will embellish in the human body is a complicated procedure. The bio-ink utilized in the organ Bioprinting procedure is very more intricate than normal Bioprinter ink. There are numerous dependent variables centered on the quantified organ and patient with regards to exclusive cells, biomaterials, and biochemical signals. The difficulty of these constituents permits for effective organ manufacture. In order for the manufactured organ to endure inside the body, biochemical and physical cues must be preserved that endorse cell endurance. Biochemical cues are related to development and adhesion factors, signaling proteins, etc. The physical characteristics are constituents outside the cell and fluid found inside the cell.

Functional Bioprinted Scaffold

There are two classifications that make up the bioink process, practical scaffold and scaffold-free. Practical scaffold utilizes biomaterials that may or may not have cells as the actual ink, while scaffold free utilizes exclusively cells. The biomaterials utilized in the usefulness of scaffold procedure come in a variation of dissimilar materials and size. From hydrogels to metal grafts, single to multiple nanometers in size – materials generated through these phases create the extracellular (outside the cell) matrix. Through these phases essentials such as dimensions, internal geometry, anti-degeneration measures, biocompatibility, as well as others must be highly considered.

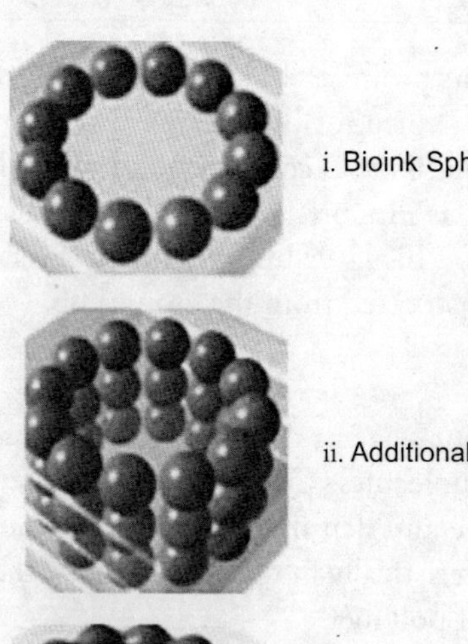

Figure 10.9: Hydrogel 3D Bioprinting Steps Forming Living Tissue

Bioprinting Approach:

Scientists in the Biological Tissue Engineering field have established methods to yield living organs that are built with the appropriate biological and mechanical characteristics. 3D bioprinting is established on three main approaches:

- Biomimicry,
- Autonomous Self-assembly and
- Mini-tissue building blocks.

The first method of bioprinting is entitled biomimicry. The main objective of this method is to create fabricated structures that are identical to the natural structure that is found in the tissues and organs in the human body. The second approach of bioprinting is autonomous self-assembly. This method depends on the physical procedure of embryonic organ growth. The third method of bioprinting is a mixture of both the biomimicry and self-assembly methods, which is entitled mini tissues. Organs and tissues are constructed from very small functional components. Mini-tissue approach takes these small pieces and manufacture and organize them into greater framework, Figure 10.9.

Successful Implants of Bioprinted Organs

Bioprinting of Human Ear

Scientists at the have effectively Bioprinted bone, muscle, and ear structures and imbedded them into rats. The novel tissues have been acknowledged by the immune system and have full-grown in size and produced their own networks of blood vessels. Human sized organs are now being Bioprinted. The completely purposeful human organs are at hand, Figure 10.10.

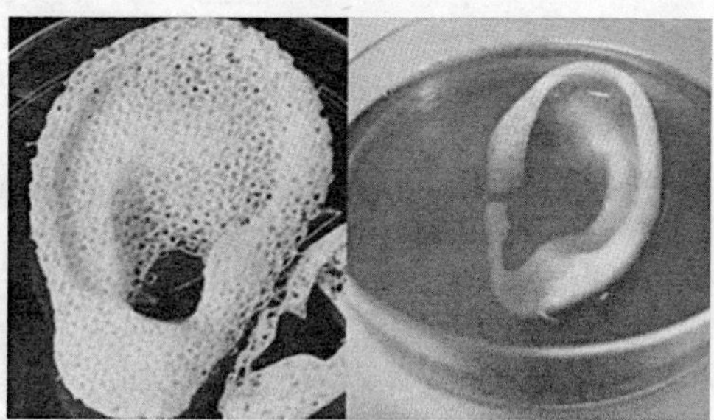

Figure 10.10: 3D Bioprinting Human Ear with Dissolvable Matrix Scaffold

Tissues Assimilate with Bone Structures

Recently, scientists successfully created a titanium jaw graft for an 83-year-old woman, Figure 10.11. Titanium was particularly most appropriate because of its light weight and non-interference with the immune system. Titanium did not fully combine with existing bone. However, Japanese scientists concentrating in biomedical 3D Bioprinting of ceramics have effectively Bioprinted tissue that has fully assimilated with current bone structures. The new 3D Bioprinting material is calcium phosphate which is the main composite in natural bone. Several months later, the grafts accomplished utilizing this novel 3D printing ceramics technology merged with the natural bonne.

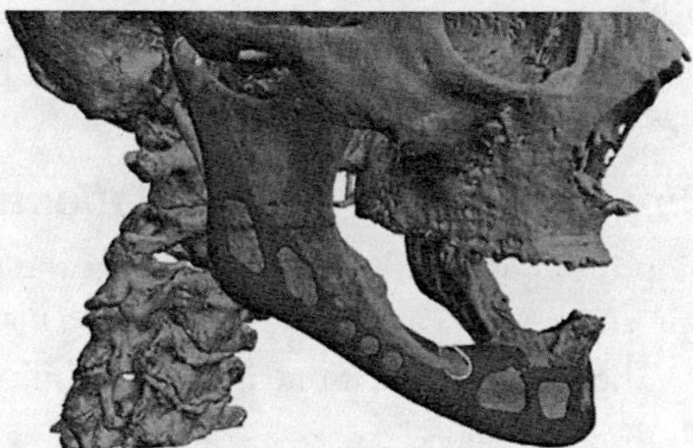

Figure 10.11: 3D Bioprinted A Jaw Implant for 83 Year Old
Courtesy: Innovation Technology

3D Bioprinting Kidneys

The kidneys are one of the toughest organs to reconstruct—if not the toughest. This is owing to the huge amount of nephrons confined in each kidney and to the nephrons' complex construction. Scientists lately advanced forwards generating a simulated kidney that could one day substitute biological donor kidneys. Utilizing 3D Bioprinting, scientists were able to reproduce the tubule element of the kidney's nephrons and provide it a vascular network for blood flow. The 3D Bioprinted tubules were capable to remain active for over two months, Figure 10.12.

Bioprinting must have more advancement time before being capable to generate a fully-efficient organ. Though, the technology has other valuable implementations. Bioprinted nephron tubules could be utilized for drug toxicity evaluation, serving govern biological kidneys' capability to filter specific chemicals, or they could be combined into existing dialysis approaches to enable the technique more comparable to the method our bodies anticipated.

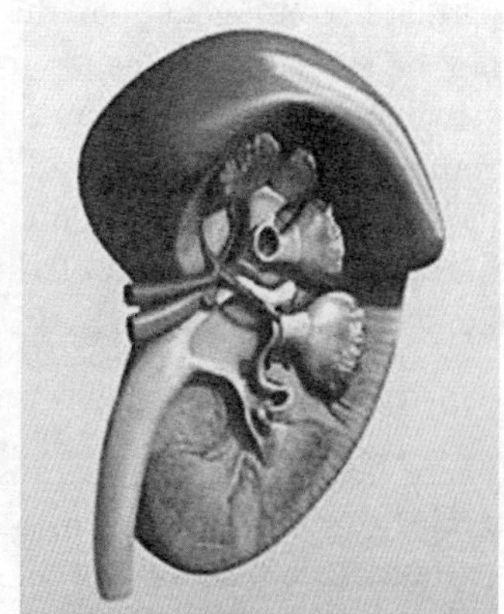

Figure 10.12: 3D Bioprinting Kidney Research
Courtesy: Organovo

3D Bioprinted Sugar Network - Artificial Liver

Scientists are a step nearer to generating a artificial liver, after a US team generated a pattern for blood vessels to propagate into, utilizing sugar. They have long been investigating with the 3D Bioprinting of cells and blood vessels, arranging up tissue structure layer by layer with synthetic cells. The synthetically engineered cells often expire before the tissue is generated, Figure 10.13. The technology, in which 3D Bioprinter utilizes sugar as its construction material will one day be utilized for transplants.

The task in comprehending how to propagate large simulated tissue is how to retain all the cells alive in the engineered tissues, because when cells are put together, they feed nutrients and oxygen from adjacent cells and become choking and dying. Sugar can be dissolved away in the existence of living tissue; it is responsive to biological tissue. Scientists presented that the utilization of 3D Bioprinter to print a random network of vessels for any tissue figure or any network of blood vessels, and then setting them with cells would generate the organ. And when the build is shaped of pipes-to-be and tissue, the sugar is dissolved away utilizing water.

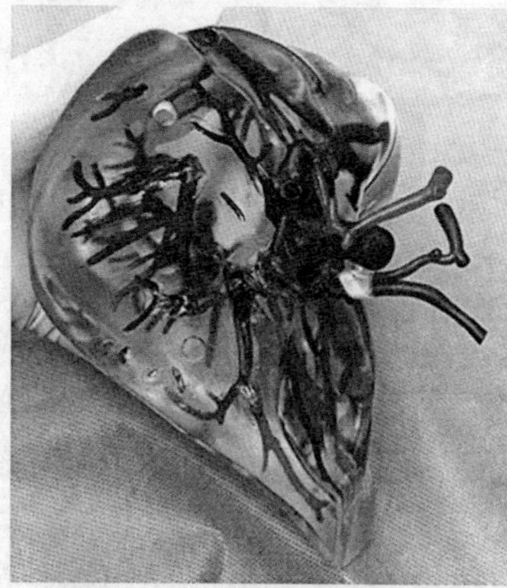

Figure 10.13: 3D Bioprinting Research - Artificial Liver Network
Courtesy: Health Essential

Bioprinting Skin – Burnt Patients

Bioprinting is mostly concentrated on the progress of 3D printed of skin and tissues. Currently, there are two main paths that are explored.

The first path: It is one is the possibility to Bioprint skin directly on the body and

The second path: It is to 3D Bioprint tissues that could actually be implemented in a human body.

For burnt patients, the skin renovation is a vital subject. The surgery takes a long time to be operative and flourish. Typically, a part of skins is detached from the body and positioned on the burnt part. Still, contingent upon the injury significance, this method might not be sufficient to shelter the wound. Not only from a physical viewpoint, it is above all a problem for their health to avert them from contamination and fluid and heat loss. Several Schools of Medicine have advanced 3D Bioprinters that can 3D print skin straight on burnt patients. The ink of the Bioprinter is prepared with diverse kinds of skin cells, Figure 10.14.

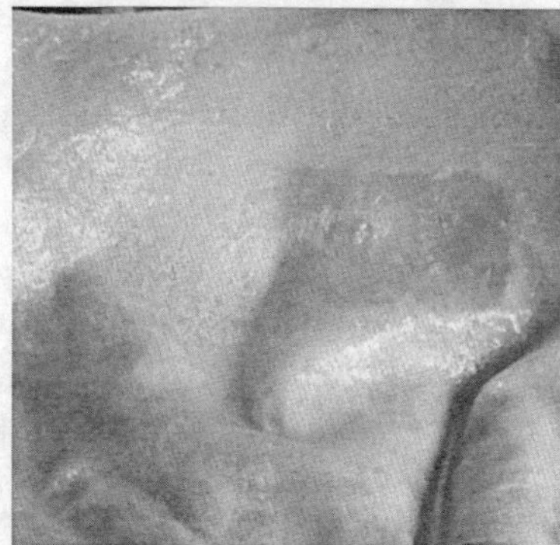

Figure 10.14: 3D Tissue Bioprinting Skin - Transplanted to Burnt Patient
Courtesy: Linköping Burns Treatment Centre – Linköping University - Sweden

Bioprinting Nerves

Every year, 250,000 Americans lose sensation or mobility because of nerve impairment from diseases such as diabetes or injuries from accidents. Scientists discovered a method to excite those nerves to regrow in lab rats. By means of 3D Bioprinter to generate divided tubes of silicon to excite and encourage nerve development. The procedure originated by scanning the rats' sciatic nerves, which regulate their back legs. From the scan, he 3D printed silicon tubes of the precise size and branching pattern. Scientists cut off a half-inch length of each rat's sciatic nerve, and then stitched the silicon guide onto the ends of the nerve endings. At first, the rats tripped as predictable. But the nerves were capable to develop. Within three months, the rats ran about normally. An artifact of Bioprinting—long parallel channels at the length of the tubes—delivered further direction to the developing nerves. It is since revealed that biomolecules can be placed down, such as proteins inside the channels as part of the Bioprinting procedure to guide nerve development. These improvements may pave the way toward stimulus of intricate nerve injuries for faster recuperation.

3D Bioprinted Robotic Skin

A technique was developed by scientists and engineers to Bioprint directly 3D print tactile sensors onto prosthetic limbs or surgical robots, Figure 10.15.

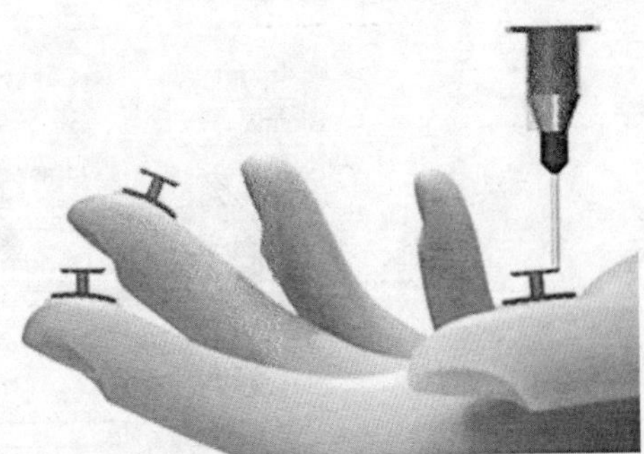

Figure 10.15: 3D Bioprinting Tactile (Touch) Sensors Directly on Robotic Skin/Hand
Courtesy: University of Minnesota

U.S. scientists have constructed a "revolutionary" procedure for 3D Bioprinting stretchable electronic sensory instruments that could provide robots the capability to sense their distance increments. The detection, is a foremost step onward in printing electronics on actual human skin.. The novel procedure is utilizing a one-of-a-kind 3D printer construct in the lab. The multifunctional 3D printer is furnished with four nozzles, to Bioprint diverse types of dedicated "ink" that make up the layers of the instrument. These flexible 3D Bioprinted sensors can expand up to three times their initial size, rendering to the investigation.

This expandable electronic textile material has functional uses. Applying this kind of "bionic/robotic skin" on surgical robots would give doctors the capability to essentially sense during minimally invasive surgeries, which would cause surgery to be easy instead of utilizing advanced cameras. These sensors could also brand it easier for other robots to stroll and interrelate with their surroundings. Scientists trust the novel 3D Bioprinting method could also be utilized to print electronics openly on actual human skin. This eventual wearable technology may ultimately be utilized for health observation or by militaries in the field to sense dangerous chemicals or explosives.

This is a fresh method to 3D printing of electronics that could take numerous instructions from health intensive care to energy production to chemical detection.

3D Bioprinted Ovaries to Give Birth

The generation of simulated ovaries for humans is a step closer after birth of vigorous pups from mice provided 'ovarian bio-prosthesis, Figure 10.16. Infertile mice have given birth to healthy pups after their fertility has been restored with ovary inserts made with a 3D Bioprinter. Scientists formed the artificial ovaries by Bioprinting porous scaffolds from a gelatin ink and providing them with follicles, the manicure, fluid-property pouches that hold immature egg cells. In tests on mice that had one ovary surgically detached, researchers established that the implantations attached to the blood supply within a week and continued on to discharge eggs naturally through the pores constructed into the gelatin structures. The work provides a step nearer making simulated ovaries for young women, whose reproductive systems have been injured by *cancer treatments*, leaving them infertile or with hormone disparities that necessitate them to take regular hormone-boosting drugs.

Figure 10.16: 3D Bioprinted Ovaries – Step Closer Aiding Pregnancy
Courtesy: Warren Centre

3D Bioprinted Vascular Networks in Ischemia

Scientists indicated they have established 3D-Bioprinted coverings permeated with cells that deliver a fresh technique to developing healthy blood vessels to treat ischemia. Their study combined 3D-Bioprinted vascular networks to guide healing angiogenesis in ischemia, Figure 10.17.

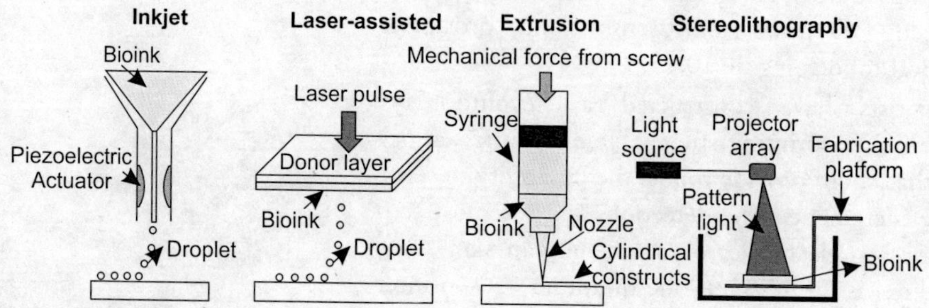

Figure 10.17: 3D Bioprinted Vascular Networks to Direct Therapeutic Angiogenesis in Ischemia

Therapeutic angiogenesis, when development factors are inoculated to boost new vessels to produce, is a hopeful trial method to treat ischemia. But in trial, the new subdivisions that sprout form a unsystematic and winding network that looks similar sort of a hairball and does not permit blood to stream efficiently through it. Researchers developed two coverings with endothelial cells—one where the cells were restructured into a precise architecture, and another where the cells were simply inoculated short of any organizational structure. *In vivo* results established the coverings with pre-organized structure reproduced a marked development in decreasing the occurrence of ischemia, while the coverings with no organization caused in the "hairball" situation. This preclinical work offers a fresh method to guide improved blood flow to precise areas of the body. The increased blood nourishment delivers valued oxygen to heal and functionally preserve essential organs such as the heart and limbs.

In Situ Bioprinting

In situ Bioprinting, in which grafts or living organs are printed in the human body during procedures, is another expected future trend, Figure 10.18. Through utilization of 3D bioprinting, cells, development factors, and biomaterial scaffolding can be deposited to restore lacerations of numerous kinds and thicknesses with exact digital control. *In situ* bioprinting for mending exterior organs, such as skin, has previously taken place. In one case, a 3D printer was utilized to seal a skin lesion with keratinocytes and fibroblasts, in stratified zones through the wound bed. This method could perhaps progress to utilize for *in situ* reparation of partly injured, diseased, or failing internal organs. A handheld 3D Bioprinter for utilization *in situ* for direct tissue restoration is an expected modernization in this area. Progressions in robotic bioprinters and robot-assisted operation may also be essential to the development of this technology.

Figure 10.18: *In Situ* Portable 3D Skin Bioprinter

Courtesy: University of Toronto - Researchers have created the first handheld 'skin Bioprinter' that forms tissues in situ for application to wounds.

Magnetic 3D Bioprinting

Magnetic 3D bioprinting is a practice that allows biocompatible magnetic ***nanoparticles*** to print cells into 3D build or 3D cell cultures. In this procedure, cells are labeled with magnetic nanoparticles, ***nano-shuttle*** that is utilized to render them magnetic. Once magnetic, these cells can be quickly Bioprinted into precise 3D designs utilizing exterior magnetic forces that mimic tissue build and performance. Magnetic 3D bioprinting can be utilized to screen for cardiovascular noxiousness, which accounts for 30% of drug extractions. Vascular plane muscle cells are magnetically Bioprinted into 3D ring configurations to mimic blood vessels that can contract and dilate. This system could possibly substitute trials utilizing ex vivo tissue, which are expensive and harvest insufficient data per trial. Furthermore, magnetic 3D bioprinting can use human cells to estimate a human *in vivo* reaction better than with animal experiments. This has been established by the bioassay which associates the advantages of 3D bioprinting in constructing tissue-like structures for investigation with the speed of magnetic printing.

The key objective users for magnetic 3D bioprinting are in the pharmaceutical and Control Research Organization, CRO industries, where this system can be incorporated initially in the drug discovery procedure as a composite screen for toxicity and effectiveness. In the upcoming years, magnetic 3D bioprinting could be implemented to the field of regenerative medicine and organogenesis.

Overall, magnetic 3D bioprinting is an operative tool to generate realistic models of native tissue, Figure 10.19.

Figure 10.19: Magnetic Levitation 3D Bioprinting Market – *Space Market Research*

3D Bio-Pen with Bio-Ink

Handheld 3D Bio-Printing Pen

It is tailored for cartilage grafts to prosper in sheep trial, Figure 10.20:

A radical tool for Bioprinting living cells *in situ* during operation inches nearer to human trials. A handheld pen that yields a bio-ink comprising a patient's own cells could soon day be utilized to mend knees injured by osteoarthritis. In a initial studies, the 3D-Bioprinting pen created excellent data when verified on sheep. The Bio-pen is an efficient new tool accessible to doctors. In the trial, cells taken from each sheep were combined with a particularly formulated Bio-ink. During surgery, the tool was then utilized to generate an exact, bespoke graft that substitute cartilage lost through osteoarthritis. The ink performs to guard the cells during the procedure, and then maintain them until they become completely incorporated into the joint. Because they are resulting from patient – or animal – being preserved, there is no risk of rejection and no need for immunosuppressive drugs.

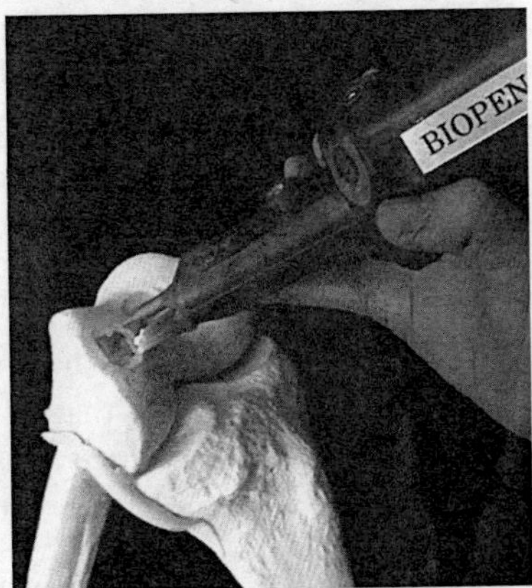

Figure 10.20: Handheld 3D Bio-pen Prints Human Cartilage
Courtesy: University of Melbourne

Scientists verified the tool on eight osteoarthritic sheep. Each had one knee cured by the Bio-pen, and another utilizing current technique. The consequences were excellent, as it was discovered that the Bio-pen achieved better in terms of quality and features of new cartilage formation.

Obstacle to Bio Printing Organs

In an ordinary organ, every cell is close only a couple of cells of a minor blood vessel and if that cell or cell group is supposed to secrete somewhat substance like insulin, a development factor or digestive enzyme, there is a channel for that as well close. These minute channels and capillary vessels are frequently lesser in diameter than a single cell which makes them exceptionally challenging to engineer by standard technologies, which can be arranged. This is where the collapse usually occurs. Samples can be made in large numbers of cells on flat sheets or a limited layers thick and can utilize abnormal mixtures to generate these vessel beds, put them together has not successful yet. Additionally, the capability to shape small beds has not been positive. Blood vessels are not that different to an oak or maple tree. If one would envisage the heart is the main trunk, and the furthest twigs are the furthest roots. It begins as a single large volume and then continues dividing and splitting and is clear that the volume in the margin is greater than in the principal core, but all the leaves are ultimately fed by a single point where those leaves signify cells.

Too many researchers are attempting to produce organs by the sameness of developing a tree by congregating a pile of leaves and hoping. Vascularization is a big difficulty in the method of 3D printing organs, since they require having a system of arteries, capillaries, and veins that support the system. They must be exit to deliver nutrients and eliminate waste generated by the cells. One choice is to vacate the space in the 3D Bioprinted tissue for veins to be included later on in the process, but scientists are now attempting to find out a method to Bioprint blood vessels as well.

Medical Devices

3D Bioprinters are utilized to fabricate a diversity of medical instruments, comprising those with intricate geometry or characteristics that equal a patient's exclusive anatomy. Some instruments are Bioprinted from a normal design to make multiple matching duplicates of the same device. Other devices, called patient-complemented or patient-specific devices, are generated from an exact patient's imaging data.

There are diverse commercially available 3D Bioprinted medical devices include the followings:

- Hearing aids and eyewear
- Instrumentation (e.g., guides to assist with proper surgical placement of a device),
- Implants (e.g., cranial plates or hip joints),
- External prostheses (e.g., hands).

Hearing Aid

3D Bioprinting has by now had a Changeable consequence on hearing aid manufacturing, Figure 10.5. Today, 99% of hearing aids that suitable into the ear are custom-made utilizing 3D Bioprinting. Everybody's ear canal is molded in a different way, and the utilization of 3D Bioprinting permits custom-molded devices to be created competently and cost-effectively.

3D Bioprinted Custom Eyewear

Tailored eyewear is more costly than the typical mass-manufactured prescription frames. This is because producing custom tailor eyewear is time-consuming and augments line items to the costs of a discrete frame. For instance, capturing facial dimensions is supplementary to an optician's responsibilities during the custom eyewear process.

Likewise, custom clay patterns augment to the material and time costs. Furthermore, errors at any phase of the procedure are costly and necessitate recalibration of patterns and measurements. 3D Bioprinting services, such as "Shapeways" and "iMaterialise," have cut down access to industrial grade printers for normal customers and allowed them to design custom frames for themselves. Though, these benefits are counterbalance by restrictions in completed products. Custom 3D printed eyewear is afflicted by the same drawbacks that distress other 3D printed objects. The final product is restricted in strength, finish and diversity. Selective Laser Sintering or SLS printers, which are characteristically employed to construction custom eyewear, are costly and restricted in their capacity to custom fabrication eyewear utilizing available materials for 3D printer.

Self-Adjustable Glasses

Scientists promoting the extensive prospective of 3D printing lately utilized a "RepRap" printer to construct self-adjustable glasses. The glasses cost just $1 to create, and the lenses can be self-adjusted, serving low-income people to evade the costs normally related to vision alteration. The project validates the humanitarian consequences of 3D-printing. With 90% of visually impaired persons living in countries where remedial eye-care is either unreasonable or impossible, the "Adspec" glasses could have a considerable global influence.

Customized Implants and Prostheses

Grafts and prostheses can be manufactured in almost any conceivable geometry via the conversion of X-ray, MRI, or CT scans into digital ".STL" 3D print files. In this method, 3D Bioprinting has been employed positively in the health care segment to allow both typical and intricate customized prosthetic limbs and surgical grafts, occasionally within 24 hours. This method has been employed to construct dental, spinal, and hip inserts. Beforehand, before grafts could be employed clinically, they had to be authenticated, which is very time-consuming.

The aptitude to rapidly yield custom grafts and prostheses resolves a clear and tenacious difficulty in orthopedics, where typical grafts are often not adequate for some patients, mainly in intricate circumstances. Beforehand, surgeons had to achieve bone graft surgeries or utilize scalpels and drills to adjust grafts by shaving pieces of metal and plastic to a wanted figure, size, and fit. This is also true in neurosurgery: Skulls have uneven figures, so it is tough to normalize a cranial graft.

In wounded of head injury, where bone is detached to provide the brain room to swell, the cranial plate that is later tailored must be precise. Although some plates are milled, more and more are shaped employing 3D printers, which enables it much easier to modify the fit and design. There have been numerous commercial and clinical achievements concerning the 3D Bioprinting of prostheses and grafts. A team of scientists successfully implanted the first 3D-printed titanium mandibular prosthesis. The implant was made by utilizing a laser to successively melt thin layers of titanium powders.

In 2013, "Oxford Performance Materials" was given FDA endorsement for a 3D-Bioprinted polyetherketoneketone, PEKK skull implant, which was first effectively grafted that year. The company manufactured 3D-printed titanium orthopedic, maxillofacial, spinal, and dental grafts. A structurally correct 3D-printed prosthetic ear able of sensing electromagnetic frequencies has been manufactured utilizing silicon, chondrocytes, and silver nanoparticles, Figure 10.21. There is a emergent trend in the direction of making 3D Bioprinted grafts out of a diversity of metals and polymers, and more lately grafts have even been Bioprinted with living cells.

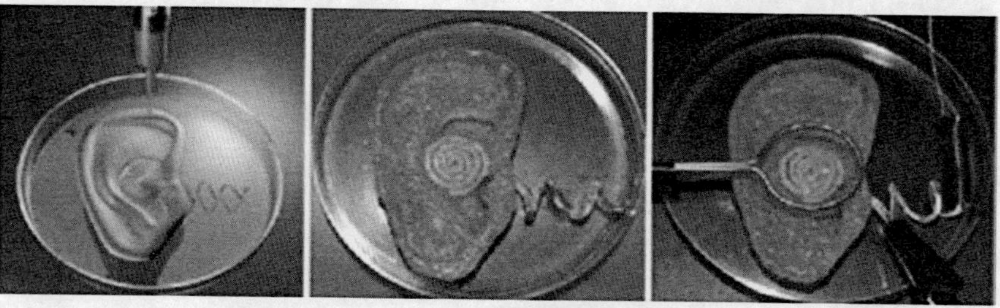

Figure 10.21: 3D Bioprinted Bionic Ear

Knee Replacement

Bioprinting technology has already transfigured joint substitutes. Knee emergency is a very common practice; there are about six diverse kinds of knees that doctors utilize, with each one the requirement to cut the bone another way. But with 3-D Bioprinting, doctors are not restricted to those six knees. They can develop one exact to each patient, Figure 10.22. Patients with tailored knees do not have to drop extra inches of bone, in its place the surgeon can amend at the optimal point, which could lead to faster heling times and better performance. Robust, flexible new knee joints imitating bone and cartilage can now be Bioprinted with nylon. These surgeries are accessible at top-tier medical services in USA.

Femur
Meniscal tear
Meniscus
Tibia

Figure 10.22: Researchers at Columbia University - Medical Center, USA Replaced Damaged Menisci Utilizing 3D Bioprinted Implant Infused with Proteins that Stimulate Stem Cells - Body Regenerated a New Functional Lining

3D Bioprinting - Imaging Data CT/MRI

Graspable (Intelligible) 3D objects bypassed the restrictions of 3D imaginings which can only be projected on flat screens. 3D workpieces can be fashioned established on CT or MRI volumetric medical imageries. By means of dedicated post-handling algorithms, a spatial model can be mined from image data sets and transferred to machine-readable data. That spatial model data is exploited by special Bioprinters for creating the final rapid prototype model. The main necessities are isotropic data, most circumstances from CT or MRI, dissection software, e.g., 3D slicer, capable to transfer a volumetric object as STL file format, slicing software, encompassed with the bioprinter, and entry-level 3D printer, most models utilize fused filament manufacture, FFM, Figure 10.23.

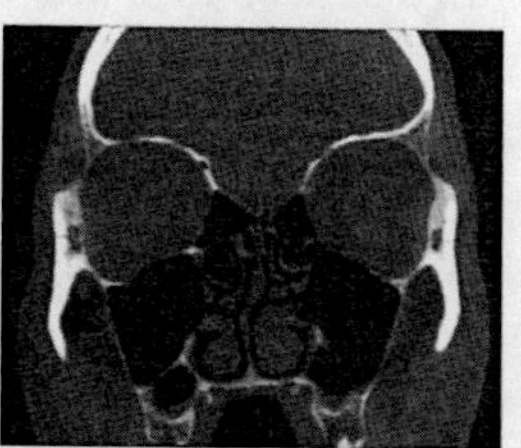

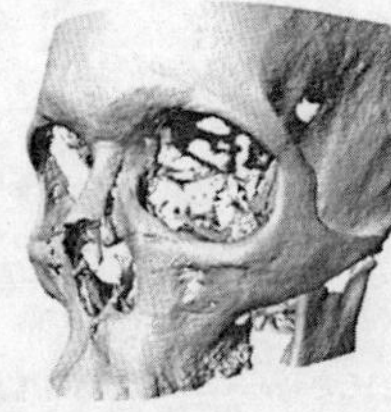

Figure 10.23: Transforming 2D Radiographic Images to 3D for 3D Bioprinting

The main implementations fields for 3D Bioprinting are urology, cardiology, cardiac surgery, vascular surgery, craniomaxillofacial surgery, neurosurgery, orthopedics, and pulmonology.

Some examples of potential applications of 3D Bioprinting in the following medicals implementations:

Cardiac surgery: In the situation of a septal metastasis of a malignant melanoma, a 3D print based on an electrocardiogram-triggered, contrast-enhanced CT scan aided visualization of the tumor, helped facilitate the preoperative planning, and was used for patient education. The segmented tumor matched the resected specimen.

Pulmonology

Anatomic relationship of a small peripheral tumor is found in regard to the bronchi. A hollow, high-resolution resin, 3D print allowed the pulmonologist to better plan the bronchoscopy and was used as a guiding reference for orientation during the procedure. It was also used for training purposes.

Urology: In a case of renal cell cancer, the 3D Bioprint-based multiphasic CT scan, hand-colored after Bioprinting, boosted visualization of the tumor and facilitated the preoperative planning for the kidney vasculature and the ureter. In another case of renal cell cancer, 3D Bioprinting facilitated preoperative planning and permitted nephron-sparing robotic surgery.

Orthopedics: 3D-Bioprint mirrored the original size of a complex fracture of the right hemi-pelvis. The mirrored Bioprint made it possible to use the non-fractured side as a template for the preoperative bending of the osteosynthesis plates, thereby saving time and increasing accuracy.

3D Bioprinted Casts for Broken Bones

It is significantly painful is a broken bone. They take a long time to recover. 3D technology can be convenient in this situation. Doctors can take X-rays or MRI scans, and then build a needed structure that is an exact match for the patient utilizing a 3D printer. This can be helpful for thousands of patients. A cortex exoskeletal cast is one such example. These casts are far better than the traditional casts used for broken bones.

Structured Anatomy

Pro-section of the hand and wrist with 3D images and 3D Bioprinted model provides the following observations:

A. Image of CT-scanned pro-section of hand and wrist;
B. The 3D computer image is constructed from the CT data, in this case, exported from the scanner workstation in DICOM format, utilizing image processing software, e.g., Amira, Avizo, Mimics, Simple ware, 3D Slicer, which creates a stereolithography file .STL;
C. Because CT scan does not provide information on color, anatomically realistic colors can be added utilizing a package such as 3D Coat;
D. The colored .STL file can then be printed in full color as a 3D copy of the original pro-section.

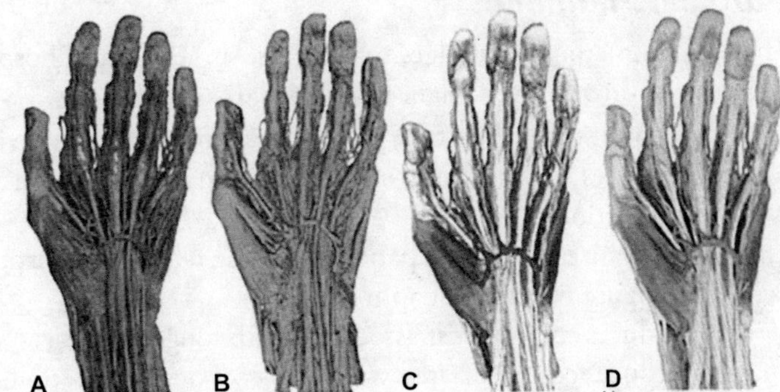

Figure 10.24: 3D Bioprinted Model of Pro-sectioned Hand

The education of anatomy has steadily been the subject of societal disagreement, particularly in the context of employing cadaveric materials in professional medical and associated health professional teaching, Figure 10.24. The reduction in dissection-based teaching in medical and allied health professional training programs has been in part due to the monetary concerns tangled in maintaining legacy programs, accessing human cadavers and concerns with health and safety attentions for students and staff showing to formalin containing embalming fluids. 3D Bioprinting permits the formation of reproductions of projected human cadaver and other anatomical examples that remove many of the above issues.

Anatomical Models for Surgical Preparation and Planning

The separate modifications and difficulties of the human body create the deployment of 3D-Bioprinted models perfect for surgical groundwork. Having a palpable model of a patient's anatomy obtainable for a physician to learn or utilize to mimic surgery is desirable to relying solely on MRI or CT scans, which aren't as informative meanwhile they are observed in 2D on a flat screen. The use of 3D-Bioprinted models for surgical exercise is also desirable to exercise on cadavers, which present problems with respect to obtainability and cost.

Cadavers also often lack the suitable pathology, so they deliver more of an experience in anatomy than an illustration of a surgical patient. Scientists at the National Library of Medicine creäte digital files from clinical data, such as CT scans, that are employed to make special 3D-Bioprinted surgical and medical replicas. 3D-Bioprinted neuroanatomical replicas can be mainly helpful to neurosurgeons by as long as a depiction of some of the most intricate configurations in the human body. The complicated, sometimes disguised relationships between cranial nerves, vessels, cerebral structures, and skull construction can be problematic to understand grounded solely on radiographic 2D images. Even a minor error in circumnavigating this complex anatomy can have actually shocking significances.

A accurate 3D model reproducing the relationship between a lesion and normal brain constructions can be helpful in deciding the safest surgical passage and can also be valuable for the neurosurgeon to practice perplexing cases. Intricate spinal malformations can also be learnt better through the utilization of a 3D model. High-quality 3D anatomical replicas with the right pathology for teaching doctors in executing colonoscopies are also essential, since colorectal cancer is the second leading reason of cancer-related demises in the U.S.

With the assistance of 3D sculpting skills, surgeon tools are effortlessly customizable in a 3D file, to be Bioprinted after. Rendering to the fact that the doctor or hospital possess a 3D Bioprinter or utilization a 3D Bioprinting service, tools can be arranged for the surgery in a day and are tailored for the patient's body. Mayo Clinic shaped around 750 surgical case models on its own 3D Bioprinters in 2018, has achieved complete cranial reconstructions, detached intricate tumors and positively detached conjoined twins with 3D Bioprinted monitors and tools.

Surgical Planning

There are numerous advantages to patients, surgeons, and hospitals in utilizing a 3D Bioprinted model to devise an optimal surgical plan. It is summarized as follows:

- Allows surgeons to try a particular approach or device in a risk-free environment;
- Provides the surgeon with practice time before performing the actual surgery, much like medical simulation;
- Minimizes the time a patient is on the table, being exposed to harmful radiation, as a surgeon tries to figure out the best approach;
- Reduces surgical cost associated with longer surgery;
- Reduces cost associated with incorrect device selection, catheters, wires, or devices types and sizes;
- Determines which are the best tools for the specific patient's case;
- Helps the surgeon choose the appropriately-sized device;
- Identifies complications so they can anticipate them in surgery.

3D Bioprinting Aorta

Figure 10.25, illustrates 3D imagining and 3D rapid prototyping of the "Aorta" in a patient with thoracic aortic aneurysm, arrow. 3D models are supportive in representing intricate pathologies. Vascular surgeons may discover models valuable for assessment of the best treatment plan as well as for exercise. Discrete patient care can help from rapid prototyping by as long as a understandable model, thus serving the patient to comprehend the pathology and to provide knowledgeable consent for surgical involvements sufficiently skilled on models, surgeons feel more self-assured while going to the operating theatre for actual surgery. Moreover, the pre-operative recreation of an exact and intricate surgery to offers a exceptional occasion to employ surgical steps in order to decide the best operating strategy, again growing the surgeon's confidence during the operation.

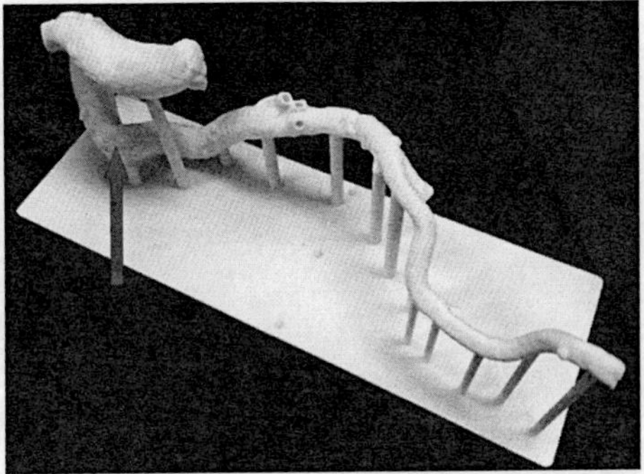

Figure 10.25: 3D Bioprinted Model Created by Computed Tomography (CT) - Scan of 89-year-old patient with Severe Aortic Stenosis - Scanned for "Transcatheter Aortic Valve"- Replacement – TAVR Evaluation
Courtesy: Stanford University – Stanford Medicine

Pharmaceutical Discovery - Drug Dosage - Delivery

3D Bioprinting technologies are even now being employed in pharmaceutical investigation and manufacture, and they capacity to be transformative. Benefits of 3D Bioprinting embrace accurate control of droplet size and amount, high reproducibility, and the capacity to yield dosage forms with intricate drug-release profiles. Intricate drug engineering procedures could also be standardized through utilizing of 3D Bioprinting to create them simpler and more feasible. 3D Bioprinting technology could be very significant in the improvement of tailored medicine, too. The first tablet, capsule, or gel fabricated by 3D Bioprinting was accepted by the FDA in August 2015. Binder-jetting into a powder bed of the drug permits very absorbent tablet, capsule, or gel to be manufactured, which allows high drug doses in a single tablet, capsule, or gel which dissolves rapidly and can be dissolved easily. This has been established for Spritam, a reformulation of levetiracetam for the treatment of epilepsy.

DENTISTRY

Already valued $2.10 Billion in the dental market, 3D Bioprinting could become a $3.1 billion industry in this sector by next year 2020, agreeing forecasts in a report released in May 2019. The prevalence of 3D Bioprinting in other markets is predictable to mature as well, but improved 3D Bioprinters and materials are now being mass-produced precisely for dentists, Figure 10.26. Also it is forecast the sale of 3D Bioprinting systems to dental labs will double by 2020. 3D Bioprinting technology is also predictable to deliver more than 60% of all dental production needs by 2025, and perhaps even more in certain areas such as dental modeling.

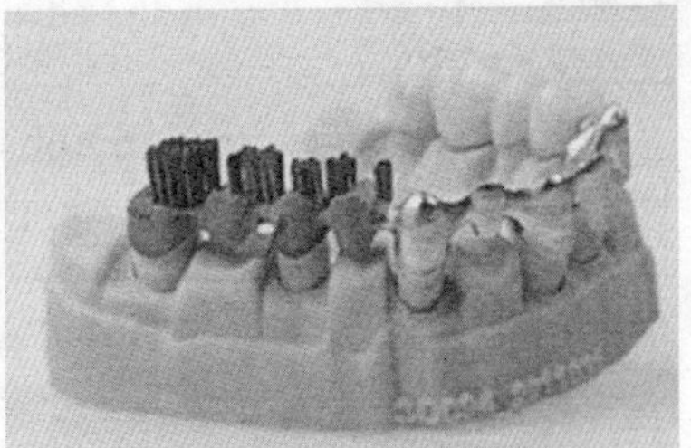

Figure 10.26: 3D Bioprinted a Patient Specific Denture

Dental Implants

Dental industry has been employing synthetic material for dentures, orthodontics, implants, crown, and bridges for numerous years. As these parts are tailor made for each person, the procedure is both time consuming and costly. Direct and indirect 3D Bioprinting, specifically Bioprinting the actual part or a mold, has been revealed to be an inexpensive and faster substitute to conventional methods.

The Use 3D Bioprinting of Dentistry Technology

Here are numerous dental applications employing 3D Bioprinting technology, below are a few a few implementations stated as follows:

Replace or Repair a Damaged Tooth: The dentist scans the patient's mouth with a small size digital wand. This generates a 3D image of the teeth and gums, which is stored as a computer file. Computer Aided Design, CAD software permits the dentist to digitally develop the tooth restoration and Bioprint the completed product on 3D Bioprinter.

Create an Orthodontic Model: Pre-3D Bioprinter technology comprises having the patient bite down on gooey, painful clay so it could toughen into a mold, which develops the preliminary model for developing a cure for braces or "Invisalign." This is not so with 3D Bioprinting. A dentist can utilize the same technology highlighted in the first example to scan the teeth, design an orthodontic appliance and Bioprint the end result in-house. "Invisalign" braces are successful commercial use of 3D printing, with 60,000 printed every day. These clear, detachable, 3D-printed orthodontic braces are tailor-made and exclusive to each patient. This product offers a good example of how 3D Bioprinting can be utilized competently and beneficially to make single, tailored, intricate items.

Produce crowns, Bridges, Caps, Dentures and More: The identical procedure defined above can be utilized to 3D Bioprint all kinds of dental implants. The only change is the exact material employed in the Bioprinting procedure.

Construct Surgical Tools: Not only can 3D Bioprinters deal with the dental implants themselves, but they can also 3D Bioprint the drill fixture required to complete specific dental procedures.

3D BIOPRINTING IN EDUCATION

The capability to yield nearly any workpiece in 3D is self-assured to transform education. As a substitute of utilizing linear, 2D teaching approaches to handover knowledge and teach ideas, one can expedite the education capability by providing hands-on 3D models. Perceptions that have been historically problematic to grasp can be directly established with 3D printed visual aids that aid students learn and recollect ideas more effortlessly. Learning through immersive experiences has long been a gratifying approach to education of all categories, and with the advent of 3D printing technology, one can implement this method virtually everywhere.

Interpreting ideas into 3D objects vicissitudes the method one teach as well as the way the students learn. From desktop printers to specialized online services employing high-grade printers, 3D technology is becoming a priceless inventive tool in many courses of study outside 3D computer design. While one may find numerous 3D printers are too expensive for utilization in a classroom environment, even the rudimentary models simplify cooperative learning. Apart from cost, another impediment to 3D printing in schools is the absence of familiarity in most mentors for utilizing this technology or the practices for implementing it to everyday educational use. Hitherto, 3D printers are a likely choice for teaching in STEM curriculum, Figure 10.27. Science, technology, engineering and mathematic, STEM is

Figure 10.27: 3D Bioprinting - University Curriculum

a term that denotes to the academic disciplines of science, technology, engineering and mathematics. The term is characteristically utilized when addressing education policy and curriculum adoptions in schools to advance competitiveness in science and technology advancement.

The implemented knowledge in designing and generating working objects prepares the students to excel in these areas of study while procuring valuable experience in chasing after prosperous technical careers.

Some of the most favorable uses for 3D printing in education embrace the followings:

- 3D printing and mathematics can work together seamlessly in order to generate precise representations of hard to comprehension mathematical ideas. Some students have struggle understanding formulas and diagrams that occur only on paper and 3D printing can be utilized to help them envision perceptions they might have trouble with.
- 3D printing permits those students who have anxiety envisioning equations, intricate graphs, and multipart mathematical arrangements to practice real-world models. Also, 3D printers convey clarity of methodological aspect to a subject which most students conventionally find out-of-date and boring.

3D printing delivers a brand new approach to procedures of design and graphics. Students no longer have to count on 2D screens and imageries to comprehend 3D models. 3D printing consents design students to discover scaled, sophisticated models they could never formulate in the lab or workshop. The overview of 3D printing to educational surroundings means students can recapitulate through distinctions of their designs and appreciate how they accomplish. Young designers can thrust their creative philosophies as far as the 3D software permits.

Implementing 3D Printing in Science – Technology - Engineering – Mathematics "STEM"

Ways 3D Printing can be used In Education are stated as follows:

1. Engineering design students is capable to print out prototypes.
2. Architecture students is capable to print out 3D models of designs.
3. History classes is capable to print out historical artifacts for consideration.
4. Graphic Design students is capable to print out 3D varieties of their creation.
5. Geography students is capable to print out landscape, demographic, or population maps.
6. Cooking students is capable to generate molds for food foodstuffs.
7. Automotive students is capable to print out auxiliary parts or adapted samples of current parts for testing.
8. Chemistry students is capable to print out 3D models of molecules.
9. Biology students is capable to print out cells, viruses, organs, and other serious biological relics.
10. Math students is capable to print out "problems" to solve in their own learning atmosphere, from scale models to city infrastructural design trials.

3D Bioprinting in Education – International Implementations

In 2019, Dimensional Research Institute executed an international survey of over 300 educational institutions concerning the implementation of 3D Bioprinting. Of those surveyed, 99% have at least one Bioprinter though direct student admission and further expansion remains to be limited due to three key obstacles:

1. **Security & Access:** Guaranteeing safe and protected right of entry to 3D printers.
2. **Cost Management:** Managing continuing costs and mitigating the Return on Investment, ROI.
3. **Incorporating into Curriculum:** Integrating 3D Bioprinting ventures into classroom curriculum.

The Benefits of 3D Bioprinting in Education

Numbers of education sectors have received this technology with open arms. This is contributed to the abundant benefits 3D Bioprinting can bring; examples of benefits are stated as follows:

- New learning possibilities – In addition to plainly feeing and viewing their designs take wings and come to life, students can touch and interrelate with the completed product and understand it the way not previously conceivable.
- Access to learning materials – If a school or an educator doesn't have right of entry to education materials, they can merely create them.
- Generates excitement and engagement – Somewhat than inactively reading from a textbook or listening to the teacher, students will have a added lively curiosity in 3D Bioprinting, as it is very much a hands-on action.
- Inspires imagination and novelty – With right to use to design software and a 3D printer, the potentials of what a student can fashion are endless, which inspires greater advancements.
- Encourages critical thinking and problem solving-skills – In ranges of education where students must workout critical thinking and problem-solving skills, a 3D printer can contribute incalculably.
- Better knowledge retaining rates – People study novel information and knowledge more efficiently if they do somewhat than just seeing or listening.
- Reasonable teaching tool – As contrasting to a few years ago when 3D printers cost thousands of dollars, they are now much more reasonable and inexpensive than alternative education means such as laptops and computers.

While some may feel as though 3D printing is merely a method for students to research and express their inspiration, this technology can be much more practical than that. In fact, education materials fashioned for subjects like mathematics and science could stimulate the next age group of engineers, architects, designers, or any other inspired yet empirical profession. It can also aid students that fight with learning theories and topics from a classic text books but are much more adept when given palpable objects to work with. Consequently, students that were previously disheartened with education could not only be revived by the prospect of learning but also reveal their hidden potential.

Revolution in Science-Technology-Engineering-Mathematics (STEM) Education

Science, Technology, Engineering & Mathematics some writers have appealed that 3D Bioprinters give an unparalleled "revolution" in STEM education. The indication for such prerogatives comes from both the low cost capability for rapid prototyping in the classroom by students, but also the manufacture of low-cost high-quality scientific equipment from open hardware designs creating open-source labs. Engineering and design philosophies are discovered as well as architectural forecasting. Students recreate replacements of museum items such as fossils and historical artifacts for education in the classroom without possibly harming sensitive collections.

Other students attentive in graphic designing can build models with intricate working parts easily. 3D Bioprinting stretches students a new viewpoint with topographic maps. Science students can investigate cross-sections of internal organs of the human body and other biological samples. And chemistry researchers can discover 3D models of molecules and the association within chemical complexes.

Architecture

Architectural models have long been a principal implementation of 3D Bioprinting procedures, for generating precise demonstration models of an architect's dream. 3D Bioprinting bids a relatively fast, easy and thriftily viable method of creating thorough models directly from 3D CAD, BIM or other digital data that architects utilize. Many effective architectural firms, now frequently utilize 3D printing (in house or as a service) as a vital part of their workflow for augmented innovation and enhanced communication. More lately some visionary architects are considering 3D printing as a direct building method. Investigation is being showed at a number of organizations on this front, most particularly Loughborough University, Contour Crafting and Universe Architecture. Iron oxide-free Portland cement powder has been utilized to generate architectural constructions up to 9 feet in height.

Automotive Industry

According to a January 2018 report by the US consultancy Frost & Sullivan, the implementation scope of 3D printing is presently limited to the production of extremely low volume parts and production tooling. Notwithstanding this, the firm predicted that the technology will produce $ 4.3 billion in the automotive industry by 2025, and achieve deeper infiltration in manufacture and the aftermarket. In cars, trucks, and aircraft, Additive Manufacturing is start to change both:

1. Unibody and fuselage design and manufacture and
2. Powertrain design and manufacture.

General Motors

When General Motors, GM was underway to build the 2014 Chevrolet Malibu, engineers at the company utilized 3D printing to save time essential in prototyping the parts for the vehicle. GM employed stereolithography, dedicated software, math data and laser sintering to construct parts out of liquid resin in order to make developments to the Malibu. General Motors said that the rapid prototyping showed particularly valuable for the floor console, which has smartphone holders for the driver and passenger. The parts were also lighter weight so that it provides the Malibu a better-quality fuel economy. GM utilized 3D printing for the front fascia design to test vehicles in wind tunnels. 3D printing was too employed to sculpt the front-seat back panels.

Ford Motor Company

Ford Motor: Company uses 3D printing to create prototypes of numerous parts in their vehicles, such as cylinder heads, brake rotors, shift knobs and vents. 3D printing was employed at "Ford's Torrence Assembly Plant" for the manufacture of the Explorer and EcoBoost engines.

Urbee 2

The team of engineers is constructing a whole vehicle with 3D printers named the "Urbee 2." The body of the original Urbee was finished through the employment of 3D printing. Though, the "Urbee 2" as a whole will be typically 3D printed, Figure 10.25. Similarly, the "Urbee 2" will also appear more similar a production-ready car associated to the original.

Urbee2 is the name of the initial car in the world car mounted employing the technology 3D printing its bodywork and car windows were "printed." The car is fashioned by collaborative work between the US engineering group and the company Stratasys, the manufacturer of Stratasys 3D printers. The Urbee2 it is a hybrid vehicle with innovative look, Figure 10.28. The engineering group will be driving the Urbee 2 from New York to San Francisco upon conclusion and is predictable to use less energy than most other production cars. Their goal is to exploit only 10 U.S. gallons of gas for the whole trip.

Figure 10.28: 3D Printing Urbee2

3D Printed Bus & Aviation

Like numerous industries, the aerospace industry is progressively implementing 3D printing and rapid prototyping technologies to advance aircraft parts in the pursuit of trimming down engineering and fabrication costs. As a matter of fact, one of the main players in the aerospace domain, "Boeing," already makes use of 3D printing technology comprehensively and printed over 22,000 parts last year. In May 2017 Airbus announced that its new "Airbus A350 XWB" incorporated over 1000 components fabricated by 3D printing, Figure 10.29.

Figure 10.29: 3D Printing Air-Bus

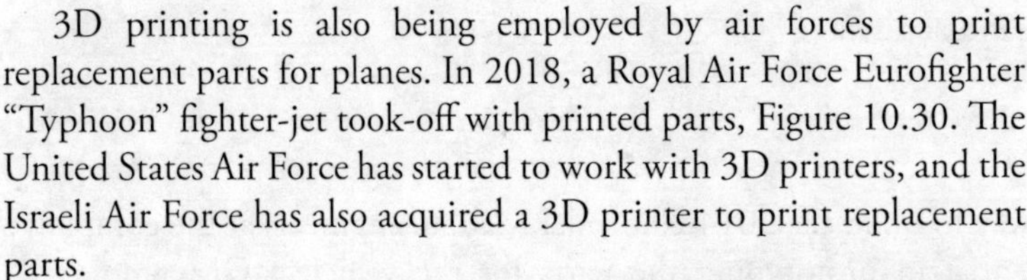

3D printing is also being employed by air forces to print replacement parts for planes. In 2018, a Royal Air Force Eurofighter "Typhoon" fighter-jet took-off with printed parts, Figure 10.30. The United States Air Force has started to work with 3D printers, and the Israeli Air Force has also acquired a 3D printer to print replacement parts.

Figure 10.30: 3D Printing Metal Airbus Wing Bracket – Lighter and Stronger Than Conventional Wing Bracket

In 2019 GE Aviation shown that it had made design for additive manufacturing to construct a helicopter engine with 16 parts instead of 900, with great possibly-impact on decreasing the intricacy of supply chains. Furthermore, General Electric (GE) recently proclaimed an investment of $80 million to 3D print fuel-nozzles for the next-generation LEAP jet engine. The unexpected predisposition of aerospace industry in the direction of additive manufacturing is mostly due to the opportunity to accomplish substantial weight decrease. According to American Airlines, for every pound of weight decreased from the aircraft, the company saves 11,000 gallons of fuel per annum.

3D Printing Advanced Metal Casting

Scientists developed a 3D design utilized to print a polymer airline seat frame. The frame was then treated in ceramic material and heated until the polymer evaporated. The subsequent ceramic mold was then utilized to metal cast the new seat frame. A proof-of-concept project could afford airlines hundreds of millions of dollars.

Autodesk has joined with a Michigan foundry in a 3D printing proof-of-concept project that caused in a novel magnesium commercial airline seat frame so light it could save an airline more than $200 million in fuel charges. Autodesk and Aristo Cast stated that if an aircraft manufacturer such as "Airbus" were to substitute all 615 seats on its "A380 jets" with the novel, lighter magnesium seat frames — across a fleet of 100 planes which characteristically have a 20-year lifespan — the interchange would save $206 million, knowing that the average jet fuel charges in 2019. Along with decreasing fuel costs, the novel seats could also ensure 126,000 fewer tons of C02 emission release if utilized on a single model aircraft.

Unmanned Aerial Systems

Recently, BAE Systems revealed 2,040 aircraft manufacturing conceptions integrating on-board 3D printing to improve Unmanned Aerial Vehicles UAVs. The concept clarifies how an aircraft scrutinizes the disaster and accounts to the mission control where the compulsory engineering data is provide to the on-board printers to print unmanned aerial vehicles rendering to the necessities of the disaster situation. Ultimately, these 3D printed UAVs will accomplish rescue operations or monitor the condition. While this idea is still on the drawing board, BAE Systems has now capitalized $250 million in research and development to guarantee that these concepts can be revolved into an actuality.

Space

NASA engineering technical staff illustrated on inventiveness and forward-thinking technology. About 70 of the parts that brand up the nomad were constructed digitally, directly from computer designs, in the heated chamber of a production-grade "Stratasys 3D Printer." For its 3D-printed components, NASA utilizes ABS, PCABS and polycarbonate materials. FDM, patented by Stratasys, is the single 3D-printing method that sustains production-grade thermoplastics, which are lightweight but tough enough for harsh end-use components.

Recently, NASA's skyrocket engine injector manufactured from a 3D printer succeeded a most important hot fire investigation. In the test, the skyrocket engine injector produced 10 times additional thrust than any injector manufactured from 3D printing in the past, Figure 10.31.

Figure 10.31: 3D Printer Manufactured Rocket Engine Injector

NASA engineering technical staff was talented to construct the injector with selective laser melting with the aim of fusing together nickel-chromium alloy powder. The examined part was the size of an injector utilized in a lesser rocket engine, but the design is comparable to what an injector appears like in a bigger engine. The injector can be scaled to a bigger size when it is fabrication-ready.

Also, NASA is looking forward to utilize 3D printing as a provision for speedy pre-prototyping. "3D printing makes it simpler to seize the dreams of the mission concepts. Utilizing3D Punters, engineers could gain peer-reviews, different design impressions, and support for final prototype. With open source design improvement, there will be a potential to incorporate multiple ideas from the outside, thereby decreasing the construct time significantly and also reducing costs.

3D Printing in Zero-Gravity

Traveling through space takes a much fuel and preparation, especially dealing with unexpected incidents in space, there may be limited equipment or spare parts available. Transporting items into space is, of course, exceptionally expensive. Astronauts may need to wait for the next mission to obtain the parts they need or resolve the failure or find out alternative solutions to their problems. With 3D printing available however, they can print out parts, tools or other items they need in the event of an emergency. One of the most logical uses for 3D printing is printing parts, tools, and other devices for astronauts while they are in space.

It can also help accelerate the building of parts for the International Space Station. The Zero-G Printer, the first 3D printer designed to operate in zero gravity, was built under a joint partnership between "NASA Marshall Space Flight Center, MSFC, and Made In Space, Inc." In September 2014, "SpaceX" delivered the zero-gravity 3D printer to the International Space Station, ISS.

On December 19, 2014, NASA emailed CAD drawings for a socket wrench to astronauts aboard the ISS, who then printed the tool utilizing its 3D printer. Applications for space offer the ability to print parts or tools on-site, as opposed to rockets to bring along pre-manufactured items for space missions to human colonies on the moon, Mars, or elsewhere.

On a related note, an engineer won a grant from NASA in 20119 to prototype a machine that will print food that is better than the freeze-dried food astronauts normally eat.

Fire Arms

AM's influence on firearms includes 2D: novel engineering methods for well-known companies, and new potentials for the making of do-it-yourself weapons. Recently the US-based group "Defense Distributed" revealed strategies to design a traditional polymers 3D printed firearm "that could be downloaded and replicated by anyone with a 3D printer." Afterward "Defense Distributed" declared their strategies, queries were upturned regarding the consequences that 3D printing and extensive consumer-level CNC machining may have on gun control efficiency.

CONSUMER PRODUCTS

Consumer Electronics

Poly-Jet technology can yield models with extraordinarily thin walls — 0.6mm or less — perfect for small tools and electronic instruments densely compacted with miniature components. Smooth finish and realistic colors make these models almost indistinguishable from the finished product.

Sporting Goods

Prototypes frequently necessitate a mixture of tough and flexible materials. A hard shell helmet, and its embellished interior, or a pair of ski goggles with tinted lenses and neoprene over molding. Only Poly-Jet technology can yield prototypes with many materials and colors in a single run, and automated construct. Consequently, it is ideal for sporting goods creators with an aesthetic charm.

"FDM Technology" operates with production-grade thermoplastics to yield components with high impact strength and great toughness. "FDM Technology" is well matched parts that essential to endure tough and frequent functional testing.

Toys

Children may develop their restricted selections of toys when buying toys. Children may imagine the precise color of their toy's hair. Children can modify their own exclusive toy, thanks to the novelties of 3D printing. The toy manufactures is among the largest recipients of the technology. Toys already incline to be small and made out of polymers, enabling it to 3D print printed easily. It is projected the total economic influence of 3D printing technology will be \$230-\$550 billion per year by 2025, \$100-\$300 billion of which will be direct consumer goods such as toys.

Jigs and Fixtures

Inkjet-based technology and FDM Technology, both available from "Stratasys," provide rapid and precise techniques to yield and fabricate tools. 3D printing jigs and fixtures with inkjet or FDM can aid decrease the backlog for an in-house machine shop and be utilized as a bridge-to-tooling answer.

ADDITIVE MANUFACTURING

Cloud-Based Additive Manufacturing

Additive manufacturing in engagement with cloud computing technologies permits decentralized and physically independent disseminated fabrication. Cloud-based additive manufacturing denotes to a service-oriented networked manufacturing entity in which service users are capable to construct workpieces through Infrastructure-as-a-Service, "IaaS," Platform-as-a-Service, "PaaS," Hardware-as-a-Service "HaaS," and Software-as-a-Service "SaaS." Distributed manufacturing is supported out by some corporations; there is also a services like "3D Hubs" that place people requiring 3D printing in connected with owners of printers. Some businesses offer on-line 3D printing services to both industrial and individual customers, working from 3D designs uploaded to the business website. 3D-printed designs are either transported to the customer or picked up from the service provider.

Mass Customization

Businesses have created services where users can customize objects utilizing simplified web based customization software, and order the subsequent items as 3D printed exclusive articles. This now permits users to generate custom cases for their mobile phones. "Nokia" has released the 3D designs for its case so that owners can modify their individual case and have it 3D printed.

Rapid Manufacturing

Progresses in RP technology have presented materials that are suitable for final manufacture, which has in turn presented the opportunity of directly manufacturing finished mechanisms. One benefit of 3D printing for rapid manufacturing lies in the comparatively low-cost production of small numbers of workpieces. Rapid manufacturing is a new method of manufacturing and many of its procedures continue unproven. 3D printing is now spreading to the field of rapid manufacturing and was acknowledged as a "next level" technology by many experts in a 2018 report. One of the most encouraging procedures appears to be the variation of selective Laser sintering (SLS), or direct metal laser sintering (DMLS) some of the better-established rapid prototyping approaches.

Rapid Prototyping

Industrial 3D printers have occurred since the early 1980s and have been expoited expansively for rapid prototyping and investigation purposes. These are commonly larger machines that utilize exclusive powdered metals, casting media, e.g. sand, plastics, paper or cartridges, and are employed for rapid prototyping by universities and commercial companies.

Agile Tooling

Agile tooling is describing the procedure of utilizing segmental means to develop tooling that is created by additive manufacturing or 3D printing approaches to permit fast prototyping and answers to tooling and fixture requirements. Agile tooling utilizes a cost effective and high quality technique to rapidly react to client and market requirements. It can be utilized in hydro-forming, stamping, injection molding and other engineering and fabrication procedures.

3D BIOPRINTED FOOD

Food is one of essential elements of life which is at the foundation of the pyramid of human essentials. Conveying the food industry to the digital era is one of the indispensable and ground-breaking implantations of 3D printing. Spread over this technology allows fast automated and repeatable procedures, freedom in design, as well as permitting large and easy changeability of the cooking process which can be tailored for each region or individual. Employing robotic layer based food printing schemes permits the recipe of the food to be digitized and saved in order to cook very repeatable and high quality dishes deprived of any margin for operator blunder. Also, the figure and beautification of the food can be tailored based on the customer or the occurrence. Additive manufacturing of food is being advanced by squeezing out food, layer by layer, into 3D substance. A large diversity of foods are applicable candidates, such as chocolate and candy, and flat foods such as crackers, pasta, and pizza.

NASA has carefully provided resourcefulness of the concept, awarding a contract to the "Systems and Materials Research Consultancy" to investigate the viability of printing food in space. One of the difficulties with food printing is the environment of the consistency of a food. For instance, foods that are not sturdy enough to be filed are not suitable for 3D printing, Figure 10.32.

Figure 10.32: 3D Bioprinting Food in Layers

CONSTRUCTION INDUSTRY

In the construction industry, Construction 3D printing can be employed to generate building modules or to 'print' entire buildings. Construction is well-suited to 3D printing as much of the information essential to produce an item will survive as a consequence of the design procedure, and the industry is already skillful in computer aided manufacturing CAD. The recent development of "building information modelling," BIM in specific may enable better exploitation of 3D printing.

Far East Decoration Design Engineering has exploited super large 3D printers to spray a combination of rapid drying cement and salvaged raw materials. This has allowed them to build 10 small demo 'houses' in less than 24 hours. They have recommended that each house can be printed for less than $5,000. Their system constructs blocks off-site by layering the cement mix in a diagonally strengthened arrangement. The blocks are then draw together on site. It is possible to use the procedure to construct larger houses or even skyscrapers in the future. In 2017, they also printed and entire villa and a five-storey apartment building. Scientists at MIT have developed a system that 3D printed the basic structure of an whole building. It is a plan concerning building that could fundamentally decrease the costs of construction, Figure 10.33.

Figure 10.33: MIT 3D Printing machine created the largest 3D printed structure to date

Courtesy: neri oxman, steven keating, julian leland, levi cai

3D Concrete Printing Robot

At the Viva Technology conference – 2017, in Paris, France, they provided registered visitors a live demonstration of the 3D Concrete Printing robot.. The host company claims the technology vividly drops the time necessary for the construction procedure by as much as a immense 75%. The concrete is truly a product called CyBe MORTAR which sets in just a matter of minutes, accomplishing its total structural strength in about an hour. Thanks to these characteristics, the robot is capable to extrude layer after layer building the structure in almost any shape physics will permit. Walls are just one of the applications, park benches and even planter boxes are also likely utilizing the 3D concrete printing technology.

The robot can transfer in as many as 6 axes and prints at a rate of 20cm per second. That means every 5 seconds it can cover a 1m. Its arm can cover a range of 2.75m and in some applications, its positioned on a height adjustable. The robot attaches through an Ethernet cable to a computer running dedicated CyBe software. This software can deal with most CAD models and does the structural authentication ahead of time.

Benefits of 3D Printing in Construction:

- Lowered materials usage
- Increasing the capacity to design a larger diversity of tailored homes and structures
- Savings of 30%-60% in construction surplus
- Decrease of manufacture time by 50%-70%
- Decrease of construction manual labor costs by 50%-80%

ENVIRONMENTAL

In Bahrain, large-scale 3D printing employing a sandstone-like material has been utilized to form unique coral-shaped configurations, which encourage coral polyps to inhabit and restore impaired reefs. These configurations have a much more natural figure than other structures utilized to build synthetic reefs, and, unlike concrete, are neither acid nor alkaline with neutral pH.

SOCIOCULTURAL APPLICATION

Fashion and Digital 3D Clothes

3D printing has arrived to the world of fashion with style designers carry out trial with 3D-printed bikinis, shoes, and dresses. In the world of extraordinary fashion courtiers such as Karl Lagerfeld designs for Chanel, Iris van Herpen and Noa Raviv experimenting with technology from "Stratasys," have engaged, and presented 3d printing in their collections. Assortments from their lines and other formation with 3D printing were displayed at the 2018 Metropolitan Museum of Art Anna Wintour Costume Center, exposition "Manus X Machina".

3D Printed Fashion Clothes

Fashion apparels are fashioned with diverse types and classes of fabrics, thread and a needle. Frequently buttons or zippers are utilized to permit people to open and close diverse portions of the clothing, Figure 10.34. To create clothes seamstresses draft clothing patterns, cut patterns and sew patterns together to fashion the final vestment. Intricate clothing patterns are often

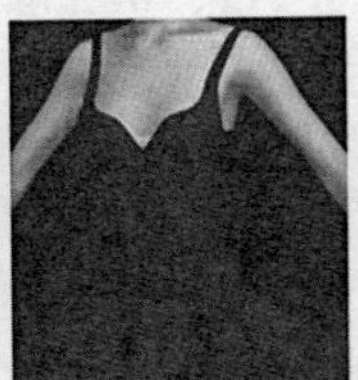
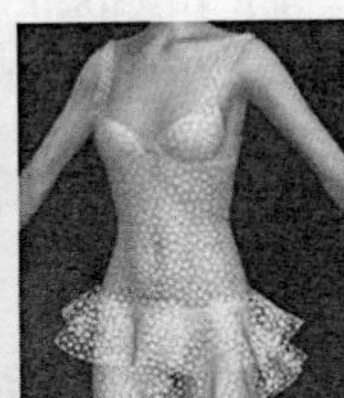

Figure 10.34: 3D Printing in US and European Fashion

drafted and restyled dozens of times. In some cases it may take a seamstress around 40 hours to cultivate a final clothing pattern, and an extra 60 hours to cut and sew it, Figure 10.34. Conventionally for thousands of years, seamstresses and tailors would sew clothing by hand.

Later in years, people invented sewing machines and transitioned to utilizing sewing machines more frequently instead of sewing by hand.

3D-Printed Fashion Shoes - On Demand

It is not merely bikinis that are being reproduced out by innovative businesses. "Kerrie Luft," a British designer fashioned a complete range of shoes exploiting the technology, which were presented at Selfridges in Oxford Street, London, England in March 2013. She originally commenced utilizing 3D printing to envisage her fashion without having to fashion an expensive mold, which is another advantage that the technology can have for fashion designers. Her outstanding fashions have established to be a ground-breaking.

In commercial production "Nike" is utilizing 3D printing to prototype and fabricating the "Vapor Laser Talon" football shoe for players of American football, and New Balance is 3D manufacturing custom-fit shoes for athletes. Also, Customers have the power to create 3D print of their own high heel designs in the store, Figure 10.35.

Figure 10.35: 3D Printing Shopper Create Custom Made Shoes
Courtesy: Retailers- Eram and Unistudio.

Art, Design & Sculpture

Artists and Sculptors are involved with 3D printing in countless of diverse means to discover form and function in ways formerly impossible. Whether purely to discover new unique expression or to study from old masters this is an exceedingly charged sector that is progressively finding novel ways of working with 3D printing and presenting the outcomes to the world. There are plentiful of artists that have now made a name for themselves by working explicitly with 3D modelling, 3D scanning and 3D printing technologies, sample of the artist as listed below:

- Joshua Harker
- Dizingof
- Jessica Rosenkrantz
- Pia Hinze
- Nick Ervinck
- Lionel Dean
- And many others.

The discipline of 3D scanning in combination with 3D printing also carries a novel element to the art world, though, in that artists and scholars now have a established procedure of reproducing the work of past masters and creating precise duplications of antique, and more recent, sculptures for close examination – works of art that they would otherwise never have been capable to interrelate with in person. The work of Cosmo Wenman is predominantly informative in this field.

Modernist Fashion Jewelry

3D printed jewelry is now on the market and it not astonishingly inexpensive. By utilizing 3D printers, jewelry creators can yield complicated designs in half the time, utilizing precious metals as well as polymers and brass. This 3D printed jewelry is an art practice in its own right, generating geometrically modern custom designs that are already rising popular amid consumers around the globe.

Museums

The popular of artworks and sculptures reserved by the Smithsonian Institute are not exposed to the public. Only 2% of its 137 million objects are being revealed at any one time as they are too historically significant or too delicate to be unprotected to the public continually. With 3D printing, museums can compromise scans of their smithereens to the public instead. In fact, not only can they envision these masterworks, they can also touch and get up close to the art or sculpture – things you can't do with the originals. On top of serving reservation great masterpieces, 3D-printed sculptures can aid educate the multitudes about old and modern art.

Cultural Heritage

In the last numbers of years 3D printing has been intensively utilized by in the cultural heritage field for conservation, renovation and dissemination objectives. Several Europeans and North American Museums have acquired 3D printers and vigorously recreate misplaced pieces of their relics. The utilization of 3D scanning technologies permits the reproduction of real objects without the utilization of molding methods that in numerous cases can be more costly, more challenging, or too invasive to be achieved, mainly for valuable artwork or fragile cultural heritage artifacts where direct interaction with the molding materials could damage the original object's surface. While one seldom think about it when visiting a museum, many of those showed pieces are very delicate and require to be handled with exceptional care. Take delicate works of art out to multiple venues, or even a national tour is endangering much to encounter the damage.

The curators from a museum in Salem, MA, selected a 3D printing answer. For a nationwide tour of an exhibition on Thomas Hart Benton, the Peabody Essex Museum generated a 3D printed reproduction of a predominantly fragile clay maquette.

The Natural History of Man, Metropolitan Museum of Art and the British Museum have employed their 3D printers to generate museum mementos that are accessible in the museum shops. Other museums, like the National Museum of Military History, and Varna Historical Museum, have gone additional extent and sell over the online platform liberating digital models of their artifacts, fashioned by utilizing "Artec 3D scanners," in 3D printing friendly file format, which everybody can 3D print at home.

Guitars

Distinct violins, whose form and sound does not wander far from the classic standard, there is a much extra liberty in guitar design and material use. 3D-printed "Gibson Les Paul" is a fully well-designed guitar exhibited at Maker Faire exhibition. Summit is already imagining a future where musicians order tailor-made guitars to yield exclusive sounds.

Gifts

3D printing is becoming widespread in the customizable gifts business, with products such as tailored models of art and dolls, in many figures: in metal or plastic, or as consumable art, such as 3D Bioprinted chocolate.

CHALLENGES OF 3D BIOPRINTING

Defending the faith of a newly born novel product has never been free from challenges, and many road blocks. The greater the challenges he greater the reward at thhe end. Below is a few of selected challenges the 3D Bioprinting technology and implementations may face:

Creating Valuable 3D Project is not an Easy Task.

Though many individuals can create a document for Bioprinting, nonetheless, developing 3D data require challenging tasks, such as the utilization of sophisticated design software and/or scanning and digitizing the data. Also, the ability to use these products necessitates training and investment, which is often beyond the reach of the average consumer.

Not all 3D is Created Equal.

Just because one may envisage a 3D image on the computer monitor, does not indicate that it may be 3D Bioprinted. 3D Bioprinters necessitate 3D data to fit in to definite geometric controls as well as to be of a precise format. This indicates there is a necessity for "middle-ware" software systems that take input 3D data and translate it to 3D data that is friendly to be 3D Bioprinted. Again, Utilization of these systems necessitates training and investment of time and resources.

Functional Challenge and Automation Challenge

There are other two types of considerable challenges; the first challenge is functional challenge, while the second is an automation challenge. It is recommended, to have the ability to use the 3D-Bioprinter to print more realistic objects than just plastic toys. The need to enable these objects to have more functions by incorporating electro-mechanical features, and electronics, such as hardware inserts and electric wires to the objects. Also, adding "InfraStructs-like" technology to make these 3D Bioprinters readable. The automation challenge is innovating the technology by speeding up the production throughput process from a small quantity in the 10's to a larger quantity in the thousands if not in the millions.

REFERENCES

1. ining, G., Kowalski, S.,"Statistical Methods for Engineers", 3rd ed. Boston:Brooks/Cole, ISBN:978-0-538-73518-6, 2011
2. Cooper, K. G. (2001). Rapid Prototyping Technology: Selection and Application, Marcel Dekker Inc., SAD, New York. ISBN 0-8247-0261-1
3. Gibson, I.L; Rosen, D.W. & Stucker, B. (2010). Additive Manufacturing Technologies: Rapid Prototyping to Direct Digital Manufacturing, Springer, ISBN 978-1-4419-1119-3, New York
4. Godec, D. (2005). Utjecaj hibridnog kalupa na svojstva injekcijski prešanog plastomernog otpreska, D.Sc. Thesis, FSB, Zagreb
5. Soloman, Sabrie "Sensors Handbook, 2nd Edition, McGraw-Hill Publishing Co., NY – ISBN 978-0-07-160570-0
6. Soloman, Sabrie "Sensors and Control Systems in Manufacturing 2nd Edition," McGraw-Hill Publishing Co., NY – ISBN 0-07-059626-3
7. Soloman, Sabrie "Affordable Automation," McGraw-Hill Publishing Co., NY – ISBN 0-07-059633-6
8. Soloman, Sabrie "Sensors Handbook, 1st Edition," McGraw-Hill Publishing Co., NY – ISBN 0-07-059630-1
9. Soloman, Sabrie "Sensors and Control Systems in Manufacturing 1st Edition," McGraw-Hill Publishing Co., NY – ISBN 978-07-10572-4

3D BIOPRINTING REVOLUTION

DR. SABRIE SOLOMAN

FEEDBACK FORM

1. Which topics from your examination point of view are inadequately or not covered in this book?

___

___

2. Is there any misprints/ mistakes/ factual inaccuracy in this book? If so, please specify.

___

___

3. What is your assessment of this book as regards the presentation of the subject-matter, expression, precision and price in relation to the other recommended books available on this subject?

___

___

4. Which competing books you regard as better than this? Please mention their author and publishers.

___

___

5. Any further suggestions/comments you would like to make for the improvement of the book?

___

___

PS: If needed, please attach a separate sheet.

Name : ____________

College : ____________

Address : ____________

Email : ____________

Mobile : ____________

Please mail the feedback form at:

KHANNA BOOK PUBLISHING CO. (P) LTD.

4C/4344, Ansari Road, Daryaganj, New Delhi-110002

You can also email this feedback form at contact@khannabooks.com or Whatsapp at +91 – 99109 09320

BEST FEEDBACK WILL BE REWARDED SUITABLY